Methods in Molecular Biology

For further volumes:
http://www.springer.com/series/7651

For over 35 years, biological scientists have come to rely on the research protocols and methodologies in the critically acclaimed *Methods in Molecular Biology* series. The series was the first to introduce the step-by-step protocols approach that has become the standard in all biomedical protocol publishing. Each protocol is provided in readily-reproducible step-by-step fashion, opening with an introductory overview, a list of the materials and reagents needed to complete the experiment, and followed by a detailed procedure that is supported with a helpful notes section offering tips and tricks of the trade as well as troubleshooting advice. These hallmark features were introduced by series editor Dr. John Walker and constitute the key ingredient in each and every volume of the *Methods in Molecular Biology* series. Tested and trusted, comprehensive and reliable, all protocols from the series are indexed in PubMed.

Engineering Natural Product Biosynthesis

Methods and Protocols

Edited by

Elizabeth Skellam

Department of Chemistry & BioDiscovery Institute, University of North Texas, Denton, USA

Humana Press

Editor
Elizabeth Skellam
Department of Chemistry & BioDiscovery Institute
University of North Texas
Denton, USA

ISSN 1064-3745 ISSN 1940-6029 (electronic)
Methods in Molecular Biology
ISBN 978-1-0716-2272-8 ISBN 978-1-0716-2273-5 (eBook)
https://doi.org/10.1007/978-1-0716-2273-5

This Humana imprint is published by the registered company Springer Science+Business Media, LLC part of Springer Nature.
The registered company address is: 1 New York Plaza, New York, NY 10004, U.S.A.

Preface

This volume collates a variety of molecular methods to study and engineer natural product biosynthesis from fungi, bacteria, and plants. As natural product producers are very diverse, general information sections highlight the importance of natural products from different perspectives and a series of detailed molecular methods to access, investigate, and engineer these small molecules are described. Chapters cover heterologous expression techniques in a variety of hosts, gene disruption methods, modification of pathway regulators, and in vitro studies. Several specialized bioinformatics chapters are also presented in addition to methods for purifying and testing small molecules of interest. Considering that natural products have heavily influenced human health and agriculture, the state-of-the-art techniques presented here provide a general reference for biological chemists and molecular biologists alike.

Denton, TX, USA *Elizabeth Skellam*

Contents

Contributors

IKURO ABE • *Graduate School of Pharmaceutical Sciences, The University of Tokyo, Tokyo, Japan; Collaborative Research Institute for Innovative Microbiology, The University of Tokyo, Tokyo, Japan*

CÉSAR AGUILAR • *Evolution of Metabolic Diversity Laboratory, Unidad de Genómica Avanzada (Langebio), Cinvestav-IPN, Guanajuato, Mexico; Department of Chemistry, Purdue University, West Lafayette, IN, USA*

FABRIZIO ALBERTI • *School of Life Sciences, University of Warwick, Coventry, UK*

TAKAYOSHI AWAKAWA • *Graduate School of Pharmaceutical Sciences, The University of Tokyo, Tokyo, Japan; Collaborative Research Institute for Innovative Microbiology, The University of Tokyo, Tokyo, Japan*

FRANCISCO BARONA-GÓMEZ • *Evolution of Metabolic Diversity Laboratory, Unidad de Genómica Avanzada (Langebio), Cinvestav-IPN, Guanajuato, Mexico*

MARTIN BAUNACH • *University of Potsdam, Karl-Liebknecht-Straße 24-25, Potsdam, Germany*

ROBERTO G. S. BERLINCK • *Instituto de Química de São Carlos, Universidade de São Paulo, São Carlos, SP, Brazil*

DARLON I. BERNARDI • *Instituto de Química de São Carlos, Universidade de São Paulo, São Carlos, SP, Brazil*

EDDER D. BUSTOS-DIAZ • *Evolution of Metabolic Diversity Laboratory, Unidad de Genómica Avanzada (Langebio), Cinvestav-IPN, Guanajuato, Mexico*

MENGHAO CAI • *State Key Laboratory of Bioreactor Engineering, East China University of Science and Technology, Shanghai, China; Shanghai Collaborative Innovation Center for Biomanufacturing, East China University of Science and Technology, Shanghai, China*

LOUISE K. CHARKOUDIAN • *Department of Chemistry, Haverford College, Haverford, PA, USA*

MARC G. CHEVRETTE • *Wisconsin Institute for Discovery and Department of Plant Pathology, University of Wisconsin-Madison, Madison, WI, USA*

AUDAM CHHUN • *Biophore – DMF, University of Lausanne, Lausanne, Switzerland*

YIT HENG CHOOI • *School of Molecular Sciences, University of Western Australia, Perth, WA, Australia*

LING CHUANG • *Institute of Botany, Leibniz University Hannover, Hannover, Germany*

JÉRÔME COLLEMARE • *Westerdijk Fungal Biodiversity Institute, Utrecht, Netherlands*

ALESCIA CULLEN • *School of Environmental and Life Sciences, The University of Newcastle, Callaghan, NSW, Australia*

KATE M. J. DE MATTOS-SHIPLEY • *School of Biological Sciences, University of Bristol, Bristol, UK*

ELKE DITTMANN • *University of Potsdam, Karl-Liebknecht-Straße 24-25, Potsdam, Germany*

MARIE A. ELLIOT • *Department of Biology and Michael G. DeGroote Institute for Infectious Disease Research, McMaster University, Hamilton, ON, Canada*

ARTEM FETTER • *Gottfried-Wilhelm-Leibniz University of Hannover, BMWZ (Zentrum für Biomolekulare Wirkstoffe), Hannover, Germany*

JAKOB FRANKE • *Institute of Botany, Leibniz University Hannover, Hannover, Germany*

Jaimie Gibbons • *Department of Biology and Microbiology, South Dakota State University, Brookings, SD, USA*

Liping Gu • *Department of Biology and Microbiology, South Dakota State University, Brookings, SD, USA*

Grayson S. Hamrick • *Department of Chemistry, Haverford College, Haverford, PA, USA*

Jo Handelsman • *Wisconsin Institute for Discovery and Department of Plant Pathology, University of Wisconsin-Madison, Madison, WI, USA*

Calvin A. Henard • *Department of Biological Sciences, University of North Texas, Denton, TX, USA; BioDiscovery Institute, University of North Texas, Denton, TX, USA*

Jessica M. Henard • *BioDiscovery Institute, University of North Texas, Denton, TX, USA*

Hindra • *Department of Biology and Michael G. DeGroote Institute for Infectious Disease Research, McMaster University, Hamilton, ON, Canada*

Christophe Jacob • *Molecular and Structural Enzymology Group, UMR 7365 CNRS-UL IMoPA, Lorraine University, Faculté de médecine, Batiment Biopôle, Vandœuvre-lès-Nancy Cedex, France*

Matthew Jordan • *School of Environmental and Life Sciences, The University of Newcastle, Callaghan, NSW, Australia*

Samba Evelyne Kabemba Kaniki • *Department of Energy Technology, Aalborg University Esbjerg, Esbjerg, Denmark*

Anusha Kishore • *Gottfried-Wilhelm-Leibniz University of Hannover, BMWZ (Zentrum für Biomolekulare Wirkstoffe), Hannover, Germany*

Kristin Labby • *Department of Chemistry, Beloit College, Beloit, WI, USA*

Colin M. Lazarus • *School of Biological Sciences, University of Bristol, Bristol, UK*

Yanran Li • *Department of Chemical and Environmental Engineering, University of California, Riverside, CA, USA*

Dany Liu • *Department of Biology and Biological Engineering, Chalmers University of Technology, Gothenburg, Sweden; Novo Nordisk Foundation Center for Biosustainability, Chalmers University of Technology, Gothenburg, Sweden*

Qi Liu • *State Key Laboratory of Bioreactor Engineering, East China University of Science and Technology, Shanghai, China; Shanghai Collaborative Innovation Center for Biomanufacturing, East China University of Science and Technology, Shanghai, China*

Casey H. Londergan • *Department of Chemistry, Haverford College, Haverford, PA, USA*

Jean-Malo Massicard • *Molecular and Structural Enzymology Group, UMR 7365 CNRS-UL IMoPA, Lorraine University, Faculté de médecine, Batiment Biopôle, Vandœuvre-lès-Nancy Cedex, France*

Lamonielli F. Michaliski • *Instituto de Química de São Carlos, Universidade de São Paulo, São Carlos, SP, Brazil*

Sreemoye Nath • *Department of Biological Sciences, University of North Texas, Denton, TX, USA; BioDiscovery Institute, University of North Texas, Denton, TX, USA*

Jorge C. Navarro-Muñoz • *Westerdijk Fungal Biodiversity Institute, Utrecht, Netherlands*

Hariharan Gopalasamudram Neelakantan • *Lichen Ecology and Bioprospecting Laboratory, Biotechnology Programme, M.S. Swaminathan Research Foundation, Chennai, India*

Brett A. Neilan • *School of Environmental and Life Sciences, The University of Newcastle, Callaghan, NSW, Australia*

Mikkel Rank Nielsen • *Department of Chemistry and Bioscience, Aalborg University Esbjerg, Esbjerg, Denmark*

MAXIMILIAN OTTO • *Department of Biology and Biological Engineering, Chalmers University of Technology, Gothenburg, Sweden; Novo Nordisk Foundation Center for Biosustainability, Chalmers University of Technology, Gothenburg, Sweden*
MEGHAN A. PEPLER • *Department of Biology and Michael G. DeGroote Institute for Infectious Disease Research, McMaster University, Hamilton, ON, Canada*
ZHILAN QIAN • *State Key Laboratory of Bioreactor Engineering, East China University of Science and Technology, Shanghai, China; Shanghai Collaborative Innovation Center for Biomanufacturing, East China University of Science and Technology, Shanghai, China*
YEYAN QIU • *Department of Biology and Microbiology, South Dakota State University, Brookings, SD, USA*
INDRA ROUX • *School of Molecular Sciences, University of Western Australia, Perth, WA, Australia*
NELLY SELEM-MOJICA • *Evolution of Metabolic Diversity Laboratory, Unidad de Genómica Avanzada (Langebio), Cinvestav-IPN, Guanajuato, Mexico; Centro de Ciencias Matemáticas, UNAM, Morelia, Michoacán, Mexico*
KARTHIK SHANMUGAM • *Lichen Ecology and Bioprospecting Laboratory, Biotechnology Programme, M.S. Swaminathan Research Foundation, Chennai, India*
VERENA SIEWERS • *Department of Biology and Biological Engineering, Chalmers University of Technology, Gothenburg, Sweden; Novo Nordisk Foundation Center for Biosustainability, Chalmers University of Technology, Gothenburg, Sweden*
JENS LAURIDS SØRENSEN • *Department of Chemistry and Bioscience, Aalborg University Esbjerg, Esbjerg, Denmark*
MUTHUKUMAR SRINIVASAN • *Lichen Ecology and Bioprospecting Laboratory, Biotechnology Programme, M.S. Swaminathan Research Foundation, Chennai, India*
LI SU • *Molecular and Structural Enzymology Group, UMR 7365 CNRS-UL IMoPA, Lorraine University, Faculté de médecine, Batiment Biopôle, Vandœuvre-lès-Nancy Cedex, France*
YI TANG • *Department of Chemical and Biomolecular Engineering, University of California, Los Angeles, CA, USA; Department of Chemistry and Biochemistry, University of California, Los Angeles, CA, USA*
JONNA TEIKARI • *University of Potsdam, Potsdam, Germany; Environmental Soil Science, Department of Agricultural Sciences, University of Helsinki, Helsinki, Finland; Institute for Atmospheric and Earth System Research, University of Helsinki, Helsinki, Finland*
KIRA J. WEISSMAN • *Molecular and Structural Enzymology Group, UMR 7365 CNRS-UL IMoPA, Lorraine University, Faculté de médecine, Batiment Biopôle, Vandœuvre-lès-Nancy Cedex, France*
KATHERINE WILLIAMS • *School of Biological Sciences, University of Bristol, Bristol, UK*
SHENG WU • *Department of Chemical and Environmental Engineering, University of California, Riverside, CA, USA*
SHANHUI XU • *Department of Chemical and Environmental Engineering, University of California, Riverside, CA, USA*
DANIELLE A. YEE • *Department of Chemical and Biomolecular Engineering, University of California, Los Angeles, CA, USA*
CARSTEN ZEILINGER • *Gottfried-Wilhelm-Leibniz University of Hannover, BMWZ (Zentrum für Biomolekulare Wirkstoffe), Hannover, Germany*

LIHAN ZHANG • *Key Laboratory of Precise Synthesis of Functional Molecules of Zhejiang Province, School of Science, Westlake University, Hangzhou, Zhejiang Province, China*
XIAFEI ZHANG • *Department of Biology and Michael G. DeGroote Institute for Infectious Disease Research, McMaster University, Hamilton, ON, Canada*
RUANBAO ZHOU • *Department of Biology and Microbiology, South Dakota State University, Brookings, SD, USA*

Chapter 1

A Bioinformatics Workflow for Investigating Fungal Biosynthetic Gene Clusters

Jorge C. Navarro-Muñoz and Jérôme Collemare

Abstract

Predicting secondary metabolite biosynthetic gene clusters is a routine analysis performed for each newly sequenced fungal genome. Yet, the usefulness of such predictions remains restricted as they provide total numbers of biosynthetic pathways with only very limited biological significance. In this chapter, we describe a workflow to predict and analyze biosynthetic gene clusters in fungal genomes. It relies on similarity networking and phylogeny to perform genetic dereplication and to prioritize candidate gene clusters that potentially produce new compounds. This basic workflow includes the generation of high-quality figures for publication.

Key words Genome mining, Natural products, Secondary metabolites, Comparative genomics, Phylogeny, Similarity networks

1 Introduction

Secondary metabolites (SMs) produced by fungi are small molecules that exhibit diverse biological activities. Some are detrimental like mycotoxins, which spoil food and feed, while others have changed human societies like the broad-spectrum antibiotic penicillin or immunosuppressive compound cyclosporin [1, 2]. The enzymes involved in the biosynthesis of these compounds are of two types: core enzymes produce the chemical backbone, while tailoring enzymes catalyze diverse reactions to modify the initial backbone (methylation, acetylation, hydroxylation, etc.). Most of these enzymes feature conserved domains, amino acid sequences responsible for specific enzymatic reactions or interactions. In fungi, core and tailoring enzymes involved in the same biosynthetic pathway are encoded by genes that often co-localize in the genome and are co-regulated, defining a biosynthetic gene cluster (BGC) [2, 3]. Sequencing fungal genomes has revealed an unexpectedly large number of BGCs, which stimulated research to link already

Elizabeth Skellam (ed.), *Engineering Natural Product Biosynthesis: Methods and Protocols*,
Methods in Molecular Biology, vol. 2489, https://doi.org/10.1007/978-1-0716-2273-5_1,

known SMs to BGCs, but also to activate uncharacterized BGCs and identify new SMs [2, 4]. From the early times in fungal genomics, bioinformatics tools to predict BGCs in fungal genomes have been developed, starting with antiSMASH [5, 6] and SMURF [7], which are mostly based on homology searches to signature conserved domains found in core enzymes. Other tools were later developed to take into account other features such as co-regulation (CASSIS [8], MIDDAS-M [9], and FunGeneClusterS [10]) or presence-absence patterns (MIPS-CG [11]). However, these more recent tools were proof of concepts that have not been maintained, in addition to not providing more accurate predictions. A new method, TOUCAN, has been recently released to predict BGCs at the protein level using features extracted from protein sequences and training datasets [12]. Nowadays, the most widely used tool is antiSMASH thanks to its user-friendly web interface and linked resources antiSMASHdb [13], MIBiG [14], and BiG-FAM [15]. Yet, the fungal version of antiSMASH—fungiSMASH—has been developed using information from bacterial BGCs and only reports genomic regions which contain a predicted core enzyme. Other tools are needed to investigate these regions further. In particular, one major question that must be answered is whether a genomic region contains a BGC that is already known. Such an option is integrated in fungiSMASH with the KnownClusterBlast module [16]. While useful, this module provides a very limited analysis of predicted BGCs in a fungal genome and is sometimes misleading as it also reports low similarities on tailoring genes only. An alternative useful analysis to perform after predicting BGCs in a fungal genome is the phylogeny of core enzymes. Considering that closely related core enzymes likely produce the same chemical backbone [17] this strategy allows for the rapid identification of known pathways. Within a given phylogenetic clade, the BGC content in tailoring genes likely varies and the comparative analysis of predicted BGCs is needed. Tools that take into account the whole BGC to perform similarity networking have recently been developed for this purpose, such as BiG-SCAPE [18]. BiG-SCAPE is becoming the reference tool for BGC networking, but it has been developed using parameters from bacteria, and it needs some tuning to be optimally used with fungal BGCs. The networking strategy also shows the advantage that it can be scaled-up to thousands of predicted BGCs, allowing large-scale analyses that include thousands of fungal genomes. Finally, it is important to visualize these BGCs to facilitate comparative analyses, but also to present results in high-quality figures. Until recently, there was no dedicated tool to create such visualization. The release of Clinker [19] filled this gap.

The workflow we suggest in this chapter (Fig. 1) is addressed to a large audience of researchers with diverse backgrounds and without in-depth bioinformatics knowledge. It allows researchers to

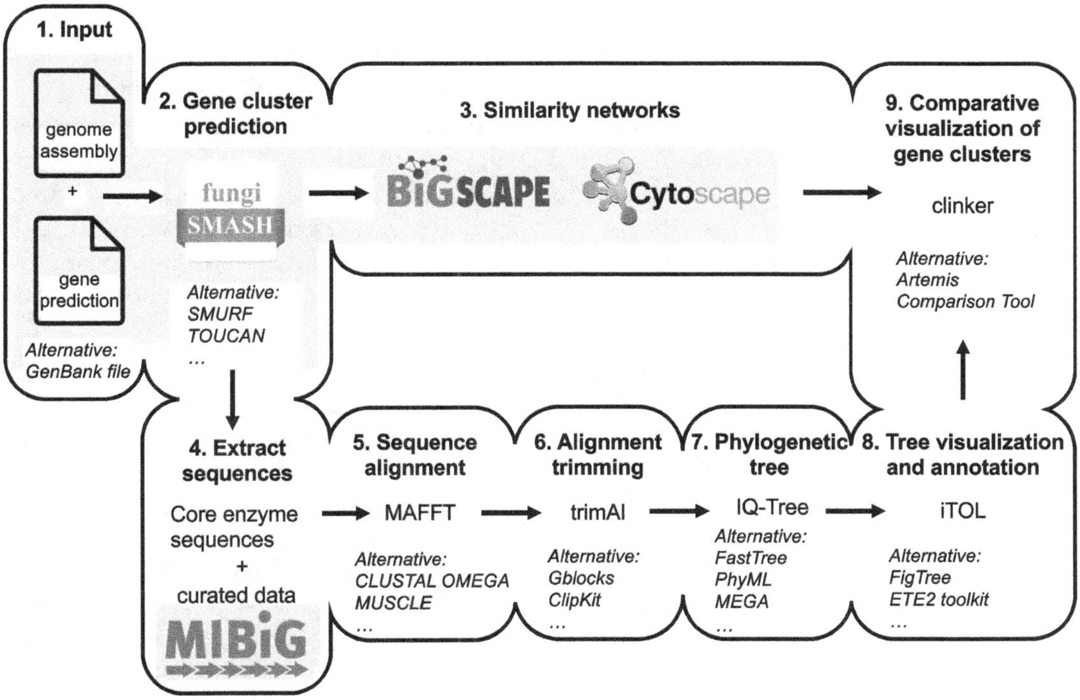

Fig. 1 Proposed workflow for the prediction and analysis of fungal biosynthetic gene clusters. Using a genome assembly and a gene prediction as a starting point, the prediction of secondary metabolite gene clusters using fungiSMASH is followed by complementary analyses. A network analysis compares predicted gene clusters as a whole and identify relationships between clusters. A phylogenetic analysis of core enzymes, together with curated sequences, provides evolutionary relationships between pathways. Both analyses allow dereplicating known pathways and selecting biosynthetic gene clusters for further functional studies. Results of each analysis and conservation of selected gene clusters across different genomes can be visualized in publication-quality figures

detect BGCs in fungal genomes and integrate relevant biological information, as well as to prioritize candidate BGCs for experimental characterization using molecular biology and synthetic biology methods.

2 Materials

2.1 Basic Requirements

1. The workflow presented in this chapter requires knowing how to use a command-line interface (terminal) with basic understanding of bash command lines. Basic knowledge of file formats (fasta, gff, GenBank, etc.) is also needed. We indicate which format to use for each tool. You can find many freely available tools on internet that are dedicated to convert files from one format to another (e.g., from gff to gff3).
2. The starting point of the presented workflow assumes that you have both a genome assembly (in fasta format) and

corresponding gene prediction (in gff3 format), or both assembly and prediction in an annotated GenBank file, from which you want to predict and analyze BGCs (Fig. 1; *see* **Note 1**).

3. All recommended software will work on any Unix-like system (MacOS, Linux). For Windows-based systems, the same instructions can be followed when using Windows Subsystem for Linux (WSL; https://docs.microsoft.com/en-us/windows/wsl/install-win10). However, we cannot guarantee the absence of conflicts when installing software because each computer configuration is different.
4. The workflow for a single or a few genomes does not require any particular requirement in regard to computer specifications. Enough storage space should be available to install all software (for example, antiSMASH will require more than 10 GB of free space) and databases, and to generate new data.

2.2 Conda

Conda is a package manager which allows users to prepare virtual environments with all the needed packages. We recommend installing the different software in their own conda environment as such environments offer important advantages: ease of installation (most packages are available from a single place, no need to compile code); reproducibility (by installing the same versions); and avoidance of conflicts (e.g. two tools in different environments need the same library, but different versions of it). Note that the same package may be installed in each environment and it is thus important to keep significant storage space on your hard drive.

1. Find and download the latest version that fits your system from https://docs.conda.io/en/latest/miniconda.html. Alternatively, once you have found the right version of conda for your system, it can be downloaded from the terminal. For example, for a Linux-based system:

```
curl -O https://repo.anaconda.com/miniconda/Miniconda3-latest-Linux-x86_64.sh
```

2. Execute installer from the terminal and follow instructions:

```
sh /path/to/Miniconda3-latest-Linux-x86_64.sh
```

3. Once conda is installed, an environment can be activated with the following command:

```
conda activate [name of environment]
```

4. To deactivate the environment, do:

```
conda deactivate
```

2.3 fungiSMASH

A webserver is available at https://fungismash.secondarymetabolites.org (*see* **Note 2**) and we do recommend using it if you need BGC prediction for a few genomes only. Because accurate gene prediction in fungal genomes is difficult, we strongly recommend to provide a gene prediction file (gff3 format) or an annotated GenBank file to fungiSMASH. Not providing such a file will yield inaccurate predictions that will not be useful for further analyses.

For a local installation when predictions are needed for large numbers of genomes, follow these instructions.

1. Setup conda channels:

```
conda config --add channels defaults
conda config --add channels bioconda
conda config --add channels conda-forge
```

2. Create an environment called "antiSMASH" in which antiSMASH and all its dependencies will be installed. The following command will install the current version of antiSMASH (6.0.0) (*see* **Note 3**):

```
conda create -n antismash antismash biopython=1.78
```

3. Activate the antiSMASH environment:

```
conda activate antismash
```

4. Setup databases for antiSMASH (this step needs to be done the first time only):

```
download-antismash-databases
```

2.4 BiG-SCAPE

These instructions will install version 1.1.2 of BiG-SCAPE.

1. If you have git in your system, clone the repository:

```
git clone https://git.wur.nl/medema-group/BiG-SCAPE.git
```

2. Alternatively, download and decompress the following file: https://git.wageningenur.nl/medema-group/BiG-SCAPE/-/archive/master/BiG-SCAPE-master.zip.
3. Create a conda environment for BiG-SCAPE using the yml file, which provides conda with a list of the required packages (including HMMER software) and defines the environment name ("bigscape"):

```
cd BiG-SCAPE
```

```
conda env create -f environment.yml
```

4. Obtain the latest version of Pfam, a database with hidden Markov models that will be used to identify protein domains. Download and decompress (into the BiG-SCAPE folder) the file Pfam-A.hmm.gz from: ftp://ftp.ebi.ac.uk/pub/databases/Pfam/current_release.
5. Setup the Pfam database (this step needs to be done the first time only):

```
conda activate bigscape
hmmpress Pfam-A.hmm
```

2.5 Other Software

All the tools are available at the time of writing this chapter. Changes in the accessibility or release of new versions may occur and will have to be updated accordingly. A user manual or readme file with installation instructions for each tool is usually available on the website where they can be downloaded (Table 1) or is part of the downloaded files. Alternatives to each suggested tool may exist, but will yield slightly different results.

1. Install clinker:

```
conda create -n clinker -c conda-forge -c bioconda clinker-py
```

2. Download and install Cytoscape (Table 1). You will need to have a working java virtual machine (e.g. OpenJDK) installed on your computer.
3. The software to build phylogenetic trees can be installed in a single conda environment ("phylogeny"):

```
conda create -n phylogeny iqtree trimal mafft
```

4. Download and install the remaining software (Table 1), following their respective instructions.

3 Methods

3.1 Prediction of Biosynthetic Regions with fungiSMASH

In this first step, the aim is to identify BGC regions in a genome using fungiSMASH. The results of this software will allow you to make a summary table with numbers for the different categories as reported by fungiSMASH, and can be used directly for the network analysis.

There are three types of input for fungiSMASH: (1) an NCBI accession number (only for the fungiSMASH webserver); (2) an annotated GenBank file containing all the contigs/scaffolds from

Table 1
Online resources to download software recommended in the workflow

Software	Download link	References
Conda environments		
conda	https://docs.conda.io/en/latest/	
BGC prediction		
fungiSMASH	https://antismash.secondarymetabolites.org/#!/download	[6]
Network analysis		
BiG-SCAPE	https://git.wageningenur.nl/medema-group/BiG-SCAPE	[18]
Cytoscape	https://cytoscape.org/	[20]
Phylogenetic analysis		
MAFFT	https://mafft.cbrc.jp/alignment/software/	[21]
trimAl	http://trimal.cgenomics.org/downloads	[22]
IQ-TREE	http://www.iqtree.org	[23]
AliView	http://ormbunkar.se/aliview/	[24]
iTOL	https://itol.embl.de	[25]
Comparative genomics figure		
Clinker	https://github.com/gamcil/clinker	[19]
Auxiliary scripts		
	https://github.com/WesterdijkInstitute/FNP_aux_scripts	

the genome; and (3) a genome assembly (fasta format) and its gene models (gff3 format).

1. Using the webserver: enter the input data in the corresponding fields. If using a combination of assembly and gene prediction files, the gene model field will appear after uploading the fasta file. Select strictness of the analysis and optionally extra features (*see* **Note 4**). Skip **Step 2** and go directly to **Step 3**.
2. Using the command-line interface, activate the antiSMASH conda environment and use the following command lines according to the type of input data (change arguments in between brackets as needed):

   GenBank file input: `antismash --cpus [number of cpus] --taxon fungi --clusterhmmer --genefinding-tool none --output-dir [folder for results] --minimal [GenBank file]`

   Pair of assembly and gene prediction files: `antismash --cpus [number of cpus] --taxon fungi --clusterhmmer`

```
--genefinding-gff3 [GFF3 file] --output-dir
[folder for results] --minimal [asssembly file]
```

In this case, there is no comparison performed to any of the antiSMASH-related databases. But more accurate comparisons will be performed in the next steps of the workflow.

3. When using the online fungiSMASH server, open the link to the result page. The page shows an overview of the identified biosynthetic regions. You will also be presented with the best hit against the MIBiG database if you have selected the "KnownClusterBlast" option. Download and decompress all results.

   When using the command-line interface, the results overview can be accessed by opening the "index.html" file inside the results folder (opens in your default web browser).

4. Browse the results (Fig. 2a): clicking on a region number will present the region subpage with more detailed information (*see* **Note 5**; detailed description of the results page are reported in the antiSMASH publications [6]). We do recommend to browse the results for manual curation of the results. For example, hybrid PKS-NRPS enzymes and regions containing both PKS and NRPS individual genes are reported as the same category of hybrid-containing regions (Fig. 2b). Browsing the results will also give you initial information that you can use to manually build a table with the number of regions with specific types of BGCs (PKS, NRPS, etc.; Fig. 2b), which is useful to indicate the production capacity of a given fungal isolate. It will also allow you to assess if some regions contain a BGC related to a characterized one. Finally, the results will also contain information about conserved domains in core enzymes, which can be useful to report (Fig. 2c).

3.2 Similarity Network Analysis

When more than one genome is being analyzed, the results from fungiSMASH are further analyzed using BiG-SCAPE [18] to build similarity networks and identify relationships between BGCs. The more genomes, the more sensible a networking analysis is. It is possible to include reference BGCs from a database like MIBiG to identify BGCs that belong to a gene cluster family (GCF) containing characterized pathways (*see* **Note 6**). Note that this approach is limited because relationships may be missed due to differences in BGC content between fungiSMASH predictions and characterized BGCs; exploring different distance cutoff values (e.g. 0.3, 0.4, 0.5) is thus recommended in such a case.

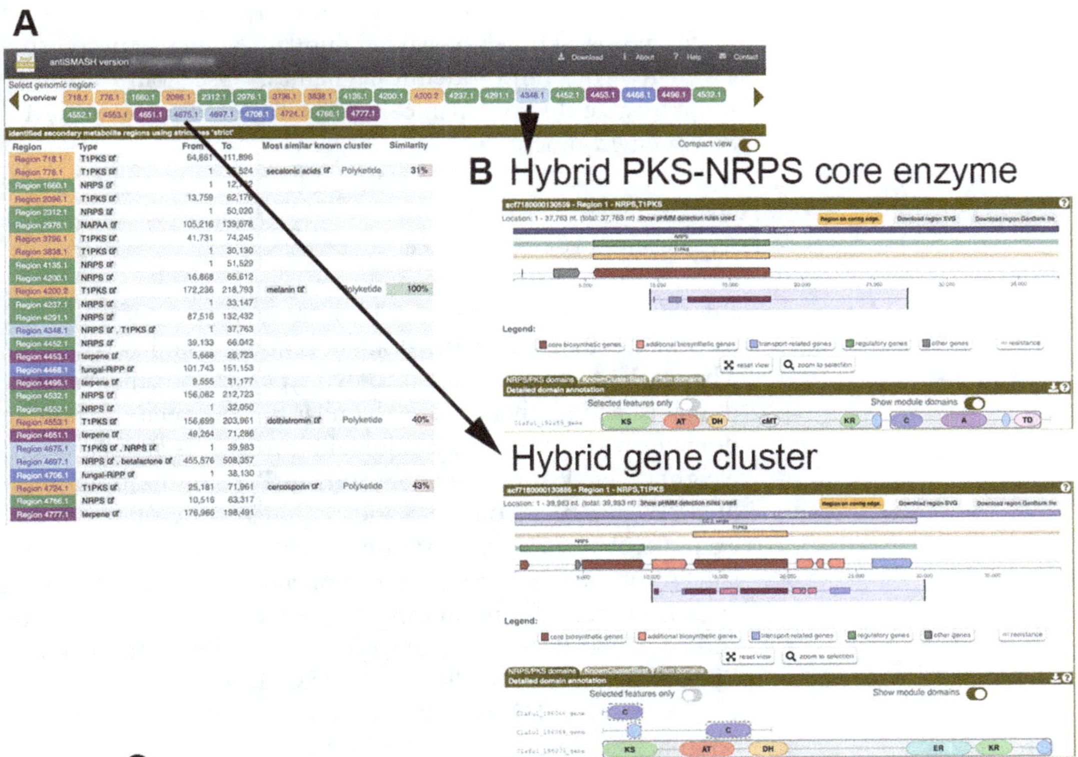

C

Table 1. Summary of secondary metabolite gene clusters as predicted by fungiSMASH.

	PKS	NRPS	Hybrid PKS-NRPS	Terpene	RiPP	Total
Clafu1	8	13	1	4	2	28

Table 2. Domain organization of polyketide synthases (PKS) and homology to known pathways.

PKS	Predicted pathway	Classification	Domain organization
776.1	emodin	non-reducing PKS	SAT-KS-AT-PT-ACP
3796.1	na	non-reducing PKS	SAT-KS-AT-PT-ACP
4200.2	DHN melanin	non-reducing PKS	SAT-KS-AT-PT-ACP-ACP-TE
4553.1	dothistromin	non-reducing PKS	SAT-KS-AT-PT-ACP-ACP-ACP-TE
4724.1	cercosporin	non-reducing PKS	SAT-KS-AT-PT-ACP-ACP-TE
718.1	na	reducing PKS	KS-AT-DH-ER-KR-ACP
2096.1	na	reducing PKS	KS-AT-DH-cMT-ER-KR-ACP
3838.1	na	reducing PKS	KS-AT-DH-cMT-ER-KR
4675.1	na	reducing PKS	KS-AT-DH-ER-KR-ACP

Fig. 2 Example of output obtained with fungiSMASH and exploitation for publication. (**a**) The overview page indicates total number of regions with predicted biosynthetic gene clusters (BGCs) in the genome of the fungus *Cladosporium fulvum* (Clafu1) and a summary of similarity to known BGCs. (**b**) Focusing on a specific region allows curating the data and reporting on the prediction of functional conserved domains found in core enzymes. In this example, two regions with similar annotation actually contains different kinds of BGCs. The upper region encodes a hybrid polyketide synthase-non-ribosomal peptide synthetase (PKS-NRPS) enzyme as shown with the typical conserved domain organizations. The lower region contains individual PKS and NRPS genes. (**c**) Example of a summary table of the results that can be immediately derived from fungiSMASH results

1. Verify that all the biosynthetic regions (GenBank files) detected by fungiSMASH do not have redundant names between species. If you find identical names, use our auxiliary script (Table 1) to rename regions within the BiG-SCAPE conda environment:

```
python rename_regions.py --inputfolder [fungiSMASH folder] --
string [unique prefix to change file names]
```

2. BiG-SCAPE uses a combination of three metrics, of which the most important one is the domain sequence similarity (DSS). The DSS is further divided into two types of domains: "anchor domains"and others. The similarity between matching domains within the "anchor domain"category is given a special weight. The list of "anchor domains" recognized in BiG-SCAPE may be incomplete and requires to be updated. For example, if you are interested in terpene cyclases or dimethylallyl tryptophan synthases, it is important to expand the list of anchor domains found in fungal core biosynthetic proteins by adding PF19086 (Terpene_syn_C_2), PF01040 (UbiA), PF06330 (TRI5), PF13249 (SQHop_cyclase_N), and PF11991 (Trp_DMAT) domains to the "anchor_domains.txt" file found in the BiG-SCAPE folder.
3. Activate the BiG-SCAPE conda environment and run the software using the following command (change the arguments in brackets according to your data). If available, include a subfolder with characterized BGCs within the input folder (*see* **Note 6**):

```
python bigscape.py --inputdir [top-level folder with GenBank
files] --outputdir [folder for results] --include_singletons
--mix --no_classify --cutoffs 0.3 0.4 0.5 --clans-off
```

4. To explore the networking results, open the "index.html" file inside your [output folder] to see an overview of the results.
5. Click on the dropdown menu on the top-right panel to select the run to visualize (Fig. 3a).
6. On the top left ("Networks"), select the BiG-SCAPE biosynthetic class to focus on ("Mixed," Fig. 3a).
7. Wait until BiG-SCAPE has organized the networks, and then start their exploration (Fig. 3b). Here you will be able to find if BGCs are related to each other or form singletons. Clicking on BGC nodes will select them; the GCF this BGC belongs to will appear in the right panel. Click on it to go to the GCF view where a basic figure can be directly exported (Fig. 3c). Identify GCFs or singletons of interest for further investigation.

Fig. 3 Using BiG-SCAPE to build similarity networks. (**a**) When the BiG-SCAPE run is finished, select the run and "mixed" class (red squares) to visualize the networks. (**b**) Example of networks generated by BiG-SCAPE with fungiSMASH predictions for six related species: *Cladosporium fulvum* (Clafu1), *Dothistroma septosporum* (Dotse1), *Zymoseptoria tritici* (Mycgr3), *Pseudocercospora fijiensis* (Mycfi2), *Mycosphaerella populorum* (Sepmu1), and *Mycosphaerella populicola* (Seppo1). In this example, the run using a cutoff of 0.6 shows several networks of BGCs that form gene cluster families (GCFs). (**c**) Detailed information about one GCF found in four out of six species, showing the similarity between the predicted BGCs in the different species. In this example, the GCF corresponds to pathways involved in the production of cladofulvin and emodin-like compounds

8. Use the results from BiG-SCAPE (https://git.wageningenur.nl/medema-group/BiG-SCAPE/-/wikis/output#results) with Cytoscape to further work with the similarity networks and generate high-quality figures (manipulate the style of the networks (colors, node shapes), export high-quality (sub)-networks for publication, include more metadata (e.g. highlight different sets of BGCs), highlight additional characterized clusters not included in the bundled MIBiG data (e.g. from a custom dataset of fungal characterized BGCs)).

   In Cytoscape, import a network using the "Import Network from File System" button on the Tool Panel (Fig. 4a). Choose a `.network` file from the BiG-SCAPE output.

9. Click on the first column name ("Clustername 1") and mark it as Source Node; mark the second column ("Clustername 2") as Target Node.
10. Select the "Raw distance" column as Edge attribute. Importing the rest of the columns is optional, but they should also be imported as Edge attributes.
11. Import annotations using the "Import Table from File" button in the Tool Panel. As an example, the called gene cluster family information can be imported directly (Fig. 4a).
12. Adjust the style of the network by changing the parameters in the "Style" menu on the left (Border width, Shape, etc.). Finally, go to File → Export → Network to Image (Fig. 4b, c) to create a figure of the network for publication.

3.3 Build a Fasta File with Amino Acid Sequences from Core Enzymes

Similarity networks are interesting to identify singletons and shared BGCs, but this kind of analysis does not provide accurate information about evolutionary relationships. For this purpose, a phylogenetic analysis is needed, which is typically performed for core enzymes of a given type (*see* **Note 7**). This analysis allows identifying evolutionary events such as gene duplications, losses, and lateral transfers. When using functionally characterized enzymes, it will also robustly confirm the close relationship to known pathways (dereplication) and identify novel enzymes that potentially produce new chemical backbones. Fasta files are needed to perform multiple sequence alignments and subsequent phylogenetic analyses. Only homologous sequences can be aligned and thus fasta files should be made for each type of core enzyme. You need to build a fasta file with characterized sequences from a repository like the MIBiG database (*see* **Note 8**), and append this file with the sequences of similar predicted core enzymes encoded in the genome(s) of interest. Depending on the repository, you may have to do it manually, or options to download or copy sequences may be offered. For this workflow, we provide a simple GenBank file parser auxiliary script

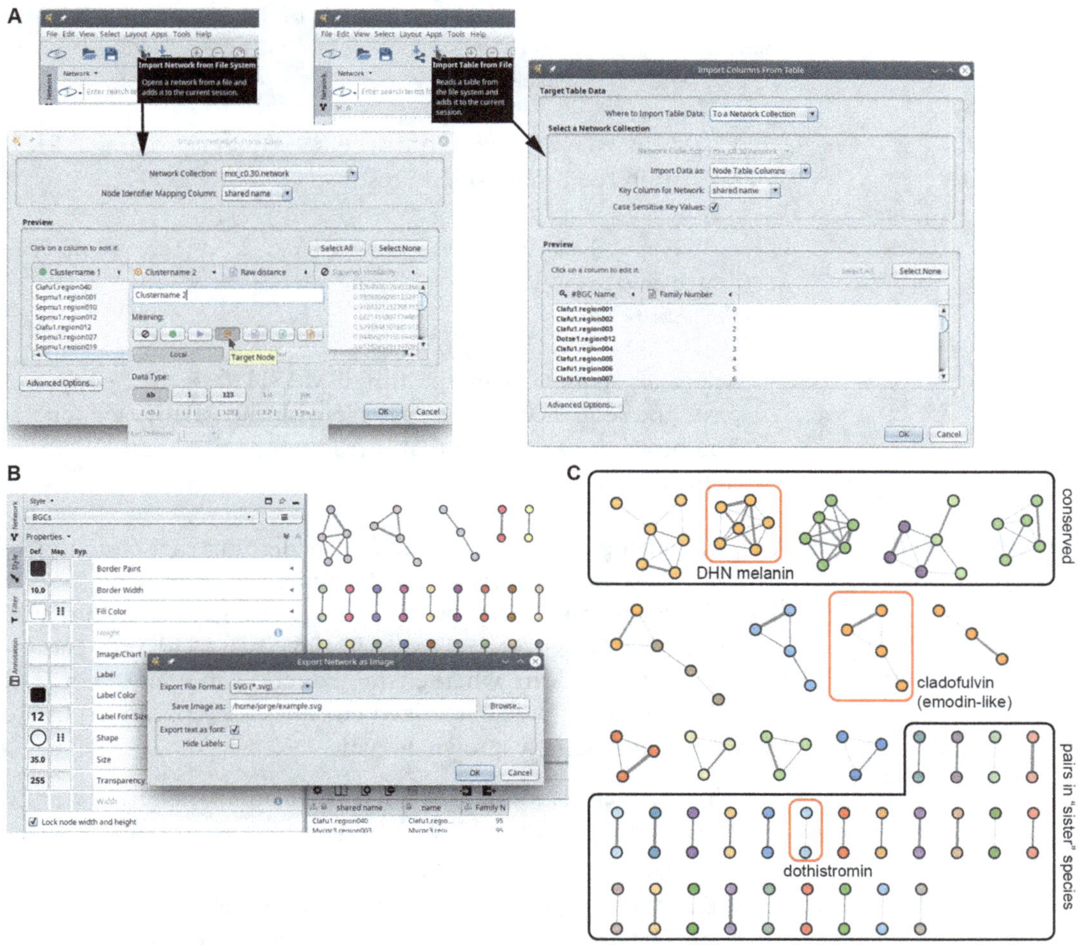

Fig. 4 Generation of figures using Cytoscape to present networks. (**a**) In Cytoscape, import a network file generated by BiG-SCAPE and import the respective annotation table. (**b**) After formatting the style of the networks, export the network representation to an image, which (**c**) you can annotate more accurately to make a figure for publication. The example shows annotated gene cluster families (singletons are not shown) in six Dothideomycetes species (BiG-SCAPE cutoff = 0.6; *see* Fig. 3). Known pathways are indicated according to the "KnownClusterBlast" results of fungiSMASH

(Table 1) which extracts sequences from specified conserved domains found in core biosynthetic proteins (*see* **Note 9**), and generates the corresponding fasta file. Below is an example method for retrieving the KS domain of polyketide synthase (PKS) sequences from the MIBiG repository and fungiSMASH results.

1. Download the zipped file with the latest version of MIBiG in GenBank format (https://mibig.secondarymetabolites.org/download) and decompress it.
2. Use the parser script (Table 1) to extract the KS sequence of fungal PKSs present in MIBiG:

```
python extract_sequences_core_enzymes.py -i [folder with MIBiG
GenBank files] -n [name of output fasta file] -t fungi -d KS --
include BGC -biosynthetic
```

3. Use the parser script to extract the KS sequence of fungal PKSs present in the fungiSMASH results:

```
python extract_sequences_core_enzymes.py -i [folder with fun-
giSMASH results] -n [name of output fasta file] -d KS -
biosynthetic
```

4. Select all sequences in the fungiSMASH fasta file, copy and paste the sequences in the fasta file with the MIBiG sequences. You now have a fasta file that is ready to be used for a phylogenetic analysis.

3.4 Phylogenetic Analysis of Core Enzymes

A typical phylogeny workflow consists in aligning the sequences of interest, removing poorly aligned sequences, removing poorly aligned regions, building a phylogenetic tree, and annotating/analyzing the tree. The obtained tree will allow you to identify monophyletic clades, indicate whether the sequences in your genome of interest are related to characterized enzymes (dereplication) or are found distant from any known enzyme.

1. Align sequences using MAFFT with default parameters: `mafft --reorder [fasta file] > [name output alignment file]` (*see* **Note 10**).
2. Open the output file in a multiple alignment viewer like AliView (Fig. 5a). If you notice that full sequences are not properly aligned or that sequences contain too large gaps, perform **Step 1** again without including those sequences which are too divergent.
3. Remove poorly aligned regions of the alignment using trimAl: `trimal -automated1 -in [alignment file] -out [name output trimmed alignment file]` (*see* **Note 11**).
4. Open the trimmed alignment in a multiple alignment viewer to assess the effect on the alignment and ascertain that the alignment is of good quality (Fig. 5a). Improving the quality of the alignment may require to perform **Step 3** again with more stringent parameters.
5. Build a phylogenetic tree using IQ-TREE and ModelFinder to find the optimal amino acid substitution model: `iqtree -s [trimmed alignment file] -m MFP -bb 1000 -alrt 1000 -T AUTO` (*see* **Note 12**).
6. The `.treefile` generated by IQ-TREE can be opened in software like iTOL, FigTree, or ETE3 toolkit (*see* **Note 13**). Root the tree using the midpoint method.

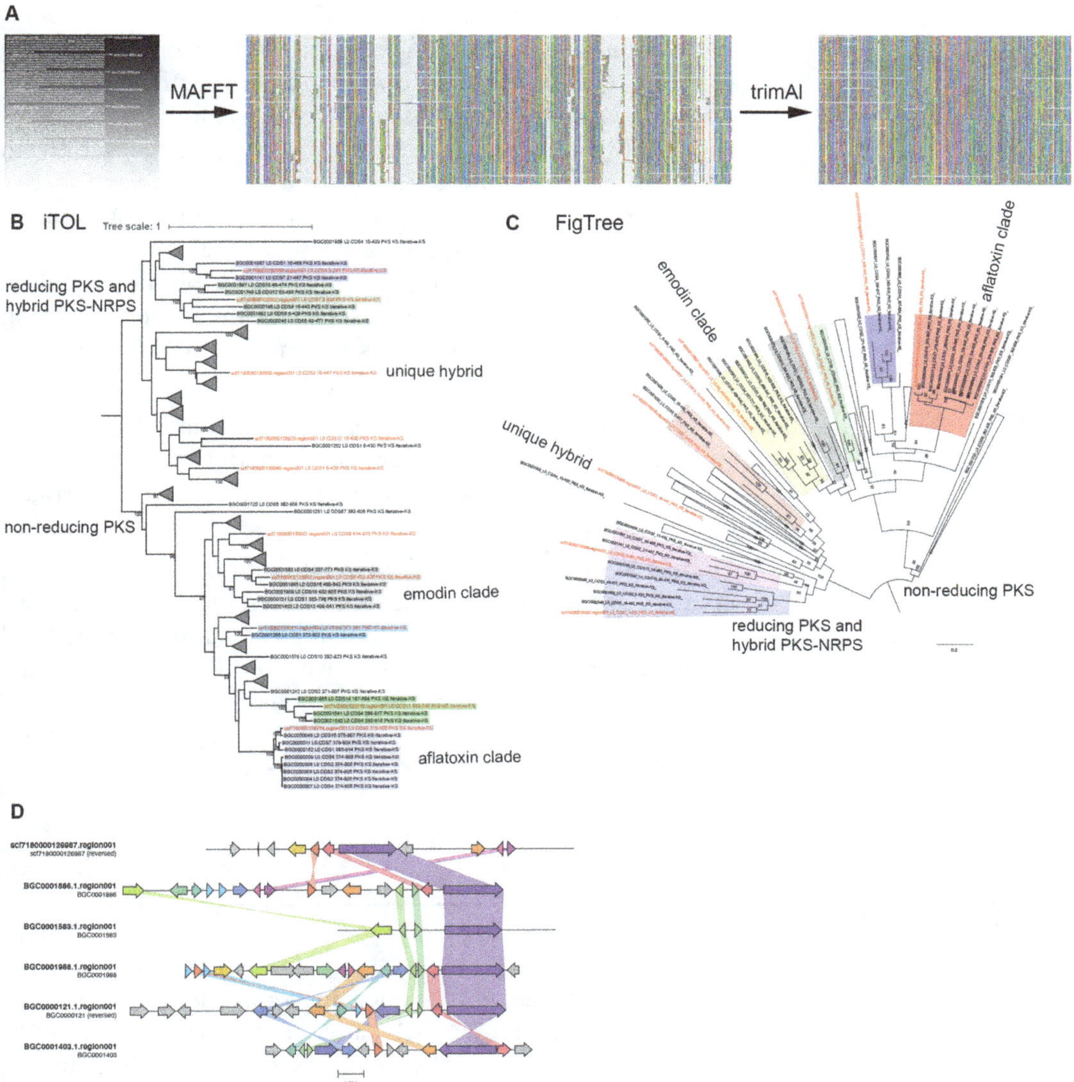

Fig. 5 Phylogenetic analysis and representation: example of polyketide synthases (PKSs) from the fungus *Cladosporium fulvum*. (**a**) KS domains from the fungiSMASH predictions and from the MIBiG databases were aligned using MAFFT. Poorly aligned regions were removed with trimAl, resulting in a good quality alignment. Alignments were inspected in AliView. A maximum-likelihood phylogenetic tree was built, midpoint-rooted, and examples of annotation with (**b**) iTOL or (**c**) FigTree are shown. These annotations highlight clades with characterized enzymes, but also sequences that appear unique. (**d**) One of the well-supported clades in this tree comprises pathways characterized to produce emodin-related anthraquinones (emodin clade). The biosynthetic gene cluster (BGC) in *C. fulvum* and related BGCs from MIBiG were visualized using Clinker. Similarity between genes is indicated and BGCs are ordered according to the phylogenetic tree. Rearrangements and differences between species are easily represented

7. Identify and highlight the clades with strong bootstrap support (typically over 95) which contain characterized core enzymes and annotate them (Fig. 5b, c). We assume that all core

enzymes within a given phylogenetic clade will produce a similar backbone [17]. List all predicted enzymes encoded in the genome of interest that belong to a clade together with characterized enzymes. This list predicts the kind of metabolites potentially produced, and thus allows genetic dereplication. In addition, if a compound is known to be produced by the species of interest, the most likely enzyme involved in its production can be deduced from the chemical structure. For example, if an emodin-derived anthraquinone is produced by the fungus of interest, a PKS enzyme that belongs to the clade containing other emodin-related synthases is a strong candidate for the production of this compound (Fig. 5b, c).

8. List all predicted enzymes unrelated to any characterized sequence. These enzymes are good candidates to prioritize for further studies as they may produce new chemical backbones.
9. To represent the loci of close homologues of a pathway of interest, use clinker [19]. Activate the conda clinker environment. The following command will create an interactive html file which you can open in your web browser to modify the layout and output SVG figures (Fig. 5d):

```
clinker -p [plot_name.html] [list of GenBank files]
```

4 Notes

1. Obtaining both files from sequencing data is not the purpose of this chapter and requires the use of dedicated software. It should be accurately performed beforehand. Alternatively, you may download such files from publicly available repositories like the National Center for Biotechnology Information (www.ncbi.nlm.nih.gov) and Joint Genome Institute MycoCosm [26]. For the former, the genome assembly (fasta) and gene prediction file (gff3), or the GenBank file containing both assembly and annotation, can be downloaded from the genome page of the species of interest. For the latter, after creating a free account, the assembly and annotation files can be obtained in the download section of each genome sequencing project, in the folders "Assembly/Assembled scaffolds (unmasked)/" (species_name.AssemblyScaffolds.fasta.gz file) and "Annotation/Filtered Models ("best")/Genes/" (choose preferentially the file in gff3 format). Both repositories offer methods to access their data, such as NCBI Datasets or JGI Globus service; read the documentation of each database for further information.

2. antiSMASH exists in three different flavors for bacterial, fungal (so-called fungiSMASH), and plant sequences. Although the differences between the fungal and bacterial antiSMASH websites seem mostly cosmetic, the bacterial version uses Prodigal as gene predictor for unannotated genomes. This gene predictor is not suitable for eukaryotic genomes and will most often create incorrect gene models which negatively impacts biosynthetic region discovery. When using the command-line version of antiSMASH, make sure to use the option [`--taxon fungi`].
3. At the time of writing, antiSMASH 6.0.0 works with biopython v1.78, but not with v1.79. This might change for future releases of antiSMASH. Verify the latest installation instructions in the official antiSMASH documentation website: https://docs.antismash.secondarymetabolites.org/install/
4. We recommend to perform a "strict" analysis as the "relaxed" or "loose" strictness tend to produce many false positives. However, if you are particularly interested in NRPS-like enzymes, it is recommended to perform a "relaxed" analysis and manually curate the results. As extra feature, we recommend to use "Cluster Pfam analysis" and "KnownClusterBlast" as these options provide useful information for enzymatic activities and dereplication, respectively. "SubClusterBlast" and "ClusterBlast" options search antiSMASH-associated databases which hold information for precursor-synthesizing subclusters and pre-calculated clusters, respectively, but currently hold no fungal data. The cassis region border detection algorithm [8] was trained using only a few available genomes at the time and is therefore not recommended.
5. If the "Cluster Pfam analysis" option was checked in the webserver (or the `--clusterhmmer` parameter was used in the command-line interface), the GenBank file corresponding to each biosynthetic region will also contain the predicted domains. This information can be used to differentiate non-reducing and reducing PKSs, for example, which contain SAT and PT domains or DH, ER, and KR domains, respectively [2].
6. Other optional parameters to consider (see full list of parameters in https://git.wageningenur.nl/medema-group/BiG-SCAPE/-/wikis/parameters): `--cores`: BiG-SCAPE uses, by default, all cores from the machine. `--mibig`: this option will include characterized BGCs from the MIBiG database (version 2.0). However, this option will not include fungal BGCs only, but all BGCs in the databases. In order to keep fungal BGCs only, we provide a simple auxiliary script (Table 1) to copy

filtered GenBank files. Use the following command in the antiSMASH conda environment:

```
python filtercopy.py -i [folder with MIBiG GenBank files] -o
[output folder] -t fungi
```

If you do this, put these fungal BGCs in your BiG-SCAPE input folder and don't use the `--mibig` option.

If you include other reference BGCs that are not included in MIBiG, they will need to be run in fungiSMASH to create the product annotations (e.g. 'TIPKS') that BiG-SCAPE uses to classify the input. Thus, when using them with BiG-SCAPE, run the following auxiliary script to process them with antiSMASH:

```
python antiSMASH_on_GenBank.py -i [folder with GenBank files]
-r [folder name for intermediate results] -o [folder name for
analyzed data]
```

7. The major types of core enzymes are polyketide synthase (PKS), non-ribosomal synthetase (NRPS and NRPS-like), terpene cyclase (TC), and dimethylallyl tryptophan synthase (DMATS). It should be noted that hybrid PKS-NRPS and NRPS-PKS enzymes exist. In this case, only the PKS or NRPS part of the enzyme should be included in the PKS and NRPS fasta files, respectively. In the case of PKSs, using the KS or AT domains provide enough information to determine evolutionary relationships [17, 27, 28]. In the case of NRPSs, these enzymes are multimodular, with a typical module consisting of three conserved domains: adenylation (A), thiolation (T), and condensation (C). Phylogeny of NRPS enzymes must be done with either the A or C domain [17, 28, 29]. If you are interested in other types of pathways that are not frequently encountered in fungal genomes, the same kind of workflow can be adapted. For easy analysis of phylogenetic trees, we recommend to use an informative name for each sequence, for example, by containing the species name, gene name, and known compound (separated by underscores; avoid special characters in the name of the sequences to prevent trimming of the names by the different tools).
8. For fungal sequences, MIBiG (https://mibig.secondarymetabolites.org/) is the major database that contains characterized enzymes. However, it is possible to use other repositories, including your own curated one. It must be noted that MIBiG still contains sequences that are only predictions and were not experimentally validated, as well as incorrect gene models.

9. Due to the modular structure of certain SM core enzymes like NRPSs, it is often better to consider using conserved domains to build phylogenetic trees. The KS (PF00109) and AT (PF00698) domains of type I PKSs have been used to build phylogenetic trees of these enzymes. The A (PF00501) and C (PF00668) domains are commonly used to build phylogenetic trees of NRPSs (only A domain for NRPS-like enzymes), while a special KS (PF00195) domain has been used to build trees of type III PKSs. The script we provide can extract KS, A, and C subsequences of core enzymes as annotated by fungiSMASH, which allows building phylogenetic trees for PKS and NRPS (-like) enzymes. Note that the models corresponding to those domains are custom and not necessarily taken from the Pfam database. For other types of core enzymes like terpene cyclases, which are more diverse, complete protein sequences could be used.
10. The `--reorder` parameter is to output the sequences according to their similarity. It allows the quick identification of divergent sequences. It is possible to modify the parameters, but our experience shows that default parameters provide good enough results. Other multiple sequence alignment software like MUSCLE [30] or CLUSTAL-OMEGA [31] can be used.
11. Instead of using `-automated1`, you can modify the parameters to trim the alignment with a different stringency. In our experience, default parameters provide good enough stringency to remove poorly aligned regions while keeping informative positions. Other trimming software such as Gblocks [32] or ClipKit [33] can be used.
12. Search for the best substitution model with ModelFinder [34] (`-m` parameter) can take a long time, especially when the alignment contains many sequences and many positions. Alternatively, you can use `-mset LG` because in our experience, the LG model is most often the best fit model, and this option only lets ModelFinder determining the best rate heterogeneity type for this substitution model. For a quick analysis, FastTree [35] could be used instead of IQ-TREE to build a phylogenetic tree. The FastTree phylogenetic tree will be less accurate, but good enough to obtain quickly an idea of the final tree. In addition to ultrafast bootstrap (`-bb`) [36], we recommend to perform an SH-like approximate likelihood ratio test (`-alrt`) [37] because these methods give slightly different results. As indicated in the IQ-TREE manual, clades are considered reliable when the SH-aLRT value is over 80% and ultrafast bootstrap over 95%. The `-T` parameter assigns automatically the number of CPUs to use.

13. The online resource iTOL is quite convenient for interactive visualization and annotation of phylogenetic trees, but annotations of the tree cannot be saved or exported without subscription. Examples of useful free alternatives, yet less user-friendly, are FigTree (http://tree.bio.ed.ac.uk/software/figtree/) and, for more advanced users, ETE3 toolkit [38]. For the latter, installation in its own conda environment can be done with the following command: `conda create -n ete -c etetoolkit ete3 ete_toolchain`. Then, a useful command line is: `ete3 mod -t [tree file] | ete3 view --mode c`, which will allow you to edit the tree in the ETE tree viewer interface.

References

1. Hyde KD, Xu J, Rapior S et al (2019) The amazing potential of fungi: 50 ways we can exploit fungi industrially. Fungal Divers 97:1–136
2. Mosunova O, Navarro-Muñoz JC, Collemare J (2020) The biosynthesis of fungal secondary metabolites: from fundamentals to biotechnological applications. In: Reference module in life sciences. Elsevier, Amsterdam
3. Keller NP, Hohn TM (1997) Metabolic pathway gene clusters in filamentous fungi. Fungal Genet Biol 21:17–29
4. Greco C, Keller NP, Rokas A (2019) Unearthing fungal chemodiversity and prospects for drug discovery. Curr Opin Microbiol 51:22–29
5. Medema MH, Blin K, Cimermancic P et al (2011) antiSMASH: rapid identification, annotation and analysis of secondary metabolite biosynthesis gene clusters in bacterial and fungal genome sequences. Nucleic Acids Res 39: W339–W346
6. Blin K, Shaw S, Steinke K et al (2019) AntiSMASH 5.0: updates to the secondary metabolite genome mining pipeline. Nucleic Acids Res 47:W81–W87
7. Khaldi N, Seifuddin FT, Turner G et al (2010) SMURF: genomic mapping of fungal secondary metabolite clusters. Fungal Genet Biol 47: 736–741
8. Wolf T, Shelest V, Nath N et al (2016) CASSIS and SMIPS: promoter-based prediction of secondary metabolite gene clusters in eukaryotic genomes. Bioinformatics 32:1138–1143
9. Umemura M, Koike H, Nagano N et al (2013) MIDDAS-M: motif-independent *de novo* detection of secondary metabolite gene clusters through the integration of genome sequencing and transcriptome data. PLoS One 8:e84028
10. Vesth TC, Brandl J, Andersen MR (2016) FunGeneClusterS: predicting fungal gene clusters from genome and transcriptome data. Synth Syst Biotechnol 1:122–129
11. Takeda I, Umemura M, Koike H et al (2014) Motif-independent prediction of a secondary metabolism gene cluster using comparative genomics: application to sequenced genomes of Aspergillus and ten other filamentous fungal species. DNA Res 21:447–457
12. Almeida H, Palys S, Tsang A et al (2020) TOUCAN: a framework for fungal biosynthetic gene cluster discovery. NAR Genom Bioinform 2:1–11
13. Blin K, Shaw S, Kautsar SA et al (2021) The antiSMASH database version 3: increased taxonomic coverage and new query features for modular enzymes. Nucleic Acids Res 49: D639–D643
14. Kautsar SA, Blin K, Shaw S et al (2019) MIBiG 2.0: a repository for biosynthetic gene clusters of known function. Nucleic Acids Res 48: D454–D458
15. Kautsar SA, Blin K, Shaw S et al (2021) BiG-FAM: the biosynthetic gene cluster families database. Nucleic Acids Res 49: D490–D497
16. Weber T, Blin K, Duddela S et al (2015) antiSMASH 3.0--a comprehensive resource for the genome mining of biosynthetic gene clusters. Nucleic Acids Res 43:1–7
17. Adamek M, Alanjary M, Ziemert N (2019) Applied evolution: phylogeny-based approaches in natural products research. Nat Prod Rep 36:1295–1312
18. Navarro-Muñoz JC, Selem-Mojica N, Mullowney MW et al (2020) A computational framework to explore large-scale biosynthetic diversity. Nat Chem Biol 16:60–68

19. Gilchrist CLM, Chooi Y-H (2021) Clinker & clustermap.js: automatic generation of gene cluster comparison figures. Bioinformatics btab007
20. Shannon P, Markiel A, Ozier O et al (2003) Cytoscape: a software environment for integrated models of biomolecular interaction networks. Genome Res 13:2498–2504
21. Katoh K, Standley DM (2013) MAFFT multiple sequence alignment software version 7: improvements in performance and usability. Mol Biol Evol 30:772–780
22. Capella-Gutierrez S, Silla-Martinez JM, Gabaldon T (2009) trimAl: a tool for automated alignment trimming in large-scale phylogenetic analyses. Bioinformatics 25:1972–1973
23. Minh BQ, Schmidt HA, Chernomor O et al (2020) IQ-TREE 2: new models and efficient methods for phylogenetic inference in the genomic era. Mol Biol Evol 37:1530–1534
24. Larsson A (2014) AliView: a fast and lightweight alignment viewer and editor for large datasets. Bioinformatics 30:3276–3278
25. Letunic I, Bork P (2021) Interactive Tree Of Life (iTOL) v5: an online tool for phylogenetic tree display and annotation. Nucleic Acids Res 49:W293–W296
26. Grigoriev IV, Nikitin R, Haridas S et al (2014) MycoCosm portal: gearing up for 1000 fungal genomes. Nucleic Acids Res 42:D699–D704
27. Kroken S, Glass NL, Taylor JW et al (2003) Phylogenomic analysis of type I polyketide synthase genes in pathogenic and saprobic ascomycetes. Proc Natl Acad Sci U S A 100: 15670–15675
28. Gallo A, Ferrara M, Perrone G (2013) Phylogenetic study of polyketide synthases and nonribosomal peptide synthetases involved in the biosynthesis of mycotoxins. Toxins (Basel) 5: 717–742
29. Bushley KE, Turgeon BG (2010) Phylogenomics reveals subfamilies of fungal nonribosomal peptide synthetases and their evolutionary relationships. BMC Evol Biol 10: 26
30. Edgar RC (2004) MUSCLE: multiple sequence alignment with high accuracy and high throughput. Nucleic Acids Res 32:1792–1797
31. Sievers F, Higgins DG (2018) Clustal omega for making accurate alignments of many protein sequences. Protein Sci 27:135–145
32. Talavera G, Castresana J (2007) Improvement of phylogenies after removing divergent and ambiguously aligned blocks from protein sequence alignments. Syst Biol 56:564–577
33. Steenwyk JL, Buida TJ, Li Y et al (2020) ClipKIT: a multiple sequence alignment trimming software for accurate phylogenomic inference. PLoS Biol 18:e3001007
34. Kalyaanamoorthy S, Minh BQ, Wong TKF et al (2017) ModelFinder: fast model selection for accurate phylogenetic estimates. Nat Methods 14:587–589
35. Price MN, Dehal PS, Arkin AP (2009) FastTree: computing large minimum evolution trees with profiles instead of a distance matrix. Mol Biol Evol 26:1641–1650
36. Minh BQ, Nguyen MAT, von Haeseler A (2013) Ultrafast approximation for phylogenetic bootstrap. Mol Biol Evol 30:1188–1195
37. Guindon S, Dufayard J-F, Lefort V et al (2010) New algorithms and methods to estimate maximum-likelihood phylogenies: assessing the performance of PhyML 3.0. Syst Biol 59: 307–321
38. Huerta-Cepas J, Serra F, Bork P (2016) ETE 3: reconstruction, analysis, and visualization of phylogenomic data. Mol Biol Evol 33:1635–1638

Chapter 2

Investigating Fungal Biosynthetic Pathways Using Heterologous Gene Expression: *Aspergillus oryzae* as a Heterologous Host

Kate M. J. de Mattos-Shipley, Colin M. Lazarus, and Katherine Williams

Abstract

A suite of molecular techniques have been developed in recent decades, which allow gene clusters coding for the biosynthesis of fungal natural products to be investigated and characterized in great detail. Many of these involve the manipulation of the native producer, for example, to increase yields of natural products or investigate the biosynthetic pathway through gene disruptions. However, an alternative and powerful means of investigating biosynthetic pathways, which does not rely on a cooperative native host, is the refactoring and heterologous expression of pathways in a suitable host strain. This protocol aims to walk the reader through the various steps required for the heterologous expression of a fungal biosynthetic gene cluster, specifically using *Aspergillus oryzae* strain NSAR1 and the pTYGS series of expression vectors. Briefly, this process involves the design and construction of up to four multigene expression vectors using yeast recombination, PEG-mediation transformation of *A. oryzae* protoplasts, and chemical extraction of the resulting transformants to screen for the presence of metabolites.

Key words *Aspergillus oryzae*, Fungal natural products, Secondary metabolism, Heterologous expression, Yeast recombination, Biosynthetic gene clusters

1 Introduction

Fungi and the secondary metabolites they produce have long been known for their bioactive properties, with the use of medicinal mushrooms dating back thousands of years [1]. The discovery of penicillin in the 1920s revolutionized modern medicine and has been followed by the discovery and development of various compounds of fungal origin, such as the immune-modulator cyclosporin [2], the antibiotic pleuromutilin [3], the antifungals griseofulvin [4] and strobilurin [5], and the cholesterol-lowering statins [6]. Fungal natural product research has previously been hampered by the relative difficulty in culturing and manipulating many fungal species, especially when compared to their bacterial

Elizabeth Skellam (ed.), *Engineering Natural Product Biosynthesis: Methods and Protocols*,
Methods in Molecular Biology, vol. 2489, https://doi.org/10.1007/978-1-0716-2273-5_2,

counterparts, as well as often vanishingly low titers of natural compounds produced under laboratory conditions. However, with the development of modern molecular techniques and the explosion of genome sequencing in the twenty-first century, fungal metabolites are accessible in a way unimaginable only decades ago. One powerful tool for the discovery and exploitation of fungal natural products is the heterologous production of compounds via the refactoring and expression of biosynthetic pathway genes in a suitable host strain.

One such host strain is the filamentous fungus *Aspergillus oryzae*. *A. oryzae* is a domesticated relative of *Aspergillus flavus*, with a long history of use in food and drink production. It has GRAS ("Generally Recognized As Safe") status and produces relatively few natural products under laboratory conditions, making it an ideal cellular factory for investigations into natural products from other species. In addition, specific strains of *A. oryzae* with corresponding molecular toolkits have been developed for just that purpose. NSAR1 is a quadruply auxotrophic (*niaD*$^-$ *sC*$^-$ Δ*argB* *adeA*$^-$) *A. oryzae* strain, which means that four separate transformation constructs can be selected for simultaneously [7]. The development of four corresponding multigene expression vectors, each of which contains four separate cassettes [8], means that 16 genes can easily be co-expressed.

This technology has already been harnessed to contribute significantly to the field of natural product research. Many complex biosynthetic pathways have been elucidated through the expression of different combinations of pathway genes, allowing the role of each encoded protein and the order of catalytic events to be determined [9–14]. Heterologous expression in *A. oryzae* has also been used to generate chemical novelty through either pathway engineering [15] or biotransformations, whereby synthetic or semi-synthetic compounds are fed into the biosynthetic pathway to produce analogues. This approach was taken by Alberti et al., who generated novel analogues of the antibiotic pleuromutilin, some of which demonstrate enhanced biological activity [12].

The following protocols aim to walk the reader through the various stages of heterologous expression (Fig. 1), including plasmid design and construction via yeast homologous recombination, transformation of *A. oryzae* with expression cassettes, and chemical analysis of the resulting strains. The protocols have been written specifically with *A. oryzae* NSAR1 and the pTYGS series of plasmids in mind, but many aspects of the protocols, as well as the overall principles, could be applied to different heterologous hosts and vector systems.

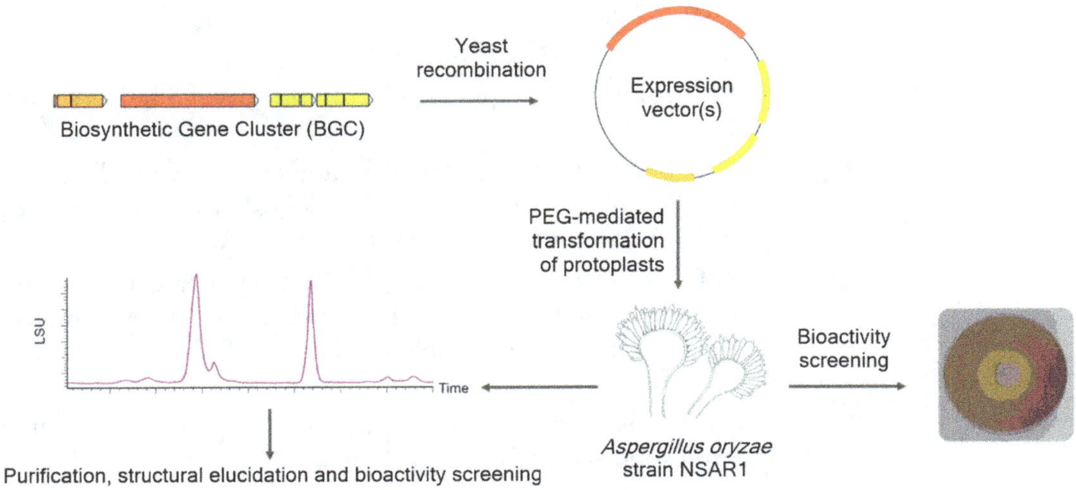

Fig. 1 An overview of the process of heterologous expression of fungal biosynthetic gene clusters in *A. oryzae*. Transformants themselves or purified compounds can be screened for desired bioactivity

2 Materials

2.1 General Laboratory Equipment

1. Benchtop microcentrifuge.
2. 1.5 ml microcentrifuge tubes.
3. 50 ml conical centrifuge tubes.
4. Freestanding centrifuge with rotor for 50 ml conical tubes.
5. Static incubator.
6. Shaking incubator.
7. Heating block or water bath for *S. cerevisiae* and *E. coli* transformations.
8. Petri dishes.
9. 250 ml conical flasks.
10. PCR thermocycler.
11. Spectrophotometer.
12. Hemocytometer.
13. Microscope.
14. Gel electrophoresis equipment.

2.2 Plasmid Design and Construction via Yeast Recombination

1. High-fidelity polymerase.
2. A uracil-auxotrophic strain of *S. cerevisiae* strain (*see* **Note 1**).
3. Restriction enzymes *Sgs*I and *Not*I.
4. YPAD: 1% (w/v) yeast extract, 2% (w/v) peptone, 2% (w/v) glucose, 0.04% (w/v) adenine sulfate, plus 1.5% (w/v) agar for solid medium.

5. Sterile H_2O.
6. 1 M lithium acetate (LiOAc) and 0.1 M LiOAc.
7. PEG solution: 50% (w/v) polyethylene glycol 3350 (*see* **Note 2**).
8. SS-DNA: 2 mg/ml denatured salmon testes DNA in TE (Tris–EDTA buffer solution: 10 mM Tris–HCl, 1 mM Na_2 EDTA, pH 8.0). *See* **Note 3** for details on SS-DNA.
9. SM-URA plates: 0.17% (w/v) yeast nitrogen base, 0.5% (w/v) ammonium sulfate, 2% (w/v) glucose, 0.077% (w/v) complete supplement mixture minus uracil (Q-biogene), 1.5% (w/v) agar.
10. pTYGS plasmid series, or other appropriate yeast-*E. coli* shuttle expression vector for use in *A. oryzae*.

2.3 Plasmid Rescue in E. coli and Screening for Correct Construction

1. Zymoprep Yeast Plasmid Miniprep Kit or similar.
2. Competent *E. coli*.
3. Suitable restriction enzymes for restriction digest analysis.
4. Polymerase.

2.4 A. oryzae Transformation

1. Malt Extract Agar (MEA).
2. GN medium: 1% (w/v) glucose, 2% (w/v) nutrient broth no. 2.
3. Sterile H_2O.
4. 0.8 M NaCl.
5. Protoplasting solution: 20 mg/ml Trichoderma lysing enzyme (Sigma) and 5 mg/ml driselase in 0.8 M NaCl—filter sterilized.
6. Solution 1: 0.8 M NaCl, 10 mM $CaCl_2$, 50 mM Tris–HCl pH 7.5.
7. Solution 2: 0.8 M NaCl, 10 mM $CaCl_2$, 50 mM Tris–HCl pH 7.5, 60% PEG 3350.
8. CZD/S top medium: 3.5% (w/v) Czapek Dox broth, 1 M sorbitol, 0.8% (w/v) agar.
9. CZD/S base medium: 3.5% (w/v) Czapek Dox broth, 1 M sorbitol, 1.5% (w/v) agar.
10. Supplements as 100× stocks:
    - (a) 1% Methionine.
    - (b) 0.5% Adenine.
    - (c) 20% Ammonium sulfate.
    - (d) 2% Arginine.

2.5 Chemical Analysis

2.5.1 Small-Scale Extraction from Plates

1. CMPA medium: 3.5% (w/v) Czapek Dox broth, 2% (w/v) maltose, 1% (w/v) peptone, 1.5% (w/v) agar.
2. Glass vials (approx. 7 ml).
3. Solvent mix: ethyl acetate, dichloromethane, and methanol (3: 2:1) with 1% acetic acid.
4. Ultrasonic bath.
5. HPLC-grade acetonitrile.

2.5.2 Liquid Cultures for Large-Scale Fermentation and Purification

1. Malt Extract Agar (MEA).
2. CMP medium: 3.5% (w/v) Czapek Dox broth, 2% (w/v) maltose, 1% (w/v) peptone.
3. Glass beaker.
4. Hand blender.
5. Buchner funnel and a side-arm flask for vacuum filtration.
6. Ethyl acetate.
7. HCl.
8. Separating funnel.
9. Anhydrous $MgSO_4$.
10. Whatmann filter paper.
11. Glass funnel.
12. Round bottom flasks and rotary evaporator.
13. HPLC-grade acetonitrile.

3 Methods

3.1 Plasmid Design and Construction via Yeast Recombination

The pTYGS series of vectors (Fig. 2) are multigene expression vectors for use in *Aspergillus oryzae* NSAR1; members of the series differ only in the selectable marker for NSAR1 transformation (either *argB*, *adeA*, *niaD*, or *sC*). Each vector contains an *amyB* expression cassette, which is strongly inducible in the presence of starch or maltose as the carbon source, as well as a further three cassettes which support strong constitutive expression in *A. oryzae* (*gpdA*, *eno*, and *adh*). The *amyB* cassette contains a Gateway® destination fragment, which means that this cassette can be used to express large genes first assembled in a Gateway entry vector, then inserted by site-specific recombination in vitro [16]. The presence of the Gateway destination fragment, which includes the *ccd*B gene (encoding the toxic protein ccdB), means that these plasmids need to be propagated in a ccdB-resistant *E. coli* strain, and subsequent recombinant plasmid propagation in a susceptible strain selects against the non-recombinant vector.

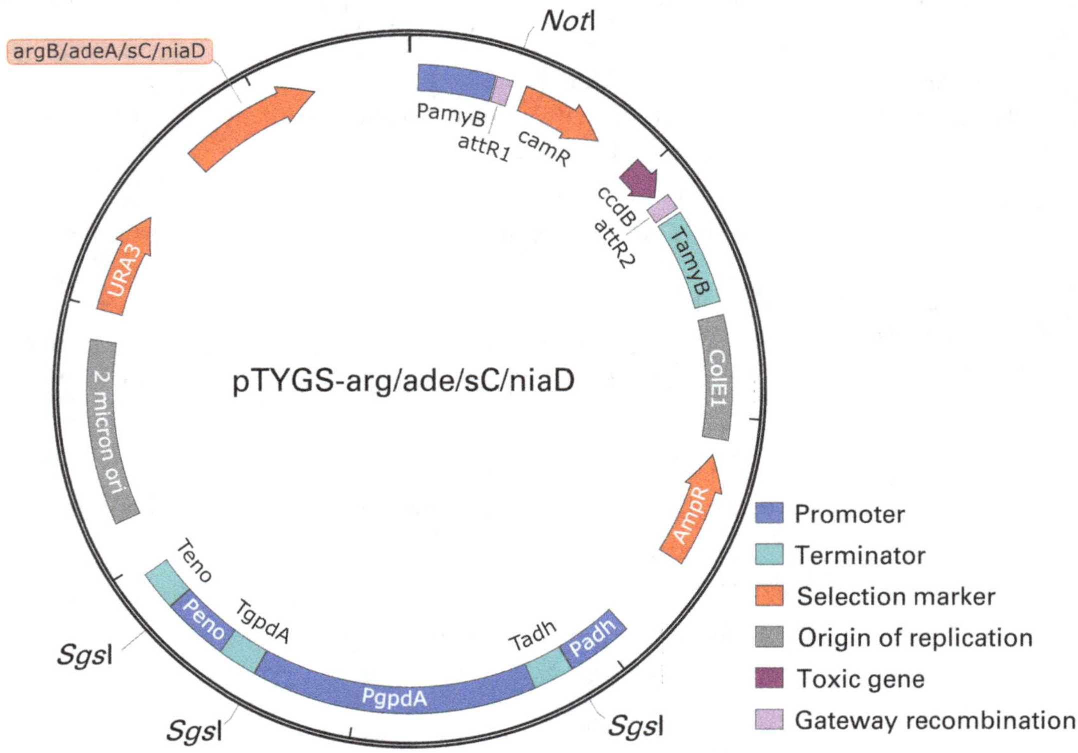

Fig. 2 Map of the pTYGS series of vectors. Each version of the vector contains a different complementation marker (*argB/adeA/sC/niaD*) for use with *A. oryzae* NSAR1

In addition to *ColE1* and *ampR* for replication and selection in *E. coli*, these vectors contain the *S. cerevisiae* 2 μm origin of replication and *URA3* gene, which allows for the construction of the expression plasmids in yeast. This process exploits the natural propensity of *S. cerevisiae* to perform homologous recombination, harnessing it to precisely join multiple DNA fragments without the need for unique restriction sites and ligation. For efficient recombination to occur, a minimum of 30 bp homology is recommended [17]. Figure 3 outlines the yeast recombination strategies which can be used to insert genes of your choice into the pTYGS vector series.

3.1.1 Plasmid and Primer Design

1. Analyze your biosynthetic gene cluster (BGC) to determine which set of genes you wish to express in *A. oryzae* (*see* **Note 5**).
2. Determine if you require more than one plasmid. Up to four genes can be expressed from a single plasmid (*see* **Note 6**).
3. Design your primers to amplify your gene(s) (*see* **Note 7**) with the following compound tails, dependent on which cassette you wish to place your gene(s) into. "N_{20}" represents your gene-specific primer sequence of approximately 20 nt:

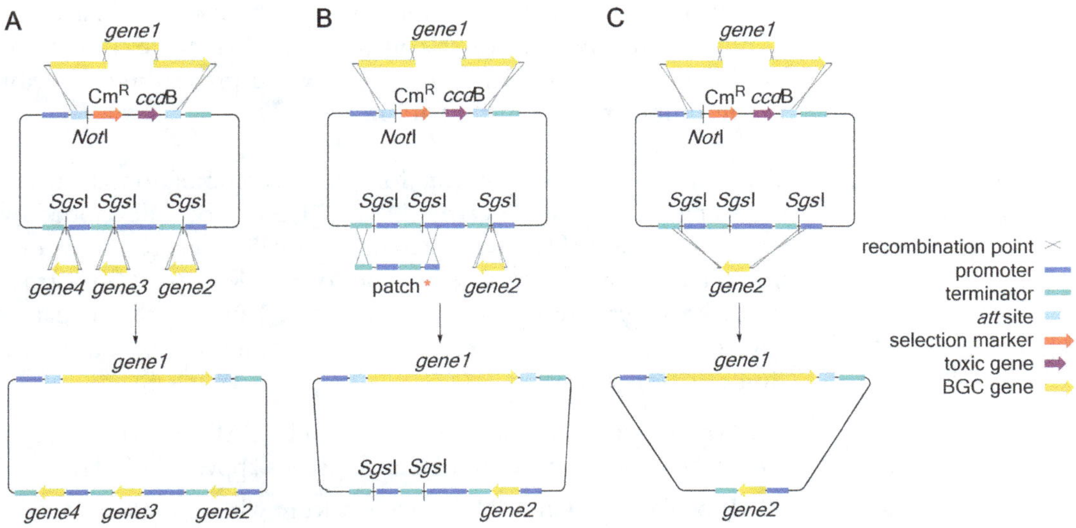

Fig. 3 Different plasmid designs using the pTYGS series, depending on the number of genes to be expressed. (**a**) shows the insertion of genes into all four expression cassettes. (**b**) shows the insertion of only two genes, with an amplified portion of the plasmid* acting as a "patch" to repair two of the cut *Sgs*I sites. To see primer sequences for amplifying a suitable patch *see* **Note 4**. This approach can be used if it is desirable to retain the two empty cassettes, which can then be used in subsequent experiments. (**c**) depicts a design where two genes are inserted, one of which is placed between the P_{adh} promoter and the T_{eno} terminator, thus removing the other cassettes. This example would require using the P_{adh} cassette compound forward primer in combination with the T_{eno} cassette compound reverse primer to amplify your gene (*see* Subheading 3.1 for primer sequences)

P_{amyB} cassette.

Forward primer: 5′-tctgaacaataaaccccacagcaagctccgN$_{20}$-3′.

Reverse primer: 5′-ctctccacccttcacgagctactacagatcN$_{20}$-3′.

P_{adh} cassette:

Forward primer: 5′- tctttcaacacaagatcccaaagtcaaaggN$_{20}$-3′.

Reverse primer: 5′- cattctatgcgttatgaacatgttccctggN$_{20}$-3′.

P_{gpdA} cassette.

Forward primer: 5′- cagctaccccgcttgagcagacatcaccggN$_{20}$-3′.

Reverse primer: 5′- gacaatgtccatatcatcaatcatgaccggN$_{20}$-3′.

P_{eno} cassette.

Forward primer: 5′- cgactgaccaattccgcagctcgtcaaaggN$_{20}$-3′.

Reverse primer: 5′- ttggctggtagacgtcatataatcatacggN$_{20}$-3′.

4. If your gene is too large to be amplified effectively as one PCR product *see* **Note 8**.
5. Amplify your genes using a high-fidelity polymerase according to the manufacturer's instructions. If possible, amplify your genes from cDNA so that the introns are removed (*see* **Note 9**).

6. Prepare your plasmid of choice by restriction digestion according to the manufacturer's instructions (*see* **Note 10**). Heat inactivate the restriction enzyme before proceeding to yeast recombination/transformation.

3.1.2 Yeast Recombination

Transformation of *S. cerevisiae* for yeast recombination is carried out using the LiOAc/SS carrier DNA/PEG protocol developed by Giezt and Woods [18]. Competent yeast cells prepared from one 50 ml culture will be sufficient to perform at least ten transformations. Each yeast recombination should be paired with a negative control where digested vector, but no PCR inserts, is added to a shot of competent yeast cells.

1. Streak out your yeast strain on solid YPAD and incubate at 30 °C (*see* **Note 11**) until clear colonies appear (2–3 days).
2. Inoculate a 10 ml YPAD starter culture with a single colony of *S. cerevisiae* and grow overnight at 30 °C with shaking at 200 rpm.
3. Add the overnight starter culture to 40 ml of YPAD in a 250 ml Erlenmeyer (conical) flask and incubate at 30 °C with shaking at 200 rpm for 4–5 h, or until the culture reaches at least 2×10^7 cells ml^{-1} (*see* **Note 12**).
4. Centrifuge the culture at 3000 × *g* for 5 min and discard the supernatant.
5. Wash the cells with 25 ml sterile H_2O and repeat the above centrifugation, then resuspend the pellet in 1 ml 0.1 M LiOAc and transfer to a 1.5 ml microcentrifuge tube.
6. Pellet the cells at 16,000 × *g* for 15 s and discard the supernatant, then resuspend the cells in 400 μl 0.1 M LiOAc (*see* **Note 13**).
7. For each transformation to be performed, transfer 50 μl of the suspension to a new 1.5 ml microfuge tube and pellet again at 16,000 × *g* for 15 s, then discard supernatant.
8. Add 240 μl of PEG solution, followed by 36 μl *1 M* LiOAc, 50 μl denatured SS-DNA (*see* **Note 3**) and up to 34 μl of DNA to the pelleted cells (*see* **Note 14**). Approximately 50–200 ng of each DNA fragment should be added (*see* **Note 15**), with linear DNA fragments to be joined containing at least 30 bp overlap.
9. Resuspend cells in the transformation mixture by vigorous vortexing, then incubate at 30 °C for 30 min and then 42 °C for 30 min.
10. Pellet the cells at *3000* × *g* for 15 s then *gently* resuspend in 1 ml of sterile H_2O (*see* **Note 16**).
11. Spread 200 μl aliquots (*see* **Note 17**) on SM-URA plates and incubate at 30 °C for 3–4 days until colonies appear.

3.1.3 Plasmid Screening and Rescue in E. coli

1. Extract plasmid directly (*see* **Note 18**) from the plate of *S. cerevisiae* transformants using a suitable method, such as the Zymoprep Yeast Plasmid Miniprep Kit (*see* **Note 19**).
2. Use 2 μl of the resulting plasmid mix transform (*see* **Note 20**) a suitable *E. coli* strain, such as DH5α or One Shot® ccdB Survival™ (*see* **Note 21**).
3. If the ratio of yeast colonies between your negative control plate and your yeast recombination plate is high (>1:10) we would recommend growing 3–4 colonies for plasmid extraction and restriction digest analysis to confirm correct construction of your plasmid.
4. If the ratio is lower, or the above method has failed to identify a correct transformant, colony PCR can be used to identify a suitable colony for subsequent plasmid preparation. *See* **Note 22** for details of colony PCR.

3.2 A. oryzae Transformation

1. Inoculate an MEA plate with *A. oryzae* NSAR1 and incubate at 28 °C until the plate is covered in sporulating mycelium (approximately 1–2 weeks).
2. Harvest conidia from a single petri-dish of *A. oryzae* NSAR1 (*see* **Note 23**) and use to inoculate 50 ml of GN medium; incubate overnight at 28 °C with shaking at 200 rpm.
3. Harvest the germinated conidia by centrifugation at 8000 × *g* for 10 min then discard supernatant. Wash the pellet once with sterile H_2O and once with 0.8 M NaCl.
4. Resuspend in 10 ml of filter-sterilized protoplasting solution and incubate at 25 °C with gentle shaking for 1.5–2 h (*see* **Notes 24** and **25**).
5. Release the protoplasts from hyphae by pipetting with a wide-bore 5 ml pipette, and filter through sterile miracloth. Centrifuge the protoplasts at 1000–3000 × *g* for 5 min and discard supernatant, then wash the pellet with solution 1 (*see* **Note 26**).
6. Resuspend pellet in 200–500 μl of solution 1 (*see* **Note 27**) depending on the number of transformations to be performed (*see* **Note 28**). For each transformation, transfer 100 μl of protoplasts to a conical 50 ml centrifuge tube on ice. Add 2–5 μg (10 μl max) of plasmid DNA to the protoplasts and mix gently.
7. Incubate the transformation mix on ice for 2 min, then add 1 ml of solution 2 and incubate at room temperature for 20 min.
8. Add 20–40 ml of molten (50 °C) CZD/S top agar containing the appropriate supplements; mix gently, but thoroughly, and pour as an overlay onto four prepared plates of CZD/S base

agar (5–10 ml per plate), again containing the appropriate supplements.

9. Incubate plates at 28 °C for 3–5 days until colonies appear.
10. To purify the resulting transformants, grow on selective media until sporulation occurs, then streak out and transfer a single spore-derived colony onto fresh selective media.

3.3 Chemical Analysis

Analysis of the resulting *A. oryzae* transformants will depend on the types of metabolites expected. Growth conditions must support the expression of the *PamyB* cassette by including maltose or starch as the carbon source, and multiple transformants must be analyzed, as not all transformants are likely to contain all cassettes due to potential fragmentation of the constructs upon integration. Transformants can be directly assayed for any desired bioactivity (e.g. antibiotic assay), and chemical extractions and subsequent purification and structural elucidation may be desirable to characterize any metabolites being produced. Some standard protocols we employ include the following:

3.3.1 Small-Scale Extraction from Plates

Extractions from agar plates are based on the protocol reported by Smedsgaard [19].

1. Grow *A. oryzae* transformants, along with a parental *A. oryzae* NSAR1 control, on CMPA plates for 1 week at 28 °C.
2. Transfer 1 cm^2 plugs of agar from growing colonies to a glass vial and add 3 ml of a solvent mix containing ethyl acetate, dichloromethane, and methanol (3:2:1) with 1% acetic acid.
3. After 1 h in an ultrasonic bath, transfer the liquid to a clean pre-weighed glass vial and dry under a nitrogen stream.
4. Once fully dry, the crude extract can be dissolved in acetonitrile or other appropriate solvent to a concentration of 10 mg/ml and analyzed by TLC / HPLC-MS/UV/ELSD.

3.3.2 Liquid Cultures for Large-Scale Fermentation and Purification

Fermentation in CMP (Fig. 4) is routinely used to scale up production of metabolites for subsequent purification and structural elucidation.

1. Inoculate liquid CMP with spores and mycelium from a mature plate of *A. oryzae* grown on MEA and incubate for 1 week at 28 °C with shaking at 200 rpm.
2. Transfer culture to a glass beaker and blend mycelium.
3. Add an equal volume of ethyl acetate and acidify with HCl, mix and leave for >1 h at room temperate.
4. Vacuum filter to separate the mycelial debris from the liquid.

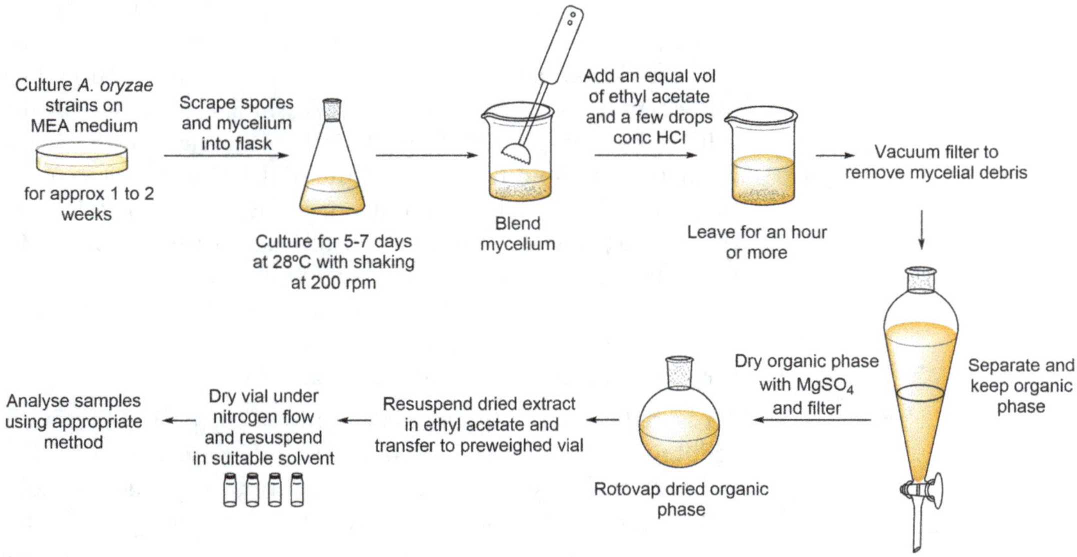

Fig. 4 An overview of the chemical extraction method typically used to screen *A. oryzae* transformants for the production of heterologously produced natural products

5. Use a separating funnel to separate the aqueous and organic layers. Discard the aqueous phase but keep the organic phase (top layer).
6. Add $MgSO_4$ to the organic phase to dry (*see* **Note 29**) and pass through filter paper placed in a glass funnel, into a round-bottomed flask.
7. Concentrate under vacuum using a rotary evaporator and resuspend in HPLC-grade acetonitrile or other appropriate solvent for subsequent analysis, i.e. TLC / HPLC-MS/UV/ ELSD.

4 Notes

1. We routinely use *S. cerevisiae* strains YPH499 or YPH500.
2. To prepare 100 ml of 50% (w/v) PEG solution, first add 50 g of PEG 3350 to approximately 30 ml of deionized H_2O then stir or incubate at 50 °C until fully dissolved. Top up to 100 ml with H_2O and autoclave. PEG must be stored in a securely capped bottle to prevent evaporation as this severely impacts transformation efficiency.
3. To prepare salmon sperm DNA (SS-DNA) for yeast recombination, dissolve 200 mg of high molecular weight salmon testes DNA (e.g. Sigma-Aldrich catalogue number D1626) in 100 ml of sterile TE using a magnetic stirrer at 4 °C. This may take some hours. Pipetting the solution gently with a wide-bore

pipette can speed up dissolution. Aliquots of SS-DNA should be stored at −20 °C. Prior to use, the carrier DNA must be denatured by boiling in a water bath for 5 min, then chilled immediately on ice before use. A similar process can be achieved by programming a PCR thermocycler to 98 °C for 5 min prior to holding at 4 °C. Unused denatured SS-DNA can be returned to the freezer and can be used immediately after thawing, as long as it has not been allowed to warm above 4 °C, without the need to denature again. It can go through 3–4 freeze-thaw cycles before further denaturing is required. If repeated denaturing is required, carrier DNA can be boiled three or four times without significant loss of activity.

4. To amplify "patch" fragments use any of the pTYGS plasmids (diluted ~1:100 and linearized) as a template and a high-fidelity polymerase. To repair just the *eno* cassette, use primers PenoF: 5′-tgctagtctccttccaacac-3′ and TenoR: 5′-gcctaatcgaatcgtcatc-3′ (product = 313 bp). To repair both the *gpdA* and *eno* cassettes (as shown in Fig. 3) use PgpdAF: 5-′-catccatactccatccttcc-3′ and TenoR (product = 1089 bp).
5. (a) Generally we find it best to express the core synthase in the cluster under the control of P_{amyB} as this promoter can be induced by the addition of malto-oligosaccharides [20].

   (b) We attempt to predict the order of biosynthesis using bioinformatics, to allow us to add the "early" genes to one plasmid and "later" genes to further plasmids.
6. We use arginine prototrophy for selection preferentially, as it gives the least background growth. Adenine prototrophy is also reliable. Methionine prototrophy can be less reliable, and it is important to check that there are no sources of methionine in your recovery medium; for example, we have found that some batches of agar are impure and will not support selection for methionine.
7. Design your forward primer-binding sequence to begin with the start codon of your gene, or if this results in a poor primer sequence, slightly upstream of your start codon. Ensure that no other ATG codons can initiate translation for an incorrect reading frame. Your reverse primer can be designed to bind to the final 20 bp of your gene or can be placed slightly downstream of your stop codon. Ensure that you reverse complement your sequence, before adding the compound tail (which has already been reverse complemented).
8. If your gene is too large to be amplified as a single fragment, primers can be designed to amplify two or more overlapping fragments which will be recombined back together in yeast as long as there is at least 30 bp of overlapping homologous sequence. Alternatively, for very large genes, the megasynthase

can be recombined from multiple PCR products into the yeast assembly vector pE-YA [21], which is a Gateway® entry vector, then transferred to the pTYGS expression vector by LR recombination in vitro. This approach is particularly beneficial if you wish to express the megasynthase from multiple different expression cassettes.

9. It is not guaranteed that *A. oryzae* will correctly splice heterologously expressed genes, especially if the genes are from a distantly related fungal species (such as a basidiomycete species). As such, it can be beneficial to express only the coding region, without introns. If the desired genes are not expressed in the native host, or it is not possible to acquire cDNA, it is possible to remove an intron through amplification and recombination of individual exons or cDNA sequences can be ordered as synthetic genes.
10. In most cases, digestion using both *Not*I and *Sgs*I will be necessary, unless you wish to leave the *amyB* cassette empty (*Not*I digest not required), or all three *Sgs*I cassettes (*adh*, *gpdA*, *eno*) empty (*Sgs*I not required).
11. Although 30 °C is the standard temperature for culturing *S. cerevisiae,* we routinely incubate at 28 °C for convenience (also applies to **steps 2**, **3**, and **11**). This does not appear to hinder the yeast recombination in any way, although growth rates may be slightly slower.
12. For strain YPH499, we find a 4-h incubation reliably results in an efficient transformation, but if a different yeast strain is being used, or transformation efficiencies are low, quantifying the yeast cells at this point may be worthwhile. This can be done using a hemocytometer or spectrophotometer. An OD600 of approximately 1.5 should correspond with a cell concentration of 2×10^7 cells ml^{-1}, but as OD600 can vary between different yeast strains and spectrophotometers, we would recommend performing an initial calibration by comparing the OD600 with cell counts using a hemocytometer.
13. Cells at this stage can be stored on ice for up to 4 weeks prior to use, although a reduction in efficiency may be observed. Alternatively, cells can be prepared for freezing and long-term storage, according to the protocol published by Gietz et al. [22].
14. Adding the PEG solution first provides a cushion to protect the yeast cells from damage by the concentrated (1 M) LiOAc; addition of other components can then be made in any order.
15. Approximately stoichiometric amounts of each fragment are required (i.e., roughly equal numbers of each fragment included in the recombination). DNA concentration can be quantified by nanodrop or estimated using agarose gel electrophoresis.

16. It is very important to be gentle with the yeast cells at this stage. The centrifugation must be performed at $\leq 3000 \times g$ and the pellet should be resuspended by gentle pipetting, not vortexing. The plating should also be done gently with efficient spreading using a sterile spreader.
17. 200 μl is normally sufficient for a large number of yeast transformants. Plating density negatively affects transformation efficiency, so it may be important to optimize the volume plated to maximize efficiency. If 200 μl is initially plated, the remaining transformation mix can be kept at 4 °C and plated out at a later date (up to 1 week).
18. We do not screen individual colonies for correct recombination at this stage; instead we scrape a mass of *S. cerevisiae* colonies from the transformation plate using a sterile toothpick or similar and resuspend the cells directly in the digestion buffer + zymolyase mix from the Zymoprep kit. Plasmid screening is done on single *E. coli* colonies following transformation with the combined yeast plasmid miniprep.
19. We favor use of the Zymoprep I kit, involving isopropanol precipitation, over the column-based Zymoprep II version for yeast plasmid recovery. This is because plasmid copy numbers in yeast cells are very low; column purification yields purer DNA but at the expense of the yield required for efficient *E. coli* transformation.
20. We find transformation of *E. coli* using electroporation to be more efficient than the heat shock method, especially as the recombinant plasmid can be large and at low concentration [23].
21. If a gene has been inserted into the *amyB* cassette, replacing the *ccdB* counter-selection gene, a standard laboratory *E. coli* strain such as DH5a should be used to propagate your plasmid. If the *ccdB* gene is still present in your plasmid design, a ccdB-resistant strain must be used.
22. Colony PCR can be used to check larger numbers of *E. coli* colonies for correct integration of your genes. For smaller genes ≤4 kb in the *adh, gpdA,* and *eno* cassettes we use vector-specific primers that will amplify over your gene from the promoter to the terminator (PadhF: 5-′-gcgtctcaccatcagcttac-3′ and TadhR: 5-′-gatgaatgctaagcacgagg-3′; PgpdAF: 5′-catccatactccatccttcc-3′ and TgpdAR: 5′-aagagagaagactcgactggg-3′; PenoF: 5′--tgctagtctccttccaacac-3′ and TenoR: 5′-gcctaatcgaatcgtcatc-3′). These will also amplify a product of ~300 bp if the gene has not been inserted, controlling for incorrect PCR set-up. For genes ≥4 kb and those in the *amyB* cassette we use one vector-specific primer (PamyBF1: 5′-ggtattgtcctgcagaatgc-3′) and one gene-

specific primer. Colonies that are positive for all recombination events can then be miniprepped for restriction analysis for final confirmation.

23. We find the easiest way to do this is to pour sterile GN from your conical flask into a petri-dish of sporulating *A. oryzae*, scrape the plate with a sterile inoculating loop to release the spores, then pour the GN and spores back into the conical flask.
24. Protoplast formation can be checked microscopically during the enzymatic cell wall digestion, although it can be difficult to identify protoplasts at this stage as they will be very dilute and can look similar to ungerminated conidia. Protoplasts should appear as perfectly round cells. We do not find it is necessary to monitor protoplast formation during the digestion, but post-filtration and centrifugation we do compare two 5 μl samples microscopically: one diluted with 5 μl H_2O and one with 5 μl Solution 1. The protoplasts in the osmotically unbalanced sample diluted with H_2O will lyse.
25. We have experienced supply issues with Trichoderma Lysing enzyme. If it cannot be obtained, we have successfully substituted Yatalase (Takara Bio) using the protocol from Tagami et al. [24].
26. Once centrifuged, immediately and gently tip the centrifuge tube upside down, then release the lid over a waste container to ensure the pellet is not disturbed.
27. Add solution 1 immediately and resuspend by gentle pipetting. If left, the pellet can become difficult to resuspend. The protoplasts can be left on ice for short periods.
28. As standard practice we simply resuspend the protoplasts in a large enough volume to conduct the desired number of transformations (up to a maximum of 500 μl). Note that it is important to perform a negative control where sterile H_2O is used instead of plasmid DNA, as this will provide an estimation of any background growth. While getting acquainted with the protocol, or if transformation efficiencies are low, the protoplasts can be quantified using a hemocytometer to check that you are getting sufficient numbers. Protoplasts should be at a concentration of approximately $1–5 \times 10^7$ ml^{-1}.
29. Magnesium sulfate ($MgSO_4$) is used as a drying agent to remove any water from your organic phase. Add $MgSO_4$, a small spatula at a time, and swirl or stir. The $MgSO_4$ should form larger clumps as water is absorbed, and the solution may also become clearer as small droplets of water are removed. If the solution is dry, then there still should be some fine drying agent visible. If not, add a little more and repeat the process. It is then important to remove the drying agent by filtering through filter paper.

Acknowledgments

This work was supported by funding from the Medical Research Council (MR/N029909/1).

References

1. Bishop KS, Kao CHJ, Xu Y, Glucina MP, Paterson RRM, Ferguson LR (2015) From 2000 years of *Ganoderma lucidum* to recent developments in nutraceuticals. Phytochemistry 114:56–65
2. Twentyman PR (1992) Cyclosporins as drug resistance modifiers. Biochem Pharmacol 43(1):109–117
3. Knauseder F, Brandl E (1976) Pleuromutilins. Fermentation, structure and biosynthesis. J Antibiot 29(2):125–131
4. Finkelstein E, Amichai B, Grunwald MH (1996) Griseofulvin and its uses. Int J Antimicrob Agents 6(4):189–194
5. Anke T, Oberwinkler F, Steglich W, Schramm G (1977) The strobilurins - new antifungal antibiotics from the basidiomycete *Strobilurus tenacellus*. J Antibiot 30(10):806–810
6. Manzoni M, Rollini M (2002) Biosynthesis and biotechnological production of statins by filamentous fungi and application of these cholesterol-lowering drugs. Appl Microbiol Biotechnol 58(5):555–564
7. Jin FH, Maruyama J-I, Juvvadi PR, Arioka M, Kitamoto K (2004) Development of a novel quadruple auxotrophic host transformation system by *argB* gene disruption using *adeA* gene and exploiting adenine auxotrophy in *Aspergillus oryzae*. FEMS Microbiol Lett 239(1):79–85
8. Lazarus CM, Williams K, Bailey AM (2014) Reconstructing fungal natural product biosynthetic pathways. Nat Prod Rep 31(10): 1339–1347
9. Heneghan MN, Yakasai AA, Halo LM, Song Z, Bailey AM, Simpson TJ, Cox RJ, Lazarus CM (2010) First heterologous reconstruction of a complete functional fungal biosynthetic multigene cluster. ChemBioChem 11(11): 1508–1512
10. Lebe KE, Cox RJ (2019) Oxidative steps during the biosynthesis of squalestatin S1. Chem Sci 10(4):1227–1231
11. Williams K, Szwalbe AJ, Mulholland NP, Vincent JL, Bailey AM, Willis CL, Simpson TJ, Cox RJ (2016) Heterologous production of fungal maleidrides reveals the cryptic cyclization involved in their biosynthesis. Angew Chem Int Ed 55(23):6784–6788
12. Alberti F, Khairudin K, Venegas ER, Davies JA, Hayes PM, Willis CL, Bailey AM, Foster GD (2017) Heterologous expression reveals the biosynthesis of the antibiotic pleuromutilin and generates bioactive semi-synthetic derivatives. Nat Commun 8(1):1831
13. Nofiani R, de Mattos-Shipley K, Lebe KE, Han LC, Iqbal Z, Bailey AM, Willis CL, Simpson TJ, Cox RJ (2018) Strobilurin biosynthesis in Basidiomycete fungi. Nat Commun 9:3940
14. de Mattos-Shipley KM, Greco C, Heard DM, Hough G, Mulholland NP, Vincent JL, Micklefield J, Simpson TJ, Willis CL, Cox RJ (2018) The cycloaspeptides: uncovering a new model for methylated nonribosomal peptide biosynthesis. Chem Sci 9(17):4109–4117
15. Fisch KM, Bakeer W, Yakasai AA, Song Z, Pedrick J, Wasil Z, Bailey AM, Lazarus CM, Simpson TJ, Cox RJ (2011) Rational domain swaps decipher programming in fungal highly reducing polyketide synthases and resurrect an extinct metabolite. J Am Chem Soc 133(41): 16635–16641
16. Katzen F (2007) Gateway® recombinational cloning: a biological operating system. Expert Opin Drug Discovery 2(4):571–589
17. Hua S-b, Qiu M, Chan E, Zhu L, Luo Y (1997) Minimum length of sequence homology required for *in vivo* cloning by homologous recombination in yeast. Plasmid 38(2): 91–96
18. Gietz DR, Woods RA (2002) Transformation of yeast by lithium acetate/single-stranded carrier DNA/polyethylene glycol method. Methods Enzymol 350:87–96
19. Smedsgaard J (1997) Micro-scale extraction procedure for standardized screening of fungal metabolite production in cultures. J Chromatogr A 760(2):264–270
20. Tada S, Gomi K, Kitamoto K, Takahashi K, Tamura G, Hara S (1991) Construction of a fusion gene comprising the Taka-amylase A promoter and the *Escherichia coli* β-glucuronidase gene and analysis of its expression in *Aspergillus oryzae*. Mol Gen Genet 229(2):301–306

21. Pahirulzaman KAK, Williams K, Lazarus CM (2012) A toolkit for heterologous expression of metabolic pathways in *Aspergillus oryzae*. Methods Enzymol 517:241–260
22. Gietz RD, Schiestl RH (2007) Frozen competent yeast cells that can be transformed with high efficiency using the LiAc/SS carrier DNA/PEG method. Nat Protoc 2(1):1–4
23. Dower WJ, Miller JF, Ragsdale CW (1988) High efficiency transformation of *E. coli* by high voltage electroporation. Nucleic Acids Res 16(13):6127–6145
24. Tagami K, Liu CW, Minami A, Noike M, Isaka T, Fueki S, Shichijo Y, Toshima H, Gomi K, Dairi T, Oikawa H (2013) Reconstitution of biosynthetic machinery for indolediterpene paxilline in *Aspergillus oryzae*. J Am Chem Soc 135(4):1260–1263

Chapter 3

Investigating Fungal Biosynthetic Pathways Using Heterologous Gene Expression: *Aspergillus nidulans* as a Heterologous Host

Danielle A. Yee and Yi Tang

Abstract

Fungal natural products encompass an important source of pharmaceutically relevant molecules. Heterologous expression of biosynthetic pathways in chassis strains enables the discovery of new secondary metabolites and characterization of pathway enzymes. In our laboratory, biosynthetic genes in a clustered pathway have been refactored in engineered heterologous hosts such as *Aspergillus nidulans*. Here we describe the assembly of heterologous expression vectors, transformation into *A. nidulans*, and detection of new compounds in the transformant strains.

Key words *Aspergillus nidulans*, Heterologous expression

1 Introduction

Secondary metabolites from fungi and their derivatives comprise a broad class of natural products with rich structural diversity and powerful biological activity [1]. Recent genome sequencing efforts have revealed many natural product gene clusters in fungi are not expressed under standard laboratory culturing conditions [2]. To access silent gene clusters, pathway genes can be reconstituted in well-established heterologous hosts such as *Aspergillus nidulans* and *Aspergillus oryzae* [3–8]. Use of these platforms has allowed for the discovery of countless natural products and characterization of their biosynthetic pathways [9, 10]. Here we describe an episomal heterologous expression system in *A. nidulans* A1145, which can accommodate up to 12 genes on three plasmids [11]. For a cleaner background during compound detection, the modified strain *A. nidulans* A1145ΔSTΔEM can be used, in which endogenous production of sterigmatocystin [12] and emericellamide [13] has been abolished [14]. The general workflow for this system is as follows: first, heterologous expression plasmids for

Elizabeth Skellam (ed.), *Engineering Natural Product Biosynthesis: Methods and Protocols*,
Methods in Molecular Biology, vol. 2489, https://doi.org/10.1007/978-1-0716-2273-5_3,

A. nidulans are assembled through yeast homologous recombination. Following this, the plasmids are transformed into *A. nidulans* protoplasts. Individual transformant colonies are then assayed for compound production. RT-PCR (reverse transcription-polymerase chain reaction) can be performed to verify genes are successfully expressed in the transformant strains.

2 Materials

2.1 Strains

1. *A. nidulans* A1145 [11] (genotype: *pyrG89; pyroA4; nkuA::argB; riboB2*) or *A. nidulans* A1145ΔSTΔEM [14].
2. *S. cerevisiae* JHY651 [15] or other yeast strain auxotrophic for uracil.
3. *E. coli* TOP10 (Invitrogen), DH10b (Thermo Scientific), etc.
4. Original fungal host strain with target gene cluster (if genomic DNA needs to be acquired).

2.2 Plasmids

1. pYTR (*see* **Note 1**).
2. pYTP (*see* **Note 2**).
3. pYTU (*see* **Note 3**).

2.3 Reagents for Cloning

Many of the commercial kits are optional. If following other protocols, additional materials are required.

1. Restriction enzymes PacI, NotI, SwaI, BamHI, PshAI.
2. Agarose gel for DNA electrophoresis.
3. Gel DNA Recovery Kit.
4. *Quick*-DNA Fungal/Bacterial Microprep Kit (Zymo Inc. USA).
5. Primers to amplify genes and promoters.
6. Proof-reading DNA polymerase.
7. Solid YPEG medium: 5 g bacto-peptone, 2.5 g yeast extract, 7.5 mL glycerol, 5 g agar, 250 mL ddH_2O. Autoclave and let solution cool to around 55 °C. Add 10 mL of 100% ethanol. Mix thoroughly and pour plates.
8. Liquid YPD medium: 5 g bacto-peptone, 2.5 g yeast extract, 5 g dextrose, 250 mL ddH_2O. Autoclave and store at room temperature.
9. Frozen-EZ Yeast Transformation II Kit (Zymo Inc. USA).
10. Solid SDC_{AA}(-U) medium: 5 g dextrose, 5 g agar, 200 mL ddH_2O. Autoclave. Prepare supplements: 1.25 g casamino acids, 1.7 g yeast nitrogen base without amino acids, 10 mg adenine, 10 mg tryptophan, 50 mL ddH_2O. Filter sterilize.

Add supplements to melted agar, mix thoroughly, and pour plates.

11. Zymoprep™ Yeast Plasmid Miniprep I (Zymo Inc. USA).
12. Isopropanol.
13. 1000× carbenicillin (50 mg/mL): Dissolve 0.5 g of carbenicillin in 10 mL ddH_2O. Filter sterilize. Prepare 1 mL aliquots and store at −20 °C.
14. 1000× ampicillin (100 mg/mL): Dissolve 1 g of ampicillin in 10 mL ddH_2O. Filter sterilize. Prepare 1 mL aliquots and store at −20 °C.
15. Solid LB medium: 6.25 g LB powder, 5 g agar, 250 mL ddH_2O. Autoclave. To prepare LB plates with carbenicillin, let melted agar cool to around 55 °C, then add 250 μL of 1000× carbenicillin. Mix thoroughly and pour plates.
16. Liquid LB medium: 6.25 g LB powder, 250 mL ddH_2O. Autoclave and store at room temperature. Aliquot and add ampicillin as needed.
17. Plasmid Miniprep Kit.
18. Yeast RNA Isolation Kit
19. SuperScript III First-Strand Synthesis System (Invitrogen).
20. Oligo-dT primer.
21. GoTaq Master Mix (Promega).

2.4 Media and Solutions for *A. nidulans* Transformation and Culturing

1. 100× uracil (500 mM): 2.24 g, 40 mL of ddH_2O. Add 10 N NaOH until fully dissolved. Filter sterilize.
2. 100× uridine (1 M): 9.76 g, 40 mL of ddH_2O. Filter sterilize.
3. 1000× pyridoxine (0.5 mg/mL): 20 mg, 40 mL of ddH_2O.-Filter sterilize.
4. 1000× riboflavin (0.125 mg/mL): 5 mg, 40 mL of ddH_2O.-Filter sterilize.
5. 20× nitrate salts: 120 g $NaNO_3$, 10.4 g KCl, 10.4 g $MgSO_4{\cdot}7H_2O$, 30.4 g KH_2PO_4. Add ddH_2O. up to 1 L, stir until dissolved, and store at room temperature.
6. Trace elements : 2.20 g $ZnSO_4{\cdot}7H_2O$, 1.10 g H_3BO_3, 0.50 g $MnCl_2{\cdot}4H_2O$, 0.16 g $FeSO_4{\cdot}7H_2O$, 0.16 g $CoCl_2{\cdot}5H_2O$, 0.16 g $CuSO_4{\cdot}5H_2O$, 0.11 g $(NH_4)_6Mo_7O_{24}{\cdot}4H_2O$. Add each to 80 mL of ddH_2O in the order shown. Add more ddH_2O to bring the total volume to 100 mL. Adjust pH to 6.5 with 1 N KOH.
7. Solid CD-sorbitol medium (1 L): 10 g glucose, 50 mL 20× nitrate salts, 1 mL trace elements, 218.6 g sorbitol (1.2 M), 20 g agar. Autoclave then pour plates as needed. Add the appropriate supplements to individual empty petri dishes

before pouring in the agar media. If not making plates immediately, solid media can be stored at room temperature and re-melted in the microwave.

8. Solid CD medium (1 L): Prepare as described above for solid CD-sorbitol medium, omitting the sorbitol.
9. Liquid CD medium (1 L): 10 g glucose, 50 mL 20× nitrate salts, 1 mL trace elements. Autoclave. Aliquot as needed and add the appropriate supplements.
10. Liquid CD-ST medium (1 L): 20 g starch, 20 g peptone (acidic digest) or casamino acids (acidic digest), 50 mL 20× nitrate salts, 1 mL trace elements. Add starch to 100 mL of ddH_2O and mix with a stir bar for 10 min. Add boiling ddH_2O to bring the total volume to 950 mL. Add the remaining ingredients. Continue to mix until everything is dissolved. Autoclave. Aliquot as needed and add the appropriate supplements.
11. Solid CD-ST medium (1 L): Prepare as described above for liquid CD-ST medium. Add 20 g of agar after everything is dissolved. Autoclave then prepare plates with the appropriate supplements.
12. Osmotic medium (500 mL): 147.9 g $MgSO_4$ (1.2 M), 10 mM sodium phosphate buffer (NaPB) (can be made from a 2 M NaPB stock: 90.9 g Na_2HPO_4 and 163.4 g NaH_2PO_4 or 187.9 g NaH_2PO_4 per liter, pH 6.5). Adjust pH to 5.8 with 1 M Na_2HPO_4 (about 25 mL). Filter sterilize and store at 4 °C. Tip: start with 450 mL of water, then the total volume will end up to be about 500 mL.
13. Trapping Buffer (1 L): 109.3 g sorbitol (0.6 M), 0.1 M Tris–HCI, pH 7.0 (can be made using 100 mL 1 M Tris). Autoclave and store at 4 °C.
14. STC buffer (1 L): 218.6 g sorbitol (1.2 M), 1.47 g $CaCl_2$ (10 mM), 10 mM Tris–HCI, pH 7.5 (can be made using 10 mL 1 M Tris). Autoclave and store at 4 °C.
15. PEG solution (100 mL): 60% PEG 4000 (BDH), 50 mM $CaCl_2$, 50 mM Tris–HCI, pH 7.5 (can be made using 50 mL 1 M Tris). Autoclave and store at room temperature.
16. Lysing enzyme, yatalase, 0.2 μm syringe filters.

2.5 Other Materials for *A. nidulans* Transformation, Culturing, and Analysis

1. Two sterile 125 mL flasks.
2. Sterile 30 mL Corex tube or sterile 50 mL Falcon tube.
3. Sterile cell strainer (Fisher, Cat No. 22363547, optional).
4. Solvents, e.g. ethyl acetate, acetone, methanol.

3 Methods

3.1 Cloning of Plasmids for A. nidulans Expression

The plasmids for heterologous expression in *A. nidulans* A1145 are denoted as pYTU, pYTP, and pYTR [5]. Their features include auxotrophic markers for uracil (*pyrG*), pyridoxine (*pyroA*), and riboflavin (*riboB*), respectively, as well as the AMA1 origin of replication for *Aspergillus* (Fig. 1). For cloning purposes, these vectors also contain the uracil auxotrophic marker (*URA3*) and 2-μm origin for *S. cerevisiae*, and the ampicillin resistance marker (*ampR*) and ColE1 origin for *E. coli*. pYTU contains the starch inducible promoter P_{glaA}, pYTP contains the starch inducible promoter P_{amyB}, and pYTR contains the constitutive promoter P_{gpdA} from *Aspergillus niger*. Additional commonly used promoters include the constitutive P_{gpdA} promoters from *Penicillium oxalicum* (P_{POgpdA}) and *Penicillium expansum* (P_{PEgpdA}), as well as the constitutive P_{coxA} promoter from *A. niger*. Typically, one to four genes expressed under different promoters can be inserted into each plasmid.

1. If using the original promoters included in the pYT vectors, digest pYTU with the restriction enzymes PacI/NotI and pYTP and pYTR with PacI/SwaI or BamHI/SwaI following the manufacturer's instructions. If using different promoters, digest pYTU with PshAI/NotI, or PshAI/PacI, and pYTP and pYTR with NotI/PacI, NotI/BamHI, or NotI/SwaI (*see* **Note 4**). Run digestion reactions on an agarose gel and recover DNA using a commercial kit such as Zymoclean™ Gel DNA Recovery Kit.
2. To obtain template DNA for amplification of target genes, culture the original fungal host on solid or liquid media such as potato dextrose agar or broth. Isolate the strain's genomic DNA using a commercial kit such as *Quick*-DNA Fungal/Bacterial Microprep Kit, phenol-chloroform extraction [16], or microwave total genomic DNA extraction [17].
3. Amplify the genes of interest and their native terminators (300–500 bp downstream from the stop codon) by PCR using the genomic DNA of the original host as the template. A high-fidelity polymerase such as Q5® High-Fidelity DNA Polymerase (NEB), AccuPrime Pfx DNA Polymerase (Invitrogen), or Phusion (NEB) should be used. If replacing the original promoters or if more than one gene is to be inserted into each plasmid, amplify the promoters to be used by PCR. Use overhang primers to introduce 25–40 bp of homology between adjacent fragments. Run PCR products on an agarose gel and recover DNA using a commercial kit such as Zymoclean™ Gel DNA Recovery Kit.

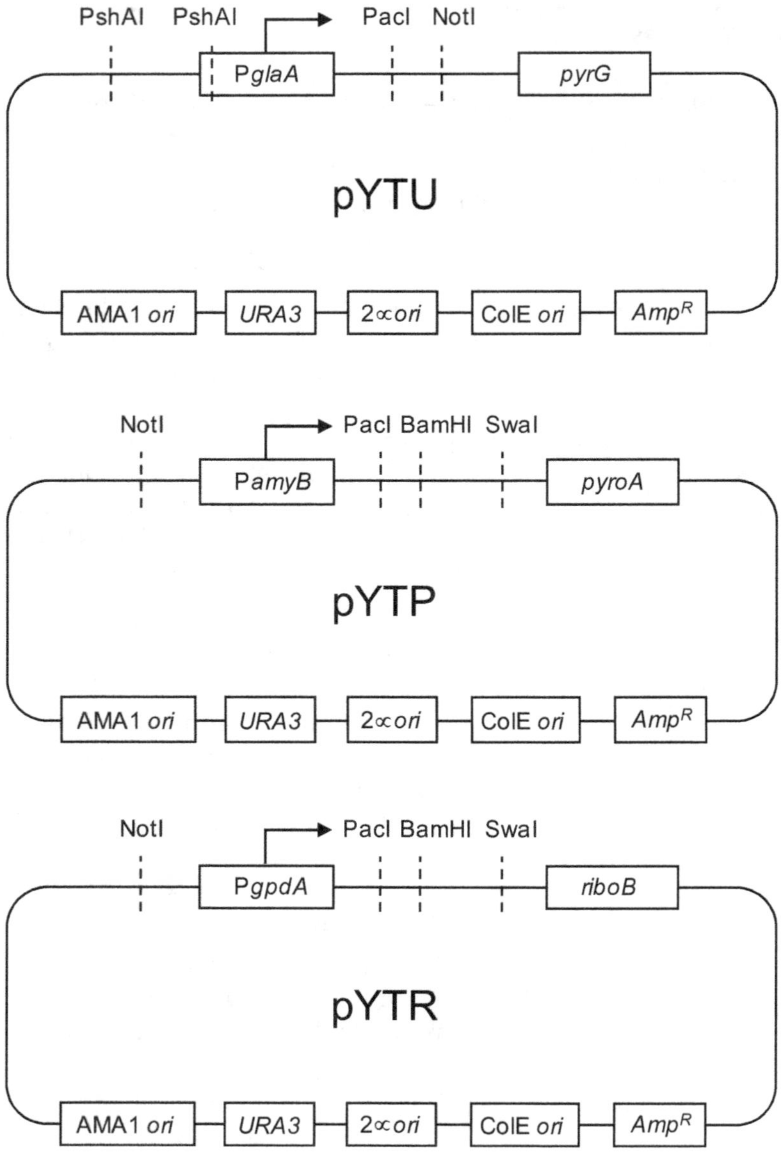

Fig. 1 Vectors for heterologous expression in *A. nidulans*: pYTU, pYTP, and pYTR

4. Co-transform the overlapping DNA fragments and their corresponding digested vectors into *S. cerevisiae* JHY651 [3] or other yeast strain auxotrophic for uracil to assemble the expression plasmids *in vivo* by yeast homologous recombination. Before preparing competent cells, yeast strains should be grown on YPEG agar plates and cultured in YPD liquid media. Yeast competent cells can be prepared using the PEG-lithium acetate method [18] or a commercial kit such as Frozen-EZ Yeast Transformation II Kit (Zymo Inc. USA). Transformants

should be plated on SDC_{AA}(-U) agar plates. Yeast colonies will appear after 1–3 days of incubation at 28 °C.

5. Extract the assembled plasmids from yeast using a commercial kit such as Zymoprep™ Yeast Plasmid Miniprep I (Zymo Inc. USA), and transform into *E. coli* TOP10, DH10b, etc. by electroporation to isolate single clones. Plate cells on LB agar plates with 50 μg/mL of carbenicillin. *E. coli* colonies will appear after 12–16 h of incubation at 37 °C.
6. Culture single *E. coli* colonies in 3–5 mL of LB medium with 100 μg/mL of ampicillin for 12–16 h, shaking at 250 rpm, 37 °C. Extract the plasmids from the overnight cultures using a commercial kit such as Zyppy™ Plasmid Miniprep Kit (Zymo Inc. USA) or QIAprep Spin Miniprep Kit (Qiagen). Confirm correct assembly by performing digestion checks with restriction enzymes followed by DNA sequencing.

3.2 Procedure for Transformation of Plasmids into A. nidulans

3.2.1 Germlings

1. Grow *A. nidulans* A1145 [11] or A1145ΔSTΔEM [14] on CD agar supplemented with uracil, uridine, pyridoxine, and riboflavin for 3–5 days at 37 °C. If using cells from a frozen protoplast stock, streak the strain on CD-sorbitol agar with the appropriate supplements.
2. When green spores appear, inoculate 25 mL of liquid CD medium containing the appropriate supplements in a sterile 125-mL flask with fresh spores (2×10^6 spores/mL, about 3 cm^2 of spores) using a sterile cotton-tipped applicator and shake at 28 °C, 250 rpm for 16–20 h. A proper culture should have an abundance of young germlings in small aggregates.

3.2.2 Digestion

1. Harvest the culture by centrifugation at 4300 × *g* for 20 min at 20 °C.
2. Remove the supernatant and resuspend the pellet in 10 mL of osmotic medium. Spin down by centrifugation at 4300 × *g* for 20 min at 20 °C.
3. Dissolve 30 mg of lysing enzymes from *Trichoderma harzianum* and 20 mg of Yatalase in 10 mL of osmotic medium and sterilize with a 0.2 μm syringe filter.
4. Remove the supernatant from the centrifuged cells and resuspend the pellet in the sterilized lysing enzyme mixture. Transfer directly into a sterile 125-mL flask.
5. Digest cells by shaking at 80 rpm for 4–6 h at 28 °C. Protoplasts are thin-walled and about two times as large as spores.

3.2.3 Harvesting Cells

1. Pour cells directly into an autoclaved 30 mL Corex tube and gently overlay with 10 mL of trapping buffer. Centrifuge in Beckmann equivalent rotor at 5200 × *g* for 20 min at 4 °C (*see* **Note 5**).

2. After centrifugation, protoplasts will accumulate in the cloudy layer at the interface of the two buffers. Remove the protoplasts from the interface with a pipet and transfer them to a sterile 15 mL falcon tube.
3. Add two volumes of STC buffer and centrifuge at 4300 × *g* for 20 min at 4 °C (*see* **Note 6**).
4. Decant the supernatant. Resuspend the protoplasts in STC buffer at a concentration of 10^8–10^9 (usually 1 mL) with minimal pipetting, which can damage the protoplasts. Aliquot 100 μL of protoplasts each in sterile 1.5 mL centrifuge tubes. If desired, store aliquots for future use at −80 °C (*see* **Note 7**).

3.2.4 Transformation

1. For each transformation, add 3 μL of each plasmid to 100 μL of protoplasts, and incubate on ice for 1 h. As a negative control, include one transformation with the appropriate empty pYT vectors. Plasmid stock DNA concentrations should be at least 100 ng/μL (*see* **Note 8**).
2. Add 600 μL of PEG solution to each tube. Mix gently by turning the tube on its side and rotating it. Incubate at room temperature for 20 min.
3. Using a pipet, drop the PEG mixture on CD-Sorbitol agar plates with the appropriate supplements if transforming only one or two plasmids. Using a spreader is not necessary and can damage the protoplasts. Incubate at 37 °C right side up to let dry. Colonies will appear after 2–4 days.

3.3 Procedure for Production of Compounds and Biotransformation

3.3.1 Production of Compounds

1. For expression of genes under starch inducible promoters, transformants must be cultured in CD-ST media. For expression of genes under constitutive promoters, CD or CD-ST media may be used (*see* **Note 9**). For liquid cultures, inoculate the spores from selected transformants in 10 mL of liquid media with appropriate supplements using a sterile cotton-tipped applicator. Shake cultures at 250 rpm at 28 °C. For solid cultures, use a sterile cotton-tipped applicator to streak selected transformants onto CD or CD-ST agar plates. Incubate at 28 °C.
2. Check cultures for compound production every other day, beginning after 2 days of growth. For liquid cultures, use a pipet with autoclaved cut tips to transfer 500 μL of culture to a 1.5 mL microcentrifuge tube (*see* **Note 10**). Centrifuge at maximum speed for 5 min. Transfer the supernatant to a new microcentrifuge tube. Add 500 μL ethyl acetate to the supernatant and vortex for 1 min. Add 500 μL acetone or methanol (*see* **Note 11**) to the cell pellet and vortex for 15 min. For solid cultures, cut out a 1 cm^2 piece of agar and transfer to a

microcentrifuge tube with 500 μL of acetone or methanol and vortex for 15 min.

3. Centrifuge the samples at max speed for 5 min. Transfer the organic layers to new microcentrifuge tubes and dry in a speed vacuum.
4. Resuspend the dried extracts in 100 μL of methanol. Centrifuge at maximum speed for 5 min and transfer 50 μL of the extract to a LC-MS vial. Inject 20 μL of extract on the LC-MS for compound detection.

3.3.2 Biotransformation

1. Use the same culturing conditions as described previously for compound production. Substrates for the biotransformation assay can be added at the beginning of culturing or after the cells have grown for a few days. Add the substrate to the culturing medium to a final concentration of 200–500 μM. If adding substrates at the beginning of a solid culture, allow melted agar to cool before adding the substrate. If feeding to a solid culture after cells have grown, the substrate can be layered on top of the plate.
2. To check cultures for biotransformation products, follow the same procedure for metabolite analysis as described in **steps 2–4** above.

3.4 Procedure for RT-PCR (Reverse Transcription-Polymerase Chain Reaction) to Verify Gene Expression

1. Use the same culturing conditions as described previously for compound production and biotransformation. Cells can be harvested for RNA extraction when compounds are expected to be produced.
2. Extract RNA from the cells using RiboPure™ Yeast RNA Isolation Kit (Ambion) following the manufacturer's instructions. Digest residual genomic DNA in the RNA extracts with DNase I (provided in the kit) at 37 °C for 4–6 h. Follow the manufacturer's instructions to inactivate the DNase.
3. Use SuperScript III First-Strand Synthesis System (Invitrogen) for cDNA synthesis with oligo-dT primers following directions from the user manual.
4. Using the synthesized cDNA as the template, set up PCR reactions to amplify fragments of the genes for heterologous expression using GoTaq Master Mix (Promega) following the manufacturer's instructions. To check for gene expression, it is not necessary to amplify the entire open reading frames of the genes. If possible, design primers flanking the introns in the target genes so the smaller cDNA product can be separated from possible bands from genomic DNA contamination. As a positive control, include a PCR reaction to amplify a region of *actA* (actin gene for *A. nidulans*) flanking its introns to assess the quality of the cDNA. Run the PCR reactions on an agarose

gel. Bands with the correct cDNA product size will indicate the genes are expressed (*see* **Note 12**).

4 Notes

1. Selection is by *riboB*, final concentration 0.125 μg/mL riboflavin.
2. Selection is by *pyroA*, final concentration 0.5 μg/mL pyridoxine HCl.
3. Selection is by *pyrG*, final concentration 10 mM uridine and 5 mM uracil.
4. If using SwaI, perform the digestions with each enzyme separately. First, digest with only SwaI using NEB buffer 3.1 at 25 °C, then column purify the reaction with a commercial kit such as DNA Clean & Concentrator (Zymo Inc. USA). Next, digest the recovered DNA with the second restriction enzyme using NEB Cutsmart® buffer at 37 °C.
5. Alternatively, pour cells into a sterile 50 mL Falcon tube and gently overlay with 10 mL of trapping buffer. Centrifuge at 4300 × *g* for 30 min at 4 °C.
6. As another alternative to **steps 1–3**, pour cells through a sterile cell strainer (Fisher, Cat No. 22363547) into a sterile 50 mL Falcon tube. Add an equal volume of STC buffer and gently invert the tube to mix. Centrifuge at 4300 × *g* for 20 min at 4 °C. Decant the supernatant and gently resuspend the pellet in 10 mL of STC buffer with minimal pipetting to wash the cells. Centrifuge at 4300 × *g* for 20 min at 4 °C.
7. Fresh protoplasts have the highest transformation efficiency. Frozen protoplasts can be used up to 3 weeks after storage but will rapidly lose competency.
8. It is optimal to add less than 10 μL of DNA solution to maintain the concentration of the STC buffer components. If plasmid DNA concentrations are low, plasmid stocks eluted in water can be lyophilized and resuspended in a smaller volume of STC buffer to increase DNA concentration.
9. Strains grow faster and to a higher cell density in CD-ST, which is a richer medium compared to CD. However, the background metabolite profile is generally cleaner in CD compared to CD-ST.
10. Uncut pipet tips may be used in early stages of culturing. It is necessary to use cut tips once the mycelium has grown too thick to be aspirated by an uncut pipet tip.
11. Depending on their solubility, some compounds may only be detected by extracting with certain solvents but not others.

Therefore, it may be advantageous to test different solvents for extraction if target compounds cannot be detected.

12. To confirm genes are properly spliced, the entire open reading frames of the genes can be amplified by PCR using a high-fidelity polymerase. The resulting PCR products can be purified and sent for DNA sequencing.

Acknowledgments

Research in related areas is supported by NIH 1R35GM118056 to YT.

References

1. Newman DJ, Cragg GM (2016) Natural products as sources of new drugs from 1981 to 2014. J Nat Prod 79(3):629–661
2. Keller NP, Turner G, Bennett JW (2005) Fungal secondary metabolism—from biochemistry to genomics. Nat Rev Microbiol 3(12): 937–947
3. Chiang YM, Oakley CE, Ahuja M, Entwistle R, Schultz A, Chang SL, Sung CT, Wang CC, Oakley BR (2013) An efficient system for heterologous expression of secondary metabolite genes in Aspergillus nidulans. J Am Chem Soc 135(20):7720–7731
4. Lubertozzi D, Keasling JD (2009) Developing *Aspergillus* as a host for heterologous expression. Biotech Adv 27(1):53–75
5. Pahirulzaman KA, Williams K, Lazarus CM (2012) A toolkit for heterologous expression of metabolic pathways in *Aspergillus oryzae*. Methods Enzymol 517:241–260
6. Anyaogu DC, Mortensen UH (2015) Heterologous production of fungal secondary metabolites in Aspergilli. Front Microbiol 6:77
7. Sakai K, Kinoshita H, Nihira T (2012) Heterologous expression system in *Aspergillus oryzae* for fungal biosynthetic gene clusters of secondary metabolites. Appl Microbiol Biotechnol 93(5):2011–2022
8. van Dijk JWA, Wang CCC (2016) Heterologous expression of fungal secondary metabolite pathways in the *Aspergillus nidulans* host system. Methods Enzymol 575:127–142
9. Qiao YM, Yu RL, Zhu P (2019) Advances in targeting and heterologous expression of genes involved in the synthesis of fungal secondary metabolites. RSC Adv 9(60):35124–35134
10. He Y, Wang B, Chen W, Cox RJ, He J, Chen F (2018) Recent advances in reconstructing microbial secondary metabolites biosynthesis in *Aspergillus spp.* Biotechnol Adv 36(3): 739–783
11. Sato M, Yagishita F, Mino T, Uchiyama N, Patel A, Chooi YH, Goda Y, Xu W, Noguchi H, Yamamoto T, Hotta K, Houk KN, Tang Y, Watanabe K (2015) Involvement of lipocalin-like CghA in decalin-forming stereoselective intramolecular [4 + 2] cycloaddition. ChemBioChem 16:2294–2298
12. Yu JH, Leonard TJJ (1995) Sterigmatocystin biosynthesis in *Aspergillus nidulans* requires a novel type I polyketide synthase. Bacteriology 177:4792
13. Chiang YM, Szewczyk E, Nayak T, Davidson AD, Sanchez JF, Lo HC, Ho WY, Simityan H, Kuo E, Praseuth A, Watanabe K, Oakley BR, Wang CCC (2008) Molecular genetic mining of the *Aspergillus* secondary metabolome: discovery of the emericellamide biosynthetic pathway. Chem Biol 15:527
14. Liu N, Hung YS, Gao SS, Hang L, Zou Y, Chooi YH, Tang Y (2017) Identification and heterologous production of a benzoyl-primed tricarboxylic acid polyketide intermediate from the zaragozic acid A biosynthetic pathway. Org Lett 19(13):3560–3563
15. Harvey CJ, Tang M, Schlecht U, Horecka J, Fischer CR, Lin HC, Li J, Naughton B, Cherry J, Miranda M, Li YF, Chu AM, Hennessy JR, Vandova GA, Inglis D, Aiyar RS, Steinmetz LM, Davis RW, Medema MH, Sattely E, Khosla C, St. Onge RP, Tang Y, Hillenmeyer ME (2018) HEx: a heterologous expression platform for the discovery of fungal natural products. Sci Adv 4(4):5459
16. Raeder U, Broda P (1985) Rapid preparation of DNA from filamentous fungi. Lett Appl Microbiol 1(1):17–20

17. Goodwin DC, Lee SB (1993) Microwave miniprep of total genomic DNA from fungi, plants, protists and animals for PCR. Biotechniques 15(3):438–441

18. Gietz RD, Schiestl RH (2007) High-efficiency yeast transformation using the LiAc/SS carrier DNA/PEG method. Nat Protoc 2(1):31

Chapter 4

Investigating Fungal Biosynthetic Pathways Using Heterologous Gene Expression: *Fusarium* sp. as a Heterologous Host

Mikkel Rank Nielsen and Jens Laurids Sørensen

Abstract

Heterologous expression of uncharacterized biosynthetic gene clusters is a popular strategy for exploring the chemical potential of filamentous fungi. Here, we describe the process of PCR-amplifying fungal gene clusters and re-assembling them in a cloning vector via target-associated recombination in *Saccharomyces cerevisiae*. The gene cluster-carrying construct is validated and used to transform protoplasts of *Fusarium graminearum*, a well-studied host that is able to express the gene cluster. Chemical analysis of transformants expressing biosynthetic genes can lead to the detection and isolation of novel compounds, such as polyketides.

Key words Gene cluster, Fusarium, Heterologous expression, Homologous recombination, TAR, Protoplast, Protoplast-mediated transformation, Fungi, Chromosomal integration

1 Introduction

Filamentous fungi possess the potential to produce an overwhelming number of bioactive secondary metabolites. The responsible genes are often located in close proximity, forming a biosynthetic gene cluster [1]. More than 1000 fungal genomes have been sequenced in the last 10–15 years and early it became obvious that fungi contain a huge unfulfilled genomic potential to produce secondary metabolites [2]. This is primarily due to tight regulation of the gene clusters, which are only activated by specific stimuli that are not present under laboratory conditions [3]. Several molecular methods can be applied to activate the silent (cryptic) gene clusters, including manipulation of local and global transcriptional regulators and epigenetic modulators [4]. Alternatively, gene clusters can be moved to another host, from where they can be heterologously expressed. This can be achieved by cloning the genes individually in

Elizabeth Skellam (ed.), *Engineering Natural Product Biosynthesis: Methods and Protocols*, Methods in Molecular Biology, vol. 2489, https://doi.org/10.1007/978-1-0716-2273-5_4,

a stepwise manner, where *Saccharomyces cerevisiae* and *Aspergillus* spp. are the preferred organisms of choice [5–7].

To accelerate the tedious cloning process, a targeted method based on in vivo homologous recombination in *S. cerevisiae* has been developed for gene clusters from bacteria [8]. However, this method has so far not been successfully applied to fungi, most likely due to their more complex and lengthy genomes. Instead, the development and application of Fungal Artificial Chromosomes (FACs) were applied to randomly clone gene clusters in *Aspergillus nidulans* followed by extensive metabolic analyses [9]. Inspired by these two approaches, we recently developed a method, which uses *S. cerevisiae* for one-step assembly of PCR amplified gene clusters [10]. As proof-of-concept, we applied this method to one gene cluster from *Fusarium solani* and two gene clusters from *Fusarium pseudograminearum*, which were subsequently introduced into the genome of the closely related *F. graminearum*.

2 Materials

Unless otherwise specified, all media and solutions are prepared using ultrapure H_2O and sterilized by autoclaving the solutions at 121 °C for 15 min. Fungal and yeast cells are washed in autoclaved and pre-chilled ultrapure H_2O.

2.1 Equipment

1. Microscope and hemacytometer.
2. PCR machine, gel electrophoresis equipment including agarose, and a spectrophotometer capable of quantifying DNA (e.g., NanoDrop 2000c).
3. Tabletop tube rocker or tube roller, vortex, water bath with temperature control (37–100 °C), and a freeze-drying system.
4. Ice.
5. Centrifuge with a capacity to process 100 mL liquid at 10,000 × *g* and 4–25 °C and a tabletop centrifuge for 1.5 mL tubes capable of spinning samples >13,000 × *g*.
6. Electroporation apparatus (e.g. MicroPulser Electroporation Apparatus, BioRad) and 0.1 mm electrocuvettes.
7. Plasmid purification kit.
8. Gel purification kit.
9. PCR purification kit.
10. DNA extraction kit suitable for filamentous fungi.
11. Miracloth cut out in squares and fitted into plastic funnels. Autoclaved wrapped in tinfoil (1 for preparation of spores, 1 for purification of genomic DNA, 1 for protoplast mediated transformation (PMT), and 1 per mutant validated).

12. Sterilized plastic syringes packed with a small ball of glass wool sealed in tin foil and sterilized by autoclaving.
13. Sterile 1.5, 15, 50 mL sterile centrifuge tubes, round-bottom test tubes (e.g. 13 × 100 mm).
14. Sterile Petri dishes, Drigalski spatulas, tweezers, scalpel, and scissors.
15. DNA ladders for standard (<10,000 bp) and high ranges (<50,000 bp).
16. 0.2 μm syringe filters for sterilizing solutions.

2.2 PCR Reagents and Primers

1. Purified genomic DNA from targeted gene cluster-harboring species.
2. High fidelity DNA polymerase and compatible reagents.
3. 20 μg/μL Bovine serum albumin (BSA).
4. 2 M NaCl: 5.84 g sodium chloride, fill H_2O to 50 mL. Mix to dissolve and store at RT (room temperature).
5. Fungal lysis buffer: To a sterile 50 mL plastic tube add 3 mL 2 M NaCl, 30 μL Triton X-100, 300 μL 1 M Tris-HCl, 300 μL 0.5 M EDTA, 3 mL 2% SDS, and 20 mL H_2O. Adjust the pH to 8.5 using a small amount of 2 M NaOH. Fill H_2O up to 30 mL, mix by vortexing and store the solution at RT.
6. pA primers for PCR-amplification and reassembly of the targeted gene cluster via target-associated recombination (TAR) [11] with the pSHUT3 vector. We usually amplify the targeted gene cluster in several (X) fragments, A_1–$A_{(X)}$ (*see* **Notes 1** and **2**), with at least 60 bp overlap to neighboring fragments [12, 13]. PCR fragments comprising the edges of the targeted gene cluster are amplified using primers pA_1-Fw and $pA_{(X)}$-Rv added 35 bp tails to facilitate assembly with the multiple cloning site of the linearized pSHUT3 vector (*see* Fig. 1a):
   (a) Fragment A_1 forward primer: 5′- AAGGGTAGCCAC GATTCGAACCACCGGTCCGGATC-[pA_1-Fw]
   (b) Fragment $A_{(X)}$ reverse primer: 5′- AGACGAGTAT TAAGTCCTCAGCGGGCCCGGCTCGA-[$pA_{(X)}$-Rv]
7. Primers for positive control PCR of constructs: C/A-Fw (5′-CTGGCCTTTTGCTCATGATT), C/A-Rv (5′-GCTCGAAT TCGTAACTTACACG).
8. pTAR primers for PCR validation of correctly assembled TAR construct: Design primer $pTAR_1$-Rv to pair with $pTAR_1$-Fw (5′- AGCAGGTGGTAAGAGGCAGA) for checking the recombination between the plasmid and fragment A_1. Design primer $pTAR_{(X+1)}$-Fw to pair with $pTAR_{(X+1)}$-Rv (5′-TTCATT GACCGTTGAGACCA) for checking the recombination between the plasmid and fragment $A_{(X)}$. Design primer pairs

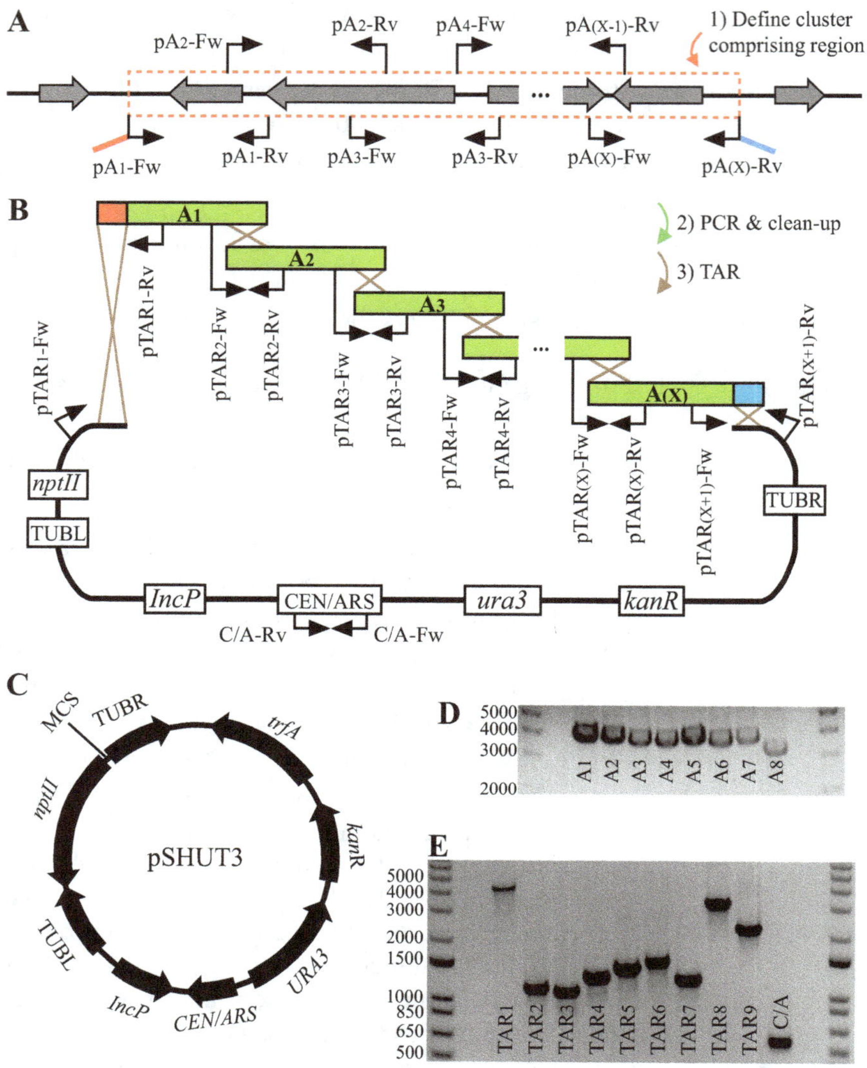

Fig. 1 Constructing and validating a vector carrying an intact copy of the targeted gene cluster. (**a**) Relative positions of pA primers (black arrows) are shown hybridizing to the gene cluster-harboring species genome for PCR-amplification of the region comprising the targeted gene cluster (dotted red square). (**b**) Resulting PCR products (green squares) with > 60 bp overlap to neighboring cluster comprising fragments carry the targeted gene cluster, collectively. The two outmost PCR products A_1 and $A_{(X)}$ are added 35 bp extra nucleotides for promoting target-associated recombination (TAR) with the linearized vector. (**c**) pSHUT3 vector map: *nptII*; neomycin phosphotransferase (G418 resistance), TUBL and TUBR; left and right borders homologous to the integration locus in *Fusarium graminearum*, MCS: multiple-cloning site including unique XhoI and BamHI recognition sites, *IncP* and *trfA*; bacterial replication, *CEN/ARS*; centromere and autonomously replicating sequence for *Saccharomyces cerevisiae*, *ura3*: uracil auxotroph selection marker, *kan*R: Aminoglycoside 3′-phosphotransferase (kanamycin resistance). (**d**) Eight PCR products A_1 to A_8 comprising the 34 kb *Fusarium solani* PKS35 gene cluster. (**e**) PCR validation with pTAR primers of an isolated TAR construct carrying the PKS35 gene cluster

for validating every homologous recombination event between cluster containing fragments, e.g. primers pTAR2-Fw/ pTAR2-Rv checking the recombination between fragments A_1 and A_2 (*see* Fig. 1b and **Note 3**).

9. Primers for validation of fungal mutants: TUB-Fw (5′-GAA TAGGTCTGCAGGGGACA), TUB-Rv (5′- ACACGTTCA GAAAGGCAAGG), nptII-Rv (5′-TTGGGTGGAGAGGCTATT).

2.3 Preparation of Fusarium Spores

1. *Fusarium graminearum* PH-1 (NRRL 31084).
2. Trace metals solution: 100 mL: 0.1 g/L $ZnSO_4 \cdot 7H_2O$, 0.05 g/L $CuSO_4 \cdot 5H_2O$.
3. 10% glycerol, sterilized. Store at 4 °C.
4. PDA: 39 g/L Potato Dextrose Agar, 1 mL/L trace metals solution, autoclaved.
5. CMC medium: To 900 mL boiling H_2O, add 15 g carboxymethylcellulose sodium salt while stirring. Add 1 g NH_4NO_3, 1 g KH_2PO_4 monobasic, 0.5 g $MgSO_4 \cdot 7H_2O$, 1 g bacto yeast extract, and fill H_2O to 1000 mL. Divide the medium in portions of 100 mL in baffled 250 mL shake flasks and autoclave (*see* **Note 4**).

2.4 Vector Assembly by TAR Cloning in Yeast

1. *Saccharomyces cerevisiae* BY4743.
2. pSHUT3 (*see* Fig. 1c and **Note 5**).
3. PCR products collectively comprising the targeted gene cluster (*see* **Note 6**).
4. Restriction endonucleases BamHI and XhoI and associated reaction buffer.
5. Solid and liquid SC–U: 1.92 g/L Yeast Synthetic Drop-Out Medium Supplement without Uracil, 6.7 g Yeast Nitrogen Base Without Amino Acids. Fill H_2O to 960 mL and adjust to pH 5.6. Autoclave the solution and add 40 mL 50% (w/v) glucose so the final concentration becomes 2%. When preparing solid medium, add 20 g/L agar before autoclaving the solution.
6. 40% (w/v) Glucose: Fill a 100 mL flask with 50 mL H_2O and slowly add 40 g D-(+) glucose while stirring. Let the glucose dissolve completely and add H_2O to 100 mL. Autoclave or filter-sterilize the solution.
7. Liquid YPAD: 10 g bacto yeast extract, 20 g bacto peptone, and 80 mg adenine hemisulfate are added to 950 mL H_2O and autoclaved. 50 mL 40% (w/v) glucose is added to a final concentration of 2%.

8. Liquid 2 × YPAD: 20 g bacto yeast extract, 40 g bacto peptone, and 80 mg adenine hemisulfate are added to 900 mL H_2O and autoclaved. 100 mL 40% (w/v) glucose is added to a final concentration of 4%.
9. LiAc 1 M: 10.2 g lithium acetate dihydrate is dissolved in 100 mL H_2O in a glass bottle and autoclaved.
10. PEG 3350 50% (w/v): Add 50 g polyethylene glycol MW 3350 to 30 mL H_2O while stirring in a beaker. Continue to stir until the PEG is completely dissolved. Control the volume by pouring the solution into a measuring glass and fill H_2O to 100 mL. Mix thoroughly and autoclave the solution in a 100 mL bottle.
11. 1 M Tris-HCl: 12.11 g Tris(hydroxymethyl)aminomethane is added to 80 mL of H_2O and stirred until completely dissolved. The pH is set to 7.5 by the addition of 7 mL concentrated HCl. The volume is set to 100 mL and the solution is autoclaved and stored at RT.
12. 0.5 M EDTA: slowly add 18.6 g ethylenediaminetetraacetic acid disodium salt dihydrate to 80 mL H_2O while stirring. Adjust the pH to 8.0 by the addition of NaOH. The EDTA will not dissolve completely before the pH is adjusted to ~8. Fill H_2O to 100 mL and store at RT.
13. 1:10 TE buffer: 10 mL/L 1 M Tris-HCl, 2 mL/L 0.5 M EDTA, adjust pH to 8.0 and autoclave.
14. Salmon ssDNA stock: Dissolve 100 mg DNA sodium salt from salmon testes in 50 mL sterile and chilled 1:10 TE buffer using a magnet stirrer. To dissolve the DNA completely the solution can be pipetted slowly up and down using a broad-nozzle tip. The dissolved DNA is distributed into 500 μL aliquots in sterile 1.5 mL tubes and stored at −20 °C.

2.5 Recovery and Validation of Cluster-Carrying Plasmids

1. Electrocompetent *E. coli* DH5α cells (or similar): Can be prepared by growing cells to OD = 0.6, chilling the cells on ice, centrifuging and washing the cells 3 times in ice-cold 10% glycerol and storing at −80 °C.
2. 50 mg/mL kanamycin stock: Dissolve 500 mg kanamycin sulfate in 10 mL H_2O and filter sterilize into sterile 1.5 mL tubes and store at −20 °C.
3. Solid and Liquid LB medium: 20 g LB Broth is mixed in 950 mL H_2O, adjust pH to 7.4. Add H_2O to 1000 mL. For solid medium, add 20 g agar before adjusting pH. Autoclave the solution. Add 500 μL kanamycin stock solution to 1 L of autoclaved medium for a final concentration of 25 μg/mL.

4. 10× salts solution: 100 mM NaCl, 25 mM KCl, 100 mM $MgCl_2$, 100 $MgSO_4$, 200 mM glucose. Autoclave solution and store at RT.
5. SOC medium, 100 mL: 2 g Bacto tryptone, 0.5 g Bacto yeast extract, 10 mL 10× salts solution. Autoclave solution and store at RT.
6. 5 M KAc: Dissolve 29.5 g potassium acetate in 75 mL of H_2O and add 11.5 mL glacial acetic acid. Adjust the volume to 100 mL and sterilize by filtration. Store at RT.
7. 2% (w/v) SDS: In a measuring cylinder, add 2 g of sodium dodecyl sulfate and fill H_2O to 100 mL. Store the solution in a flask at RT.
8. SPE buffer: to 400 mL H_2O add 91 g D-sorbitol, 10 mL 0.5 M EDTA, 1.04 g $Na_2HPO_4 \cdot 7H_2O$, 0.16 g $NaH_2PO_4 \cdot 1H_2O$, 500 μL β-mercaptoethanol. Mix to dissolve completely and set the volume to 500 mL before sterilizing by filtration. The solution can be stored at RT.
9. Zymolyase stock: 100 mg Zymolyase 20T from *Arthrobacter luteus*, 0.5 mL 1 M Tris-HCl, 5 mL sterile 50% glycerol, and 4.5 mL H_2O is mixed in a sterile 50 mL centrifuge tube by vortexing. Aliquot the enzyme stock into 500 μL portions in sterile 1.5 mL tubes and store at −20 °C for up to a year.
10. 50% (w/v) glycerol solution, sterilized.
11. 99% ethanol.
12. 2-propanol.

2.6 Protoplast-Mediated Transformation

1. 5×10^6 *F. graminearum* spores, fresh (maximum 2 weeks old).
2. 80 mg/mL G418 stock: Add 400 mg Geneticin G418 sulfate to a 15 mL tube and fill sterile H_2O up to 5 mL. Vortex the solution to mix. Sterilize the stock solution by filtrating it through a 0.2 μm syringe filter and store at −20 °C in 1 mL aliquots.
3. Sucrose (55% w/v): slowly add 275 g table sugar to 200 mL H_2O while stirring. Fill H_2O to 500 mL. The sucrose has to be completely dissolved before autoclaving, autoclave and store at RT.
4. Regeneration and selective-overlay medium: in a 1000 mL flask, add 10 g Agarose (SERVA 11404, *see* **Note 7**), 1 g Bacto yeast extract, and 1 g peptone from casein, fill H_2O to the 500 mL mark and autoclave. Immediately after autoclaving, add 500 mL sterile 55% sucrose and mix. For making selection medium, add 1 mL G418 stock per liter of selective-overlay medium to a final concentration of 80 μg/mL (*see* **Note 8**).

5. Czapek Dox agar: 35 g/L Czapek Dox broth, 20 g/L agar, 1 mL/L Trace metal solution, autoclaved. Add 3.75 mL G418 stock to a final concentration of 300 μg/mL, and pour plates.
6. Liquid YEPD for cultivation of fungi: 3 g bacto yeast extract and 10 g bacto peptone is added to 950 mL H_2O and autoclaved. 50 mL sterile 40% (w/v) glucose is added to a final concentration of 2%.
7. 1 M sorbitol: Add 18.2 D-sorbitol and fill H_2O up to 100 mL. Stir the solution to dissolve the sorbitol and autoclave. Store the solution at RT.
8. 1 M $CaCl_2$: Add 14.7 g $CaCl_2 \cdot 2H_2O$, fill H_2O to 100 mL. Filter sterilize and store at RT.
9. STC buffer: 43.7 g D-sorbitol, fill H_2O to 188 mL, autoclave. Add 2 mL 1 M Tris-HCl, 10 mL 1 M $CaCl_2$, H_2O to 200 mL. Autoclave.
10. PEG 8000 30% (w/v): Dissolve 60 g polyethylene glycol MW 8000 in up to 195 mL H_2O and autoclave. Add 2 mL 1 M Tris-HCl and 100 μL 1 M $CaCl_2$ and fill with sterilized H_2O to 200 mL.
11. 1.2 M KCl, 100 mL: Dissolve 8.95 g potassium chloride in 80 mL of H_2O. Mix to dissolve and adjust the volume to 100 mL. The solution can be stored at RT.
12. 10 μg pSHUT3 in a concentration of approximately 250 ng/μL (*see* **Note 9**).
13. Protoplasting solution: 500 mg Driselase from *Basidiomycetes* sp., 1 mg Chitinase from *Streptomyces griseus*, 100 mg Lysing Enzymes from *Trichoderma harzianum* is added to 20 mL 1.2 M KCl and mixed in a centrifuge tube on a tube rocker/roller for 30 min before filter sterilizing through a 0.2 μm filter into a sterile centrifuge tube. The protoplasting solutions can be made in advance and stored at −20 °C (*see* **Note 10**).
14. 100 μm cell strainer.
15. 1 mL plastic tips with broad nozzles for the micropipette. These can be prepared by cutting the tip of regular 1000 μL plastic tips with scissors and autoclaving the tips in a closed container.

3 Methods

3.1 Production of *Fusarium* Spores

1. Inoculate *F. graminearum* on PDA medium and incubate at 25 °C for 1–2 weeks.
2. Using a sterile scalpel cut out 0.5 × 0.5 cm agar cubes overgrown with mycelium. Under sterile conditions, transfer

five agar cubes to 100 mL autoclaved CMC medium (*see* **Note 11**) in a baffled 250 mL shake flask. Incubate the culture at 20 °C, 150 rpm for 4–7 days in total darkness (*see* **Note 12**).

3. Harvest the spores by pouring the culture through a layer of sterile Miracloth removing hyphae and agar cubes. Centrifuge the flow through at 10,000 × *g* 4 °C for 15 min and decant off the medium keeping as much of the pellet as possible. Wash the pellet in 100 mL cold sterile H_2O, centrifuge and remove the H_2O by decanting. Repeat the wash with 50 mL cold sterile H_2O. Remove all liquid using a micropipette and resuspend all pellets in a total of 5.0 mL H_2O and calculate the spore concentration using a hemacytometer. Store the spores at 4 °C (*see* **Note 13**).

3.2 Target Associated Recombination (TAR)-Mediated Cloning of Targeted Gene Cluster

3.2.1 PCR-Amplification of Cluster Comprising Fragments A_1–$A_{(X)}$

1. Cultivate the targeted gene cluster-harboring species and extract genomic DNA for PCR. There are several methods to extract fungal DNA. In our lab, *Fusarium* genomic DNA is purified using the FastDNA SPIN Kit for Soil from 100 mg lyophilized mycelium harvested from a 5 day old 20 mL YPAD medium.
2. Using pA primers shown in Fig. 1a, PCR-amplify gene cluster-comprising fragments A_1–$A_{(X)}$ using a high fidelity DNA polymerase following the manufacturer's guidelines and compatible reagents. As an example, we amplified the 34 kb *PKS35* gene cluster of *F. solani* this way in eight fragments A_1–A_8 (*see* Fig. 1d).
3. Evaluate 1 μL of the PCR product with gel electrophoresis. If the gel displays unspecific PCR products, proceed to **step 4**, otherwise proceed to **step 5**.
4. Unspecific PCR products are undesirable as they contaminate the TAR cloning reaction and can result in unwanted recombination. Load the total PCR reaction volume on a 0.8% agarose gel, run electrophoresis, and isolate the PCR product using the QIAquick Gel Extraction kit. Proceed to **step 6**.
5. Clean PCR fragments using a PCR purification kit and quantify the DNA concentration of the product using the Nanodrop instrument.
6. To enhance concentration of PCR products, pool a minimum of 200 fmol of every A_1–$A_{(X)}$ fragment (~180–250 ng for a 3–4 kb dsDNA molecule). Mix the PCR products with 5× volume PB buffer (QIAGEN) and proceed following the manufacturers' guidelines for the QIAquick PCR Purification Kit. Elute the concentrated DNA pool in 30 μL pre-warmed 55 °C elution buffer. Alternatively, a standard precipitation with phenol:chloroform:isoamyl alchohol (25:24:1) can be used to concentrate the PCR products.

3.2.2 Plasmid Backbone Preparation

1. Grow *E. coli* carrying pSHUT3 in 25 mL LB medium containing 25 μg/mL kanamycin at 37 °C shaking 200 rpm for 2 days and isolate plasmid DNA using the QIAprep Spin Miniprep Kit (*see* **Note 14**).
2. Linearize 0.5–2.0 μg pSHUT3 plasmid miniprep with BamHI and XhoI applying the supplier's protocol and reagents, and load the entire digest reaction on a 0.8% agarose gel. Perform electrophoresis and purify the 8.628 bp backbone using the QIAquick Gel Extraction Kit (*see* **Note 15**).

3.2.3 Yeast Transformation [14]

1. Inoculate a colony *S. cerevisiae* in 5 mL YPAD medium and incubate 30 °C 200 rpm for 12–16 h.
2. Determine the cell titer of the overnight culture by pipetting 10 μL culture into 1000 μL H_2O and mix by inversion, measure the OD_{600} using a spectrophotometer, and calculate the cell titer in the culture assuming an OD_{600} of 0.1 equals 1×10^6 cells/mL. Remember to multiply by the dilution factor. Inoculate 2.5×10^8 cells from the overnight culture in 50 mL 2 × YPAD medium (the titer should be 5×10^6 cells/mL) in a sterile baffled 250 mL cultivation flask and incubate at 30 °C, 200 rpm until the cell titer reaches 2×10^7 cells/mL (two generations, typically 4 h).
3. Denature 500 μL of salmon ssDNA solution in a 100 °C H_2O bath for 5 min and cool immediately on ice.
4. Prepare 1630 μL transformation mix from 250 μL denatured salmon ssDNA solution, 180 μL 1 M LiAc, and 1.2 mL PEG 3350 50% (w/v). Mix in a sterile 15 mL tube and keep on ice.
5. Harvest the yeast cells centrifugation at 3000 × *g* in a 50 mL centrifuge tube for 5 min. Discard the medium and wash the pellet twice in 25 mL sterile H_2O by centrifuging and discarding the H_2O. Resuspend the cell pellet in 1.0 mL of sterile H_2O and transfer the suspension to a sterile 1.5 mL tube. Centrifuge the cell suspension at 13,000 × *g* for 30 s and remove the supernatant with a micropipette.
6. Resuspend the cell pellet in 1.0 mL of sterile H_2O and distribute 100 μL (1×10^8 cells) into a 1.5 mL tube per transformation reaction (*see* **step 8**). Centrifuge at 13,000 × *g* for 30 s and discard the supernatant.
7. Resuspend the cell pellets in 326 μL transformation mix by carefully pipetting up and down 10–20 times.
8. Setup the following transformation reactions:
   (a) Negative growth control: add 34 μL H_2O to a tube of 326 μL cells in transformation mix and vortex.
   (b) Negative assembly control: add 10–50 fmol linearized plasmid backbone extracted from an agarose gel plus

H_2O in a total volume of 34 μL to a tube of 326 μL cells in transformation mix and vortex.

(c) Positive control: Add 1 μL pSHUT3 miniprep and 34 μL H_2O to a tube of 326 μL cells in transformation mix and vortex.

(d) TAR reaction: Add the pool of gene cluster comprising PCR products containing 200 fmol of each fragment A_1–$A_{(X)}$, 10–50 fmol linearized pSHUT3 extracted from an agarose gel (important: use the same amount of linearized plasmid as in the negative assembly control) plus H_2O in a total volume of 34 μL to a tube of 326 μL cells in transformation mix and vortex.

9. Incubate transformation reactions in a 42 °C H_2O bath for 40 min.
10. Centrifuge the tubes at 13,000 × *g* for 30 s and remove the supernatant using a micropipette.
11. Add 1000 μL of sterile H_2O and stir the pellet with a sterile micropipette tip. Resuspend the cells completely by vortexing.
12. For each transformation reaction, plate 20, 200, and 700 μL of the cells on SC–U agar. The 20 μL volume can be delivered to the plate in a puddle of 100 μL sterile H_2O. Spread the cells using a sterilized Drigalski spatula and incubate the plates at room temperature until the liquid has been absorbed into the agar plates.
13. Incubate the plates for 3–4 days at 30 °C and count the colonies.
14. Consider the colony count of the different transformation reactions to determine the success of the TAR cloning. To interpret the result, consult Table 1.

3.3 Recovery and Validation of TAR Constructs

3.3.1 Extraction of Plasmid DNA from Yeast Transformants [15]

1. Streak individual colonies from the TAR reaction plate onto fresh SC–U plates and incubate at 30 °C for 1–2 days.
2. Harvest the cells by washing the plate in 5 mL sterile H_2O by running a spatula over the colonies several times. Transfer the liquid to a 50 mL tube and pellet the cells by centrifugation at 1000 × *g* for 5 min. Remove the liquid by decanting.
3. Resuspend the cells in 5 mL 1 M sorbitol by vortexing. Pipette 1 mL of the cell suspension into a 1.5 mL tube and pellet the cells in a tabletop centrifuge for 13,000 × *g* for 30 s. Discard the supernatant.
4. Resuspend the cells in 500 μL SPE solution.
5. Add 40 μL zymolyase 20T solution, mix carefully by pipetting up and down, and digest the cells' walls for 1 h at 30 °C.

Table 1
Target-associated recombination cloning outcomes and suggested actions

	Transformation reaction					
	a	b	c	d		
Circular plasmid	–	–	+	–		
Linearized plasmid	–	+	–	+		
A1-A$_{(x)}$	–	–	–	+	Result	Action
Scenario 1	0 colonies	0 colonies	100–1000 colonies	100 colonies	Correct assembly	Proceed to Subheading 3.3. Recovery and Validation of TAR Constructs
Scenario 2	0 colonies	10 colonies	100–1000 colonies	100 colonies	Plasmid digest incomplete. The TAR reaction contained a low amount of native circular plasmid yielding prototrophic colonies with a low frequency.	Proceed to Subheading 3.3. Recovery and Validation of TAR Constructs. Increase the number of screened colonies from reaction D.
Scenario 3	0 colonies	100 colonies	100–1000 colonies	100 colonies	Plasmid digest incomplete. The TAR reaction contained a large amount of native circular plasmid yielding prototrophic colonies with a high frequency	Discard transformants. Optimize plasmid digest and gel extraction
Scenario 4	Any growth	–	–	–	Contamination	Discard transformants. Make sure all reagents are sterilized and the experiment is carried out under sterile conditions.

6. Harvest yeast spheroplasts by centrifugation at 5000 × *g* for 5 min, discard the supernatant, and resuspend the spheroplasts thoroughly in 500 μL 0.5 M EDTA by pipetting up and down several times, slowly.
7. Lyse the cells by mixing in 50 μL 2% SDS solution and incubate the reaction at 70 °C for 15 min.
8. Add 50 μL 5 M KAc and incubate the reaction on ice for 30 min.
9. Pellet cell debris at full speed in a tabletop centrifuge for 15 min.
10. Avoiding any precipitate, pipette the DNA-containing supernatant to a clean 1.5 mL tube and record the volume.
11. Add an equal amount of RT 2-propanol, mix by inverting the tube and precipitate the DNA by centrifuging 5 min at full speed in a tabletop centrifuge. Remove the supernatant by pipetting. Avoid dislocating the white/transparent pellet with the micropipette.
12. Wash the pellet in 500 μL 99% ethanol. Remove the ethanol using a micropipette. Wipe the opening of the tube with kitchen paper and let the tube sit open under sterile conditions to evaporate remaining ethanol.
13. Resuspend the dry DNA pellet in 300 μL sterile H_2O and store the isolated DNA from yeast at −20 °C (*see* **Note 16**).
14. (Optional) Use 1 μL of isolated DNA from yeast in diagnostic PCRs (*see* Subheading 3.3.3).

3.3.2 Recovery and Proliferation of Constructs in E. coli

1. Thaw electrocompetent *E. coli* cells in a bucket of ice.
2. In a pre-chilled electroporation cuvette, mix 1 μL isolated DNA from yeast with 20 μL electrocompetent *E. coli* cells. Let the DNA and cells sit on ice to mix for 1 min.
3. Electroshock the cells in an electroporation apparatus set to 1.8 kV and immediately resuspend the cells in 1 mL SOC medium. Transfer the cell suspension into a sterile round-bottom test tube and incubate at 37 °C, 250 rpm for 1 h.
4. Spread 500 μL of the transformation mix on LB agar containing 25 μg/mL kanamycin. Incubate the plates at 37 °C overnight.

3.3.3 Validation of TAR Constructs

1. Proliferate *E. coli* and purify TAR constructs using a miniprep kit.
2. Validate plasmid constructs with PCR using pTAR primers (*see* Fig. 1b) amplifying across sections where recombination occurred between DNA fragments. We validated constructs carrying the PKS35 gene cluster in this way with nine TAR checks (*see* Fig. 1e). As a control use C/A–Fw/Rv primers to

test for the presence of the plasmid backbone. As a control, we recommend performing all of the above reactions on genomic DNA of the gene cluster-harboring species.

3. (Optional) Validate the TAR constructs using suitable restriction enzymes. Evaluate the digest patterns via gel electrophoresis next to a high-range DNA ladder.
4. (Optional) Full-length plasmid sequencing can be used to validate the constructs thoroughly and enables detection of potential errors introduced in the recombination.
5. Freeze and store *E. coli* strains carrying validated TAR constructs in 10% glycerol.

3.4 Transformation of the Targeted Gene Cluster into F. graminearum

3.4.1 Preparation of TAR Construct Carrying the Targeted Gene Cluster DNA

1. Awaken *E. coli* carrying the validated TAR construct carrying the targeted gene cluster on LB containing 25 μg/mL kanamycin.
2. Inoculate liquid LB cultures and prepare 10 μg of a validated TAR construct carrying an intact in maximum volume of 40 μL of TE buffer (*see* **Note 9**).
3. (Optional) Linearize the vector construct with restriction endonuclease treatment prior to fungal transformation (*see* **Note 17**).

3.4.2 Protoplast-Mediated Transformation [16]

1. Vortex the spore solution and inoculate 5 × 10^6 *F. graminearum* spores to 100 mL of YEPD medium. Incubate the fungal culture at 29 °C for 14–16 h with agitation.
2. Prepare 165 mL regeneration medium per transformation reaction. Autoclave the medium and mix in the sugar. Place the medium in a pre-heated water bath set to 55 °C to prevent the agarose from solidifying.
3. Prepare 20 mL filtered protoplasting solution in a 50 mL centrifuge tube.
4. In the microscope, control the fungal culture contain germinated hyphae and is not contaminated (Fig. 2a). Harvest the mycelium by pouring it through a layer of sterile Miracloth in a funnel. Rinse the mycelium with 50–100 mL 1 M Sorbitol. Cut a square in the Miracloth carrying the mycelium using sterilized scissors and resuspend it in 20 mL protoplasting solution by rolling the tube, remove the Miracloth using sterile tweezers. Incubate the protoplasting reaction at 29 °C on a tube roller set to 60 rpm for 1–3 h, while monitoring the protoplasting efficiency in the microscope (*see* Fig. 2b and **Note 18**).
5. When an abundance of protoplasts are released (*see* Fig. 2c), immediately harvest the protoplasts by pouring the solution through a sterile syringe packed with glass wool to remove undigested mycelium and hyphae and collect the flow through (*see* Fig. 2d). Pour the solution through a 100 μm cell strainer

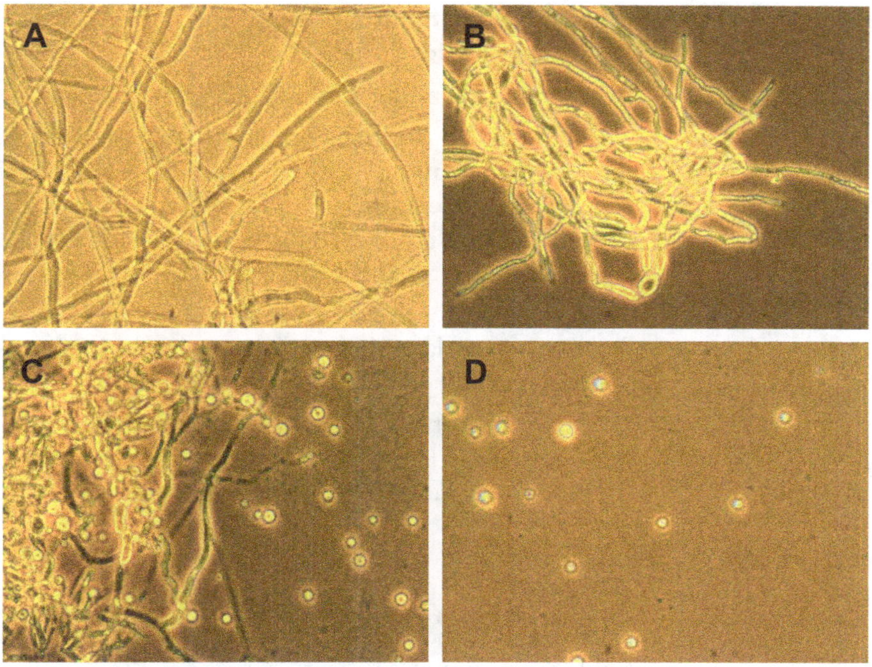

Fig. 2 Protoplasting mycelium observed under the microscope. **A**: Freshly germinated hyphae of *Fusarium graminearum*. **B**: 5 minutes after submerging the mycelium into the protoplasting solution. **C**: Protoplasts were released after 60 to 90 minutes. **D**. Filtered protoplasts free of undigested hyphae and cell debris

into a clean 50 mL centrifuge tube. From this point onwards, keep the protoplasts chilled on ice.

6. Centrifuge protoplasts at 1000 × *g* for 15 min at 4 °C and remove the supernatant by decanting carefully.
7. Resuspend the protoplasts in 10 mL STC buffer by pipetting up and down slowly with a micropipette equipped with a sterile broad-nozzle tip.
8. Centrifuge the protoplasts at 1000 × *g* for 15 min at 4 °C and discard the supernatant. Gently resuspend the pellet in 1000 μL STC buffer without vortexing.
9. Transfer the protoplasts to a 1.5 mL tube and count the protoplast concentration using a hemacytometer.
10. Adjust the concentration to 1×10^8 protoplasts/mL in STC buffer.
11. Prepare the transformation reaction by adding the following to a 50 mL centrifuge tube on ice: 100 μL STC buffer, 50 μL PEG 8000 30% (w/v), 10 μg linearized or circular vector DNA dissolved in maximum 40 μL TE buffer, 100 μL protoplasts in STC (1×10^7 protoplasts). Mix the reagents by swirling the tube gently and incubate at RT for 20 min.

12. Add 2 mL PEG 8000 30% (w/v) and mix gently by inversion. Incubate at RT for 5 min.
13. Add 4 mL STC buffer and mix gently by inversion.
14. Mix 600 μL of the transformation reaction with 15 mL of molten regeneration medium (equilibrated at 55 °C) in a sterile tube and cast the contents into a petri dish. Repeat this process to cast a total of 10 plates per transformation reaction. Incubate the plates at RT for 2 h.
15. Prepare 165 mL selective-overlay medium per transformation reaction. Autoclave the medium and mix in the sugar and antibiotics. Place the medium in a water bath set to 50 °C.
16. Overlay each petri dish with 15 mL selective-overlay medium (*see* **Note 19**). Incubate the plates at RT in the dark until transformants emerge (typically 4–8 days).

3.4.3 Selecting Mutants

1. Isolate mutants on Czapek Dox agar containing 300 μg/mL G418.
2. Sporulate isolated transformants in CMC medium and isolate single-spore colonies twice to ensure genetically stable isolates.

3.4.4 Screening Fungal Transformants with Colony PCR

1. Transfer a minute amount of 3–5 day old mycelium into 300 μL fungal lysis buffer using a sterile toothpick or pipette tip. Vortex the lysis solution for 1 min. Pellet cell debris for 2 min in a tabletop centrifuge. The DNA can be stored at −20 °C for several months.
2. Prepare PCR master mix with 1 μL fungal lysate and 2 μg/μL BSA. Setup the flowing reactions: primers TUB-FW/TUB-Rv to check if the insertion site is empty. Apply primers TUB-Fw/nptII-Rv and pTAR$_{(X+1)}$-Fw/TUB-Rv to validate in locus integration. As a control, we recommend to carry out the PCR reactions on isolated DNA from *F. graminearum* and the targeted gene cluster-harboring species.

3.5 Mutant Genome Validation

1. Cultivate transformants in 20 mL 2 × YPAD for 5–7 days.
2. Harvest the mycelium through a layer of sterile Miracloth.
3. Rinse the mycelium in 50 mL sterile H_2O and squeeze the Miracloth to remove liquid from the mycelium. Purify genomic DNA from lyophilized tissue of transformants using the FastDNA SPIN Kit for Soil.
4. Perform Southern blot or whole genome sequencing of transformants to determine cluster integrity and copy number (*see* **Note 20**).

3.6 Identification of Target Compounds Produced by Targeted Gene Cluster

1. Inoculate validated mutants alongside wild types of *F. graminearum* and the targeted gene cluster-harboring species on different media (e.g. YES, CYA, MEA). Grow the fungi for 14 days to stimulate the accumulation of secondary metabolites.
2. Extract metabolites from plates as described by Smedsgaard [17].
3. Perform liquid-chromatography mass-spectrometry using extracts from *F. graminearum* as controls in search for novel compounds [18–22].

4 Notes

1. The genes that should be included in the targeted gene cluster region can be difficult to predict for uncharacterized gene clusters. A thorough genetic analysis is in order to associate putative members of the gene cluster's protein function to secondary metabolism. Be particularly careful when defining the beginning and end of the targeted gene cluster: remember to include regulatory elements necessary for expression (promoters, terminators). In addition, look out for regions conserved in *F. graminearum*. We observed the extremes of the transferred *PKS40–NRPS32* cluster were conserved in the chromosome 2 of *F. graminearum* resulting in unintended integrations at this exact position in all analyzed transformants.
2. We recommend amplifying the targeted gene cluster in multiple sections and thus the number of primers necessary will depend on the targeted gene cluster's size. Shorter PCR products are easier to obtain in high concentration and longer fragments are tedious to purify from an agarose gel, if applicable. We usually amplify the targeted gene cluster in (overlapping) sections of 3–4 kb. In addition, when using TAR for cloning biosynthetic genes, it is a popular choice designing primers with melting temperatures around 66 °C [7, 13] in order to eliminate the risk of unspecific product formation and in turn eliminate the necessity for gel purification of fragments prior to assembly.
3. The required total number of pTAR primer pairs for validating correct recombination between plasmid and PCR fragments can be expressed as $X + 1$, where X is the number of PCR fragments comprising the targeted biosynthetic genes. As a positive control, the pTAR primers can be used in combination with genomic DNA from the targeted gene cluster-harboring species.

4. CMC takes 2–3 h to dissolve completely. Add the CMC to the boiling H_2O in smaller portions and wait for it to dissolve. We always use CMC from Sigma (C4888) as other variants can be insoluble.
5. The pSHUT3 vector comprises two border sequences TUBL and TUBR for promoting genomic integration [10, 23] via homologous recombination in *F. graminearum* (obtainable by request).
6. The DNA concentration of every PCR product should be at least 30 fmol/μL and ideally above 40 fmol/μL. The molar concentration of a dsDNA molecule can be estimated from the measured concentration [ng/μL] using an online calculation tool like the NEBioCalculator (New England Biolabs).
7. Different brands of agarose can affect the survival rate of fungal protoplasts and therefore we recommend using agarose from SERVA.
8. The selective overlay medium is prepared with a concentration of 80 μg/mL G418. Since the regeneration medium contains no antibiotics, the effective concentration of G418 becomes 40 μg/mL medium in overlaid plates (15 mL regeneration medium + 15 mL selective-overlay medium).
9. The plasmid backbone comprises the bacterial origin of replication *IncP* (*oriV*) which is compatible with both *Agrobacterium tumefaciens* and *E. coli*. We found it is necessary to modify the plasmid purification procedure (i.e. miniprep) when working with a low-copy origin such as *IncP* in order to ensure sufficient material for TAR cloning: (1) Increase the preculture volume to at least 25 mL, (2) Pool and concentrate the yield from several plasmid purification runs, (3) Elute by pipetting pre-warmed elution buffer (55 °C) to the center of the column filter and let it incubate at RT for 10 min before centrifuging. For protoplast-mediated transformation, 10 μg of plasmid DNA is required. To extract plasmid DNA in high concentration, will high–yield plasmid recovery systems such as Maxiprep (various providers) or the Miraprep procedure [24] be feasible.
10. Many enzyme batches will contain insoluble resin particles, which will obfuscate monitoring the protoplasting process in the microscope. It is critical to let the enzymes dissolve in the KCl solution for 30 min with agitation and remove the insoluble particles via filtration (0.2 μm) before introducing the mycelium.
11. Mung bean extract is another popular medium for inducing sporulation in *Fusarium* spp. and will work for *F. graminearum*.

12. Monitor the sporulation process daily by using a microscope on a small sample pipetted from the culture beginning 4 days post-inoculation. Ideally, the culture is harvested when a high density of conidia is observed, before germination.
13. We recommend using *Fusarium* spores no more than 2 weeks old, which have not been frozen.
14. In the development of pSHUT3, we tested different origins of replication for both yeast and bacteria. We found the high-copy *2μ* yeast origin of replication was suitable for capturing a 10 kb gene cluster [25] and alleviated the plasmid purification process from yeast significantly. However, when attempting to clone larger gene clusters we obtained zero yeast clones, and we replaced the yeast origin of replication with *CEN6/ARSH4*. The low-copy bacterial replication elements, *IncP* and *trfA*, can be substituted for the popular high-copy pUC origin of replication when cloning large fungal gene clusters [8, 26]. However, readers of this protocol should notice this modification eliminates the option to use the vector in *Agrobacterium*-mediated transformation [14].
15. In our experience, one of the bottlenecks in this workflow is getting enough DNA for the TAR cloning. The required amount of DNA in H_2O can take up to 34 μL of the transformation solution volume, i.e. the concentration of both PCR products and backbone should be relatively high. We suggest optimizing the DNA concentration of PCR products and gel purified linearized plasmid backbone before attempting yeast transformation. (1) To ensure sufficient amounts of PCR product, prepare 150 μL PCR mix for every fragment. (2) Always elute DNA in the smallest possible volume when using the purification kits. (3) Pre-warm elution buffer or H_2O to 55 °C, pipette the elution buffer to the column and let incubate at RT for 5 min before spinning, pipette the elute back into the same column and repeat the process to increase the nucleotide yield. (4) Keep in mind most gel extraction kits recover at most 60–80% of the applied DNA. When experiencing poor DNA yield from gel purification we suggest increasing the input of DNA (plasmid DNA in digest) two- to fourfold the required amount.
16. Pipetting up and down 25–30 times improves solubility significantly. Although vigorous agitation can result in sheering of large DNA molecules, in our experience, dissolving the DNA completely is key to recovering the plasmids in *E. coli*. In addition, DNA samples can be stored at 4 °C overnight to improve solubility.
17. We have experienced both circular and linearized plasmid DNA will yield transformants. However, we observe a small increase

in the number of transformants obtained when digesting constructs with a unique restriction endonuclease cutting the plasmid backbone but not inside the cluster, prior to transformation. Undesired plasmid backbone integration can occur in *F. graminearum* through non-homologous end joining. Alternatively, the gene cluster can be excised from the plasmid by restriction endonuclease digestion, purified from the backbone using gel electrophoresis and applied to transform protoplasts.

18. The protoplasting reaction must be stopped at an optimal time point: too early; can result in a low yield of protoplasts, too long; can significantly decrease the viability of the fragile protoplasts. The exact timing is difficult to predict due to activity differences in enzyme batches. We suggest monitoring the reaction in 15–20 min intervals. Keeping the enzymes in contact with the protoplasts for a prolonged period of time can significantly decrease their viability and thus the transformation efficiency.
19. Selective-overlay medium contains 80 μg/mL G418. By overlaying 15 mL regeneration medium with 15 mL selective-overlay medium, the effective G418 concentration is consequently halved to 40 μg/mL. Alternative selection markers for transforming protoplasts of *F. graminearum* comprise *hph*: 200 μg/mL hygromycin B in selective-overlay medium, *nat1*: 100 μg/mL nourseothricin in selective-overlay medium.
20. When transforming filamentous fungi such as *F. graminearum* it is relevant to analyze the genome for any additional or unintended recombination events, as the introduced DNA can recombine unspecifically with the genome through non-homologous end joining resulting in disruption of the native genetic integrity. Colony PCR utilizing primers binding outside the insertion site combined with primers annealing inside the introduced DNA can be applied to identify false positive colonies [23]. In addition, we recommend analyzing the entire genome for additional integration events (that is, integration of recombinant DNA outside the targeted locus in addition to integration in the desired locus) with either Southern blot or whole genome sequencing [10], since PCR cannot discriminate those cases.

Acknowledgments

This work was supported by the Novo Nordisk Foundation grants NNF15OC0016028 and NNF18OC0034952.

References

1. Keller NP, Hohn TM (1997) Metabolic pathway gene clusters in filamentous fungi. Fungal Genet Biol 21(1):17–29. https://doi.org/10.1006/fgbi.1997.0970
2. Hertweck C (2009) Hidden biosynthetic treasures brought to light. Nat Chem Biol 5(7): 450–452. https://doi.org/10.1038/nchembio0709-450
3. Brakhage AA (2013) Regulation of fungal secondary metabolism. Nat Rev Microbiol 11(1): 21–32. https://doi.org/10.1038/nrmicro2916
4. Keller NP (2019) Fungal secondary metabolism: regulation, function and drug discovery. Nat Rev Microbiol 17(3):167–180. https://doi.org/10.1038/s41579-018-0121-1
5. Billingsley JM, DeNicola AB, Tang Y (2016) Technology development for natural product biosynthesis in *Saccharomyces cerevisiae*. Curr Opin Biotechnol 42:74–83. https://doi.org/10.1016/j.copbio.2016.02.033
6. He Y, Wang B, Chen W, Cox RJ, He J, Chen F (2018) Recent advances in reconstructing microbial secondary metabolites biosynthesis in *Aspergillus* spp. Biotechnol Adv 36(3): 739–783. https://doi.org/10.1016/j.biotechadv.2018.02.001
7. Yap HYY, Muria-Gonzalez MJ, Kong BH et al (2017) Heterologous expression of cytotoxic sesquiterpenoids from the medicinal mushroom *Lignosus rhinocerotis* in yeast. Microb Cell Factories 16:103. https://doi.org/10.1186/s12934-017-0713-x
8. Yamanaka K, Reynolds KA, Kersten RD, Ryan KS, Gonzalez DJ, Nizet V, Dorrestein PC, Moore BS (2014) Direct cloning and refactoring of a silent lipopeptide biosynthetic gene cluster yields the antibiotic taromycin A. PNAS 111(5):1957–1962. https://doi.org/10.1073/pnas.1319584111
9. Clevenger K, Bok J, Ye R et al (2017) A scalable platform to identify fungal secondary metabolites and their gene clusters. Nat Chem Biol 13: 895–901. https://doi.org/10.1038/nchembio.2408
10. Nielsen MR, Wollenberg RD, Westphal KR, Sondergaard TE, Wimmer R, Gardiner DM, Sørensen JL (2019) Heterologous expression of intact biosynthetic gene clusters in *Fusarium graminearum*. Fungal Genet Biol 132: 103248. https://doi.org/10.1016/j.fgb.2019.103248
11. Kouprina N, Larionov V (2016) Transformation–associated recombination (TAR) cloning for genomics studies and synthetic biology. Chromosoma 125:621–632. https://doi.org/10.1007/s00412-016-0588-3
12. Noskov VN, Koriabine M, Solomon G, Randolph M, Barrett JC, Leem S-H, Stubbs L, Kouprina N, Larionov V (2001) Defining the minimal length of sequence homology required for selective gene isolation by TAR cloning. Nucleic Acids Res 29(6):e32. https://doi.org/10.1093/nar/29.6.e32
13. Kuijpers NG, Solis-Escalante D, Bosman L, van den Broek M, Pronk JT, Daran JM, Daran-Lapujade P (2013) A versatile, efficient strategy for assembly of multi–fragment expression vectors in *Saccharomyces cerevisiae* using 60 bp synthetic recombination sequences. Microb Cell Factories 12:47. https://doi.org/10.1186/1475-2859-12-47
14. Frandsen RJ, Frandsen M, Giese H (2012) Targeted gene replacement in fungal pathogens via *Agrobacterium tumefaciens*–mediated transformation. Methods Mol Biol 835:17–45. https://doi.org/10.1007/978-1-61779-501-5_2
15. Gietz RD, Schiestl RH (2007) High–efficiency yeast transformation using the LiAc/SS carrier DNA/PEG method. Nat Protoc 2(1):31–34. https://doi.org/10.1038/nprot.2007.13
16. Noskov VN, Lee NC, Larionov V et al (2011) Rapid generation of long tandem DNA repeat arrays by homologous recombination in yeast to study their function in mammalian genomes. Biol Proced 13:8. https://doi.org/10.1186/1480-9222-13-8
17. Varga E, Wiesenberger G, Hametner C, Ward TJ, Dong Y, Schöfbeck D, McCormick S, Broz K, Stückler R, Schuhmacher R, Krska R, Kistler HC, Berthiller F, Adam G (2015) Discovery of a novel type A trichothecene. Environ Microbiol 17:2588–2600. https://doi.org/10.1111/1462-2920.12718
18. Smedsgaard J (1997) Micro–scale extraction procedure for standardized screening of fungal metabolite production in cultures. J Chromatogr A 760(2):264–270. https://doi.org/10.1016/s0021-9673(96)00803-5
19. Jørgensen SH, Frandsen RJ, Nielsen KF, Lysøe E, Sondergaard TE, Wimmer R, Giese H, Sørensen JL (2014) *Fusarium graminearum* PKS14 is involved in orsellinic acid and orcinol synthesis. Fungal Genet Biol 70:24–31. https://doi.org/10.1016/j.fgb.2014.06.008
20. Klitgaard A, Iversen A, Andersen MR et al (2014) Aggressive dereplication using UHPLC–DAD–QTOF: screening extracts for

up to 3000 fungal secondary metabolites. Anal Bioanal Chem 406:1933–1943. https://doi.org/10.1007/s00216-013-7582-x

21. Sørensen JL, Hansen FT, Sondergaard TE, Staerk D, Lee TV, Wimmer R, Klitgaard LG, Purup S, Giese H, Frandsen RJN (2012) Production of novel fusarielins by ectopic activation of the polyketide synthase 9 cluster in *Fusarium graminearum*. Environ Microbiol 14:1159–1170. https://doi.org/10.1111/j.1462-2920.2011.02696.x
22. Westphal KR, Nielsen KAH, Wollenberg RD, Møllehøj MB, Bachleitner S, Studt L, Lysøe E, Giese H, Wimmer R, Sørensen JL, Sondergaard TE (2019) Fusaoctaxin A, an example of a two-step mechanism for non-ribosomal peptide assembly and maturation in fungi. Toxins 11(5):277. https://doi.org/10.3390/toxins11050277
23. Wollenberg RD, Saei W, Westphal KR, Klitgaard CS, Nielsen KL, Lysøe E, Gardiner DM, Wimmer R, Sondergaard TE, Sørensen JL (2017) Chrysogine biosynthesis is mediated by a two-module nonribosomal peptide synthetase. J Nat Prod 80(7):2131–2135. https://doi.org/10.1021/acs.jnatprod.6b00822
24. Pronobis MI, Deuitch N, Peifer M (2016) The Miraprep: a protocol that uses a Miniprep kit and provides Maxiprep yields. PLoS One 11(8):e0160509. https://doi.org/10.1371/journal.pone.0160509
25. Sørensen JL, Benfield AH, Wollenberg RD, Westphal K, Wimmer R, Nielsen MR, Nielsen KF, Carere J, Covarelli L, Beccari G, Powell J, Yamashino T, Kogler H, Sondergaard TE, Gardiner DM (2018) The cereal pathogen *Fusarium pseudograminearum* produces a new class of active cytokinins during infection. Mol Plant Pathol 19:1140–1154. https://doi.org/10.1111/mpp.12593
26. Yin WB, Chooi YH, Smith AR, Cacho RA, Hu Y, White TC, Tang Y (2013) Discovery of cryptic polyketide metabolites from dermatophytes using heterologous expression in *Aspergillus nidulans*. ACS Synth Biol 2(11):629–634. https://doi.org/10.1021/sb400048b

Chapter 5

Heterologous Expression of Fungal Biosynthetic Pathways in *Aspergillus nidulans* Using Episomal Vectors

Indra Roux and Yit Heng Chooi

Abstract

Filamentous fungi produce a wide diversity of secondary metabolites, whose biosynthesis is encoded in biosynthetic gene clusters (BGCs). As novel BGCs are often found in fungal species that are genetically intractable or difficult to cultivate, heterologous expression is increasingly being used for compound discovery. In addition, heterologous expression is a useful strategy to elucidate the function of the genes within a BGC and shed light on their enzymatic mechanisms. Here, we describe a method for BGC elucidation using multi-marker AMA1-based pYFAC vectors for episomal expression in the fungal host *Aspergillus nidulans*. The pYFAC vectors have the advantage of high transformation efficiency and support high compound production. In addition, different pathway intermediates can be easily evaluated by testing different vector combinations. This protocol encompasses different AMA1-based strategies for BGC expression such as cloning of a BGC native sequence, promoter exchange or transcription factor overexpression. We also describe procedures for *A. nidulans* protoplasting, transformation, and small-scale culture analysis of strains containing AMA1 vectors.

Key words *Aspergillus nidulans*, Heterologous expression, AMA1, Polyketide synthase, Non-ribosomal peptide synthetase, Filamentous fungi, Biosynthetic gene clusters, Secondary metabolites, Synthetic biology

1 Introduction

Fungal secondary metabolites (SMs), such as polyketides and non-ribosomal peptides, display diverse structures and bioactivities. The pathways responsible for the biosynthesis of fungal SMs are encoded on biosynthetic gene clusters (BGCs) [1]. The increasing amount of BGCs identified in the genomes of filamentous fungi represents an almost untapped resource for drug discovery. However, most BGCs remain silent or lowly expressed under laboratory culture conditions and their encoded compounds are not detected. Several strategies can be used to manipulate silent BGCs for increased SM production in fungal genome mining [2]. Heterologous expression of BGCs in well established host is becoming a

Elizabeth Skellam (ed.), *Engineering Natural Product Biosynthesis: Methods and Protocols*,
Methods in Molecular Biology, vol. 2489, https://doi.org/10.1007/978-1-0716-2273-5_5,

popular strategy for accessing the chemistry encoded by cryptic BGCs, especially when they are identified in fungal species that are difficult to cultivate or genetically intractable [3]. Additionally, heterologous expression via step-wise pathway reconstitution is a powerful approach for elucidation of SM pathways and biosynthetic mechanisms, which could inform the discovery of related pathways and expand the enzymatic toolbox for engineering of novel pathways [2, 4, 5].

Filamentous fungal hosts are often preferred for the expression of BGCs from filamentous fungi, due to compatibility with mRNA splicing, the presence of specialized cellular compartments and secretion systems, and overall better genetic compatibility than bacterial or yeast hosts [3, 6, 7]. Thus, BGC sequences can be obtained by PCR amplification [8] or genome capture [9] bypassing the need for intron removal or codon optimization. The fungus *Aspergillus nidulans* has been successfully used as a heterologous host for the expression of BGCs from diverse fungi [10]. Strains used for heterologous expression have been modified for auxotrophic selection (e.g. *pyroA4*, *riboB2*, *pyrG89*), reduced random integrations (*nkuA*Δ) and a cleaner metabolic background (e.g. *ST*Δ) [11, 12]. Additionally, *A. nidulans* has a well-established genetic toolbox, that includes autonomously replicating episomal vectors based on the AMA1 sequence [13]. AMA1 vectors maintained by auxotrophic pressures are increasingly being used for heterologous expression [10]. Additionally, many genetic parts, such as promoters and terminators, have been shown to be effective for the episomal expression of BGCs [8, 14] . Here, we describe strategies for AMA1-based expression of heterologous BGCs in *A. nidulans* using multi marker pYFAC vectors. The backbone of pYFAC vectors was derived from the pKW20088 carrying *pyrG* marker developed by Watanabe lab [15]. Our lab subsequently expanded the vectors to *riboB* and *pyroA* markers, and additional vectors with multiple alcohol responsive promoters for inducible expression [8]. The vector set also facilitates pathway elucidation by easily evaluating different biosynthetic gene combinations in a modular manner (Fig. 1). In our hands, this system enables robust and high production of heterologous secondary metabolites. Additionally, the AMA1 sequence increases transformation efficiency, which minimizes the number of protoplasts needed per transformation reaction [13].

The protocol below can be used to explore BGCs from diverse fungi from the Pezizomycotina subphylum [8, 16, 17]. BGCs can be identified with bioinformatic tools using annotated fungal genomes obtained de novo or sequences available on databases such as FUNGIDB [19–20]. A popular algorithm to identify BGCs is antiSMASH [21], and candidate clusters can be compared by homology to known clusters in MIBIG [22] or to other BGC databases with cblaster [23]. Various strategies to prioritize BGCs

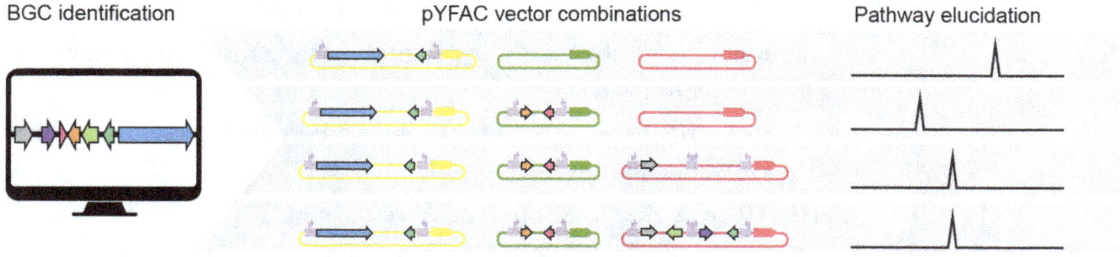

Fig. 1 Strategy for heterologous pathway elucidation in *Aspergillus nidulans* by evaluating AMA1-based pYFAC vector combinations. In the schematic, the candidate genes identified in a silent or lowly expressed biosynthetic gene cluster (BGC) are cloned under strong promoters in pYFAC vectors with the markers *pyrG*, *riboB*, and *pyroA* (yellow, green, and magenta). The modular and combinatorial nature of pYFAC vectors facilitates creating strains for heterologous expression of pathway intermediates

in genome mining have been discussed previously and will increase the odds for the discovery of novel SMs with desired biological activities or properties [24, 25].

2 Materials

2.1 Cloning of Biosynthetic Genes on AMA1 Vectors

1. Vectors with the AMA1 sequence, auxotrophic markers (*pyrG*, *pyroA*, *riboB*), and components for replication in *Escherichia coli* and *Saccharomyces cerevisiae*, such as the pYFAC vectors. Empty pYFAC vectors with the markers *pyrG*, *riboB*, and *pyroA* are available via Addgene under IDs #168982, #168975, #168976, respectively [8, 26]. The vectors pYFAC-CH2/3/4 containing the empty four alcohol inducible promoter cloning site are available via Addgene under IDs #168978, #168979, #168980 [8]. Alternatively, other vectors containing AMA1 sequence and auxotrophic markers are available via FGSC (Fungal Genetics Stock Centre) [27]. Other AMA1-based vectors such as pYTU, pYTP and pYTR by Tang lab and pKW20088 by Watanabe lab have also been demonstrated for BGC expression [14, 15].
2. Software for DNA sequence visualization, editing, and primer design.
3. PCR primers for BGC amplification, colony PCR, and sequencing (*see* **Note 1**).
4. PCR reagents and enzymes of choice. We use high-fidelity DNA polymerase such as PfU, Q5 High-Fidelity (New England Biolabs) or PrimeSTAR GXL (Takara Bio).
5. Genomic DNA (gDNA) from the fungal strain containing the BGC of interest (*see* **Note 2**).
6. PCR thermocycler with the option to decrease temperatures inside a loop and programmable ramps, and PCR tubes.

7. Stained agarose gel and gel purification kit.
8. Restriction enzymes for plasmid linearization and analysis of digestion pattern.
9. Reagents for in vitro homology-based cloning (also known as Gibson assembly).
10. 10-beta electrocompetent *E. coli*.
11. *S. cerevisiae URA3* deficient strain, e.g., strain BJ5464.
12. Yeast Transformation Kit.
13. Yeast Plasmid MiniPrep Kit.
14. Liquid and solid Luria-Bertani (LB) media and 50 mg/mL ampicillin solution to use at final concentration 50 μg/mL.
15. Drop-out ura medium lacking uracil: For 1 L of media dissolve 6.7 g of yeast nitrogen base without amino acids, 20 g of glucose, 1.4 g of yeast synthetic drop-out medium supplements, 76 mg of histidine, 76 mg of tryptophan, and 380 mg of leucine. To prepare the solid version of the media add 20 g of agar per liter.

2.2 *Aspergillus nidulans* Strain Construction and Analysis

1. *Aspergillus nidulans* engineered triple auxotrophic strain with the mutations *pyrG89* (pyrimidine auxotrophy), *riboB2* (riboflavin auxotrophy), and *pyroA4* (pyridoxine auxotrophy), e.g. *A. nidulans* FGSC A1145 (TN02A7) from the Fungal Genetics Stock Center [12]. For a cleaner metabolic background, strains with the sterigmatocystin gene cluster deleted such as *A. nidulans* LO4389 [28] or additional BGCs deleted such as *A. nidulans* A1145*ΔSTΔEM* [29], LO7890 or LO8030 [11] are preferred.
2. Glucose minimal media (GMM): For 1 L dissolve in ultrapure water 10 g glucose, add 20 mL of 50× nitrate salts solution, and 1 mL of trace elements solution, and adjust the pH to 6.5 with NaOH solution. To prepare 300 mL of 50× nitrate salts solution, dissolve 90 g $NaNO_3$, 7.8 g KCl, 7.8 g $MgSO_4 \cdot 7H_2O$, and 22.8 g KH_2PO_4, autoclave and store at room temperature. To prepare 100 mL of trace elements solution dissolve 2.20 g of $ZnSO_4 \cdot 7H_2O$, 1.10 g of H_3BO_3, 0.50 g of $MnCl_2 \cdot 4H_2O$, 0.50 g of $FeSO_4 \cdot 7H_2O$, 0.16 g of $CoCl_2 \cdot 5H_2O$, 0.16 g of $CuSO_4 \cdot 5H_2O$, 0.11 g of $(NH_4)_6Mo_7O_{24} \cdot 4H_2O$, and 6.00 g of Na_4EDTA (*see* **Note 3**). To prepare solid GMM, also add 16 g of agar per liter. If supplements are needed, add 10 mL/L of riboflavin 0.025% w/v stock (filter-sterilized added after autoclaving, stored in the dark at 4 °C), 1 mL/L of pyridoxine-HCl stock solution of 0.2% w/v (filter-sterilized added after autoclaving, stored at 4 °C), or uridine 1.26 g/L and uracil 0.56 g/L (added to the media before autoclaving).

3. Stabilized minimal media (SMM): Solid GMM with an additional 218.6 g of sorbitol per liter.
4. Mycelium washing solution: For 1 L, dissolve in ultrapure water 147.9 g of $MgSO_4 \cdot 7H_2O$ (0.6 M). Autoclave and store at 4 °C.
5. Osmotic Buffer: For 500 mL final volume, first completely dissolve 147.9 g of $MgSO_4 \cdot 7H_2O$ (1.2 M) in 10 mM equimolar sodium phosphate buffer (*see* **Note 4**). Adjust the pH to 5.8 with 1 M Na_2HPO_4 (*see* **Note 4**). Filter-sterilize and store at 4 °C.
6. Trapping Buffer: For 1 L of solution dissolve in ultrapure water 109.3 g of sorbitol (0.6 M) adding Tris-HCI to a final concentration of 0.1 M. Adjust pH to 7.0, autoclave and store at 4 °C.
7. STC buffer: For 1 L of ST solution dissolve 218.6 g of sorbitol (1.2 M) adding Tris-HCI to 10 mM (10 mL of 1 M Tris-HCl, pH 7.5). Autoclave and store at 4 °C. Separately, prepare 1 M $CaCl_2$, filter-sterilize and store at −20 °C. Add $CaCl_2$ to a final concentration of 10 mM to prepare STC buffer.
8. Polyethylene glycol (PEG) solution with $CaCl_2$: This solution contains PEG 4000 60% w/v, $CaCl_2$ 50 mM, Tris-HCl 50 mM, pH 7.5. To prepare 100 mL solution, we add the PEG directly into the bottle, where the exact level of 95 mL is previously marked, and dissolve it in 95 mL of ultrapure water with Tris-HCl. Warming facilitates dissolving. Before autoclaving place a magnetic stirrer bar inside the bottle. Add 5 mL of filtered 1 M $CaCl_2$ into the autoclaved PEG solution and mix by stirring. Maintain the PEG solution at room temperature to avoid precipitation and stir before using.
9. 0.1% (w/v) Tween-80 solution in ultrapure water, filter sterilized.
10. Sterile 15 and 50 mL conical centrifuge tubes (Falcon tubes).
11. Sterile 5 mL syringe, sterile cotton, and tweezers.
12. Extralyse enzyme mix (Laffort).
13. Centrifuge with a swinging bucket and cold temperature option (4 °C).
14. 250 mL conical flasks with cotton plugs (sterile).
15. Incubator shaker with clamps for 250 mL conical flasks.
16. Sterile glass test tubes with cap.
17. Syringe and 0.22 μm sterile syringe filters.
18. Miracloth.
19. Sterile pipette tips (10, 200, and 1000 μL) and sterile 1.5 and 2 mL microcentrifuge tubes.

20. Sterile 90 mm petri dishes.
21. Phase contrast microscope (40× magnification objective) and cell counting chamber.
22. Sterile wooden cotton buds and Drigalski glass spatula.
23. 10 mL tips for solvent manipulation.
24. Ethyl acetate, methanol, acetic acid, acetone.
25. Sonicator.
26. Liquid chromatography coupled to a diode array detector and mass spectrometer system (LC-DAD-MS), vials and inserts.

3 Methods

3.1 Cloning of Biosynthetic Genes on AMA1 Vectors

There are multiple strategies to express a BGC using AMA1 vectors. The choice of strategy depends on several variables such as number of genes in the cluster, presence of cluster-specific transcription factors (TFs), interest on BGC elucidation, ease of cloning, etc. For example, if the BGC is actively expressed in the host strain, complete reconstitution of the gene cluster under their native regulatory regions on pYFAC AMA1-based vectors often leads to compound production [30]. In some cases, cloning of BGCs that are silent on their native context on AMA1 vectors can also lead to detectable compound production [9, 27, 31]. However, BGC overexpression is often needed to achieve high titers, especially for BGCs silent on their native host. If the BGC contains a putative TF gene, we often opt for TF overexpression and cloning of the rest of the cluster genes under their native promoter. TF overexpression can be achieved by cloning the TF under a strong promoter such as *gpdA* (P_{gpdA}) or alcohol inducible promoters in an AMA1 vector [8, 30].

For BGCs with no cluster-specific TF available, or when TF overexpression results in low titers, expression of all the genes from a BGC under strong promoters is a robust alternative. The vectors pYFAC-CH2/3/4 contain four alcohol inducible promoters per vector to test up to 12 genes simultaneously using the three auxotrophic markers (Fig. 2) [8]. Using pYFAC-CH2/3/4 vectors for expression is also robust strategy to obtain high compound titers across diverse BGCs, facilitating downstream purification. By evaluating different gene combinations or truncated pathway versions, different pathway intermediates can be obtained, which also facilitates the elucidation of the biosynthetic pathway [31].

All pYFACs AMA1-based vectors contain components for replication and selection on both *E. coli* and *S. cerevisiae* (Fig. 2). Thus, cloning can be performed using in vitro isothermal homology-based assembly (also known as Gibson assembly) or in vivo transformation assisted recombination (TAR) in *S. cerevisiae*. While the

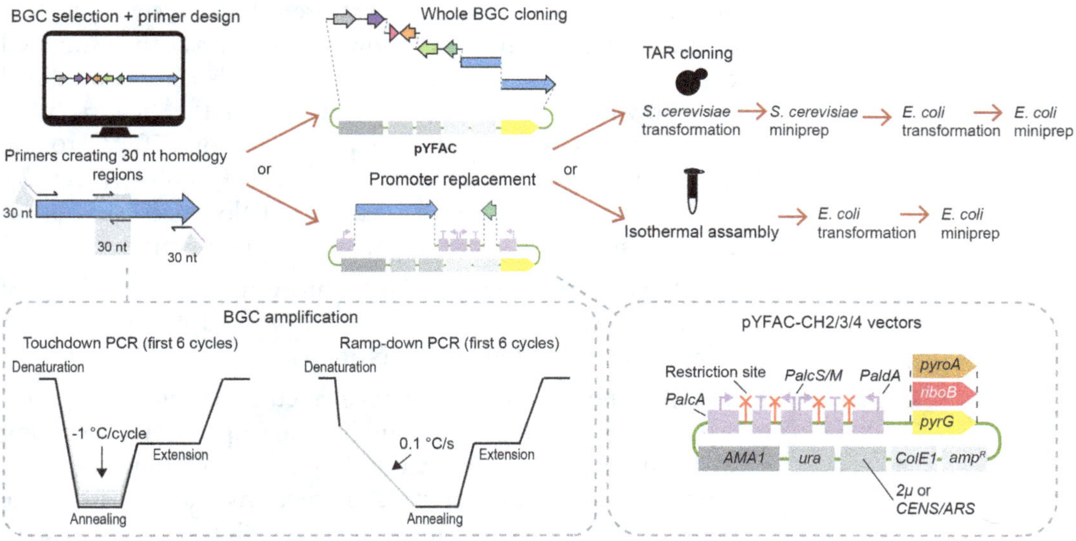

Fig. 2 Overview of cloning strategies for AMA1-based expression in *Aspergillus nidulans*. We often clone the whole biosynthetic gene cluster (BGC) under their native promoter sequence or replace just the promoter of the cluster-specific transcription factor gene for BGC overexpression. Alternatively, we employ promoter replacement of all the genes in the BGC, using pYFAC-CH2/3/4 vectors. To amplify BGC fragments, we design PCR primers with 30 nt homology overlaps that can be used for either in vitro isothermal assembly or in vivo yeast cloning. To increase PCR success, we add a loop of touchdown or ramp-PCR in the first 6 cycles of the reactions (total of 34 cycles). The components or pYFAC-CH2/3/4 vectors are indicated in the scheme

overall time of cloning by isothermal assembly is faster (~3-days cloning), TAR cloning is a reliable alternative for the assembly of large vectors (~6-days cloning). We usually design primers that create 30-nt homology overlaps between fragments, so the same set of PCR fragments can be used for isothermal assembly or TAR cloning, increasing the chances of successful cloning.

1. Design primers to create overlapping PCR fragments of 4–5 kb spanning across the BGC for whole cluster reconstruction or the coding region of the gene of interest if using promoter replacement (Fig. 2). Verify that the region of the primer that anneals to the genomic DNA (gDNA) template has a melting temperature of 60 °C. For the fragments containing the 5′ and 3′ ends for assembly on the vector, design primers that contain a 30-nt overhang with homology to the ends of the digested vector. For the primers amplifying junctions between genomic fragments, 30-nt of homology can be created by customizing the distance between the Fw and Rv primers of adjacent fragments (*see* Fig. 2).
2. Set up 50–100 μL PCR reactions using high-fidelity polymerase. For successful PCR amplification is important to have good quality gDNA template (*see* **Note 1**). We usually set a variation of touchdown PCR [32], starting 6 °C above the estimated annealing temperature (T_m −3 °C) and sequentially

decreasing the annealing temperature −1 °C/cycle in the first 6 cycles (Fig. 2). Then we follow by 28 cycles at the estimated annealing temperature. Alternatively, we also use a modified rampdown-PCR protocol [33], where in the first 6 cycles the annealing temperature ramps down from 70 °C to the final annealing temperature at a rate of 0.1 °C/s (*see* Fig. 2). Both strategies favor specificity and can help troubleshoot problems arising from secondary structure in long primers.

3. Run the PCR product on a 0.8% agarose gel. Excise the band of the expected amplicon size from the gel and purify the DNA using a gel extraction kit (*see* **Note 5**).
4. The vector assembly can be performed by in vitro isothermal assembly or in vivo yeast recombination (Fig. 2). When cloning by isothermal assembly we use the master mix following proportions as per manufacturer's instructions, but we often scale the reaction down to ~3 μL volume in PCR tubes (*see* **Note 6**). When the reaction is completed, use 1.5 μL to transform by electroporation 10-beta electrocompetent *E. coli* as per manufacturer instructions (*see* **Note 7**). Screen colonies for the desired transformation event by colony PCR or by proceeding to miniprep and digestion pattern analysis. Standard alkaline lysis miniprep usually yields 1–4 μg/μL of pYFAC-derived vector as measured by NanoDrop. If cloning by TAR, transform *S. cerevisiae* with the PCR fragments and digested vector using Frozen-EZ Yeast Transformation II Kit (ZymoResearch) following the manufacturer's instructions (*see* **Note 8**). Grow the cells for 2 days on dropout medium lacking uracil (or relevant auxotrophic marker). Select colonies for yeast miniprep (*see* **Note 9**). Yeast miniprep can be performed with a kit, such as ZymoPrep Yeast Plasmid MiniPrep I Kit (*see* **Note 10**). Amplify the vector by electroporating ~1.2 μL of the yeast miniprep to *E. coli*, and proceeding to *E. coli* miniprep (Fig. 2) (*see* **Note 11**).
5. Verify the vectors by observing digestion pattern and sequencing of the cloning junctions or whole vector.

3.2 Preparation of *Aspergillus nidulans* Protoplasts

One of the advantages of using AMA1 vectors is that they support high transformation efficiency in *A. nidulans*. As the number of protoplasts used per reaction can be scaled down, the following procedure yields protoplasts for 15–25 transformation events with AMA1-based vectors. Additionally, protoplasts can be aliquoted and stored at −70 °C for a couple of months with reliable transformation efficiency. The protoplasting procedure is derived from Yelton et al. and Lim et al. with modifications [34, 35].

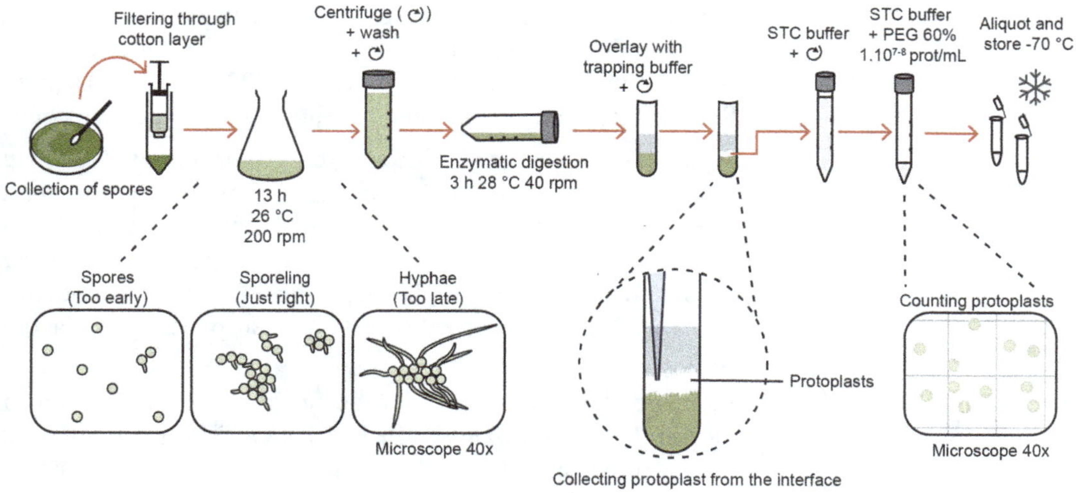

Fig. 3 Overview of *Aspergillus nidulans* protoplast preparation. After 13 h of overnight incubation, the protoplast are prepared from sporelings. From sporeling collection to protoplast storage the procedure can be completed in $\leq$6 h

1. Before starting, prepare washing solution, osmotic buffer, trapping buffer and STC buffer (*see* Subheading 2). Verify having all the sterile materials needed.
2. Inoculate the triple auxotrophic *A. nidulans* strain across the surface of 1–2 solid GMM plates supplemented with riboflavin, pyridoxin, uracil, and uridine. Incubate at 37 °C for 3 days.
3. Collect the conidia from the plate of *A. nidulans* by adding ~3 mL of filter sterilized 0.1% Tween-80 solution or media, and gently scraping the plate with a sterile cotton bud to create a spore suspension. With sterile tweezers, grab autoclaved cotton and place it inside a 5 mL sterile syringe inserted into a 15 mL conical centrifuge tube (*see* Fig. 3), creating a layer of ~2 cm of cotton inside the syringe. The spore suspension is gently collected with a pipette and passed through the cotton to filter residual hyphae. Inoculate with the spore suspension 100 mL of GMM with all supplements and ampicillin at 50 μg/mL in a sterile 250 mL flask capped with a cotton plug.
4. Incubate the flask for 13 h at 26 °C and 200 rpm shaking, until most of the spores have started germinating to sporelings. This can be assessed with a microscope using a 40× objective and observing individual sporelings with small hyphal elongation or by seeing small aggregates of spores and sporelings (*see* Fig. 3). For high protoplasting efficiency, it is important to stop the culture in the sporeling stage before long hyphae are formed (*see* Fig. 3).
5. Transfer the culture to two 50 mL conical centrifuge tubes, and centrifuge at 4 °C and 3220 × *g* for 15 min. Remove the

supernatant from each tube and resuspend in 50 mL of washing solution. In this step the spores can be pooled in one of the conical tubes. Centrifuge at 4 °C and 3220 × *g* for 15 min. Remove the supernatant gently without disrupting the pellet.

6. Separately prepare the protoplasting solution by dissolving ~0.6 g of Extralyse in 10 mL of osmotic buffer (*see* **Note 12**). Filter-sterilize the protoplasting solution with a 0.22 μm syringe filter (low protein binding recommended). Transfer the 10 mL of sterile protoplasting solution to the tube containing the fungal pellet, thoroughly resuspending to avoid clumps.
7. Incubate the closed conical tube horizontally in a shaker at 40 rpm for 2.5 h at 28 °C (*see* Fig. 3). Optionally, the state of the protoplasting can be observed by analyzing a small sample by microscopy, as the protoplasts are observed as thin-walled and double the size of the spores.
8. Transfer 3 mL of the cell suspension into sterile 15 mL glass test tubes (3–4 tubes). Gently overlay with 3 mL of chilled trapping buffer by slowly ejecting the trapping buffer on the side of the glass tubes (*see* Fig. 3). It is important to be gentle in the manipulation of the tubes to avoid disturbance in the interface between layers. Centrifuge the closed tubes for 15 min at 4 °C and 3220 × *g* using slow braking and slow acceleration.
9. Remove the protoplasts from the interface and collect them in a sterile 15 mL centrifuge tube hold on ice (~1 mL per glass tube). The protoplasts are observed as a cloudy layer in the interface between solutions and are collected by pipetting through the trapping buffer layer with a P1000 pipette (*see* Fig. 3).
10. Add 2 volumes of cold STC buffer to the protoplasts and centrifuge at 4 °C at 3220 × *g*. Gently pour the supernatant without disrupting the protoplast pellet.
11. Resuspend the pellet in 1 mL cold STC. Take 2 μL of the protoplasts and dilute 1/10 in STC to count with a hemocytometer. Once the protoplast concentration is known, dilute the protoplast solution to 1×10^7–1×10^8 protoplasts/mL with STC and ¼ final volume of PEG solution (*see* **Note 13**) (usually a final volume of ~2 mL).
12. Aliquot 100 μL of the protoplast mix in 2 mL microcentrifuge tubes, wrap the tubes in paper for slow freezing, and store at −70 °C (or −80 °C) for cryopreservation (*see* **Note 14**).

3.3 Transformation of *Aspergillus nidulans*

Transformation of *A. nidulans* with AMA1-based vectors is a reliable procedure that usually yields 10–100 colonies (Fig. 4). For novel SM identification, strains containing the BGCs encoded on

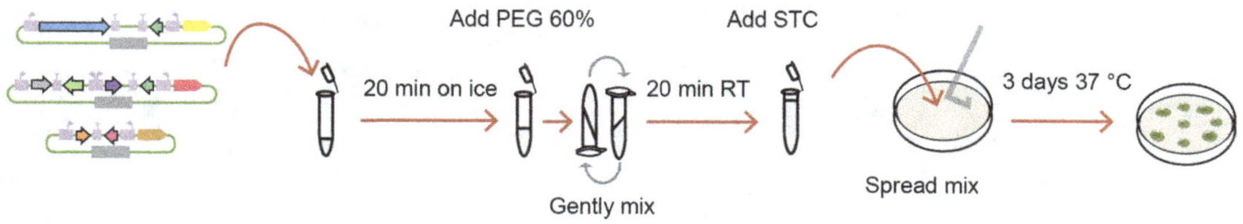

Fig. 4 Overview of polyethylene glycol (PEG)-mediated transformation of *Aspergillus nidulans* with AMA1-vectors. Room temperature (RT). Stabilized minimal media (SMM)

AMA1 vectors are compared with control strains containing the empty AMA1 vectors.

1. Thaw on ice the 2 mL tubes containing 100 μL of cryopreserved protoplasts. Add 40 μL of STC buffer and 2–3 μL of each vector (~3μg of vector DNA as measured by Nanodrop). Gently mix by tapping the tube. Incubate on ice for 20 min. While incubating, place the closed PEG solution in the magnetic stirrer to mix it before the next step (*see* **Note 13**).
2. Add 450 μL PEG solution to the tube, mix gently by inversion (Fig. 4). Optionally, the PEG solution can be added incrementally (225 μL and mix, and then add 225 μL and mix again). Incubate at room temperature for 20 min.
3. Add ~1.5 mL STC buffer to a total volume of 2 mL. Mix by inversion and transfer 1 mL per plate to two plates of selective media, spreading gently with a sterile Drigalski spatula (Fig. 4) (*see* **Note 15**).
4. Incubate for 3 days at 37 °C. Individual colonies should be observed.

3.4 Small-Scale Culture and Metabolic Profile Analysis

Extracts from small-scale liquid cultures can be screened for new compounds by chromatographic techniques such as LC-DAD-MS. Candidate peaks are identified by comparing the metabolic profile of strains expressing the BGC of interest to control strains harboring empty vectors.

1. Per strain, select three individual colonies for biological replicates and spread each onto the surface of an individual GMM plate. Grow the strains for 3 days at 37 °C.
2. Under sterile conditions, inoculate 50 mL of LMM in a 250 mL conical flask to a final concentration of ~2×10^6 spores/mL (*see* **Note 16**). This is usually equivalent to a third of a plate of spores collected by scraping with 0.1% Tween-80. Prepare a ~20% glycerol stock with the rest of the spores and store at −70 °C to preserve the strains for large-scale cultures.
3. If alcohol inducible promoters are used for expression, incubate the flasks at 37 °C and shaking at 200 rpm for 18 h. After

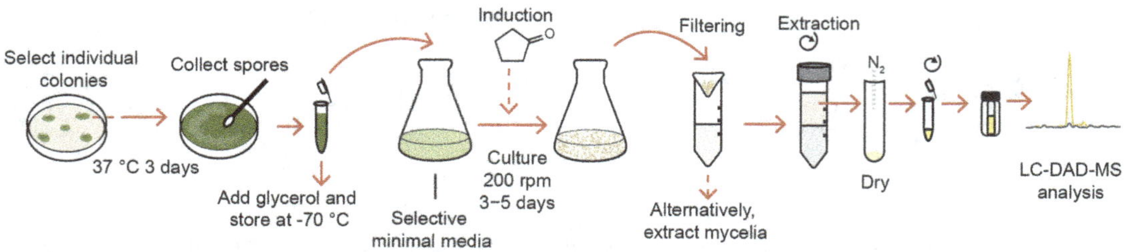

Fig. 5 Overview of small-scale liquid culture of *Aspergillus nidulans* harboring AMA1-based vectors and downstream secondary metabolite analysis by liquid chromatography coupled to a diode array detector and mass spectrometry (LC-DAD-MS). If alcohol inducible promoters are used, cyclopentanone is added for induction 18 h after the start of the culture. New peaks are identified by comparing the metabolic profile of the strains expressing the heterologous BGC (yellow) with the control strains harboring empty vectors (grey)

that time add 133 μL of cyclopentanone for induction and lower the temperature to 25 °C until the end of the culture [35]. If culturing for 4 days, after 2 days add an additional 133 μL of cyclopentanone (*see* **Note 17**). If the genes are expressed under a constitutive promoter, it is recommended to cultivate 3–5 days at a single temperature that reflects the optimal of the organism from which the BGC is originated, e.g., 25 °C when expressing BGCs from fungi that grow in temperate weather.

4. At the end of the culture, mycelia are usually observed as small pellets (*see* **Note 18**). For analysis of SM in the culture media, filter the culture through Miracloth to collect 20 mL of the media on a 50 mL conical centrifuge tube (Fig. 5).
5. Add 20 mL of ethyl acetate:methanol:acetic acid 89.5:10:0.5 to the tubes, and mix by inversion of the closed tubes in a fume hood, opening the lid between mixing to release pressure (*see* **Note 19**). Centrifuge to separate the phases.
6. With a 10 mL pipette transfer the organic phase to a glass tube. Dry the organic layer under a stream of nitrogen gas or by vacuum drying.
7. Generally, SMs are secreted to the media, but in some cases mycelia extraction is necessary [36]. In those cases, collect mycelia by vacuum filtering with a Büchner-Kitasato flask and funnel covered with a filter paper. This minimizes the amount of water in the mycelia. Transfer the mycelia to a 50 mL conical centrifuge tube, add 15 mL of acetone, and sonicate for 15 min twice. Centrifuge and transfer 5–10 mL into a clean glass tube. Dry the organic layer under a stream of nitrogen gas or vacuum drying.
8. Resuspend the dried crude extract in 300 μL MeOH by sonication and transfer to 1.5 mL microcentrifuge tubes. Centrifuge at maximum speed (~20,238 × *g*) for 10 min. Without

disturbing the pellet transfer the extract to a vial with a new insert. Optionally, clean the samples with HPLC filters.

9. Run samples in a LC-DAD-MS. We usually use a linear gradient of 5–95% acetonitrile–water (containing 0.1% v/v formic acid) in 10 min followed by 95% acetonitrile for 3 min, with a flow rate of 0.70 mL/min, detecting ions in a *m/z* range 100–1000 in both positive and negative ion mode and analyzing DAD signals 220–440 nm.

10. Candidate peaks are identified by comparing DAD chromatogram or total ion chromatogram between strains containing the BGC of interest to control strains. Verify consistency between biological replicates when identifying new peaks. Samples from small-scale cultures can also be analyzed by sensitive techniques such as LC-DAD-MS/MS. For scaling up for compound purification, use the glycerol stock from the strains with novel SM production to inoculate plates for 2 L of large-scale culture (*see* **Note 20**). After culturing and purification, proceed with NMR and X-ray crystallography analysis of the compounds of interest. Methods for compound purification and analysis is beyond the scope of this protocol and can vary from compound to compound (*see* **Note 21**).

4 Notes

1. We often design primers that can be used for both Sanger sequencing and for colony PCR. If available, next generation whole plasmid sequencing can be used instead of Sanger sequencing to confirm correct vector assembly.

2. Good quality gDNA can be visualized on a gel as a single band. We usually obtain gDNA from ground frozen mycelium incubated in 10 mM Tris, 10 mM EDTA, and 1% SDS at 70 °C for 30 min and 400 rpm. After precipitation of the supernatant with equal volume of 2.8 M potassium acetate, samples are cleaned up extracting with chloroform:isoamyl alcohol (24: 1). The aqueous phase is precipitated with 0.7 volume of isopropanol and centrifugation and washed with 70% ethanol. The DNA Pellet is resuspended in TE buffer. gDNA is also RNAse treated.

3. When preparing 100 mL of trace elements solution, first dissolve in 80 mL of MQ water following the order of reagents as indicated, making sure each is dissolved and then make up the volume to 100 mL. Adding KOH can help to fully dissolve. Store at 4 °C.

4. The 10 mM sodium phosphate buffer can be prepared from a 1 M NaH_2PO_4 and 1 M Na_2HPO_4 stocks to create an

equimolar solution with pH 6.8. Once the $MgSO_4 \cdot 7H_2O$ is fully dissolved, adjusting the pH with Na_2HPO_4 can take time. Do not adjust the pH with NaOH or use heat to dissolve, as it will form precipitate.

5. If the amounts of DNA recovered are low, use a vacuum concentrator before cloning.
6. We generally use NEBuilder HiFi DNA Assembly Master Mix (New England Biolabs, MA, USA) for in vitro isothermal assembly. Alternatively, we use the Hot Fusion mix developed by Fue et al. [37].
7. Using cells that support high transformation efficiency for large vectors is key to cloning success with AMA1 vectors, which are usually ≥14 kb. Nevertheless, we obtain good results using diluted commercial 10-beta electrocompetent *E. coli* (New England Biolabs). We keep 5 μL aliquots that, once thawed, we dilute 1/10 with cold sterile glycerol 10% to a final volume of 50 μL before use.
8. Alternatively, transform yeast by the lithium acetate/single-stranded carrier DNA/PEG method from Gietz and Schiestl [38].
9. For TAR, we particularly encourage to also transform a negative control with the digested backbone only. Visual comparison of the number of colonies on the TAR transformants and the control is a good predictor of successful cloning. Alternatively, do a yeast colony PCR on the TAR transformants. For yeast colony PCR, first aliquot 15 μL of 20 mM NaOH into PCR tubes, inoculate with a part of the colony and incubate at 100 °C for 10 min. Centrifuge and use 1 μL of supernatant as template in a 10 μL PCR reaction for screening.
10. Alternatively, the yeast cells can be protoplasted before lysis using the same solutions as for preparing *A. nidulans* protoplasts (*see* Subheading 3.2). For that, we grow the yeast colonies overnight in 3 mL of liquid drop-out medium, and we resuspend the pellet in 2 mL of osmotic buffer with 100 mg of Extralyse (no need to sterilize). After 1.5 h of digestion, we proceed with an overlay of trapping buffer and collection of the protoplasts (similar to Fig. 3), and centrifuge after mixing with 2 volumes ST buffer. The miniprep can then be carried out with the protoplast pellet using standard alkaline lysis with the same solutions as for *E. coli* miniprep.
11. To avoid arcing during electroporation when using DNA from yeast miniprep, we often clean up the yeast miniprep by passing it through a Sephadex G-25 spin column or by repeating the washing step of the miniprep.

12. Alternatively, 30 mg Lysing Enzymes from *Trichoderma* (Sigma) and 20 mg Yatalase (Takara) can be used. Higher concentrations of Extralyse can also be evaluated.
13. Mix the closed PEG solution on the magnetic stirrer for ~20 min before using to avoid crystallization and facilitate pipetting. Maintain the PEG solution at room temperature.
14. If it is the first time using a new protoplast batch, also include a negative control without DNA. Transformation efficiency can be lower for vectors $\geq$30 kb, but $\geq$10 colonies per plate are usually obtained. If the transformation efficiency is high once the protoplast batch is tested, the protoplast solution can be further diluted and aliquoted before use in next transformation rounds.
15. If less vectors are needed, auxotrophies can be supplemented in the media or empty vectors can be used to transform *A. nidulans* and complement all auxotrophies. Seal plates with parafilm to avoid evaporation and cover with aluminum foil if supplementing with riboflavin as it is light sensitive.
16. Flasks are closed with cotton wool plugs and protected with aluminum foil. If the cultures contain riboflavin, ensure that dark conditions are maintained throughout incubation. Optionally, ampicillin can be added to the medium at 50 μg/mL.
17. The production can be evaluated by taking in sterility 10 mL of the culture on the second day for analysis by LC-DAD-MS.
18. Variation in the fungal growth morphologies between biological replicates growing on liquid culture could affect compound production titers [39].
19. Acidified ethyl acetate is important to make some compounds extractable by neutralizing charges. For example, we observed that dehydromicroperfuranone was only extractable with an acidified ethyl acetate mix but not with ethyl acetate alone [27].
20. The volume needed for large-scale culture varies depending on compound titer and the efficiency of compound isolation. Based on the experience in our lab, the range of culture volume required to obtain sufficient pure compound for structural characterization can vary from 2 to 10 L and up to 30 L (rarely) for trace compounds.
21. In general, our lab uses a series of chromatography separations involving columns with silica gel or Sephadex LH-20 as stationary phase followed by semi-preparative or preparative chromatography using C18 column. The mobile phase choices (forward or reverse phase) vary and depend on the properties of the compounds.

Acknowledgments

This work was supported by the Australian Research Council (FT160100233 and DP210102180). We thank Prof Kenji Watanabe (University of Shizuoka) for the gift of pKW20088, which served as a base for the pYFAC system. We also are grateful to Prof Berl Oakley for the gift of A. nidulans strains LO7890 and LO8030 and Prof Yi Tang for the A. nidulans A1145ΔSTΔEM, that form the basis of this and many of our works.

References

1. Keller NP (2019) Fungal secondary metabolism: regulation, function and drug discovery. Nat Rev Microbiol 17:167–180. https://doi.org/10.1038/s41579-018-0121-1
2. Kjærbølling I, Mortensen UH, Vesth T, Andersen MR (2019) Strategies to establish the link between biosynthetic gene clusters and secondary metabolites. Fungal Genet Biol 130: 107–121
3. Anyaogu DC, Mortensen UH (2015) Heterologous production of fungal secondary metabolites in *Aspergilli*. Front Microbiol 6:77. https://doi.org/10.3389/fmicb.2015.00077
4. Pickens LB, Tang Y, Chooi Y-H (2011) Metabolic engineering for the production of natural products. Annu Rev Chem Biomol Eng 2: 211–236. https://doi.org/10.1146/annurev-chembioeng-061010-114209
5. Lazarus CM, Williams K, Bailey AM, Prod N, Lazarus CM, Williams K, Bailey AM (2014) Reconstructing fungal natural product biosynthetic pathways. Nat Prod Rep 31:1339–1347. https://doi.org/10.1039/C4NP00084F
6. He Y, Wang B, Chen W, Cox RJ, He J, Chen F (2018) Recent advances in reconstructing microbial secondary metabolites biosynthesis in *Aspergillus* spp. Biotechnol Adv 36: 739–783. https://doi.org/10.1016/j.biotechadv.2018.02.001
7. Zhang W, Du L, Qu Z, Zhang X, Li F, Li Z, Qi F, Wang X, Jiang Y, Men P, Sun J, Cao S, Geng C, Qi F, Wan X, Liu C, Li S (2019) Compartmentalized biosynthesis of mycophenolic acid. Proc Natl Acad Sci 116: 13305–13310. https://doi.org/10.1073/pnas.1821932116
8. Hu J, Sarrami F, Li H, Zhang G, Stubbs KA, Lacey E, Stewart SG, Karton A, Piggott AMM, Chooi Y-HH (2019) Heterologous biosynthesis of elsinochrome A sheds light on the formation of the photosensitive perylenequinone system. Chem Sci 10:1457–1465. https://doi.org/10.1039/c8sc02870b
9. Clevenger KD, Bok JW, Ye R, Miley GP, Verdan MH, Velk T, Chen C, Yang KH, Robey MT, Gao P, Lamprecht M, Thomas PM, Islam MN, Palmer JM, Wu CC, Keller NP, Kelleher NL (2017) A scalable platform to identify fungal secondary metabolites and their gene clusters. Nat Chem Biol 13:895–901. https://doi.org/10.1038/nchembio.2408
10. Caesar LK, Kelleher NL, Keller NP (2020) In the fungus where it happens: history and future propelling *Aspergillus nidulans* as the archetype of natural products research. Fungal Genet Biol 144:103477. https://doi.org/10.1016/j.fgb.2020.103477
11. Chiang YM, Ahuja M, Oakley CE, Entwistle R, Asokan A, Zutz C, Wang CCC, Oakley BR (2016) Development of genetic dereplication strains in *Aspergillus nidulans* results in the discovery of aspercryptin. Angew Chem 55: 1662–1665. https://doi.org/10.1002/anie.201507097
12. Nayak T, Szewczyk E, Oakley CE, Osmani A, Ukil L, Murray SL, Hynes MJ, Osmani SA, Oakley BR (2006) A versatile and efficient gene-targeting system for *Aspergillus nidulans.* Genetics 172:1557–1566. https://doi.org/10.1534/genetics.105.052563
13. Aleksenko A, Clutterbuck AJ (1997) Autonomous plasmid replication in *Aspergillus nidulans*: AMA1 and MATE elements. Fungal Genet Biol 21:373–387. https://doi.org/10.1006/fgbi.1997.0980
14. Ling, Liu Man-Cheng, Tang Yi, Tang (2019) Fungal Highly Reducing Polyketide Synthases Biosynthesize Salicylaldehydes That Are Precursors to Epoxycyclohexenol Natural Products. Journal of the American Chemical Society 141(50) 19538-19541 https://doi.org/10.1021/jacs.9b09669
15. Yuta, Tsunematsu Noriyasu, Ishikawa Daigo, Wakana Yukihiro, Goda Hiroshi, Noguchi Hisao, Moriya Kinya, Hotta Kenji,Watanabe (2013) Distinct mechanisms for spiro-carbon

formation reveal biosynthetic pathway cross-talk. Nature Chemical Biology 9(12) 818-825 https://doi.org/10.1038/nchembio.1366

16. Li H, Gilchrist CLM, Phan CS, Lacey HJ, Vuong D, Moggach SA, Lacey E, Piggott AM, Chooi YH (2020) Biosynthesis of a new benzazepine alkaloid nanangelenin A from *Aspergillus nanangensis* involves an unusual l-kynurenine-incorporating NRPS catalyzing regioselective lactamization. J Am Chem Soc 142:7145–7152. https://doi.org/10.1021/jacs.0c01605
17. Kessler SC, Zhang X, McDonald MC, Gilchrist CLM, Lin Z, Rightmyer A, Solomon PS, Gillian Turgeon B, Chooi YH (2020) Victorin, the host-selective cyclic peptide toxin from the oat pathogen *Cochliobolus victoriae*, is ribosomally encoded. Proc Natl Acad Sci U S A 117:24243–24250. https://doi.org/10.1073/pnas.2010573117
18. Cacho RA, Tang Y, Chooi YH (2014) Next-generation sequencing approach for connecting secondary metabolites to biosynthetic gene clusters in fungi. Front Microbiol 5:774. https://doi.org/10.3389/fmicb.2014.00774
19. Basenko EY, Pulman JA, Shanmugasundram A, Harb OS, Crouch K, Starns D, Warrenfeltz S, Aurrecoechea C, Stoeckert CJ, Kissinger JC, Roos DS, Hertz-Fowler C (2018) FungiDB: an integrated bioinformatic resource for fungi and oomycetes. J Fungi 4:39. https://doi.org/10.3390/jof4010039
20. Medema MH (2021) The year 2020 in natural product bioinformatics: an overview of the latest tools and databases. Nat Prod Rep 4: 301–306. https://doi.org/10.1039/d0np00090f
21. Blin K, Shaw S, Steinke K, Villebro R, Ziemert N, Lee SY, Medema MH, Weber T (2019) AntiSMASH 5.0: updates to the secondary metabolite genome mining pipeline. Nucleic Acids Res 47:W81–W87. https://doi.org/10.1093/nar/gkz310
22. Medema MH, Kottmann R, Yilmaz P, Cummings M, Biggins JB, Blin K, De Bruijn I, Chooi YH, Claesen J, Coates RC, Cruz-Morales P, Duddela S, Düsterhus S, Edwards DJ, Fewer DP, Garg N, Geiger C, Gomez-Escribano JP, Greule A, Hadjithomas M, Haines AS, Helfrich EJN, Hillwig ML, Ishida K, Jones AC, Jones CS, Jungmann K, Kegler C, Kim HU, Kötter P, Krug D, Masschelein J, Melnik AV, Mantovani SM, Monroe EA, Moore M, Moss N, Nützmann HW, Pan G, Pati A, Petras D, Reen FJ, Rosconi F, Rui Z, Tian Z, Tobias NJ, Tsunematsu Y, Wiemann P, Wyckoff E, Yan X, Yim G, Yu F, Xie Y, Aigle B, Apel AK, Balibar CJ, Balskus EP, Barona-Gómez F, Bechthold A, Bode HB, Borriss R, Brady SF, Brakhage AA, Caffrey P, Cheng YQ, Clardy J, Cox RJ, De Mot R, Donadio S, Donia MS, Van Der Donk WA, Dorrestein PC, Doyle S, Driessen AJM, Ehling-Schulz M, Entian KD, Fischbach MA, Gerwick L, Gerwick WH, Gross H, Gust B, Hertweck C, Höfte M, Jensen SE, Ju J, Katz L, Kaysser L, Klassen JL, Keller NP, Kormanec J, Kuipers OP, Kuzuyama T, Kyrpides NC, Kwon HJ, Lautru S, Lavigne R, Lee CY, Linquan B, Liu X, Liu W, Luzhetskyy A, Mahmud T, Mast Y, Méndez C, Metsä-Ketelä M, Micklefield J, Mitchell DA, Moore BS, Moreira LM, Müller R, Neilan BA, Nett M, Nielsen J, O'Gara F, Oikawa H, Osbourn A, Osburne MS, Ostash B, Payne SM, Pernodet JL, Petricek M, Piel J, Ploux O, Raaijmakers JM, Salas JA, Schmitt EK, Scott B, Seipke RF, Shen B, Sherman DH, Sivonen K, Smanski MJ, Sosio M, Stegmann E, Süssmuth RD, Tahlan K, Thomas CM, Tang Y, Truman AW, Viaud M, Walton JD, Walsh CT, Weber T, Van Wezel GP, Wilkinson B, Willey JM, Wohlleben W, Wright GD, Ziemert N, Zhang C, Zotchev SB, Breitling R, Takano E, Glöckner FO (2015) Minimum information about a biosynthetic gene cluster. Nat Chem Biol 11:625–631. https://doi.org/10.1038/nchembio.1890
23. Gilchrist CLM, Booth TJ, Chooi YH (2020) cblaster: a remote search tool for rapid identification and visualisation of homologous gene clusters. bioRxiv. https://doi.org/10.1101/2020.11.08.370601
24. Gilchrist CLM, Li H, Chooi YH (2018) Panning for gold in mould: can we increase the odds for fungal genome mining? Org Biomol Chem 16:1620–1626. https://doi.org/10.1039/c7ob03127k
25. Ziemert N, Alanjary M, Weber T (2016) The evolution of genome mining in microbes – a review. Nat Prod Rep 33:988–1005. https://doi.org/10.1039/C6NP00025H
26. Chooi YH, Zhang G, Hu J, Muria-Gonzalez MJ, Tran PN, Pettitt A, Maier AG, Barrow RA, Solomon PS (2017) Functional genomics-guided discovery of a light- activated phytotoxin in the wheat pathogen *Parastagonospora nodorum* via pathway activation. Environ Microbiol 19:1975–1986. https://doi.org/10.1111/1462-2920.13711
27. Roux I, Woodcraft C, Hu J, Wolters R, Gilchrist CLM, Chooi YH (2020) CRISPR-mediated activation of biosynthetic gene clusters for bioactive molecule discovery in filamentous fungi.

ACS Synth Biol 9:1843–1854. https://doi.org/10.1021/acssynbio.0c00197

28. Ahuja M, Chiang YM, Chang SL, Praseuth MB, Entwistle R, Sanchez JF, Lo HC, Yeh HH, Oakley BR, Wang CCC (2012) Illuminating the diversity of aromatic polyketide synthases in *Aspergillus nidulans*. J Am Chem Soc 134:8212–8221. https://doi.org/10.1021/ja3016395
29. Liu N, Hung YS, Gao SS, Hang L, Zou Y, Chooi YH, Tang Y (2017) Identification and heterologous production of a benzoyl-primed tricarboxylic acid polyketide intermediate from the zaragozic acid a biosynthetic pathway. Org Lett 19:3560–3563. https://doi.org/10.1021/acs.orglett.7b01534
30. Roux I, Bowles S, Kalaitzis JA, Vuong D, Lacey E, Chooi Y-H, Piggott AM (2021) Characterisation and heterologous biosynthesis of burnettiene A, a new polyene-decalin polyketide from Aspergillus burnettii. Organic & Biomolecular Chemistry 19(43):9506–9513. https://doi.org/10.1039/D1OB01766G
31. Li H, Lacey AE, Shu S, Kalaitzis JA, Vuong D, Crombie A, Hu J, Gilchrist CLM, Lacey E, Piggott AM, Chooi Y-H (2021) Hancockiamides: phenylpropanoid piperazines from *Aspergillus hancockii* are biosynthesised by a versatile dual single-module NRPS pathway. Org Biomol Chem 19:587–595. https://doi.org/10.1039/d0ob02243h
32. Korbie DJ, Mattick JS (2008) Touchdown PCR for increased specificity and sensitivity in PCR amplification. Nat Protoc 3:1452–1456. https://doi.org/10.1038/nprot.2008.133
33. Oakley CE, Edgerton-Morgan H, Oakley BR (2012) Tools for manipulation of secondary metabolism pathways: rapid promoter replacements and gene deletions in *Aspergillus nidulans*. Methods Mol Biol 944:143–161. https://doi.org/10.1007/978-1-62703-122-6_10
34. Yelton MM, Hamer JE, Timberlake WE (1984) Transformation of *Aspergillus nidulans* by using a *trpC* plasmid. Proc Natl Acad Sci 81:1470–1474. https://doi.org/10.1073/pnas.81.5.1470
35. Lim FY, Sanchez JF, Wang CCC, Keller NP (2012) Toward awakening cryptic secondary metabolite gene clusters in filamentous fungi. Methods Enzymol 517:303–324. https://doi.org/10.1016/B978-0-12-404634-4.00015-2
36. Li H, Wei H, Hu J, Lacey E, Sobolev AN, Stubbs KA, Solomon PS, Chooi YH (2020) Genomics-driven discovery of phytotoxic cytochalasans involved in the virulence of the wheat pathogen *Parastagonospora nodorum*. ACS Chem Biol 15:226–233. https://doi.org/10.1021/acschembio.9b00791
37. Fu C, Donovan WP, Shikapwashya-Hasser O, Ye X, Cole RH (2014) Hot fusion: an efficient method to clone multiple DNA fragments as well as inverted repeats without ligase. PLoS One 9:e115318. https://doi.org/10.1371/journal.pone.0115318
38. Gietz RD, Schiestl RH (2007) Frozen competent yeast cells that can be transformed with high efficiency using the LiAc/SS carrier DNA/PEG method. Nat Protoc 2:1–4. https://doi.org/10.1038/nprot.2007.17
39. Veiter L, Rajamanickam V, Herwig C (2018) The filamentous fungal pellet—relationship between morphology and productivity. Appl Microbiol Biotechnol 102:2997–3006. https://doi.org/10.1007/s00253-018-8818-7

Chapter 6

Targeted Genetic Engineering via *Agrobacterium*-Mediated Transformation in *Fusarium solani*

Mikkel Rank Nielsen, Samba Evelyne Kabemba Kaniki, and Jens Laurids Sørensen

Abstract

Members of the *Fusarium solani* species complex are filamentous fungi that can act as pathogens to many crops and animals. Although relevant, a robust molecular toolbox is missing for the investigation of gene function and metabolism. In this chapter, we describe how *Agrobacterium*-mediated transformation can be used to facilitate gene targeting. A flexible vector system, based on in vivo recombination in *Saccharomyces cerevisiae*, is utilized to achieve overexpression and gene deletion of targeted biosynthetic genes in *F. solani* f. sp. *pisi*.

Key words Agrobacterium-mediated transformation, *Fusarium solani*, *Fusarium*, Gene targeting, Knockout, Overexpression, Genomic integration, Filamentous fungi, AMT, ATMT

1 Introduction

The *Fusarium solani* species complex (FSSC) encompasses more than 90 species [1–4]. Collectively, they are ubiquitously present in the environment as decomposers, but they can also act as damaging pathogens of important crops, such as legumes, fruits, and solanaceous crops [5]. Furthermore, members of the FSSC are implicated in mycotic infections of humans and other animals, where they are responsible for two-thirds of the reported fusarioses [4, 6]. In addition to the pathogenic relevance, the FSSC also contain a heavy arsenal of bioactive secondary metabolites, including 13 polyketides and 13 non-ribosomal peptide synthetases [7, 8].

Despite the huge agricultural impact and clinical relevance, only few functional genetics papers have been published, primarily due to a limited molecular toolbox. Early transformation studies in *F. solani* were based on protoplast transformation (e.g. [9, 10]), although this approach has not gained wide usage due to difficulties in preparing high quality protoplasts. To overcome this problem,

Elizabeth Skellam (ed.), *Engineering Natural Product Biosynthesis: Methods and Protocols*,
Methods in Molecular Biology, vol. 2489, https://doi.org/10.1007/978-1-0716-2273-5_6,

an *Agrobacterium*-mediated transformation (AMT) protocol was applied by Liu and Bao [11]. The AMT protocol was further improved in our lab by optimizing incubation temperature, medium and by switching to *A. tumefaciens* strain AGL-1 instead of LBA4404 [12], which has been the preferred strain for other *Fusarium* species [13–15].

Besides transformation efficacy, one of the important obstacles in fungal transformation is the generation of transformation vectors. AMT vectors for *Fusarium* transformation can be assembled by various means including enzyme- and ligase-dependent, fusion PCR, Xi, In-Fusion, and USER friendly cloning methods [16]. We recently developed an additional method, where the vectors are assembled in vivo in *Saccharomyces cerevisiae* by target-associated recombinational (TAR) cloning [17]. This approach allows fast and reliable generation of vectors and thereby provides a valuable tool for improving our understanding of *F. solani*.

2 Materials

All solutions are prepared from ultrapure demineralized (Type 1) water. When H_2O is used for washing cells, it is first sterilized by autoclaving at 121 °C for 15 min and stored at 4 °C.

2.1 Instruments

1. DNA measurement instrument such as the Qubit HS DNA Assay or the Nanodrop 2000c spectrophotometer (Thermo Scientific).
2. Spectrophotometer for measure optical density (OD_{600}) of cell cultures.
3. PCR thermocycler.
4. Agarose gel electrophoresis equipment, including running buffer, a DNA standard, and a DNA stain.
5. Lyophilizing system (freeze-drying equipment).
6. Bead beater, for instance, a Bead Mill 24 Homogenizer (Thermo Scientific).
7. Water bath 37–100 °C.
8. Microscope and hemacytometer for determination of spore quality and concentration.
9. Incubator with temperature control 19–37 °C with the option for orbital shacking.
10. Tube rocker, alternatively use a tube roller.
11. Microcentrifuge.
12. Electroporator.

2.2 Sterile Equipment

1. Miracloth cut to fit in plastic funnels, wrapped in tinfoil, and autoclaved.
2. Tweezers.
3. Drigalski spatula.
4. Tooth picks and inoculation loops.
5. Black filter papers, cut to fit a petri dish, placed in a glass bowl or glass beaker, covered with tinfoil, and autoclaved at 121 °C for 15 min to sterilize (*see* **Note 1**).
6. Plastic syringes packed with glass wool, wrapped in tinfoil, and autoclaved.
7. Autoclaved 100 mL, 250 mL (baffled), and 1000 mL Erlenmeyer cultivation flasks.
8. Disposable sterile plastic syringes.
9. Disposable 0.22 μm sterile syringe filters.
10. Bottle-top 0.45 μm vacuum filters for sterilizing solutions.
11. 55 and 90 mm petri dishes.
12. 1.5, 15, and 50 mL centrifuge tubes.
13. Cryogenic tubes.
14. Disposable sterile scalpels.
15. Parafilm.

2.3 Strains

1. *Escherichia coli* DH5α.
2. *Agrobacterium tumefaciens* AGL-1.
3. *Fusarium solani* f.sp. *pisi* 77-13-4 mpVI (FGSC 9596).
4. *Saccharomyces cerevisiae* BY4743.

2.4 Plasmids

1. pSHUT4-*eYFP*, Addgene repository #133094 [17].
2. pSHUT3-32 (available upon reasonable request) [18].
3. pRF-HU2 (*see* **Note 2**) [19].

2.5 Biological Kits

1. FastDNA SPIN Kit for soil supplied with 2 mL tubes of Lysing Matrix E (MP Biomedicals).
2. PCR purification kit, e.g. the QIAquick PCR Purification Kit.
3. Gel extraction kit, e.g. the QIAquick Gel Extraction Kit.
4. Miniprep plasmid purification kit, e.g. the QIAprep Spin Miniprep Kit.

2.6 Enzymes

1. BamHI.
2. XhoI.
3. DpnI.

4. Zymolyase stock: 100 mg Zymolyase 20T from *Arthrobacter luteus* dissolved in 10 mL of 25% glycerol, 0.05 M Tris-HCl. Store at −20 °C in 500 μL aliquots.
5. High fidelity DNA polymerase, e.g. the Phusion Hot Start II DNA Polymerase and compatible PCR reagents.

2.7 Primers

All primer positions are shown in Figs. 1, 2, and 3a, b.

2.7.1 Primers for Construction of the Overexpression Vector pSHUT4::geneX

Design primers for amplification of targeted *geneX* comprising tails for TAR assembly in *S. cerevisiae* (*see* Fig. 1a):

OE_geneX_Fwd 5′-AGTTCGCACCACCTTCAAAG-[*geneX* forward sequence]-3′

OE_geneX_Rev 5′-CAAATGTTTGAACGATCGGC-[*geneX* reverse sequence]-3′

2.7.2 Primers for Validation of the pSHUT4::geneX TAR Construct

Primers for PCR validation and sequencing across the multiple-cloning site (*see* Fig. 1b):

E022 5′-TTTCCCCCATACCTCCTTTC-3′

E023 5′-AAGACCGGCAACAGGATTC-3′

2.7.3 Primers for Construction of the Gene Deletion Vector pKO-geneY

Primers for amplification of the vector backbone (*see* Fig. 2a):

F073 5′-AACGTCGTGACTGGGAAAAC-3′

F074 5′-CGCTCATGATCAGATTGTCG-3′

Primers for amplification of the *hph* gene:

F071 5′-AATTAAGTCCTCAGCGGGCC-3′

F072 5′-CATCGGTGATGTCGGCGATA-3′

1. Design primers for amplification of 800–1200 bp border sequences left (UP) and right (DOWN) of the targeted *geneY* added tails for TAR assembly in *S. cerevisiae* (*see* Fig. 2b):

   KO_geneY_UpFwd 5′-CGACAATCTGATCATGAGCG-[UP forward sequence]-3′

   KO_geneY_UpRev 5′-GGCCCGCTGAGGACTTAATT-[UP reverse sequence]-3′

   KO_geneY_DwFwd 5′-TATCGCCGACATCACCGATG-[DOWN forward sequence]-3′

   KO_geneY_DwRev 5′-GTTTTCCCAGTCACGACGTT-[DOWN reverse sequence]-3′

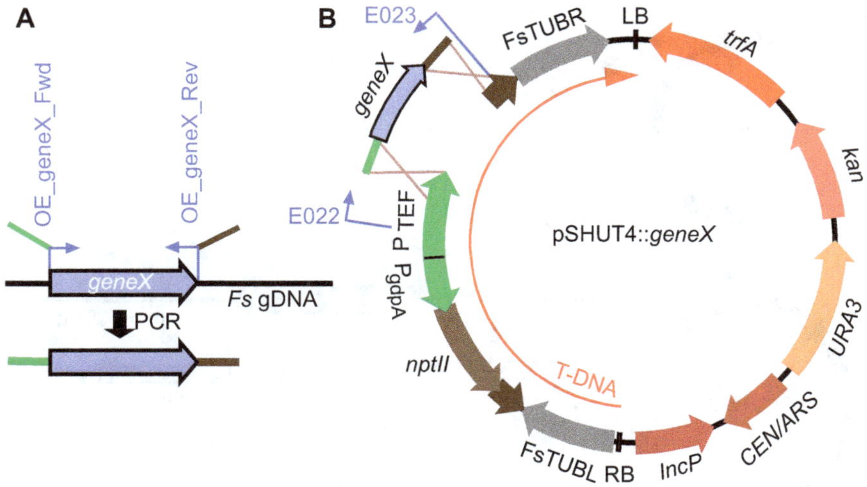

Fig. 1 Construction of the AMT overexpression vector pSHUT4::*geneX*. (**a**) The targeted gene, *geneX* (light blue), desired to be overexpressed is PCR-amplified using primers (dark blue) containing tails (green, brown) with homology to the linearized vector pSHUT4′. (**b**) Target-associated recombinational cloning in vivo in *Saccharomyces cerevisiae*. Homologous recombination (X's) between primer tails and the linearized plasmid results in fusion between *geneX* and the constitutive promoter $P_{TEF\text{-}1\alpha}$ from *Aspergillus nidulans*. Primers used in construct validation are shown as dark blue arrows. The T-DNA cassette is marked as a red arrow

*2.7.4 Primers for Validation of the pKO-*geneY *TAR Construct (*See *Fig. 2c)*

1. Primers for PCR validation and sequencing across the UP border sequence:

   C090 5′-CAAGTTCAAGGAGCGAAAGG-3′

   G058 5′-GCCTGGACGACTAAACCAAA-3′

2. Primers for PCR validation and sequencing across the DOWN border sequence:

   G059 5′-TCGTTTACCCAGAATGCACA-3′

   C093 5′-CCGACGGATGTTCGACTATT-3′

2.7.5 Primers for Colony PCR Screening of F. solani *OE::*geneX *Transformants (*See *Fig. 3a)*

D096 5′-AGGCAAAGCAAGGTAAAGCA-3′ in combination with C091 (see below).

C091 5′-CTGCCTCTTACCACCTGCTC-3′

D097 5′-GCAGTTTCTCCACAGCCTTC-3′ in combination with E022 (see above).

2.7.6 Primers for Colony PCR Screening of F. solani *Δ*geneY *Transformants (*See *Fig. 3b)*

G056 5′-AATTAAGTCCTCAGCGGGCC-3′

G057 5′-CATCGGTGATGTCGGCGATA-3′

1. Design a primer locus_geneY_Fwd to pair with G058 (see above) for validating recombination between UP and the targeted locus.

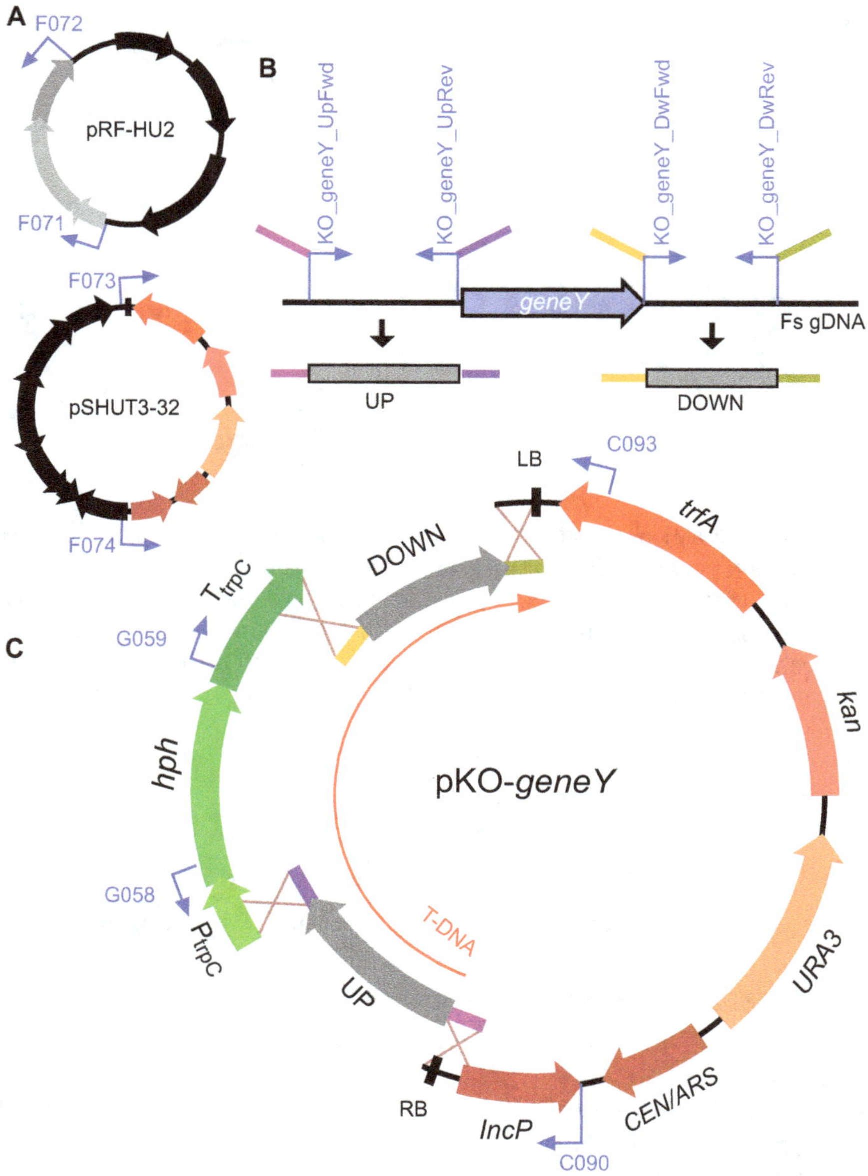

Fig. 2 Construction of the AMT gene deletion vector pKO-*geneY*. (**a**) Vectors pRF-HU2 and pSHUT3-32 are used as PCR templates for amplification of the *hph* selection marker (green arrows) and the vector backbone (red arrows) used in assembly of pKO-*geneY*. (**b**) The genomic locus comprising *geneY* targeted for gene deletion. Two 800–1200 bp border sequences are PCR-amplified into UP and DOWN (grey boxes) using the primers shown (dark blue) with tails (pink, purple, orange, light green) containing homology to the vector backbone and *hph* selection marker. (**c**) Target-associated recombinational cloning in vivo in *Saccharomyces cerevisiae*. Homologous recombination (X's) between primer tails and PCR fragments results in fusion of a T-DNA cassette comprising UP and DOWN sequences separated by *hph*. Primers used in construct validation are shown as dark blue arrows. The T-DNA cassette is marked as a red arrow

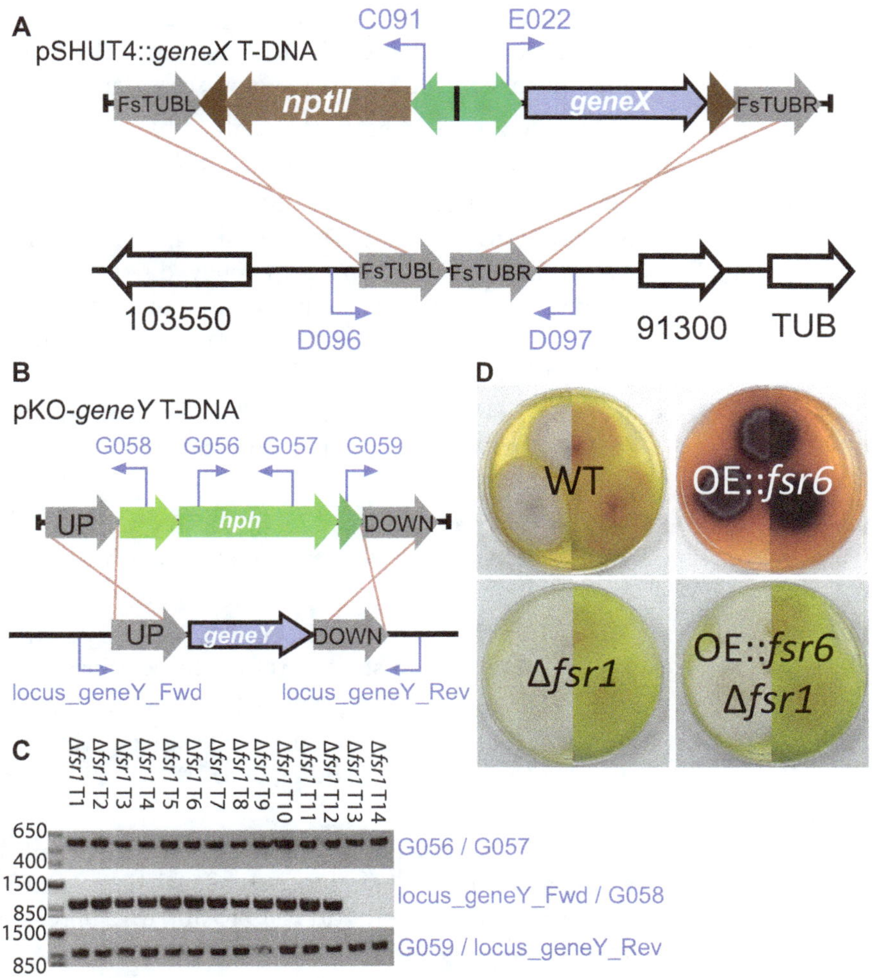

Fig. 3 Genetic engineering by targeted genomic integration of T-DNA cassettes in the genome of *F. solani*. (**a**) The overexpression vector pSHUT4::*geneX* T-DNA recombines through homologous recombination with the FsTUBL and FsTUBR sequences resulting in genomic integration of the *nptII* selection marker together with the targeted *geneX* fused to $P_{TEF-1\alpha}$. (**b**) The gene deletion vector pKO-*geneY* T-DNA recombines with the genomic DNA up and downstream of *geneY* resulting in deletion of *geneY* and gene replacement with *hph*. Primers used in fungal colony PCR are shown as dark blue arrows. (**c**) Fungal colony PCR screening of 14 *F. solani* Δ*fsr1* transformants reveals efficient gene replacement in ~80% of isolates. (**d**) Photographs (top and bottom of agar plates) of validated *F. solani* strains including the wild type (WT), OE::*fsr6*, Δ*fsr1*, and OE::*fsr6* Δ*fsr1*. OE::*fsr6* displays an overproduction of mycelial pigments, while Δ*fsr1* and OE::*fsr6* Δ*fsr1* both display an albino phenotype in comparison to the wild type

2. Design a primer locus_geneY_Rev to pair with G059 (see above) for validating recombination between DOWN and the targeted locus.

2.8 Solutions

1. 2 mg/mL salmon sperm DNA (ssDNA): Dissolve 100 mg DNA sodium salt from salmon testes in 50 mL 1:10 TE buffer using a magnet stirrer. To dissolve the DNA completely the

solution can be pipetted slowly up and down using a broad-nozzle tip. The dissolved DNA is distributed into 500 μL aliquots in sterile 1.5 mL tubes and stored at −20 °C.

2. 1 M LiAc: 10.2 g lithium acetate dihydrate is dissolved in 100 mL H_2O in a glass bottle and autoclaved.
3. 50% (w/v) PEG 3350: Add 50 g polyethylene glycol (PEG) MW 3350 to 30 mL H_2O while stirring in a beaker. Continue to stir until the PEG is completely dissolved. Control the volume by pouring the solution into a measuring glass and fill H_2O to 100 mL. Mix thoroughly and autoclave the solution in a 100 mL bottle.
4. 1 mM HEPES buffer: 0.238 g/L HEPES buffer (MW = 238.3 g/mol), pH adjusted to 7.4. Filter sterilize and store at 4 °C.
5. 10% (w/v) glycerol: Mix 115 mL 87% glycerol with 885 mL H_2O. Autoclave and store at room temperature or store at 4 °C for washing cells.
6. 20% (w/v) glycerol: Mix 22.9 mL 87% glycerol with 77.1 mL H_2O. Autoclave and store at room temperature.
7. 20% (w/v) glucose: Fill a 100 mL flask with 50 mL H_2O and slowly add 50 g D-(+) glucose while stirring. Let the glucose dissolve completely and add H_2O to 100 mL. Filter sterilize the solution.
8. 40% (w/v) glucose: Fill a 500 mL flask with 250 mL H_2O and slowly add 200 g D-(+) glucose while stirring. Let the glucose dissolve completely and add H_2O to 500 mL. Filter sterilize the solution.
9. Trace metals for PDA: Dissolve 0.1 g/L $ZnSO_4 \cdot 7H_2O$ and 0.05 g/L $CuSO_4 \cdot 5H_2O$ in 100 mL H_2O. Autoclave and store at room temperature.
10. 2.5× salts solution: To boiling H_2O, add the following salts while stirring to dissolve: 3.625 g/L KH_2PO_4, 5.125 g/L K_2HPO_4, 0.375 g/L NaCl, 1.250 g/L $MgSO_4 \cdot 7H_2O$, 1.250 g/L $(NH_4)_2SO_4$, 0.0062 g/L $FeSO_4 \cdot 7H_2O$. Filter-sterilize the solution and store at room temperature (*see* **Note 3**).
11. 5 M KOH. Dissolve 7.013 g KOH pellet in 25 mL H_2O in a volumetric flask. WARNING: KOH a strong base. Use protective equipment.
12. 10 mM acetosyringone: Dissolve 98.1 mg acetosyringone in 45 mL H_2O and stir to dissolve. Adjust the pH to 8.0 using 5 M KOH. Fill H_2O to 50 mL and filter sterilize. Use immediately or store at 4 °C for 24 h.

13. 1 M MES buffer: dissolve 19.52 g MES buffer (2-[*N*-morpholino]ethanesulfonic acid, MW = 195.24 g/mol) in 80 mL H_2O. Adjust the pH to 5.3 using 5 M KOH. Filter sterilize and use immediately or freeze at −20 °C (*see* **Note 4**).
14. BSA: Bovine serum albumin PCR grade. Alternatively, 20 mg/mL dissolved in H_2O and stored at −20 °C.
15. Fungal lysis buffer: To a sterile 50 mL plastic tube add 3 mL 2 M NaCl, 30 μL Triton X-100, 300 μL 1 M Tris-HCl, 300 μL 0.5 M EDTA, 3 mL 2% SDS, and 20 mL H_2O. Adjust the pH to 8.5. Fill H_2O up to 30 mL, mix by vortexing and store the solution at room temperature.

2.9 Antibiotics Stocks

WARNING: antibiotics are toxic chemicals and can cause harm. Make sure to use appropriate protective equipment.

1. 50 mg/mL kanamycin: dissolve 500 mg kanamycin sulfate in 10 mL H_2O. Stir to dissolve and filter (0.22 μm) to sterilize. Store at −20 °C in 1 mL aliquots.
2. 120 mg/mL hygromycin B: Dissolve 1.2 g hygromycin B in 10 mL H_2O by stirring. Stir to dissolve and filter (0.22 μm) to sterilize. Store at −20 °C in 1 mL aliquots.
3. 300 mg/mL geneticin G418: Dissolve 3 g Geneticin G418 sulfate in 10 mL H_2O by stirring. Stir to dissolve and filter (0.22 μm) to sterilize. Store at −20 °C in 1 mL aliquots.
4. 300 mg/mL cefoxitin: Dissolve 3 g cefoxitin sodium salt in 10 mL H_2O by stirring. Stir to dissolve and filter (0.22 μm) to sterilize. Store at −20 °C in 1 mL aliquots.
5. 10 mg/mL rifampicin: Dissolve 100 mg rifampicin in 10 mL 99% ethanol by stirring. Stir to dissolve and filter (0.22 μm) to sterilize. Store at −20 °C in 1 mL aliquots.

2.10 Media

1. PDA: 39 g/L Potato Dextrose Agar, 1 mL/L trace metals solution, autoclaved. When the agar has cooled down below 50 °C, add antibiotics, mix well, and cast plates.
2. PDB: 24 g/L Potato Dextrose Broth, 1 mL/L trace metals solution, autoclaved.
3. YPD broth: 10 g/L bacto yeast extract, 20 g/L bacto peptone. Dissolve in H_2O and remember to leave room for sugar. Autoclave and add 50 mL/L 40% (w/v) glucose to a final concentration of 2% (w/v).
4. 2 × YPAD broth: 20 g/L bacto yeast extract, 40 g/L bacto peptone, 80 mg adenine hemisulfate. Dissolve in H_2O and remember to leave room for sugar. Autoclave and add 100 mL/L 40% (w/v) glucose to a final concentration of 4% (w/v).

5. SC-U agar: 1.92 g/L Yeast Synthetic Drop-out Medium Supplement without Uracil, 6.7 g/L Yeast Nitrogen Base Without Amino Acids, 20 g/L agar. Dissolve completely in H_2O while leaving room for sugar. Autoclave the solution and immediately add 50 mL/L 40% (w/v) glucose. Mix well and cast plates.
6. LB broth: 20 g/L LB Broth Powder. Adjust pH to 7.4 and autoclave.
7. LB agar: 20 g/L LB Broth Powder, 20 g/L agar. Adjust pH to 7.4 and autoclave. When the agar has cooled down, add antibiotics, mix well, and cast plates.
8. ½LB pH 7.7: 10 g/L LB Broth Powder, Adjust the pH to 7.7 and autoclave.
9. CMC: To 900 mL boiling H_2O, add 15 g carboxymethylcellulose sodium salt while stirring. Add 1 g NH_4NO_3, 1 g KH_2PO_4 monobasic, 0.5 g $MgSO_4 \cdot 7H_2O$, 1 g bacto yeast extract and fill H_2O to 1000 mL. Distribute 70 mL in 250 mL baffled cultivation flasks and autoclave.
10. YM broth: 0.4 g/L bacto yeast extract, 10 g/L mannitol, 0.1 g/L NaCl, 0.1 g/L $MgSO_4$, 0.5 g/L $K_2HPO_4 \cdot 3H_2O$. Adjust pH to 7.0 and autoclave.
11. IMAS, 250 mL: Add the following to an autoclaved 250 mL flask; 100 mL 2.5× salts solution, 2.5 mL 20% (w/v) glucose, 6.25 20% (w/v) glycerol, 126.25 mL H_2O, 10 mL 1 M MES buffer, and 5 mL 10 mM acetosyringone (fresh).
12. IMAS agar, 500 mL: To a 500 mL flask add 5 mL 20% (w/v) glucose, 12.5 20% (w/v) glycerol, 20 g agar, 252.5 mL H_2O. Autoclave. Immediately add 200 mL pre-heated 2.5× salts solution and let the agar cool down below 50 °C before adding 20 mL 1 M MES buffer and 10 mL 10 mM acetosyringone (fresh) under sterile conditions. Mix and immediately cast plates.
13. V8 agar: 175 mL/L V8 juice (Campbell Soup Company), 3 g/L $CaCO_3$, 20 g/L agar. Adjust pH to 6.4 and autoclave. Equilibrate the molten agar in a 55 °C water bath. Add antibiotics and mix. Immediately cast plates (*see* **Note 5**).

3 Methods

3.1 Purification of Fungal DNA

1. Cultivate *F. solani* in 20 mL PDB for 4 days at 25 °C before harvesting the mycelium by pouring the culture through a layer of sterile Miracloth and washing residual medium away with 50 mL H_2O.

2. Transfer the fresh mycelium to a sterile tube and puncture the lid using scissors. Lyophilize the mycelium pellet by placing the tube in the freeze dryer over night.
3. Grind the mycelium in a Lysing Matrix E on the Bead Mill 24 for 30 s at 5.0 m/s.
4. Purify genomic DNA using the FastDNA SPIN Kit for Soil.

3.2 Preparation of Fragments for Construction of Overexpression Vector (See Fig. 1)

For overexpression in *F. solani*, the targeted "*geneX*" can be fused to the constitutive promoter $P_{TEF\text{-}1\alpha}$ and terminator NOS in the vector pSHUT4-*eYFP*. The vector comprises a transfer-DNA (T-DNA) cassette comprising the neomycin phosphotransferase (*nptII*) two border sequences (FsTUBL, FsTUBR) that can facilitate genomic integration in a non-coding region of chromosome 2 via homologous recombination (*see* **Note 6**).

1. PCR-amplify the targeted gene coding sequence using primers OE_geneX_Fwd/OE_geneX_Rev comprising 20 bp tails for TAR assembly in *S. cerevisiae* (*see* Fig. 1a and **Note 7**). Purify the PCR product using the QIAquick PCR purification kit.
2. In order to splice out the *eYFP* gene, digest 1 μg pSHUT4-*eYFP* using FD BamHI and FD XhoI in 1× FastDigest Buffer at 37 °C for 4 h, inactivate the enzymes at 65 °C for 15 min. Load and run the entire digest volume on a 1% agarose gel electrophoresis and purify the 9736 bp linearized vector molecule using the QIAquick gel recovery kit.
3. Control the purified DNA molecules using gel electrophoresis and determine the nucleotide concentration.

3.3 Preparation of Fragments for Construction of Deletion Vector (See Fig. 2)

For gene deletion a vector is prepared comprising nucleotide sequences upstream and downstream of the targeted *geneY*. The upstream and downstream regions, labeled UP and DOWN, are amplified through PCR and fused into a vector construct separated by a *hph* selection marker for facilitating gene replacement and selection.

1. PCR the vector backbone (*trfA-kan-URA3-CEN6/ARSH4-IncP*) using primers F073/F074 using 1–10 ng pSHUT3-32 as the PCR template (*see* Fig. 2a).
2. PCR-amplify the *hph* gene (P_{trpC}:*hph*:T_{trpC}) from pRF-HU2 using primers F071/F072 (*see* Fig. 2a).
3. Treat the vector backbone and *hph* PCR products by adding 1 μL DpnI to 50 μL PCR reaction, mix and incubate at 37 °C for 1 h (*see* **Notes 8** and **9**). Purify the product using the QIAquick PCR purification kit.
4. Amplify UP and DOWN fragments using primers KO_geneY_UpFwd/KO_geneY_UpRev and KO_geneY_DwFwd/KO_geneY_DwRev comprising 20 bp tails for TAR cloning in

S. cerevisiae (*see* Fig. 2b). Purify the UP and DOWN fragments using the PCR purification kit.

5. Control the purified DNA molecules using gel electrophoresis and determine the nucleotide concentration.

3.4 Assembly of Vector Constructs

This sub-protocol utilizes in vivo TAR assembly by transforming DNA fragments into *S. cerevisiae* using the LiAc/ssDNA/PEG method [20].

1. Inoculate 5 mL 2 × YPAD with a single colony of *S. cerevisiae* freshly grown on YPD. Incubate the culture at 30 °C, 200 rpm, for 12–16 h.
2. Determine the cell titer and from the preculture inoculate 50 mL 2 × YPAD with 2.5×10^8 cells in a baffled 250 mL shake flask (initial titer: 5×10^6 cells/mL). Incubate the culture at 30 °C 200 rpm for 4 h.
3. Denature 500 μL ssDNA in a 100 °C water bath for 5 min and place in ice to cool.
4. Prepare the transformation reagents in a 15 mL tube, add and mix 500 μL denatured ssDNA, 360 μL 1 M LiAc, 2400 μL 50% (w/v) PEG 3350 (*see* **Note 10**).
5. Harvest *S. cerevisiae* cells when the culture reaches the titer 2×10^7 cells/mL by centrifugation at 3000 × *g* for 5 min at 4 °C. Wash the cell pellet in 25 mL H_2O by centrifugation and discarding the supernatant. Repeat this wash once.
6. Resuspend the cell pellet in 1 mL H_2O and transfer to a clean 1.5 mL tube. Centrifuge and discard the supernatant.
7. Resuspend the cell pellet in 1 mL H_2O and divide it into clean tubes in 100 μL portions. Centrifuge and remove the H_2O by pipetting carefully. Keep the cells on ice.
8. Resuspend cell pellets in 326 μL of the transformation reagents solution by pipetting up and down to dissolve the cell pellet.
9. Setup the transformation reaction(s) for assembly of an overexpression vector construct (pSHUT4:*geneX*, *see* Fig. 1b) by adding DNA fragments dissolved in a total volume of 34 μL H_2O or buffer EB to cells suspended in transformation reagents solution: 200 fmol *geneX* PCR product, 40–100 fmol linearized pSHUT4′. Mix the DNA and cells by carefully pipetting up and down (*see* **Note 11**).
10. Setup the transformation reaction(s) for assembly of a gene deletion vector construct (pKO-*geneY*, *see* Fig. 2c) by adding DNA fragments dissolved in a total volume of 34 μL H_2O or buffer EB to cells suspended in transformation reagents solution: 200 fmol UP, 200 fmol DOWN, 200 fmol *hph*, 100 fmol

pSHUT3-32 backbone. Mix the DNA and cells by carefully pipetting up and down (*see* **Note 12**).

11. Incubate the transformation reactions in a water bath at 42 °C for 40 min.
12. Centrifuge and discard the supernatant. Resuspend the cells in 500 μL H_2O and spread the cells on SC-U agar using a disinfected Drigalski spatula.
13. Incubate the plates for 3–4 days at 30 °C.

3.5 Isolation of TAR Constructs

In order to recover the vector constructs from yeast, plasmid DNA is extracted from freshly grown cells and transformed into *E. coli* cells for proliferation. This method utilizes the reagents supplied with the QIAprep spin miniprep kit (QIAGEN).

1. Pick and streak a single transformant colony on SC-U agar and incubate for 2–3 days at 30 °C.
2. With a sterile inoculation loop, resuspend 10 μL of freshly grown cells in a clean 1.5 mL tube with 250 μL Resuspension Buffer P1 added 100 μg/mL RNase. Add 5 μL Zymolyase stock and incubate at 37 °C for 1 h.
3. Warm up Elution Buffer EB in a 55 °C water bath.
4. Add 250 μL Lysis Buffer P2 and mix immediately by inverting the closed tube slowly 4–6 times. Do not let this reaction exceed 5 min.
5. Add 350 μL Neutralization Buffer N3 and mix immediately by inverting the closed tube slowly 4–6 times.
6. Centrifuge the sample for 10 min at 13,000 rpm.
7. Transfer 800 μL supernatant into a QIAprep 2.0 spin column and centrifuge for 60 s. Discard the flow-through.
8. Wash the DNA by adding 750 μL Wash Buffer PE to the spin column and centrifuge for 60 s. Discard the flow-through.
9. Centrifuge the column again for 1 min to remove residual wash buffer.
10. Place the spin column in a clean 1.5 mL tube and add 20 μL 55 °C Buffer EB to the center of the column filter. Incubate the sample for 5 min at room temperature.
11. Elute the DNA by centrifugation at 13,000 rpm for 1 min. Store the yeast plasmid DNA at −20 °C.
12. Use 1 μL of isolated yeast plasmid DNA to transform chemically competent DH5α cells using standard heat shock procedure and select the cells on LB agar containing 50 μg/mL kanamycin. Incubate transformed *E. coli* over night at 37 °C.

3.6 Validation of Vector Constructs

1. Inoculate *E. coli* strains carrying vector constructs in liquid LB and purify plasmid DNA applying the QIAprep Spin Miniprep Kit (*see* **Note 13**). Store plasmid DNA samples at −20 °C.
2. PCR validation of overexpression vector pSHUT4::*geneX* is carried out using primers E022/E023.
3. PCR validation of gene deletion vector pKO-*geneY* can be performed using primers C090/G058 across the UP insert and G059/C093 across the DOWN insert.
4. Purify PCR products applying the QIAquick PCR purification kit.
5. Validate the diagnostic PCR products by sequencing, e.g. Sanger sequencing service.
6. (optional) Digest vector minipreps using eligible restriction endonucleases and analyze the result via gel electrophoresis.

3.7 Transforming Validated Vectors into A. tumefaciens

3.7.1 Preparation of Competent Agrobacterium tumefaciens Cells

1. In a 1 L shake flask, inoculate 500 mL LB broth containing 10 μg/mL rifampicin with 400 μL of a fresh overnight culture of *A. tumefaciens*. Incubate the culture at 28 °C 150 rpm until the culture reaches an OD_{600} value of 0.5.
2. Chill the shake flask on ice for 30 min.
3. Chill 1 mM HEPES buffer and 10% glycerol on ice.
4. Transfer the culture to centrifuge bottles and centrifuge the cells at 4000 × *g* 4 °C for 15 min. From this point on, the cells should be kept cold throughout the preparation.
5. Remove as much of the cultivation medium as possible by decanting and resuspend the cell pellet in 500 mL chilled 1 mM HEPES buffer. Centrifuge as above.
6. Resuspend in 250 mL 1 mM chilled HEPES buffer. Centrifuge as above.
7. Resuspend in 10 mL 1 mM chilled HEPES buffer. If applicable, transfer the cells to a 15 or 50 mL centrifuge tube. Centrifuge as above.
8. Remove as much of the supernatant as possible by pipetting. Resuspend in 2 mL chilled 10% glycerol and make aliquots in 1.5 mL tubes (e.g. 55 μL per tube) and freeze on dry ice for 2 min. Alternatively, snap-freeze in liquid nitrogen for 30 s. Immediately store competent cells at −80 °C.

3.7.2 Electroporation of A. tumefaciens

1. Chill 0.1 mm electroporation cuvettes on ice (*see* **Note 14**).
2. Thaw competent *A. tumefaciens* cells at room temperature and immediately place in a bucket of ice. Transfer 50 μL cells to a chilled 0.1 mm electroporation cuvette on ice.
3. Add 1–2 μL plasmid in buffer EB (2–100 ng DNA) and mix. Tap the cuvette to remove air bubbles. Keep on ice.

4. Set the electroporation apparatus to 2.2 kV. Dry the surface of the cuvette with kitchen paper, insert the cuvette in the electroporation chamber, and shock the cells. The pulse length should be approximately 5 ms (*see* **Note 15**).
5. Immediately add 0.5 mL of YM broth transfer to a 1.5 mL tube and incubate at 28 °C 200 rpm for 60–90 min.
6. Plate the cells on LB agar containing 20 μg/mL rifampicin and 50 μg/mL kanamycin. Incubate the plates at 28 °C for 2–3 days.
7. (Optional) Validate *A. tumefaciens* strains carrying plasmids with colony PCR by suspending a minute amount of freshly grown cells form a colony into 50 μL PCR mix and prolong the initial denaturation step at 99 °C to 10 min. Use plasmid specific primers as previously. Alternatively, the QIAprep spin miniprep kit can be applied to overnight cultures of Agrobacterium inoculated with single colony isolates for re-validating vector constructs.
8. Streak single colony isolates on LB containing 20 μg/mL rifampicin and 50 μg/mL kanamycin.

3.8 Production of F. solani Spores

For AMT, adequate amounts of macroconidia (spores) can be prepared in advance and stored at −20 °C. 1.2×10^7 spores are required for each transformation experiment not counting controls. We recommend pipetting spores using broad-nozzle pipette tips to reduce stress.

1. Inoculate *F. solani* on PDA agar and cultivate for 7–28 days.
2. Inoculate 3 × 70 mL CMC medium in baffled 250 mL shake flasks with agar cubes cut from the PDA plate and induce sporulation at 20 °C, 100 rpm for 4–6 days in darkness.
3. Observe the sporulation process in a microscope. The conidia should not germinate before harvesting.
4. Chill H_2O and 10% glycerol in a bucket of ice.
5. Harvest the spores by pouring the culture through a syringe packed with glass wool.
6. Centrifuge the flow-through at 10,000 × *g* for 15 min at 4 °C. Carefully discard the supernatant by decanting.
7. Resuspend the spore pellet in 25 mL chilled H_2O and centrifuge as before. Discard the supernatant and repeat this wash in H_2O twice.
8. Resuspend the spore pellet in 5 mL chilled 10% glycerol. Determine the spore concentration using a hemacytometer. Freeze on dry ice for 2 min or snap-freeze in liquid nitrogen for 30 s and immediately store at −80 °C in cryogenic tubes.

3.9 Targeted Genetic Engineering via Agrobacterium-Mediated Transformation [12, 17]

In this subprotocol, spores of *F. solani* are transformed using the AMT protocol. Please note some media must be prepared fresh, i.e., on the same day they are used. In addition, we recommend rinsing all glassware in demineralized H_2O twice before preparing media and solutions in order to remove residual soap.

3.9.1 Day 1

1. Inoculate a single *A. tumefaciens* colony carrying pSHUT4::*geneX* or pKO-*geneY* in 20 mL of ½LB pH 7.7 containing 10 μg/mL rifampicin and 50 μg/mL kanamycin in a sterile and capped 100 mL Erlenmeyer flask. Incubate the culture at 28 °C and 100 rpm for 12–16 h (*see* **Note 16**).

3.9.2 Day 2

1. Add acetosyringone to water and start dissolving it using a magnet stirrer. Remember to adjust the pH (*see* **Note 17**).
2. Prepare 35 mL liquid IMAS medium per transformation experiment.
3. Prepare 10 IMAS agar plates per transformation experiment. In addition, prepare two plates for fungal negative controls and one plate for *Agrobacterium* negative control.
4. Place black filter papers using sterile tweezers onto 11 IMAS plates including one fungal negative control plate. Soak the filter papers in 100–400 μL liquid IMAS (depending on the moisture of the plate) to help filters stick to the agar. Rub the filters with a Drigalski spatula to eliminate air bubbles beneath the filters. Store all IMAS agar plates, with and without filters, at 4 °C, overnight.
5. Add 10 mL IMAS medium added 50 μg/mL kanamycin to three sterilized 100 mL Erlenmeyer flasks. Inoculate with 100 μL, 200 μL, and 500 μL of *Agrobacterium* preculture. Place the IMAS cultures at 28 °C 100 rpm until the OD_{600} reaches 0.4–0.6 the next day (timing: 12–20 h).
6. Save remaining liquid IMAS for Day 3 at 4 °C.

3.9.3 Day 3

1. Monitor OD_{600} of *Agrobacterium* cultures (*see* **Note 18**).
2. Thaw frozen spores and dilute them to the concentration 6×10^6 spores/mL in liquid IMAS. Prepare 2.2 mL of fungal spores in IMAS per transformation experiment, and keep on ice.
3. When the *Agrobacterium* culture reaches the $OD_{600} = 0.4–0.6$, mix *F. solani* spores and *Agrobacterium* culture 1:1 in a total volume of 4.0 mL in a sterile centrifuge tube. Incubate the co-cultivation for 1 h at 28 °C 100 rpm at a 45° angle on a tube rocker.

4. Spread 400 μL of the co-culture onto 10 IMAS plates fitted with a black filter paper.
5. For negative growth controls: spread 100 μL diluted spores or 100 μL *Agrobacterium* culture onto plates without filters. In addition, spread 100 μL of diluted spores to an IMAS plate holding a black filter paper.
6. Incubate IMAS plates with the co-culture at 21–22 °C for 3 days in total darkness (*see* **Note 19**).

3.9.4 Day 6

1. Prepare and filter sterilize cefoxitin. For selection of cells transformed with the pSHUT4::*geneX*, dissolve and filter sterilize geneticin. For selection of cells transformed with pKO-*geneY*, dissolve and filter sterilize hygromycin.
2. Cast 11 V8 agar plates per transformation experiment containing 300 μg/mL cefoxitin and either 120 μg/mL hygromycin or 150 μg/mL geneticin. Label the plates with the corresponding vector name, antibiotic and "S1."
3. Using sterile tweezers carefully transfer the black filter papers from S1 IMAS to V8 plates containing antibiotics. Make sure to eliminate air bubbles under the filters by rolling the filter on in one smooth motion. Discard the IMAS plates.
4. Incubate the plates at 21–22 °C in darkness again (*see* **Note 20**).

3.9.5 Day 10–14

1. Cast V8 agar plates containing the appropriate combination of 150 μg/mL geneticin or 120 μg/mL hygromycin B. Mark these plates with respective plasmid name, selection and "S2."
2. Move the black filter papers from S1 V8 to the S2 V8 agar plates while being mindful of eliminating air bubbles below the filters.
3. Incubate at 21–22 °C in darkness until colonies emerge.

3.10 Isolation of Transformants

1. Cast PDA plates in small petri dishes (Ø = 55 mm) containing 150 μg/mL geneticin or 120 μg/mL hygromycin.
2. Look for colonies on the S1 and S2 plates and isolate individual colonies by transferring a minute amount of mycelium on the tip of a toothpick to PDA containing antibiotics. Grow fungal isolates at 28 °C in darkness for 1 week.
3. When the isolates grow on PDA, streak isolates onto new PDA plates containing 200 μg/mL geneticin or 120 μg/mL hygromycin and incubate at 28 °C.
4. (Optional): In order to secure genomic stability, we recommend sporulating isolates and isolate single-spore colonies.

5. (Optional): When isolates have grown on agar for 7–14 days the petri dishes can be wrapped in Parafilm and stored at 4 °C for up to 2 months.
6. Cut out agar cubes using a sterile scalpel and transfer them under sterile conditions to a cryogenic tube. Submerge the agar cubes in 10% glycerol and freeze isolated strains at −80 °C.

3.11 Screening Fungal Transformants with Colony PCR

3.11.1 Preparation of Colony PCR Template DNA

From a PDA plate, transfer a minute amount of 3–5 day old mycelium into 300 μL fungal lysis buffer using a sterile toothpick or pipette tip. Vortex the lysis solution for 1 min. Pellet cell debris for 2 min in a tabletop centrifuge. The lysate can be stored at −20 °C for several months.

3.11.2 Screening OE::geneX Overexpression Mutants (See Fig. 3a)

Genomic integration of the pSHUT4::*geneX* vector T-DNA (FsTUBL-P_{gdpA}:*nptII*:T_{NC}-$P_{TEF\text{-}1\alpha}$:*GeneX*:T_{NOS}-FsTUBR) is guided by two border sequences identical to a non-coding locus via homologous recombination PCR validation includes primers binding within the T-DNA cassette paired with primers outside the integration locus.

1. Prepare PCR master mix comprising 2 μg/μL BSA and use 1 μL of the fungal lysate as the template DNA. Setup the flowing reactions: Use primers D096/D097 to check if the insertion site is empty (will yield a 2715 bp band in the unmodified locus). Use the primer pairs D096/C091 and E022/D097 to validate recombination between the T-DNA and genome locus. As a control, we recommend to carry out the PCR reactions on isolated DNA from *F. solani*.

3.11.3 Screening ΔgeneY Knockout Mutants (See Fig. 3b)

Successful gene replacement with the pKO-*geneY* T-DNA [UP-P_{trpC}:*hph*:T_{trpC}-DOWN] is guided by homologous recombination between the cloned UP and DOWN segments of the T-DNA and the targeted *geneY*. PCR validation includes primers binding within the T-DNA cassette paired with primers hybridizing the genome surrounding *geneY*.

1. Prepare PCR master mix comprising 2 μg/μL BSA and use 1 μL of the fungal lysate as the template DNA. Setup the flowing reactions: Use primers G056/G057 to amplify a section of the *hph* selection marker. Use primer locus_geneY_Fwd/G058 and G059/locus_geneY_Rev to check for gene replacement with the T-DNA. As a control, we recommend to carry out the PCR reactions on isolated DNA from wild type *F. solani*. We transformed *F. solani* with *A. tumefaciens* pKO-*fsr1* for gene deletion of *fsr1*, the polyketide synthase responsible for mycelial pigmentation, and screened 14 mutants this way (Fig. 3c).

3.12 Preparation of OE::geneX ΔgeneY Double Mutants

Double mutants, i.e., a combination between a knockout and an overexpression modification, are prepared by performing the first modification as described above, validating a mutant and performing another iteration of AMT and selection for the second modification. This is possible due to the overexpression vector T-DNA comprises *nptII* (G418^R) while the gene deletion T-DNA carries *hph* (Hygromycin^R). In this case, both antibiotics are added to the selective V8 and PDA.

For instance, first we overexpressed the transcription factor *fsr6* resulting in overproduction of mycelial pigmentation [17]. Next, spores of the OE::*fsr6* mutant was used in AMT with *A. tumefaciens* pKO-*fsr1*, targeting the core polyketide synthase *fsr1* (PKS3), responsible for mycelial pigmentation. The OE::*fsr6* Δ*fsr1* double mutant showed an albino phenotype, comparable to that of the deletion mutant Δ*fsr1* (*see* Fig. 3d).

4 Notes

1. Never autoclave the same black filter papers twice as they become too soft and will break when transferred between agar plates.
2. Other plasmids carrying *hph* fused to the *A. nidulans* P_{trpC}/T_{trpC} are available. Those are pRF-HU, pRF-HUE, pRF-HU2E [19].
3. Earlier, the *Fusarium* AMT protocol recipe described IMAS containing $CaCl_2$ that we have chosen to omit. The addition of this salt causes heavy precipitation and can clog the filtration system. Remember to shake the solution well before mixing IMAS medium. When preparing IMAS agar, the 2.5× salts solution can be pre-heated in a 55 °C water bath.
4. MES can precipitate and form crystals when stored at −20 °C. To re-dissolve, vortex vigorously for 10–15 min.
5. In order to prevent the V8 and IMAS agar from solidifying in the flask, equilibrate the medium in a 55 °C water bath immediately after autoclaving to keep the agar molten.
6. The pSHUT4-*eYFP* vector targets a non-coding position in the genome between two genes annotated NECHADRAFT_103550 and NECHADRAFT_91300 found upstream of NECHADRAFT_66759 which encodes the β-tubulin gene [21]. This site was chosen inspired by the approach applied in *F. graminearum* [22].
7. Other cloning systems such as the Gibson cloning approach will work with the primers described in this method.

8. DpnI will work directly in some polymerase reaction buffers (post-PCR cycling) such as the HF buffer supplied with the Phusion Hot Start II DNA Polymerase.
9. Remaining PCR template, in this case pSHUT-*eYFP*, will contaminate the PCR product in the TAR cloning as circular plasmid can result in false-positive prototroph colonies on drop-out substrate. Thus, we apply DpnI to eliminate remaining plasmid molecules from the PCR product.
10. PEG is viscous and can be difficult to pipette accurately. Prepare sterile broad-nozzle pipette tips by cutting the ends off 1000 μL micropipette tips and autoclave.
11. In addition, a negative control reaction can be setup where the *geneX* insert is replaced with H_2O.
12. In addition, two negative control reactions can be performed: (1) *hph* fragment and H_2O and (2) pSHUT3-32 vector backbone and H_2O.
13. The plasmid backbones comprise the low-copy origin of replication *IncP* (*oriV*). In order to obtain sufficient amounts of plasmid DNA for validation experiments increase the cultivation volume to 10–25 mL in the miniprep procedure.
14. Electroporation cuvettes can be recycled: Rinse with 70% ethanol, then with H_2O and finally with 99% ethanol. Discard all liquid and dry the cuvette using kitchen paper. Air dry the cuvette in the UV box to degrade any remaining DNA and store under sterile conditions.
15. If a loud pop is heard during delivery of the electroporation shock, the system has short-circuited and the cells are dead. Short circuits can be caused by a high salt concentration in the sample or from high concentrations of DNA; attempt to electroporate using 0.5 μL DNA or 0.1 μL in a fresh cuvette.
16. It is difficult to predict the timing for the *Agrobacterium* IMAS culture since the cultures must reach optimal OD and then used immediately. The growth can be affected by several parameters such as plasmid size, preculture conditions, aeration, etc. Therefore, we always use three flasks of IMAS per *Agrobacterium* strain inoculated with different volumes of preculture. When one of the three cultures reaches the desired OD, the co-cultivation is initiated and the remaining IMAS cultures are discarded.
17. The acetosyringone powder can take up to 3 h to dissolve. Therefore, we recommend covering the beaker with parafilm or cling film to prevent evaporation. Adjusting the pH will aid dissolving the compound faster. Acetosyringone is an unstable molecule and the solution should be made fresh every time. Do not heat the solution. We usually wrap the beaker in tinfoil to

prevent degradation from sunlight. At pH 8.0, the solution will turn light yellow if the compound is dissolved. The compound will deteriorate in contact with air. If possible, store under inert gas and never freeze the chemical powder.

18. *Agrobacterium* cultures can become lumpy and this can obscure optical density measurements. Break the lumps apart by pipetting cultures up and down with a micropipette.
19. 2–4 mm white aerial mycelium should appear from the black filters. Do not let the aerial mycelium become too dense before continuing to the next step.
20. The following days, most of the aerial mycelium will disappear due to the selection pressure, and the filters may appear completely blank until colonies arise. The mycelium on the fungal growth control should collapse and produce no colonies. However, spontaneous resistance to G418 can develop at a low rate in *F. solani* and result in small colonies displaying inhibited growth. These false-positives are unlikely to proliferate when continuously streaked unto selective agar.

Acknowledgements

This work was supported by the Novo Nordisk Foundation grants NNF15OC0016028 and NNF18OC0034952.

References

1. Nalim FA, Samuels GJ, Wijesundera RL, Geiser DM (2011) New species from the *Fusarium solani* species complex derived from perithecia and soil in the Old World tropics. Mycologia 103:1302–1330
2. O'Donnell K (2000) Molecular phylogeny of the *Nectria haematococca-Fusarium solani* species complex. Mycologia 92:919–938
3. Šišić A, Al-Hatmi AMS, Baćanović-Šišić J, Ahmed SA, Dennenmoser D, de Hoog GS, Finckh MR (2018) Two new species of the *Fusarium solani* species complex isolated from compost and hibiscus (*Hibiscus* sp.). Antonie Van Leeuwenhoek 111:1785–1805
4. O'Donnell K, Al-Hatmi AMS, Aoki T, Brankovics B, Cano-Lira JF, Coleman JJ, de Hoog GS, Di Pietro A, Frandsen RJN, Geiser DM, Gibas CFC, Guarro J, Kim H-S, Kistler HC, Laraba I, Leslie JF, López-Berges MS, Lysøe E, Meis JF, Monod M, Proctor RH, Rep M, Ruiz-Roldán C, Šišić A, Stajich JE, Steenkamp ET, Summerell BA, van der Lee TAJ, van Diepeningen AD, Verweij PE, Waalwijk C, Ward TJ, Wickes BL, Wiederhold NP, Wingfield MJ, Zhang N, Zhang SX (2020) No to *Neocosmospora*: phylogenomic and practical reasons for continued inclusion of the *Fusarium solani* species complex in the genus *Fusarium*. mSphere 5:e00810–e00820
5. Zhang N, O'Donnell K, Sutton DA, Nalim FA, Summerbell RC, Padhye AA, Geiser DM (2006) Members of the *Fusarium solani* species complex that cause infections in both humans and plants are common in the environment. J Clin Microbiol 44:2186–2190
6. O'Donnell K, Sutton DA, Wiederhold N, Robert V, Crous PW, Geiser DM (2016) Veterinary fusarioses within the United States. J Clin Microbiol 54:2813–2819
7. Hansen FT, Gardiner DM, Lysøe E, Fuertes PR, Tudzynski B, Wiemann P, Sondergaard TE, Giese H, Brodersen DE, Sørensen JL (2015) An update to polyketide synthase and non-ribosomal synthetase genes and nomenclature in *Fusarium*. Fungal Genet Biol 75: 20–29
8. Hansen FT, Sørensen JL, Giese H, Sondergaard TE, Frandsen RJ (2012) Quick guide

to polyketide synthase and nonribosomal synthetase genes in *Fusarium*. Int J Food Microbiol 155:128–136

9. Dickman MB, Kolattukudy PE (1987) Transformation of *Fusarium solani pisi* using the cutinase promoter. Phytopathology 77:1740–1740
10. Wasmann CC, VanEtten HD (1996) Transformation-mediated chromosome loss and disruption of a gene for pisatin demethylase decrease the virulence of *Nectria haematococca* on pea. Mol Plant-Microbe Interact 9: 793–803
11. Liu HW, Bao XM (2009) Overexpression of the chitosanase gene in *Fusarium solani* via *Agrobacterium tumefaciens*-mediated transformation. Curr Microbiol 58:279–282
12. Romans-Fuertes P, Sondergaard TE, Sandmann MI, Wollenberg RD, Nielsen KF, Hansen FT, Giese H, Brodersen DE, Sørensen JL (2016) Identification of the non-ribosomal peptide synthetase responsible for biosynthesis of the potential anti-cancer drug sansalvamide in *Fusarium solani*. Curr Genet 62:799–807
13. Malonek S, Rojas MC, Hedden P, Gaskin P, Hopkins P, Tudzynski B (2004) The NADPH-cytochrome P450 reductase gene from *Gibberella fujikuroi* is essential for gibberellin biosynthesis. J Biol Chem 279:25075–25084
14. Song ZS, Cox RJ, Lazarus CM, Simpson TJ (2004) Fusarin C biosynthesis in *Fusarium moniliforme* and *Fusarium venenatum*. ChemBioChem 5:1196–1203
15. Sørensen LQ, Lysøe E, Larsen JE, Khorsand-Jamal P, Nielsen KF, Frandsen RJN (2014) Genetic transformation of *Fusarium avenaceum* by *Agrobacterium tumefaciens* mediated transformation and the development of a USER-Brick vector construction system. BMC Mol Biol 15:15
16. Frandsen RJN (2011) A guide to binary vectors and strategies for targeted genome modification in fungi using *Agrobacterium tumefaciens*-mediated transformation. J Microbiol Methods 87:247–262
17. Nielsen MR, Holzwarth AKR, Brew E, Chrapkova N, Kaniki SEK, Kastaniegaard K, Sorensen T, Westphal KR, Wimmer R, Sondergaard TE, Sorensen JL (2019) A new vector system for targeted integration and overexpression of genes in the crop pathogen *Fusarium solani*. Fungal Biol Biotechnol 6:25
18. Nielsen MR, Wollenberg RD, Westphal KR, Sondergaard TE, Wimmer R, Gardiner DM, Sørensen JL (2019) Heterologous expression of intact biosynthetic gene clusters in *Fusarium graminearum*. Fungal Genet Biol 132:103248
19. Frandsen RJ, Andersson JA, Kristensen MB et al (2008) Efficient four fragment cloning for the construction of vectors for targeted gene replacement in filamentous fungi. BMC Mol Biol 9:70
20. Gietz RD, Schiestl RH (2007) High-efficiency yeast transformation using the LiAc/SS carrier DNA/PEG method. Nat Protoc 2(1):31–34
21. Coleman JJ, Rounsley SD, Rodriguez-Carres-M, Kuo A, Wasmann CC, Grimwood J et al (2009) The genome of *Nectria haematococca*: contribution of supernumerary chromosomes to gene expansion. PLoS Genet 5(8): e1000618
22. Westphal KR, Wollenberg RD, Herbst FA, Sørensen JL, Sondergaard TE, Wimmer R (2018) Enhancing the production of the fungal pigment Aurofusarin in *Fusarium graminearum*. Toxins 10(11):485

Chapter 7

Investigating Fungal Biosynthetic Pathways Using *Pichia pastoris* as a Heterologous Host

Zhilan Qian, Qi Liu, and Menghao Cai

Abstract

Fungal natural products have extensive biological activities, and thus have been largely commercialized in the pharmaceutical, agricultural, and food industries. Recently, heterologous expression has become an irreplaceable technique to functionalize fungal biosynthetic gene clusters and synthesize fungal natural products in various chassis organisms. This chapter describes the general method of using *Pichia pastoris* as a chassis host to investigate fungal biosynthetic pathways.

Key words Fungal natural products, Biosynthetic pathways, *Pichia pastoris*, Heterologous expression, Chassis host

1 Introduction

Fungi are well-known as a vast treasure trove of natural products. Fungal natural products have broad biological activities, such as antitumor, antibacterial, immunosuppressive, *etc.* They can also biologically control weeds and be used as herbicides, and even color food, allowing their use in various fields [1–6]. Penicillin [7], the first broad-spectrum antibiotic discovered, is the most famous representative of fungal natural products used for medical treatment. Fingolimod (Gilenya) [7–9], derived from myriocin, was approved by FDA to treat multiple sclerosis. Caspofungin [10], an ester peptide derivative from echinocandin, is the first choice for the clinical treatment of fungal infections. The statin family [11, 12], consisting of compounds of fungal origin such as compactin and lovastatin, are the most common cholesterol-lowering agents. Fungal natural products are also widely applied in the agricultural and food industries. Strobilurin fungicides [13] produced by mushrooms have been used worldwide as pesticides

Zhilan Qian and Qi Liu contributed equally with all other contributors.

Elizabeth Skellam (ed.), *Engineering Natural Product Biosynthesis: Methods and Protocols*, Methods in Molecular Biology, vol. 2489, https://doi.org/10.1007/978-1-0716-2273-5_7,

since the 1990s. *Monascus* red [14], containing a variety of azaphilone derivatives, is the traditional natural food coloring agent in southeast Asia. These compounds show extensive biological activities due to their structural diversity. Based on their biosynthetic origin, fungal natural products can be divided into terpenoids, polyketides, alkaloids, non-ribosomal peptides, hybrid compounds, *etc.* [15].

Fungal genome sequencing reveals that the number of silent gene clusters in fungi far exceeds the number of natural products that have been discovered. It indicates that there is still a large number of natural compounds with unknown activity to be explored [16]. However, many fungi grow slowly, and their production process is susceptible to environmental change. Also, most fungi have not yet been developed with a mature genetic manipulation system. These bring great difficulties in functional identification of biosynthetic gene clusters and their encoded compounds in fungi. Heterologous expression has been proved effective in validating fungal gene clusters and exploring natural products [17–19]. Recently, the rapid development of synthetic biology and bioinformatics technologies has allowed heterologous expression to be a versatile way for the clarification of biosynthetic pathways and discovery of new compounds [15, 20–22].

Currently, heterologous hosts used for the production of fungal natural products mainly include *Escherichia coli*, yeast, and filamentous fungi [19, 23, 24]. *E. coli* has a perfect genetic operating system, but fungal proteins expressed in this prokaryotic host are often inactive, which restricts its use in expression of fungal gene clusters. Filamentous fungi are used with the advantage of expressing fungal genes without removing introns in most cases. Nevertheless, the immature genetic operating system severely limits the use of fungal hosts, and the background secondary metabolic pathways sometimes affect functional identification of heterologous gene clusters [25]. Yeast has become an alternative chassis organism for expression of fungal gene clusters despite requiring the removal of introns in fungal genes. Various yeast hosts and their related genetic editing tools have been reported. The downstream processing technologies of yeast expression are also well-developed [26, 27]. Until now, different kinds of fungal metabolites have been successfully produced in yeast, *e.g.*, polyketides, terpenoids, alkaloids, flavonoids, *etc.*, many of which have achieved high-level production [28–32].

This chapter introduces the general methods for the investigation of fungal biosynthetic pathways in a methylotrophic yeast, *Pichia pastoris* (syn. *Komagataella phaffii*). *P. pastoris* is a generally recognized as safe (GRAS) strain identified by the FDA and permitted for use in the pharmaceutical and food industries [33–35]. The advantages, including: strong expression capacity; fast-growth and high cell-density; simple and low-cost culture medium; and mild protein glycosylation; have made it a preferred and versatile protein expression host both in lab and in industry [33–35]. These merits also make it a chassis host candidate for synthesis of natural compounds and value-added chemicals. Recently, *P. pastoris* has been successfully applied to heterologous production of various natural products such as 6-MSA [36], citrinin [37], terreic acid [38], lovastatin [39–41], and other compounds [35]. Here we introduce two methods that were successfully applied in assembly and investigation of fungal biosynthetic pathways in *P. pastoris.*

2 Materials

2.1 Plasmid Construction

1. Plasmids required are listed in Table 1.
2. Restriction endonucleases : *Eco*RI; *Sal*I; *Bam*HI; *Apa*I; *Xho*I.
3. PCR MasterMix, DNA Polymerase.
4. Competent *Escherichia coli* TOP10 cells (Invitrogen).
5. Medium (w/v): LB medium (1% NaCl, 0.5% yeast extract, 1% tryptone); LLB medium (0.5% NaCl, 0.5% yeast extract, 1% tryptone). To prepare solid medium, 2% (w/v) agar is added. Antibiotics are used at the following concentrations: zeocin (50 μg/ml); ampicillin (100 μg/ml).
6. DNA purification kit.
7. Mini plasmid kit.
8. ClonExpress™II one step cloning kit (Vazyme Biotech).
9. Primers required for plasmid construction are listed in Table 2.

2.2 P. pastoris Transformation

1. *P. pastoris* strain GS115 (Invitrogen), Δ*ku70* [42].
2. Restriction endonucleases: *Bln*I.
3. Medium (w/v): YPD (1% yeast extract, 2% tryptone, 2% glucose); MM (1.34% yeast nitrogen base without amino acids, 0.5% (v/v) methanol); YND (0.67% yeast nitrogen base without amino acids, 2% glucose); MGY (0.67% yeast nitrogen base without amino acids, 1% glycerol). Zeocin is used at 100 μg/ml. D-Biotin is used at 0.4 μg/ml. L-Histidine is used at 50 μg/ml.
4. Sorbitol-LiAc solution: 0.6 M sorbitol, 0.001 M Tris, 0.15 M LiAc, 0.01 M DTT, pH 7.5.
5. Yeast genomic DNA extraction kit.
6. Electroporator

Table 1
List of parent plasmids used in this work

Plasmid	Description	Source
pPICZ B	Zeocin^R; P_{AOX1}-based expression vector	Invitrogen
pGAPZαA	Zeocin^R; P_{GAP}-based expression vector	Invitrogen
pDTgl	pUC18 derivative carrying homologous arm of P_{TEF1}UP-gl target	[42]
pDFgl	pUC18 derivative carrying homologous arm of P_{FLD1}UP-gl target	[42]
3.5k-PFgl+PTgl	pPIC3.5K derivative carrying *PARS*, *CAS9*, P_{HXT1}, P_{FLD1}UP-gl and P_{TEF1}UP-gl	[42]

Table 2
List of primers used in this work

Primer name	sequence
TT~5AOX-F	5′ TTAAGTGAGACCTTCGTTTGTGCGGATCAGATCTAACATCCAAAGAC 3′
BamHI~TT-R	5′ AAGCTATGGTGTGTGGGGGATCCGCACAAACGAAGGTCTC 3′
TT~GAP-F	5′ TTAAGTGAGACCTTCGTTTGTGCGGATCTTTTTTGTA GAAATGTCTTGG 3′
BamHI~GAP-R	5′ AAGCTATGGTGTGTGGGGGATCCATAGTTGTTCAATTGATTGA 3′
Tglup~AOX-F	5′ CCTCCTTCAACTAGTCTCGACAGATCTAACATCCAAAGACG 3′
Tgldown~TT-R	5′ GATGGTTAAAAGGACGGGCCTCTCACTTAATCTTCTGTACTCT 3′
Fglup~AOX-F	5′ AATTGATATACTAGTCTCGACAGATCTAACATCCAAAGACG 3′
Fgldown~TT-R	5′ GACCTACCCTACGACGGGCCTCTCACTTAATCTTCTGTACTCT 3′
HAPTglUP-F	5′ CTATGACCATGATTACGAATTCGAGCT 3′
HAPTglDOWN-R	5′ TGCCTGCAGGTCGACTCTAG 3′
HAPFglUP-F	5′ CATGATTACGAATTCGAGCT 3′
HAPFglDOWN-R	5′ GTCGACTCTAGACGGACTCA 3′
inCas9 R1-F	5′ CCGAGTGACAGGGCGATAAGA 3′
3AOX1-R	5′ GCAAATGGCATTCTGACATCC 3′

2.3 Extraction and Detection of Products

1. Filter membrane (0.22 μm, Millipore).
2. Anhydrous methanol; ethyl acetate; acetonitrile (chromatographic grade); formic acid (chromatographic grade); acetic acid (chromatographic grade).

3. High-performance liquid chromatography equipped with suitable reverse column for products analysis.

2.4 Transcriptional Level Analysis

1. Yeast RNA extraction kit.
2. Reverse transcription kit.
3. RT-qPCR master mix solution.

2.5 Protein Expression Level Analysis

1. Potassium phosphate buffer (50 mM, pH 7.0); PMSF (1 mM, Beyotime); anhydrous methanol.
2. Loading buffer (5×): 1.5 M Tris–HCl, pH 6.8, 50% (v/v) glycerol, 10% (w/v) SDS, 0.5% (w/v) bromophenol blue, 5% (v/v) beta-mercaptoethanol.
3. SmartPAGE™ precast protein gel (Smart Lifesciences, #SLE008); MOPS running buffer (Smart Lifesciences, #SLB009).
4. His-tag antibody (1:2000 dilution, Beyotime); secondary antibody: horseradish peroxidase-labeled goat anti-rabbit IgG (H +L) (1:1000 dilution, Beyotime).
5. PVDF membrane (Beyotime).
6. PBS buffer, pH 7.0; PBST buffer (1 l PBS buffer, 500 μl Tween 20).
7. Blocking solution: 10% (w/v) skimmed milk powder in PBST buffer.
8. ECL luminescent kit (Beyotime).

3 Methods

3.1 Pathway Assembly by Big Plasmid Carrying Multiple Biosynthetic Genes

3.1.1 Construction of Expression Plasmids

Construction of pPICZ-GeneX

1. Design and synthesize primers to amplify DNA fragments of the target gene X (*see* **Notes 1** and **2**). Purify PCR products by gel electrophoresis and recover the DNA fragments from the agarose gel using DNA purification kit.
2. Digest the plasmid pPICZ B with *Eco*RI and *Sal*I, and recover the plasmid backbone from enzyme digestions using DNA purification kit.
3. Assemble the DNA fragments recovered in **steps 1** and **2** together by ClonExpress™II one step cloning kit. Add the reaction mix to 80 μl competent *E. coli* TOP10 cells and incubate on ice for 10 min. Heat pulse the mix at 42 °C for 90 s. Recover cells on ice for 5 min. Add the mix to 1 ml of LB medium and incubate at 37 °C for 1 h at 200 rpm. Spread the cells in LLB-zeocin plates for colony selection. Incubate the plates overnight at 37 °C. Pick single colonies and inoculate

0.4 ml LLB-zeocin medium at 37 °C and 200 rpm for 2 h. Verify the positive colonies harboring pPICZ-GeneX by colony PCR and DNA sequencing.

Co-expression of two genes (e.g. construction of pPICZ-GeneA+GeneB)

4. Cultivate the correct transformants in LLB-zeocin medium overnight at 37 °C and 200 rpm. Isolate the plasmids pPICZ-GeneA and pPICZ-GeneB (*see* **Note 3**) from *E.coli* using mini plasmid kit.
5. Digest the pPICZ-GeneA with *BamH*I and recover the plasmid backbone from enzyme digestions using DNA purification kit.
6. Amplify the P_{AOX1}-*GeneB*-TT fragments with primers TT~5AOX-F and BamHI~TT-R (*see* **Note 4**) using the pPICZ-GeneB as the template. Purify PCR products by gel electrophoresis and recover the DNA fragments from the agarose gel using DNA purification kit.
7. Assemble the DNA fragments recovered in **steps 5** and **6** together by ClonExpress™II one step cloning kit. The transformation of competent *E. coli* TOP10 cells and the selection of positive clones harboring the plasmid pPICZ-GeneA +GeneB follow the protocol described in **step 3**.
8. Cultivate the correct *E. coli* in LLB-zeocin medium overnight at 37 °C and 200 rpm. Isolate the pPICZ-GeneA+GeneB from *E.coli* using mini plasmid kit. If there are multiple expression cassettes in the follow-up, repeat **steps 4–7** to add the expression cassettes one by one, with the same endonucleases and primers (*see* **Note 5**).

Construction of pPICZ-GeneA+GeneB+P_{GAP} (*see* **Note 6**)

9. Digest the plasmid pPICZ-GeneA+GeneB with *BamH*I, and recover the plasmid backbone from enzyme digestions using DNA purification kit.
10. Amplify the P_{GAP} fragments with primers TT~GAP-F and BamHI~GAP-R using the plasmid pGAPZαA as the template. Purify PCR products by gel electrophoresis and recover the DNA fragments from the agarose gel using DNA purification kit.
11. Assemble the DNA fragments recovered in **steps 9** and **10** together by ClonExpress™II one step cloning kit. The transformation of competent *E. coli* TOP10 cells and the selection of positive clones follow the protocol described in **step 3**.

3.1.2 P. pastoris Transformation

1. Digest the plasmid pPICZ-GeneA+GeneB+P_{GAP} with *Bln*I and recover the plasmid backbone from enzyme digestions using DNA purification kit.
2. Preparation of *P. pastoris* GS115 or Δ*ku70* cells for transformation: Grow the cells in 5 ml YPD medium overnight to the logarithmic phase (OD_{600} = 4–8). Then transfer all the seed into 100 ml of YPD medium and grow the cells until OD_{600} = 1.3–1.5. Centrifuge the sample at 5000 × *g* for 5 min and discard the supernatant. Resuspend the cells in 20 ml sorbitol-LiAc solution and incubate at 30 °C and 200 rpm for 30 min. Centrifuge the sample at 5000 × *g* for 5 min at 4 °C and discard the supernatant. Resuspend the cells in 20 ml ice-cold sorbitol solution (1 M) and then centrifuge at 4 °C and 5000 × *g* for 5 min. Repeat this wash one more time. Discard the supernatant, resuspend the cells in 1 ml ice-cold sorbitol solution (1 M), and store the cells at −80 °C.
3. Add 1 μg of linearized plasmid into 80 μl of competent yeast cells from **step 1**. Transfer the mix to an ice-cold electroporation cuvette (2 mm). Use a 2.00 kV pulse to transform the yeast by electroporation. Resuspend the cells immediately and softly in 1 ml ice-cold sorbitol solution (1 M) and transfer to a sterile tube with 1 ml YPD medium. Incubate the cells at 30 °C and 200 rpm for 2 h.
4. Spread the cells on YPD agar plates with zeocin, and incubate at 30 °C for 2–3 days. Pick single colonies and incubate each colony in YPD liquid medium with zeocin in 2 ml sterile eppendorf tubes at 30 °C and 200 rpm for 2 days. Isolate genomic DNA using Yeast DNA kit. Verify positive colonies by PCR using genomic DNA as the template.

3.1.3 Cultivation and Sampling of P. pastoris

1. Cultivate the cells in 50 ml YPD medium overnight to the logarithmic phase. Centrifuge the cell cultures at 5000 × *g* for 5 min, and wash twice with sterile distilled water. Resuspend the cells with MM medium and inoculate cells into MM medium (adding biotin and histidine) to an OD_{600} of 1.0. Add anhydrous methanol by 0.5% (v/v) every 24 h.
2. Harvest 2 ml cell cultures by centrifugation at 12,000 × *g* for 2 min and discard the supernatant. Store the cells at −80 °C for subsequent RNA transcriptional analysis and protein expression analysis (*see* **Note 7**).
3. Collect 10 ml cell cultures in centrifuge tubes and add 20 ml ethyl acetate. Vortex the mix for 5 min and centrifuge at 5000 × *g* for 5 min. Evaporate the organic phase (*see* **Note 8**) under reduced pressure and dissolve the dry residue in 1 ml methanol. Store the sample at 4 °C for subsequent HPLC analysis.

3.1.4 Analysis of Production

HPLC Analysis

1. Filter the sample across 0.22 μm membrane and transfer it to HPLC auto-sampler vials.
2. Equilibrate the analytical column with the mobile phase of the initial composition over 10 min.
3. Maintain column temperature at the required levels. Commonly, use the injection sample volume of 20 μl and the flow rate of 1 ml/min. The detection wavelength depends on the maximum absorption wavelength of each specific compound.
4. To determine the molecular weight of the product, it could be further analyzed by LC-MS under the same conditions.

RNA Transcriptional Analysis

1. Extract total RNA from yeast cells by Yeast RNA extraction kit.
2. Obtain cDNA by reverse transcription.
3. Design specific primers to amplify the fragments by PCR using cDNA as the template to check the transcription of the target gene.
4. For comparison of gene transcription levels, conduct reverse transcription-quantitative PCR (RT-qPCR) with RT-qPCR master mix solution.

Protein Expression Analysis

1. Extraction of intracellular proteins from *P. pastoris*: Harvest cell cultures by centrifugation at 12,000 × *g* for 2 min and discard the supernatant. Wash the cells with ice-cold 50 mM potassium phosphate buffer twice and then resuspend it in 1 ml buffer added with 1 mM PMSF. Transfer cell sample to a 2-ml screw-cap tube with 1.8 g zirconium beads, and lyse in a bead mill eight times (vortex at high speed for 1 min and keep on ice for 1 min each time). Centrifuge cell lysates at 4 °C and 12,000 × *g* for 5 min, and collect the supernatant.
2. SDS-PAGE: Mix 40 μl protein supernatant with 10 μl loading buffer (5×) and incubate at 95 °C for 10 min. Centrifuge the mix at 12,000 × *g* for 2 min and collect the supernatant. Load 10 μl of the sample per well of the precast protein gel. Run in the gel at a constant pressure of 130 V for 80 min.
3. Western-blotting: Electro transfer the sample from SDS-PAGE gel to an equilibrated PVDF membrane at a constant pressure of 100 V. Block the PVDF membrane in blocking solution and incubate on a rotary shaker (75 rpm) for 2 h at room temperature or overnight in the refrigerator at 4 °C. Add first antibody (His-tag antibody, 5 μl in 10 ml blocking solution) and incubate at a rotary shaker (75 rpm) for 3 h at room temperature or overnight in refrigerator at 4 °C. Wash membrane three times

(10 min for each time) with PBST buffer. Add second antibody (horseradish peroxidase-labeled goat anti-rabbit IgG [H+L], 10 μl in 10 ml blocking solution) and incubate at a rotary shaker for 1.5 h. Wash membrane three times (10 min for each time) with PBST buffer. Dry the PVDF membrane with filter paper and detect the proteins in the membrane with ECL luminescent kit.

3.2 Pathway Assembly by CRISPR-Cas9 Mediated Multiple-Gene Integration

For construction of pPICZ-GeneX, refer to Subheading 3.1.1 (*see* **Note 9**).

3.2.1 Construction of Expression Vectors

Construction of Donor Cassette Plasmids (Double-gene integration as an example)

1. Digest the plasmids pDTg1with endonucleases *Xho*I and *Apa*I, and so does the pDFg1. Recover the plasmid backbones from enzyme digestions using DNA Purification Kit.
2. Amplify the P_{AOX1}-GeneA-TT fragment with primers Tg1up~AOX-F and Tg1down~TT-R using the plasmid pPICZ-GeneA as the template. Amplify the P_{AOX1}-GeneB-TT fragment with primers Fg1up~AOX-F and Fg1down~TT-R using the plasmid pPICZ-GeneB as the template. Purify PCR products by gel electrophoresis and recover the DNA fragments from the agarose gel using DNA purification kit.
3. Assemble the DNA fragments recovered in **steps 1** and **2** together by ClonExpress™II one step cloning kit, which produces two plasmids pDTg1-GeneA and pDFg1-GeneB separately. The transformation of competent *E. coli* TOP10 cells and the selection of positive clones using ampicillin follow the protocol described in Subheading 3.1.1, **step 3**.
4. Repeat **steps 1–3** above with other backbone-cloning vectors [42] if there are a number of cassettes to be constructed. Alternatively, combine with the method of constructing large plasmids to integrate multiple expression cassettes on the same donor plasmid, which could be used flexibly when necessary.

3.2.2 P. pastoris Transformation

1. Amplify the Tg1HAUP-P_{AOX1}-GeneA-TT-Tg1HADO fragment with primers HAPTg1UP-F and HAPTg1DOWN-R using the plasmid pDTg1-GeneA as the template. Amplify the Fg1HAUP-P_{AOX1}-GeneB-TT-Fg1HADO fragment with primers HAPFg1UP-F and HAPFg1DOWN-R using the plasmid pDFg1-GeneB as the template. Purify PCR products by gel electrophoresis and recover the DNA fragments from the agarose gel using DNA purification kit.
2. Preparation of *P. pastoris* Δ*ku70* cells for transformation: Refer to Subheading 3.1.2, **step 2**.

3. Add 100 ng plasmid 3.5k-PFg1+PTg1 (*see* **Note 10**) and donor DNA (1 μg Tg1HAUP-P_{AOX1}-GeneA-TT-Tg1HADO, 1 μg Fg1HAUP-P_{AOX1}-GeneB-TT-Fg1HADO) (*see* **Note 11**) into 80 μl of Δ*ku70* competent cells from **step 1**. Transformation by electroporation was as same as **step 3** in Subheading 3.1.2. Incubate at 30 °C and 200 rpm over 3 h.
5. Wash the cells twice with sterile water, spread the cells on YND agar plates, and incubate at 30 °C for 4–5 days. Pick single colonies and inoculate each colony into YPD liquid medium in 2 ml sterile Eppendorf tubes at 30 °C and 200 rpm for 2 days. Isolate genomic DNA using TIANamp yeast DNA kit. Verify positive transformants by PCR using genomic DNA as the template.
4. Streak positive transformants on YPD agar plates and incubate at 30 °C for 2 days. Pick single colonies and incubate each colony in YPD liquid medium in 2 ml sterile Eppendorf tubes at 30 °C and 200 rpm for 2 days. Isolate genomic DNA using TIANamp yeast DNA kit. The strains losing the plasmid 3.5k-PFg1+PTg1 was verified by PCR with primers in Cas9 R1-F and 3AOX1-R using genomic DNA as the template (*see* **Note 12**).

3.2.3 Cultivation and Sampling of P. pastoris

Refer to Subheading 3.1.3.

3.2.4 Analysis of Production

Refer to Subheading 3.1.4.

4 Notes

1. The introns within the target gene fragments need to be removed in advance. Possible methods are as follows: Synthesize the complete gene without introns directly (i.e. extract fungal RNA, perform reverse transcription to obtain cDNA, and amplify gene fragments using cDNA as a template) or extract fungal genomic DNA as a template, design primers to amplify each exon fragments separately, and splice the fragments together by overlap PCR.
2. For the construction of large plasmids carrying multiple genes, *BamH*I and *Bln*I sites within the gene sequences need to be mutated in advance. Design specific primers to introduce single base substitutions via overlap PCR.
3. Any number of plasmids for single gene expression cassettes can be constructed through the three steps in 3.1.1, with the only differences being in primer design.

4. The downstream primer BamH~TT-R includes a *Bam*HI site, which introduces specific restriction sites for subsequent insertion of other expression cassettes.
5. Assembly of multiple DNA fragments at a time can be achieved with ClonExpress™ multiS one step cloning kit. The primers need to be designed separately.
6. The number of expression cassettes that can be added to a single plasmid is limited. In this section, *Shble* (zeocin resistance gene) was used as the selection marker, and P_{GAP} was used as the integration site. Some other resistance and integration sites can be selected when multiple genes are involved and the construction strategy keeps the same. For plasmids carrying multiple P_{GAP} sequence, the *Bln*I restriction site should be mutated to ensure that only one P_{GAP} can be linearized by *Bln*I for genome integration.
7. The samples used for transcriptional analysis should not be stored for a long time. Preferably, extract RNA immediately after sampling.
8. When the stratification is not obvious, extend the centrifugal time or increase the speed appropriately.
9. For plasmid construction here, the target gene does not need to be mutated for the specific restriction site.
10. The gRNA plasmid needs to match the donor fragments, and the combination of different sites can be selected when necessary.
11. When the donor fragments are large, the amount of donor fragments needs to be increased to ensure transformation efficiency.
12. In order to ensure that the gRNA plasmids are not retained, wash the cells twice with sterile water and spread the cells on the MGY or YND agar plate (histidine deficient). If no colony grows, it illustrates that the plasmids (harboring histidine synthetic gene cassette) have been completely lost.

References

1. Butcher MG, Meyer PA, Hallen RT, Albrecht KO, Clayton CK, Polikarpov E et al (2018) Fungal metabolites as precursors to renewable transportation fuels. Fuel 215:123–141. https://doi.org/10.1016/j.fuel.2017.10.052
2. Gupta S, Chaturvedi P, Kulkarni MG, Van Staden J (2020) A critical review on exploiting the pharmaceutical potential of plant endophytic fungi. Biotechnol Adv 39:107462. https://doi.org/10.1016/j.biotechadv.2019.107462
3. Hyde KD, Xu J, Rapior S, Jeewon R, Lumyong S, Niego AGT et al (2019) The amazing potential of fungi: 50 ways we can exploit fungi industrially. Fungal Divers 97:1–136. https://doi.org/10.1007/s13225-019-00430-9
4. Terkar A, Borde M (2021) Endophytic fungi: novel source of bioactive fungal metabolites. In: New and future developments in microbial biotechnologyand

bioengineering, pp 95–105. https://doi.org/10.1016/b978-0-12-821005-5.00006-5

5. Vijayalakshmi S, Karthik K, Winny Fred Crossia A, Subashini G, Bhuvaneswari S, Panneerselvam A et al (2020) Fungal secondary metabolites: a potential source of anticancer compounds. In: New and future developments in microbial biotechnology and bioengineering, pp 81–93. https://doi.org/10.1016/b978-0-12-821006-2.00007-8
6. Yuan S, Gopal JV, Ren S, Chen L, Liu L, Gao Z (2020) Anticancer fungal natural products: mechanisms of action and biosynthesis. Eur J Med Chem 202:112502. https://doi.org/10.1016/j.ejmech.2020.112502
7. Beekman AM, Barrow RA (2014) Fungal metabolites as pharmaceuticals. Aust J Chem 67:827–843. https://doi.org/10.1071/ch13639
8. Schueffler A, Anke T (2014) Fungal natural products in research and development. Nat Prod Rep 31:1425–1448. https://doi.org/10.1039/c4np00060a
9. Strader CR, Pearce CJ, Oberlies NH (2011) Fingolimod (FTY720): a recently approved multiple sclerosis drug based on a fungal secondary metabolite. J Nat Prod 74:900–907. https://doi.org/10.1021/np2000528
10. Balkovec JM, Hughes DL, Masurekar PS, Sable CA, Schwartz RE, Singh SB (2014) Discovery and development of first in class antifungal caspofungin (CANCIDAS(R))--a case study. Nat Prod Rep 31:15–34. https://doi.org/10.1039/c3np70070d
11. Mulder KC, Mulinari F, Franco OL, Soares MS, Magalhaes BS, Parachin NS (2015) Lovastatin production: from molecular basis to industrial process optimization. Biotechnol Adv 33:648–665. https://doi.org/10.1016/j.biotechadv.2015.04.001
12. Xiong Z, Cao X, Wen Q, Chen Z, Cheng Z, Huang X et al (2019) An overview of the bioactivity of monacolin K/lovastatin. Food Chem Toxicol 131:110585. https://doi.org/10.1016/j.fct.2019.110585
13. Feng Y, Huang Y, Zhan H, Bhatt P, Chen S (2020) An overview of strobilurin fungicide degradation:current status and future perspective. Front Microbiol 11:389. https://doi.org/10.3389/fmicb.2020.00389
14. Agboyibor C, Kong WB, Chen D, Zhang AM, Niu SQ (2018) *Monascus* pigments production, composition, bioactivity and its application: a review. Biocatal Agric Biotechnol 16:433–447. https://doi.org/10.1016/j.bcab.2018.09.012
15. Hillman ET, Readnour LR, Solomon KV (2017) Exploiting the natural product potential of fungi with integrated -omics and synthetic biology approaches. Curr Opin Syst Biol 5:50–56. https://doi.org/10.1016/j.coisb.2017.07.010
16. Li YF, Tsai KJS, Harvey CJB, Li JJ, Ary BE, Berlew EE et al (2016) Comprehensive curation and analysis of fungal biosynthetic gene clusters of published natural products. Fungal Genet Biol 89:18–28. https://doi.org/10.1016/j.fgb.2016.01.012
17. Harvey CJB, Tang M, Schlecht U, Horecka J, Fischer CR, Lin HC et al (2018) HEx: a heterologous expression platform for the discovery of fungal natural products. Sci Adv 4:eaar5459. https://doi.org/10.1126/sciadv.aar5459
18. Skellam E (2019) Strategies for engineering natural product biosynthesis in fungi. Trends Biotechnol 37:416–427. https://doi.org/10.1016/j.tibtech.2018.09.003
19. Zhang X, Hindra, Elliot MA (2019) Unlocking the trove of metabolic treasures: activating silent biosynthetic gene clusters in bacteria and fungi. Curr Opin Microbiol 51:9–15. https://doi.org/10.1016/j.mib.2019.03.003
20. Lyu HN, Liu HW, Keller NP, Yin WB (2020) Harnessing diverse transcriptional regulators for natural product discovery in fungi. Nat Prod Rep 37:6–16. https://doi.org/10.1039/c8np00027a
21. Mosunova O, Navarro-Muñoz JC, Collemare J (2020) The biosynthesis of fungal secondary metabolites: from fundamentals to biotechnological applications. In: Reference module in life sciences. https://doi.org/10.1016/b978-0-12-809633-8.21072-8
22. Ren H, Shi C, Zhao H (2020) Computational tools for discovering and engineering natural product biosynthetic pathways. iScience 23:100795. https://doi.org/10.1016/j.isci.2019.100795
23. Alberti F, Foster GD, Bailey AM (2017) Natural products from filamentous fungi and production by heterologous expression. Appl Microbiol Biotechnol 101:493–500. https://doi.org/10.1007/s00253-016-8034-2
24. Kjaerbolling I, Mortensen UH, Vesth T, Andersen MR (2019) Strategies to establish the link between biosynthetic gene clusters and secondary metabolites. Fungal Genet Biol 130:107–121. https://doi.org/10.1016/j.fgb.2019.06.001
25. Vassaux A, Meunier L, Vandenbol M, Baurain D, Fickers P, Jacques P et al (2019) Nonribosomal peptides in fungal cell factories: from genome mining to optimized

heterologous production. Biotechnol Adv 37: 107449. https://doi.org/10.1016/j.biotechadv.2019.107449

26. Baghban R, Farajnia S, Rajabibazl M, Ghasemi Y, Mafi A, Hoseinpoor R et al (2019) Yeast expression systems: overview and recent advances. Mol Biotechnol 61:365–384. https://doi.org/10.1007/s12033-019-00164-8
27. Guirimand G, Kulagina N, Papon N, Hasunuma T, Courdavault V (2021) Innovative tools and strategies for optimizing yeast cell factories. Trends Biotechnol 39:488. https://doi.org/10.1016/j.tibtech.2020.08.010
28. Ignea C, Raadam MH, Motawia MS, Makris AM, Vickers CE, Kampranis SC (2019) Orthogonal monoterpenoid biosynthesis in yeast constructed on an isomeric substrate. Nat Commun 10:3799. https://doi.org/10.1038/s41467-019-11290-x
29. Ma T, Shi B, Ye Z, Li X, Liu M, Chen Y et al (2019) Lipid engineering combined with systematic metabolic engineering of *Saccharomyces cerevisiae* for high-yield production of lycopene. Metab Eng 52:134–142. https://doi.org/10.1016/j.ymben.2018.11.009
30. Srinivasan P, Smolke CD (2019) Engineering a microbial biosynthesis platform for de novo production of tropane alkaloids. Nat Commun 10:3634. https://doi.org/10.1038/s41467-019-11588-w
31. Chen R, Yang S, Zhang L, Zhou YJ (2020) Advanced strategies for production of natural products in yeast. iScience 23:100879. https://doi.org/10.1016/j.isci.2020.100879
32. Palmer CM, Miller KK, Nguyen A, Alper HS (2020) Engineering 4-coumaroyl-CoA derived polyketide production in *Yarrowia lipolytica* through a beta-oxidation mediated strategy. Metab Eng 57:174–181. https://doi.org/10.1016/j.ymben.2019.11.006
33. Schwarzhans JP, Luttermann T, Geier M, Kalinowski J, Friehs K (2017) Towards systems metabolic engineering in *Pichia pastoris*. Biotechnol Adv 35:681–710. https://doi.org/10.1016/j.biotechadv.2017.07.009
34. Wang J, Wang X, Shi L, Qi F, Zhang P, Zhang Y et al (2017) Methanol-independent protein expression by *AOX1*promoter with *trans*-acting elements engineering and glucose-glycerol-shift induction in *Pichia pastoris*. Sci Rep 7: 41850. https://doi.org/10.1038/srep41850
35. Pena DA, Gasser B, Zanghellini J, Steiger MG, Mattanovich D (2018) Metabolic engineering of *Pichia pastoris*. Metab Eng 50:2–15. https://doi.org/10.1016/j.ymben.2018.04.017
36. Gao L, Cai M, Shen W, Xiao S, Zhou X, Zhang Y (2013) Engineered fungal polyketide biosynthesis in *Pichia pastoris*: a potential excellent host for polyketide production. Microb Cell Fact 12:77. https://doi.org/10.1186/1475-2859-12-77
37. Xue Y, Kong C, Shen W, Bai C, Ren Y, Zhou X et al (2017) Methylotrophic yeast *Pichia pastoris* as a chassis organism for polyketide synthesis via the full citrinin biosynthetic pathway. J Biotechnol 242:64–72. https://doi.org/10.1016/j.jbiotec.2016.11.031
38. Kong C, Huang H, Xue Y, Liu Y, Peng Q, Liu Q et al (2018) Heterologous pathway assembly reveals molecular steps of fungal terreic acid biosynthesis. Sci Rep 8:2116. https://doi.org/10.1038/s41598-018-20514-x
39. Liu Y, Bai C, Liu Q, Xu Q, Qian Z, Peng Q et al (2019) Engineered ethanol-driven biosynthetic system for improving production of acetyl-CoA derived drugs in Crabtree-negative yeast. Metab Eng 54:275–284. https://doi.org/10.1016/j.ymben.2019.05.001
40. Liu Y, Bai C, Xu Q, Yu J, Zhou X, Zhang Y et al (2018) Improved methanol-derived lovastatin production through enhancement of the biosynthetic pathway and intracellular lovastatin efflux in methylotrophic yeast. Bioresour Bioprocess 5:22. https://doi.org/10.1186/s40643-018-0202-z
41. Liu Y, Tu X, Xu Q, Bai C, Kong C, Liu Q et al (2018) Engineered monoculture and co-culture of methylotrophic yeast for de novo production of monacolin J and lovastatin from methanol. Metab Eng 45:189–199. https://doi.org/10.1016/j.ymben.2017.12.009
42. Liu Q, Shi X, Song L, Liu H, Zhou X, Wang Q et al (2019) CRISPR-Cas9-mediated genomic multiloci integration in *Pichia pastoris*. Microb Cell Fact 18:144. https://doi.org/10.1186/s12934-019-1194-x

Chapter 8

Evolutionary Genome Mining for the Discovery and Engineering of Natural Product Biosynthesis

Marc G. Chevrette, Nelly Selem-Mojica, César Aguilar, Kristin Labby, Edder D. Bustos-Diaz, Jo Handelsman, and Francisco Barona-Gómez

Abstract

Genome mining has become an invaluable tool in natural products research to quickly identify and characterize the biosynthetic pathways that assemble secondary or specialized metabolites. Recently, evolutionary principles have been incorporated into genome mining strategies in an effort to better assess and prioritize novelty and understand their chemical diversification for engineering purposes. Here, we provide an introduction to the principles underlying evolutionary genome mining, including bioinformatic strategies and natural product biosynthetic databases. We introduce workflows for traditional genome mining, focusing on the popular pipeline antiSMASH, and methods to predict enzyme substrate specificity from genomic information. We then provide an in-depth discussion of evolutionary genome mining workflows, including EvoMining, CORASON, ARTS, and others, as adopted by our group for the discovery and prioritization of natural products biosynthetic gene clusters and their products.

Key words Evolutionary genome mining, Natural products evolution, Specialized metabolism, EvoMining, CORASON, ARTS

1 Introduction to Secondary Metabolism and Evolutionary Genome Mining

Historically, the evolutionary underpinnings of natural products (NPs; also called secondary metabolites) have largely been neglected; bacteria and fungi are often grown in the laboratory under controlled conditions where their expressed secondary metabolites undoubtedly do not influence fitness in the same way they do in their natural settings. NPs were largely thought to confer non-essential metabolic functions (hence, "secondary") and thus be governed by mostly neutral (rather than adaptive) evolutionary processes. With the arrival of the genomic era, a different reality was unveiled. As the first genomes of NP-producing microorganisms

Marc G. Chevrette and Nelly Selem-Mojica contributed equally to this work.

Elizabeth Skellam (ed.), *Engineering Natural Product Biosynthesis: Methods and Protocols*, Methods in Molecular Biology, vol. 2489, https://doi.org/10.1007/978-1-0716-2273-5_8,

were analyzed through the lens of functional annotation, it became apparent that they dedicate large portions of their genomes to the biosynthesis of secondary metabolites [1, 2]. As more and more genomes continue to be sequenced, we can now look at NP pathways in the context of populations, where the chemical logic dictated by biosynthetic genes is shaped by both neutral and adaptive forces, including natural selection. This offers a contrasting view to the NP research of the past. In nature, secondary metabolism is "secondary" in name only. Its products have been selected for both biologically and ecologically relevant functions through complex evolutionary processes. Secondary metabolism is sometimes referred to as "specialized" metabolism to reflect that its metabolic products may be "specialized" through evolution to modulate fitness in specific environments (e.g. as antibiotics, siderophores, signals, etc.).

This philosophical shift has begun to shed light on the evolutionary principles and rules that influence how NP biosynthetic pathways assemble and change over time [3]. Just as biochemical principles govern the synthesis of NPs, evolutionary forces shape their conservation and diversification across the tree of life. Together, biochemical and evolutionary principles provide the foundations to what can be called "evolutionary genome mining" of NPs. In this paper we aim to provide a workflow to leverage functional annotations of NP-producing organisms to construct evolutionary predictions of novel biosynthetic pathways [4–6] or to direct engineering strategies for the biosynthesis of NPs of interest [7, 8].

Before providing a typical workflow for evolutionary genome mining, a word of caution should be mentioned. The main difference between standard genome mining and evolutionary approaches is that strictly speaking, for evolutionary genome mining there is less need for empirically validated data, which is the foundation of rule-based approaches such as antiSMASH [9, 10]. In these more common approaches, we find what we look for: nonribosomal peptides (NRPs), polyketides (PKs), terpenes, phosphonates, etc., the known biosynthetic classes with common domains and genes with well-established functions. Predictions are based on previous experimental data and empirical knowledge, often obtained after laborious, interdisciplinary efforts. In contrast, evolutionary genome mining uses relative divergence between genes and domains (some of which may have yet unknown functions) to deduce evolutionary patterns. Based on the intrinsic predictive power of evolutionary theory, one can construct completely unprecedented predictions from evolutionary genome mining. Uncovering this "biosynthetic dark matter" otherwise unseen by rules-based approaches accelerates the discovery of truly novel NPs and their biosynthetic enzymes and pathways. Thus, in the following subsection, we briefly describe the main evolutionary principles that have emerged during the last decade to better understand the development and use of evolutionary genome mining tools.

1.1 Principles of Evolutionary Genome Mining

Although the theory of evolution is well-established and has served as the main conceptual framework for studying biological systems, it was not until recently that evolutionary thinking has been integrated into applied natural products research. Moreover, to migrate from speculative evolutionary hypotheses into quantitative studies, large datasets are needed. In recent years, large genomic datasets have enabled us to pursue such studies in the NP community. Similar to the enzymatic principles that shape our understanding of NP biosynthesis (such as adenylation [A] domain specificity of nonribosomal peptide-synthetases [NRPSs], or chain elongation along polyketide synthases [PKSs]), the following broad evolutionary principles can be listed.

1. Enzyme promiscuity drives pathway evolution through genetic expansion-and-recruitment events that provide the building blocks for assembly into NP biosynthetic pathways.
2. Once enzymes (or domains) are recruited into NP biosynthesis, they tend to cluster together as multidomain megasynthases and/or biosynthetic gene clusters (BGCs). It is widely appreciated that this rapid evolution occurs under strong positive selection, but it is also clear that negative selection and neutral forces can influence BGC dynamics.

Once genetic events cluster biosynthetic enzymes together, the resulting pathways can recruit other auxiliary elements, such as regulators, domain–domain interactors, transporters, and resistance genes. Empirical evidence established these principles in the last decade or so, and these were exploited by researchers for the development of the three main evolutionary genome mining tools that will be addressed in this paper: (1) EvoMining, (2) CORASON, and (3) ARTS. Other tools (such as BiG-SCAPE and DeepBGC) are also briefly highlighted; while not explicitly rooted in evolutionary principles, these tools may be useful in performing comparative analyses from which evolutionary hypotheses can be formed. In the following sections, we will provide the basics of each one of these tools, but first we will briefly comment on the functional annotation, which is essential for successful evolutionary genome mining. Also important are the repositories or databases that feed into evolutionary genome mining algorithms, and these will be introduced as well. In line with this, we will describe and define below antiSMASH, Prokka, and RAST as the main tools used for preparing the raw data (materials) for evolutionary genome mining, but many others should be considered before embarking on an evolutionary genome mining study. Likewise, we will comment on tools available to predict the substrate specificities and overall structure of nonribosomal peptides and polyketides. Again, there are many comparable tools, but we will focus on SANDPUMA and transPACT because they leverage evolutionary principles and phylogenetics to make their predictions.

2 Functional Annotation and Databases for NP Research

To identify and interpret evolutionary events for genome mining, first the genes and gene clusters within the genome must be annotated. Here, we assume that the raw materials available to the researcher are an assembled prokaryotic genome and the workflows and databases are described below. Many command-line algorithms can both assemble and evaluate the quality of the draft genome and have been thoroughly reviewed elsewhere [11, 12]. There are also web-based free and automatic assembly services such as PATRIC (available at https://www.patricbrc.org) [13]. It is important to emphasize that genomic assemblies should be as contiguous as possible, but this is often not practical due to the repetitive nature of some BGCs, making them susceptible to fragmentation across many contigs. Short-read sequencing technologies, such as Illumina, are, thus, generally not the best suited for genome mining when the assembly is not done carefully, and long-read sequencing technologies (e.g. Pacific Biosciences, Oxford Nanopore, etc.) or hybrid long-and-short read assemblies are preferred when possible. However, for the majority of use cases, high coverage Illumina-only assemblies are sufficient for genome mining. Regardless of the read length, coverage, or technology used it is important to be mindful of putative BGCs that fall on contig edges, as fragmented assemblies that shatter BGCs into multiple contigs can lead to over counting. Once genomic data is assembled, workflows are typically the same, but it is important to be aware of the type of sequencing used and assess the quality of the assembly to properly interpret downstream analyses. This is especially relevant when dealing with metagenomic assemblies. Even though it might be tempting to apply all workflows directly to metagenomic assemblies or even to metagenomic bins, the underlying assumptions made by metagenome-specific programs are quite different from the ones made by regular assemblers, and consequently, a more refined pre-processing of the data is often needed before employing the genome mining workflows described here.

2.1 Functional Annotation of Microbial Genomes

Two popular tools to annotate genomic assemblies are the Rapid Annotation using Subsystem Technology or RAST [14], and Prokka [15]. Both are workflows that automatically run a series of specialized algorithms on any given genome to find and annotate coding sequences (or CDS), tRNAs, rRNAs, and other genomic features. To annotate the coding sequences that form the bulk of the genes explored in the subsequent genome mining, RAST uses GLIMMER2 [16] and Prokka uses Prodigal [17] to identify open reading frames to be refined by subsequent steps. Typically, these consist of searches against curated protein databases obtained from phylogenetically similar organisms: RAST against its own online databases and Prokka against UniProt. Here, it is important to note that regardless of their differences, both workflows perform better when the assembly is of good quality, i.e., the resulting genome is not fragmented, and the contigs, or scaffolds, are long enough to contain the multiple genes found in BGCs.

The main difference between RAST and Prokka is in their presentation. RAST is generally used as an automated web service, which means that for annotation a genome must be uploaded through the RAST web page so that processing can proceed remotely using RAST's web servers. After the workflow finishes, the results are stored and made accessible online and can be visualized through the SEED viewer, which was developed by the same team [18] or downloaded to a local computer. While the graphic interface is user-friendly, a more programmatic approach is needed to annotate dozens or hundreds of genomes. For these cases, RAST is also available as a command-line API through the myRAST' docker container. Prokka, on the other hand, does not have a graphic interface; it runs entirely on the command line making it easy to run with many genomes and integrate into custom post-assembly workflows.

2.1.1 myRAST Genome Annotation

After genome assembly, the annotation can be standardized by the RAST platform. DNA contigs can be uploaded to RAST for annotation and the file can be downloaded, as follows:

1. Access RAST either by its web interface following directions at http://rast.nmpdr.org/rast.cgi or by myRAST command-line docker image (github nselem/myrast).
2. To use myRAST docker image for the first time download the docker image from the dockerhub typing the following at the command line: "docker pull nselem/myrast."
3. To annotate a genome once the myRAST docker image is installed, run the image on the same folder where the genome fasta file is located: "docker run -i -t -v $(pwd):/home nselem/myrast /bin/bash."

To upload the genome using your RAST account, employ the command: svr_submit_RAST_job -user <user> -passwd <pass> -fasta <file> domain Bacteria -bioname "Organism name" -genetic_code 11 -gene_caller rast.

There are also other programs and workflows to annotate genomes that might provide an extra level of flexibility and control for more experienced users. Some popular gene-calling algorithms (that as discussed above are used as part of other pipelines) can be used independently: for example, Prodigal [17] and GlimmerHMM [19] are scalable algorithms for calling prokaryotic genes. Likewise, downstream comparison either through automatic pipelines or best BLASTP/HMMER searches against gene ontologies (e.g., COG, eggNOG, KO) and many other databases, can be used to assign putative function. In these more manual approaches, it is important to evaluate which functional databases and levels of annotation make the most sense for the scientific question. If it is not clear, then automatic workflows like RAST and Prokka are recommended, as they will provide functional information at many levels.

2.2 Databases of Known BGCs

Since genome sequencing has become economically viable for research groups around the world, available genomic data of NP-producing organisms has been increasing rapidly. Thus, the development of well-structured and curated databases has become crucial for NP research. While there are other databases reviewed elsewhere [20, 21], here we will limit ourselves to provide an overview of the two databases most commonly used in MIBiG and the antiSMASH database, referred to as antiSMASH-db.

Minimal Information about a Biosynthetic Gene Cluster (MIBiG) is a freely available, public database created in 2015 [22]. MIBiG is a curated repository of BGCs, intended to aggregate the annotated genomic sequence responsible for the biosynthesis of a particular natural product. While the genomic sequence is a principal feature of MIBiG, it is not the only one. Through incremental additions on each new version of the database, chemical data about the NP, functional roles of the genes in the clusters, and graphical breakdowns of well-known biosynthetic modules have been added. All featured BGCs have some level of validation linking the BGC to the production of its NP, and the genomic annotations are manually curated to check for misannotations or other artifacts introduced by automated tools. This high level of curation, along with an easy-to-use graphical user interface, has made MIBiG a very reliable database typically used as the starting point for genome mining.

However, MIBiG is not without blind spots. It links biosynthesis to entire BGCs, but does not validate the function of each gene, sometimes resulting in genes with hypothetical or unknown functions. A related challenge is that BGC boundaries are often difficult to computationally predict, so it is hard to know when the BGC truly begins and ends, which can result in incomplete BGCs (or BGCs with "extra" genes at the boundaries) reported in the database. Finally, due to the nature of MIBiG's indexing system, which takes into account the compound and the organism that produces it, the same pathway can be entered several times, since the same NP can be produced by multiple organisms. This, however, can also be considered a valuable feature, depending on the user's goal, as long as the genome sequences of the cognate organisms are available and can be readily identified.

Another popular tool used by genome mining workflows is the antibiotics and secondary metabolites analysis shell database [23] coupled with the antiSMASH software, which is aimed at exploring the biosynthetic potential of a genome (discussed in more detail later in this chapter). antiSMASH-db was introduced in 2016 and has been updated in 2019 and 2021. The latest version of this database, in contrast to MIBiG, does not rely on third-party submissions. Rather, it is compiled by the developers: all of the genomes in the NCBI Refseq database with a high-quality assembly level (scaffold or higher) were downloaded and de-replicated to weed out all of the polymorphic assemblies of commonly sequenced organisms to select one representative assembly per organism. All of these were passed through the whole antiSMASH workflow, and the predicted BGCs were indexed and logged. Therefore, these antiSMASH outputs comprise the bulk of the database. As opposed to the MiBIG database, no experimental validation is carried out for any of the BGCs. The lack of such stringency allowed the antiSMASH-db team to log information of a more diverse array of organisms, making this DB more taxonomically comprehensive with entries for taxonomic groups absent from MIBiG. Nevertheless, it is important to reiterate that this is a database of predictions, and as such, it cannot be taken as a complete portrait of biological reality since the limitations and inherent biases of the workflow used to obtain the predictions can influence the database entries. One of the most important limitations of the rules-based nature of the antiSMASH algorithm is that it can only characterize BGCs with previously reported NP biosynthetic logic, which in turn can limit the biosynthetic novelty it can uncover.

3 Genome Mining Programs

3.1 antiSMASH

antiSMASH is an integrated pipeline for the automated mining of genomic data capable of identifying BGCs that fall into many biosynthetic classes, such as polyketides (PKs), nonribosomally synthesized peptides (NRPs), terpenes, aminoglycosides, aminocoumarins, indolocarbazoles, lantibiotics, bacteriocins, nucleosides, beta-lactams, butyrolactones, siderophores, and melanins [9, 24]. Its versatility and easy-to-use graphical interface have made antiSMASH the most widely used software for NP genome mining.

antiSMASH finds BGCs in a target genome by seeking genes and domains that are homologous to those in carefully trained models. All genes of the query genome are scored against a database of highly conserved hidden Markov model (HMM) profiles, which are representative of specific BGC families. Once all the homologous genes in the target genome are identified, predefined cluster rules are employed to define potential individual clusters encoded in each genomic region, according to the particular BGC family to which it was found to be most similar. These potential clusters are then extended by adding more up and downstream genes following family-specific rules of expansion. This is done to ensure that the predicted BGC will be as complete as possible. Since there is not an automated method to accurately predict gene cluster borders (with exception of the exclusive CASSIS algorithm [25] for fungal genomes that can detect co-regulated genes), the BGCs predicted by antiSMASH are shown, and labeled, as regions, to reflect the fact that the BGC borders are just offsets defined in the cluster detection rules, and conserved parts of it might have been left out. Along with the BGC detection process, the antiSMASH pipeline integrates several specialized modules, such as the ones used to predict the AT domains in PKSs, and the A domains in NRPSs, to provide a more in-depth analysis of each predicted BGC. Once BGC regions are identified, antiSMASH compares the biosynthetic regions of the genome to the rest of the BGCs in the antiSMASH-db.

The antiSMASH output is very informative and gives a schematic profile of each predicted BGC, its genes (and their corresponding sequences), the family of BGCs to which it belongs, and homologous BGCs found in antiSMASH-db, MIBiG, and other databases as well as information about the predicted role of each gene. For some biosynthetic classes, like PKSs and NRPSs, core chemical structures are predicted and reported to the user.

3.1.1 General Use of antiSMASH to Identify BGCs

As described above, input for antiSMASH should be a good quality bacterial genome assembly. antiSMASH documentation recommends submission of annotated genomes, in which gene starts and stops are identified, or use of Genbank, EMBL, NCBI accession number, or plain FASTA format. If a FASTA file (commonly extension *.fasta or *.fna) containing no annotations is submitted, Prodigal [17] will be used to annotate the genome. If an NCBI accession number is used, the user needs to submit a genomic nucleotide accession rather than an NCBI assembly. Note NCBI re-annotations and/or reconfiguration of its programmatic access can interrupt the direct workflow between antiSMASH and NCBI; this happens every few months or so. If antiSMASH cannot directly connect to NCBI via an accession number, an alternative option is to download the nucleotide sequence file for the genome ["Genbank (full)" or "FASTA (text)"; note that "Genbank (full)," which includes the entirety of the sequence information, is not the same as "Genbank"'which only includes annotations] and then upload this file to antiSMASH.

3.1.2 Job Submission

A job may be submitted to the antiSMASH webpage which sends a job into a queue for running remotely. Alternatively, antiSMASH may be used as a standalone, command-line process if running many jobs is preferred. The default for detection strictness "relaxed" will detect well-defined BGCs as well as BGCs missing one or more functional parts. For most jobs, the parameter defaults, KnownClusterBlast, ClusterBlast, SubClusterBlast, and ActiveSiteFinder are recommended, while Cluster Pfam analysis and Pfam-based GO term annotation are recommended to be left unselected to keep runtime low, especially with a genome that has already been functionally annotated via RAST or otherwise. Currently, antiSMASH 6 beta is available, which additionally allows for MIBiG/antiSMASH-db comparisons, class-independent RiPP identification with RRE-Finder [26], and TIGRFam analysis [27].

3.1.3 Stringency Levels

antiSMASH5+ can detect BGCs at three stringency levels: strict, relaxed (default), and loose. For a full description of the rules, see the antiSMASH github repository.

Strict: This will follow strict rules for finding a BGC. For most biosynthetic classes, this will be the same as relaxed, but for some (mostly multimodular biosynthesis), this will have subtle differences between strict and relaxed. For example, KS, AT, and ACP domains must all be present for a T1PKS to be identified under strict rules. Similarly, C, A, and PCP domains must all be present for NRPSs.

Relaxed: Same as strict, but not all hallmark domains are necessary to call a cluster. If only one or a few NRPS domains are found and some are not seen, these will be designated "NRPS-like" clusters.

Loose: Same as relaxed, but new classes are added. These are left out of the default as they are often part of primary metabolism instead of secondary metabolism ("false positives"). These classes include saccharides, fatty acids, and halogen-containing regions.

For almost all use cases, the default of relaxed is recommended.

3.1.4 Extended Parameters

The following extended parameters are described in more detail. For further information, refer to the antiSMASH 5.0 User Manual (accessed 2021-02-04).

KnownClusterBlast: BGCs identified in query genomes are compared to BGCs from the experimentally characterized MIBiG repository.

ClusterBlast: BGCs identified in query genomes are searched against antiSMASH-db, a comprehensive gene cluster database composed of antiSMASH-predicted BGCs. Thus, these BGCs may or may not be experimentally characterized.

SubClusterBlast: Genes in identified BGCs are searched against common biosynthetic operons (e.g., the biosynthesis of non-proteinogenic amino acids; for the full list, *see*: https://bitbucket.org/antismash/antismash/src/bfa2f1c0fe97de910c19ee2be241b415e6fa97bd/antismash/generic_modules/subclusterblast/subclusters.txt?at=master&fileviewer=file-view-default).

ActiveSiteFinder: It predicts the active site residues of several highly conserved biosynthetic enzymes.

Cluster Pfam analysis: Pfam is used to annotate each gene in the BGC.

Pfam-based GO term annotation: Genes are further annotated with GO terms based on their Pfam annotation.

For most exercises, KnownClusterBlast, ClusterBlast, SubClusterBlast, and ActiveSiteFinder are recommended to be selected, while Cluster Pfam analysis and Pfam-based GO term annotation are recommended to be left unselected. Depending on the antiSMASH queue, this will (usually) allow for the analysis to finish within 10–30 min.

3.1.5 Interpreting Results

Results will be forwarded via email if an email address was provided. They will also appear on the web page from where the job was submitted. The results will first depict a cluster overview table. This will include an overview map at the top of the screen that shows sequentially the clusters identified. Below, a list of gene cluster regions includes more detailed info about each cluster including the type (they will also be color-coded by type), coordinates (location within the sequence), and most similar known clusters (compared to MIBiG). From this overview page, clicking on each identified cluster ("Region") will reveal more information by opening a new page with all the collected information of the region where the BGC was identified. This results page is very friendly and intuitive and contains links to the results obtained using the databases incorporated to antiSMASH analysis for the user to dive into some aspect of the output which facilitates the navigation and the interpretation of the data (Fig. 1).

3.2 Algorithms Predicting Enzyme Substrate Specificity

3.2.1 SEARCHGTr (Glycosyltransferase Specificity)

Among the wide diversity of natural products, there are some (typically those with antibiotic activity) that are glycosylated in the late stages of their biosynthesis. Among the various classes of glycosyltransferases (GTrs), one group distinguishes to differentially glycosylate the secondary metabolite aglycone. The nature of the sugar, the site of glycosylation, and the quantity of sugars added affect the ability of the antibiotic. Correspondingly this great variety of modifications results from variations in donor substrates. Likewise, there can be a vast repertoire of acceptor substrates, as these aglycones can be PK or NRP molecules with all the diversity which they imply.

SEARCHGTr is a software for the analysis of GTrs. This tool correlates sequences of GTrs to the chemical structures of their

Fig. 1 Results page generated by antiSMASH. (1) Version of antiSMASH used. (2) Regions of the genome where potential BGCs were located. (3) Analysis mode. (4) Percentage of similarity with known BGCs. (5) Region identified by antiSMASH can be explored by clicking on the desired region, which displays a new page with the following structure: (**a**) Location of the region within the genome. (**b**) Individual information about each of the genes that composite the possible cluster within that region. To change the gene, click on the gene of interest. (**c**) General features about the genes in the region. (**d**) Specific annotations about all the genes in the region. Depending on features selected, up to five tabs with detailed information are presented: Domains of the proteins, Pfam data, MIBiG comparison, KnownClusterBlast, and SubClusterBlast analysis. (**e**) Gene cluster prediction details are presented in a few tabs: Classification of the putative cluster, possible products, and RiPP recognition elements identified by the RRE-Finder tool (if applicable). This last feature (**e**) is presented only if the input data is enough to make these predictions

corresponding substrates and can predict the donor/acceptor specificity, as well as identify putative substrate-binding residues. The software also provides interfaces to other public databases like GenBank, SWISS-PROT, CAZY, PDB, PDBSum, and PubMed, to provide the maximum power for your analysis [28]. In summary, SEARCHGTr, assists in the identification of biosynthetic products of BGCs found in bacterial genomes and is considered an extraordinary accessory tool after massive genome mining analysis, providing a deep comprehensive computational exploration of proteins involved in biosynthesis of aglycone core and its downstream glycosylations.

3.2.2 NRPS and PKS Substrate Specificity Prediction

Algorithms have been developed to predict the substrate specificities of NRPS adenylation (A) domains and T1PKS acyltransferase (AT) domains. The ability to identify the substrate specificity of the A and AT domains is necessary to rationally infer or engineer new products. Importantly, both A and AT domains have a phylogenetic signal that corresponds to their substrate specificity: where these domains fall on a phylogenetic tree can inform which building blocks they incorporate [3].

On the one hand, the modular enzymes NRPSs are responsible for the synthesis of NRPs; they typically have at least three domains: an A domain that confers specificity for an amino acid substrate, a peptidyl carrier protein (PCP) that tethers the substrate to the NRPS, and a condensation (C) domain that catalyzes the formation of peptide bonds between the amino acid substrate and the synthesized peptide chain. In nature, many NRPSs have more than ten modules and many hundreds of unique NRPS substrates have been described [29]. Thus there is tremendous combinatorial potential and structural diversity within NRPS chemistry. Genomic mining of natural products has been used to predict NRPS substrate specificity directly from the sequence of A domains. Initially, this was done by comparing the so-called Stachelhaus codes [30], a 10-residue barcode of sorts that corresponded to the A domain residues that are part of the active site and thus are predicted to drive substrate specificity. For some specificities, these codes work well, for others, especially more exotic, non-proteinogenic amino acids, predictive power is limited.

A more expanded approach for both A domains in NRPSs and AT domains in T1PKSs was developed by Minowa [31]. These authors simplified PKs and NRPs as carbonyl/peptidyl units to extract the structural motifs efficiently and analyzed the AT and A domain sequences of its synthases as well as their physical interactions to predict which substrates are selected and assembled into highly ordered chemical structures. Also, the structural motifs of known compounds were analyzed to infer possible functions turning this methodology into an integrative analysis that links genomics and chemical information to predict biosynthetic pathways, structures, and bioactivity of PKs/NRPs. Although this methodology was not released as a formal pipeline its development allowed the birth of other methodologies based on the hidden Markov model (HMM) for the analysis of protein sequences.

Another algorithm based on HMM to predict the substrate specificity was proposed in 2013 by Khayatt and coworkers [32]. Like all algorithms that use this method, the objective is to determine the unknown parameters of the subject of study, from observable parameters. The extracted parameters can be used to carry out successively run analyses, in this case for pattern recognition. One of the biggest advantages of implementing HMMs is that it eliminates the need for multiple sequence alignment and residue selection that is hampering the alignment-based clustering methods, which at the same time results in a high prediction quality for the substrate specificity of the A domains in NRPS enzymes. This algorithm could be used until a while ago with a simple web-based tool, however, the web server is no longer publicly hosted, so it is suggested to retrieve the HMMs from the manuscript's materials and/or contact the authors to use this pipeline.

NRPSPredictor2 then expanded on the Stachelhaus code to those A domain residues within an 8 Angstrom distance from the active site and used these to develop support vector machines to classify A domain specificity [33], but this is limited to specificities for which a lot of validated training data exists. In an alternative approach, the phylogenetically driven algorithm prediCAT addresses the problem of predicting possible NRPS substrates, calculating a confidence score for each A domain based on patterns of A domain monophyly. Again, this method is only as good as its training information.

These initial approaches were combined with previous training data to retrain all high-performing classes of algorithms, which later were combined into a single ensemble method called SANDPUMA [34]. This method was designed for automatic retraining to ensure its accuracy increases as more experimentally data is entered through MIBiG or other repositories. SANDPUMA is a high confidence method to predict the specificity of NRPSs through many orthogonal methods and was employed in v4 of the antiSMASH analysis pipeline. In antiSMASH v5+, this has been replaced by NRPSPredictor2 to minimize runtime. Nevertheless, processing of A domains with SANDPUMA can be a valuable follow-up to basic antiSMASH analysis, especially in taxonomically diverse organisms, and is freely available at https://bitbucket.org/chevrm/sandpuma and as a docker image at https://hub.docker.com/r/chevrm/sandpuma/under the GNU Public License 3 (GPL3).

On the other hand, bacterial multimodular enzymes called *trans*-acyltransferase polyketide synthases (*trans*-AT PKSs) possess the ability to synthesize a great diversity of polyketides. Recently, the transPACT algorithm has implemented a phylogenomics-based approach to automate the classification of *trans*-AT PKS modules from bacteria based on their ketosynthase (KS) domains, intending to delve into the biosynthetic logic and evolutionary patterns of this enzyme family [35]. The logic of this tool is similar to prediCAT for NRPS A domains, making predictions for substrate specificity based on the phylogenetic placement of query domains. The development of this tool has proven useful in identifying relationships between cryptic PK BGCs, optimizing its genome mining as well as the discovery of chemical variants with specific characteristics. Natural hybrids are rare among the most widely studied *cis*-AT PKSs, so leveraging the evolutionary patterns revealed in trans-AT PKSs may allow for a more direct approach to PKS engineering. The transPACT source code is available freely under an open-source AGPL-3.0 license from https://github.com/chevrm/transPACT [36].

4 NP Specialty Databases

There is a long history of efforts to systematize the knowledge regarding NP BGCs [22, 37, 38] and specialized genome mining databases now allow for evolutionary explorations of secondary metabolism. Evolutionary information in these repositories can be used by laboratories even before genomes are available. For example, once a microorganism has been isolated and taxonomically identified, the BGCs of representatives of related species can be consulted. With this information, once a cluster has been prioritized, it is possible to direct efforts to sequence its amplicons or to investigate the expression only of the regions of interest. Search by species is available at antiSMASH-db, these outputs can feed other evolutionary genomic mining algorithms such as EvoMining [39]. It is important to remember, however, that taxonomy often underestimates biosynthetic diversity and strains with high organism-level phylogenetic similarity may have many unshared, unique metabolic genes and products [40, 41].

In addition to MIBiG [22] and antiSMASH-db [23], which have been described previously, there are other specialty databases of note for NP genome mining [37, 42–44]. Analogous to MIBiG in the sense that it contains clusters with experimental evidence those interested in carbohydrate-active enzymes (CAZymes) can use dbCAN-PUL [42]. Furthermore, information about a species of interest can be obtained from databases that combine experimental and bioinformatics BGCs evidence. Predicted and experimentally validated BGCs have been combined in IMG-ABC [44] that additionally provide environmental metadata and a collection of BGCs from metagenomic sources. To annotate antibiotic resistance, CARD [43] and RESFINder [45] provide comprehensive catalogs. When needed for molecule structures, the Natural Products Atlas (NP Atlas) offers a collection of 24,594 compounds [46].

Additionally, there are specialty databases that attempt to define BGC families within known natural products classes. This is of utmost importance as each natural products class contains multiple families of BGCs. Families are made up of BGCs that share core genes and vary in ornamental and/or precursor-supply genes. One of the first programs that were used to give a distance between pairs of clusters was MultiGeneBlast [47]. Since then, the identification of families has continued to improve by developing approaches to address the increasing number of BGCs [48, 49]. BiG-FAM [50] is a database that synthesizes current knowledge about BGC families. This database is indispensable in evolutionary genome mining because it provides the distribution of BGCs families distributed throughout taxonomic lineages. Once a family of interest has been chosen in BiG-FAM, the evolutionary dynamics can be studied in detail with visualization tools such as CORASON [48].

5 Evolutionary Genome Mining of NPs

Once genomes are annotated and organized within ad hoc databases, it is desirable to place target genomes within their taxonomic context. For this, phylogenomic analyses based on conserved proteins of related organisms, which vary amongst different bacterial lineages, instead of single markers, is desirable. Thus, groups of conserved orthologs are typically obtained de novo, for which many tools are available. This includes automatic pipelines for this purpose, such as AutoMLST [51]. Although establishing the phylogenetic relationships of the target genomes is not strictly needed for evolutionary genome mining, phylogenetic context increases the quality of analysis and eases the interpretation of the results obtained [3, 52].

It should be noted that although the selected tools described next rely on the construction of phylogenetic trees, these are gene trees rather than species trees obtained after the suggested phylogenomic analyses. The main difference between species and gene trees is that the former provide phylogenetic relationships believed to portray the natural history of a species, as they use conserved, but phylogenetically informative genes (e.g. 16S *rRNA*, *rpoB*, *dnaE*, etc.) where divergence is assumed to be driven by mostly neutral forces. In contrast, gene trees used for the analysis of BGCs may reflect on the species relationship but not necessarily, and may be more or less conserved throughout the compared genomes. This implies that the use of biosynthetic genes for reconstructing phylogeny can, and often does, affect the output.

5.1 EvoMining

EvoMining [39] is a pipeline that aims to find new BGCs by detecting enzyme gene families with extra copies that have diverge from their conserved metabolism homologs. To find these divergent enzyme genes, EvoMining integrates three different datasets: (1) the annotated genomes of a taxonomic lineage to be mined; (2) the seed enzyme genes of the conserved families, obtained from the core pangenome of a targeted taxonomic lineage and/or enzyme genes included in representative genome-scale metabolic model [4, 53]. Whereas (1) and (2) datasets are better if compiled by the user, precompiled databases for Actinobacteria, Cyanobacteria, *Pseudomonas,* and Archaea are available. Based on these datasets, EvoMining seeks genes from the conserved enzyme families that exist in extra copies in the genomes of interest. The identified diverging enzyme genes are then used to search for homologs in MIBiG, which represent the final dataset. This allows for the establishment of genetic relationships between expanded enzymes with known BGCs (Fig. 2).

EvoMining differs from other BGC discovery algorithms, like antiSMASH, in that it is blind to previous knowledge about

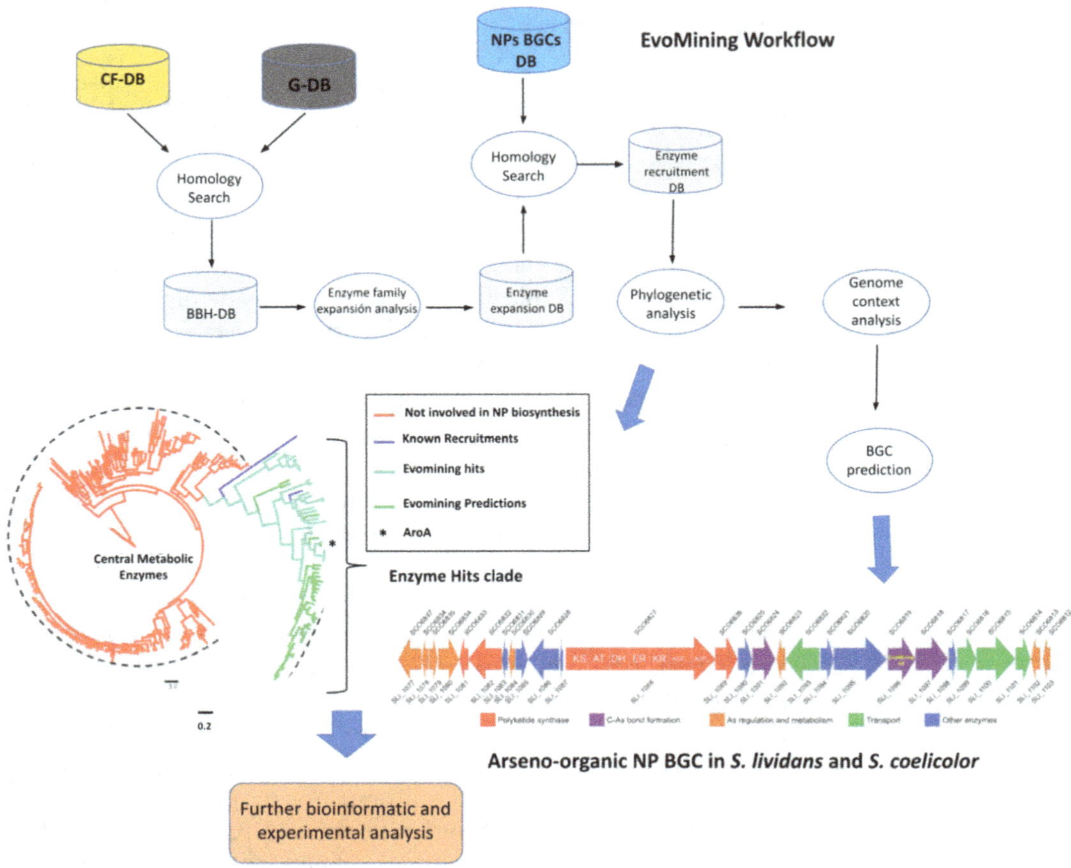

Fig. 2 EvoMining workflow for the identification of novel BGCs. The three input databases are shown in yellow (CF-DB), gray (G-DB), and blue (NP-DB). Internal databases are shown in light grey (BBH-DB/Enzyme expansion DB/Enzyme recruitment DB). Ovals depict processes. A phylogenetic tree recapitulates the evolution of one NP biosynthetic enzyme. The case of 3-phosphoshikimate-1-carboxyvinyl transferase family is shown as an example [4]. The topology of the tree shows a major clade associated with central metabolism (red), and a divergent clade with at least three subclades, in which divergent AroA enzymes from *S. lividans* and *S. coelicolor* are located. Color code is shown in the key. After the visual inspection of the phylogenetic tree it is possible to determine recruited enzymes. Genomic context analysis of the cognate genes can then be performed, which can be used to delimit the putative BGC

biosynthetic enzymes of NPs. This is achieved by directing the searches to genes that adhere to the evolutionary principle of gene expansion-and-recruitment, or functional specialization, assuming that the diverging genes are recruited by a BGC. Because of this, EvoMining can find truly novel NP BGCs families, although it also requires a more in-depth revision of the predictions generated, since it can give a higher rate of false positives. For instance, recruitments into other metabolic fates rather than NP biosynthesis are possible (e.g. xenobiotic degradation, catabolism, etc.). Another limitation of EvoMining is that it requires that the taxonomic lineage that is being mined has significant genomic information (i.e. taxonomic coverage) since the quality of the predictions depends greatly on the accuracy of evolutionary relationships. Thus, EvoMining has been referred to as "low-confidence/high-novelty" by Medema and Fischbach, 2015 [54].

In practical terms, EvoMining reconstructs a phylogenetic tree with all the ortholog sequences found in the G-DB (including central and expanded enzymes), which allows visualizing the evolutionary history of a determined enzyme family. EvoMining then marks in a color code throughout the tree the different enzymes, as follows: central metabolic enzymes (red); seed central enzymes contained in the CF-DB (orange); expanded enzymes recruited into NP biosynthesis according to MIBiG (blue); expanded enzymes potentially recruited into NP biosynthesis according to antiSMASH database (cyan); and the expanded enzymes that are phylogenetically close to the blue branches, absent in MIBiG or antiSMASH database (green). The last two results are referred to as EvoMining hits and EvoMining predictions, respectively. Note that both are predictions made independently of previous knowledge, but EvoMing hits coincide with antiSMASH whereas EvoMining predictions are completely novel and potentially unprecedented BGCs.

5.1.1 Preparation of EvoMining Databases

EvoMining requires three databases: a genome database (G-DB), a conserved metabolism enzyme families database (CF-DB), and a natural product database (NP-DB), which are integrated as shown in Fig. 2, and resulting in several other internal DBs. The G-DB consists of all those genomes or sequences in which the search will be performed. The CF-DB will direct the course of the searches, since central metabolic enzymes are the seed sequences to seek for enzymatic expansions, which include paralogs or homologs obtained after gene duplications or horizontal gene transfer, respectively. Thus, the quality of the CF-DB, to be curated manually if needed in order to include only central metabolic enzymes, is a sine qua non condition for EvoMining success. Likewise, the more diverse the G-DB is, the better, as central metabolism orthologs diverge according to the evolutionary history of the organisms themselves, i.e. within genomic lineages. The NP-DB or MIBiG database helps identify enzymatic expansions in known BGC, but whose functions are typically unknown but predicted to be associated with NP biosynthesis by association. Analyses within a known and limited universe helps expand EvoMining predictions to unknown areas of the phylogeny. To assist in this process, antiSMASH is used to annotate the genomic vicinities of EvoMining hits.

To assemble and integrate these DBs, follow the EvoMining tutorial, which is exemplified on the operative system Ubuntu 14.04 with Docker version 18.01.0-ce, build 03596f5. In this demonstration, a G-DB of 1246 Actinobacteria genomes was gathered, consisting of 25 private and 1221 public Actinobacteria genomes with at least three genes on average per contig (available at NCBI in 2016). The CF-DB consists of 295 Actinobacteria central enzymes. Sequences are organized in 93 families involved in nine metabolic pathways, including amino acid biosynthesis, glycolysis, pentose phosphate pathway, and tricarboxylic acids cycle. The seed sequences are those from *Streptomyces coelicolor*, *Mycobacterium tuberculosis,* and *Corynebacterium glutamicum*. Any machine with a docker engine is also suitable for the protocol. To install docker on other operative systems, such as Mac, cloud, Windows, and Linux servers, follow docker directions at https://docs.docker.com/engine/installation. For DB assembly follow these instructions:

Genome Database (G-DB)

1. Create a folder GENOMES and download the amino acid and annotation files from the RAST server by typing:

```
"svr_retrieve_RAST_job <user> <password> <jobId> table_txt >
$<jobId>.txt"
"svr_retrieve_RAST_job <user> <password> <jobId> amino_acid >
$<jobId>.faa"
```

   (Note: To optimize uploading/downloading time when dealing with several genomes on batch follow directions at https://github.com/nselem/myrast.)

2. Gather on a directory the genome database including the FASTA and annotation files considering the number and the diversity of the bacterial genomes.
3. Write a tab separated file where each line should contain three columns: the RAST ID, RAST genome ID and the organism names, which is equivalent to the RAST IDs file.

Central Families Database (CF-DB)

1. Set C_DB enzymes on a fasta file with the central enzyme sequences organized with the following special fasta format:

   >Metabolic Subsystem|Step number on this subsystem| Function_Seed_Number|Organism code

   Example:

   >E4P_AMINO_ACIDS|6|3_phosphoshikimate_1carboxyvinyltransferase_2|Scoe

   On the central pathway devoted to amino acid biosynthesis, starting with Erythrose 4-phosphate (E4P_AMINO_ACIDS) as the precursor, the sixth step of the subsystem is catalyzed by the enzyme 3-phosphoshikimate-1-carboxyvinyl transferase. There are three seeds in this enzyme family (*S. coelicolor*, *M. tuberculosis,* and *C. glutamicum*), In this example, the header corresponds to the seed number two, i.e. the *S. coelicolor* (Scoe) ortholog.

Natural Products Database (NP-DB)

The NP-DB used, as we mentioned before, is MIBiG-DB (accessed in 2017). To link this DB to the EvoMining pipeline, all sequences that belong to a BGC on MIBiG repository are retrieved. Metadata, such as producer organisms and compounds class, are also integrated. MIBiG is ready to use and included on the docker image by default. To include another NP-DB, set a fasta file with all the desired Natural products coding genes.

Internal EvoMining Databases

After the user completes compilation of the starting databases, phylogenetic histories of each of the enzyme families from central metabolism are reconstructed by EvoMining pipeline, as follows.

1. Copies devoted to central metabolism are identified by Best Bidirectional Hits (BBH) after a BLAST search conducted between each enzyme family from the CF-DB against all genes in the G-DB.
2. Recruited families are temporarily stored after filtering expanded families with not known recruitment within NP biosynthesis. This is done when at least one hit on the NP-DB is obtained when using the expanded family sequences as a query. BLAST with an e-value threshold of 0.001 is used.
3. The sequences of known recruitments from NP-DB are added to the expanded family, leading to a new internal database of expanded-and-recruited families, which is aligned with muscle and curated with Gblocks.
4. The aligned informative regions are submitted for reconstruction of a phylogenetic tree by the neighbor-joining software FastTree.
5. The SVG Perl module and Newick utilities kit provides an interactive visualization of the tree, which is color-coded as explained before. The tree visualization allows some degree of user interaction including access to the genomic context or predicted BGC.

5.1.2 EvoMining Job Submission

1. To run EvoMining for the first time download EvoMining docker image from the docker hub.

```
"docker pull nselem/newevomining:latest".
```

2. Run EvoMining docker image on a directory that contains the G-DB and the RastIds file

```
"docker run -i -t -v $(pwd):/var/www/html/EvoMining/exchange
-p 80:80 nselem/newevomining:latest /bin/bash"
```

3. If you are running EvoMining in a new organism for the first time, EvoMining will automatically run the FaatoEvo.pl script

and create the EvoMining organism fasta file. The header format contains pipe separated camps as follows:

```
>gi|gen Id|Organism Id|NCBI Id|Function|Organism Name
```

4. Initialize EvoMining pipeline Inside EvoMining docker image. This step runs the blast search of the CF-DB enzymes against the G-DB. This process may take a few hours depending on the size of the genomic database.

```
"perl startevomining -g mygenomes -r myRastIds -c centralDB -n
Natural products".
```

where -g -c -n modifiers stand for the genomic, central, and natural product databases, respectively, and -r is the tab separated Rast Ids file. To run an example, use default data and run "startevomining" without modifiers:

```
"perl startevomining"
```

5. Visualize expansion heatmap after the startevomining.pl script is finished.
6. Open a browser with the following address http://localhost/EvoMining/html/index.html. Follow directions of the user interface to start the expansion heatmap view, this should be done in minutes.
7. To search for recruitments, proceed to the recruitment search by pressing the next button on the heatmap visualization. BLAST searches at this stage may take a few hours according to the size of the G-DB.
8. For phylogenetic reconstruction, select the expanded-and-recruited enzyme families and oversee the alignment, curation, and phylogenetic reconstruction process. After selection, submit one tree at a time. It may take hours or days according to the number of trees selected and the size of the G-DB.
9. When reconstruction of trees is over, the web browser displays the option to visualize the tree. Select the desired tree and visualize it. Files *.nwk and *.csv will become available in the tree directory as outputs. These files are compatible with MicroReact, which provides a better visualization interface.
10. The tree is visually examined in the search for EvoMining predictions. Once a hit is chosen, observe if its genomic context is conserved, suggesting a BGC, which should be different to the genomic context of enzyme genes devoted to central metabolism.

A Note on Docker Images

A docker image is a way to wrap a piece of software along with all its dependencies, assuring that the software will stay functional resisting the time pass. During this tutorial several docker images were requested. This subsection explains in detail the general structure of the commands frequently used.

The following is a Docker image running as an example:

```
docker run -i -t -v $(pwd):<Docker directory> -p 80:80 <image name> /bin/bash
```

where docker run means that the docker engine will run an image, the one indicated on the image name. In this tutorial, images used are nselem/newevomining, nselem/orthocores, and nselem/myrast.

The modifiers **-i** -**t** -**v** mean interactive mode, with a terminal and with a volume system. A volume is a way to share files between the computer and the docker image. The path of a local directory and a fixed path of the docker image separated by a colon are specified after the volume modifier -**v**. The docker image will check if files exist in the local directory, and every file produced will be stored in the local directory and will persist when closing the docker image. The local directory used in this tutorial is $(pwd), an abbreviation of print working directory. Examples of fixed docker directory paths are /home and /usr/src/CORE.

The /bin/bash at the end of the instructions means that the image should start with an interactive bash session, where other pieces of software can be run like perl or python scripts. Finally, the -p 80:80 instruction in EvoMining stands for the interaction with a web browser on port 80. If this port is not available the user should select another port such as 87.

5.2 CORASON

CORASON (CORe Analysis of Synthetic Orthologs to prioritize Natural product biosynthetic gene clusters) is a tool to find conserved homologs of a query BGC (or of any genetic locus) in a user-friendly supplied database of genomes. All genes of a query BGC can be systematically investigated until the conserved gene core is identified. Regardless of its physical position within the query BGC the conserved genes are used to generate a whole-BGC phylogenomic or CORASON tree. It is therefore important to note that genomes or sequences lacking one gene of the conserved gene core defined by the user, will be missed. Because of this, choosing the conserved gene core is crucial, which is determined by the window of phylogenomic analysis: what is conserved between some organisms may not be conserved for other, closely related, organisms. Moreover, it is recommended to avoid PKS or NRPS genes, even if they are conserved, since their associated enzyme families are over-represented in bacterial genomes, including their known modular multiplicity, leading to confounding results (for instance, a false

BGC duplicity within a genome). Instead, single-domain enzyme genes, from enzyme families with relatively limited memberships, should be used, such as those identified by EvoMining (e.g. [48],).

To use CORASON, the user needs to provide the query BGC, pre-define the conserved gene core, and assemble a database of annotated genomes. CORASON then searches for the conserved genes in every genome, and once an homologue is found, the search is expanded to its vicinity, to find existing homologs of the rest of the genes in the query BGC. If there is at least one additional hit, the putative BGC is considered to be a CORASON hit, and the program saves it. Otherwise, it is discarded. Once every hit is found, all of the homologs of the conserved genes are aligned and a whole-BGC phylogenetic tree is built. The output of CORASON consists of such phylogeny, and at every tip of the tree, the genomic vicinity is drawn. Genes that are homologous throughout the BGCs identified are colored similarly, facilitating visual inspection of the output.

5.3 BiG-SCAPE

Once BGCs are identified by antiSMASH (or otherwise), it is often useful to compare them to BGCs from other databases and organisms (Fig. 3). These global comparisons can identify putative evolutionary events and reveal gene and/or domain content to known BGCs, aiding in chemical predictions. BiG-SCAPE is a useful tool that uses an all-by-all comparison of HMMER-annotated Pfam domains to compute a distance score between BGCs. Three major comparisons are calculated: Jaccard dissimilarity, domain duplication score, and adjacency index; all in different weights finely tuned to specific biosynthetic classes (e.g., T1PKSs have one set of weights, RiPPs another, terpenes another, and so on). BGCs from related genomes, or from various databases (including MIBiG and antiSMASH-db), can be included to provide context.

BiG-SCAPE itself comes packaged with MIBiG, so comparisons to known BGCs can be performed automatically. From the all-by-all distance matrix, a network can be created at various cutoffs to identify and explore gene cluster families (GCF). Cytoscape and networkX are useful network exploration tools to assess and visualize the BiG-SCAPE gene cluster network with a graphical user interface or programmatically, respectively. For large-scale analyses of BGCs, BiG-SLICE distills the Pfam information into a single vector to allow for rapid calculations of (dis)similarity between BGCs. Although some resolution is lost with this abstraction, it enables very large analyses between thousands of BGCs at a time.

5.4 ARTS

The Antibiotic Resistance Target Seeker (ARTS) is based on the principle that antibiotic-producing organisms must contain a resistance mechanism to protect themselves from their own antibiotic production [55, 56]. Some of these resistance mechanisms are genes included within the BGC for antibiotic synthesis. It has

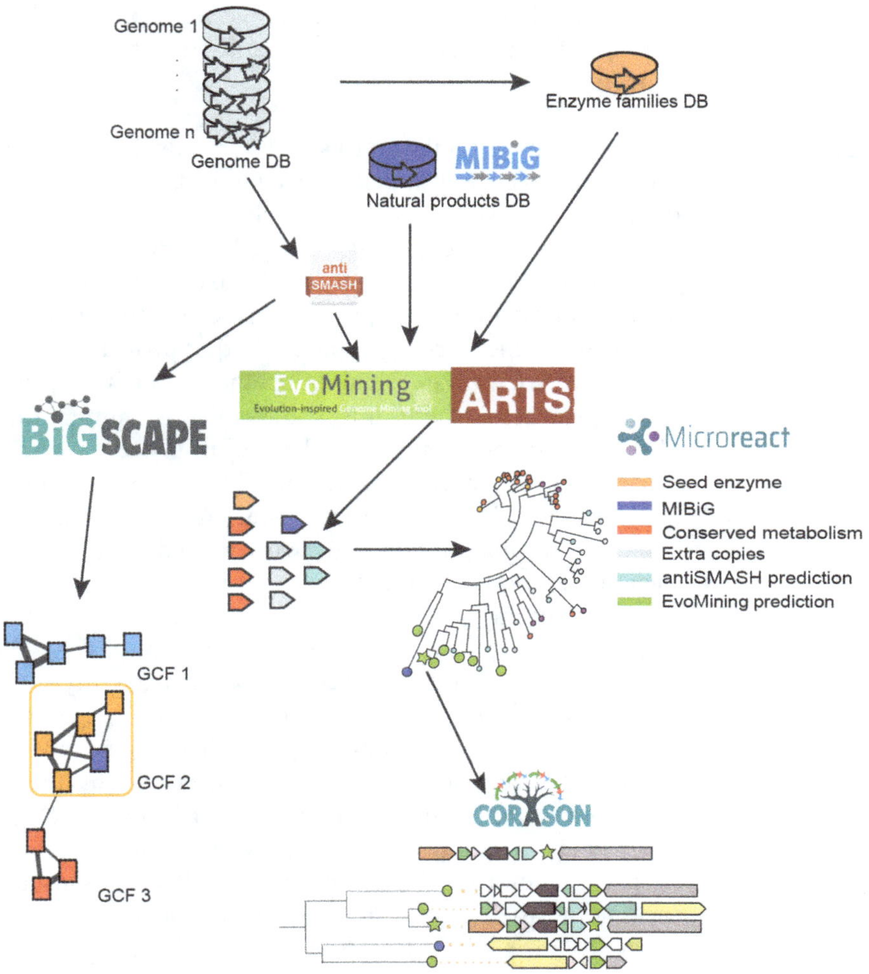

Fig. 3 General workflow for evolutionary genome mining of natural products. BiG-SCAPE classifies BGCs into gene families according to gene content and synteny. The BGCs can be provided by antiSMASH or other means of annotation and curation. ARTS and EvoMining are genome mining methods based on comparative (phylo)genomics of phylogenetically related lineages. Extra copies of selected enzyme families are localized and the distance to genes that belong to BGCs from MIBiG is measured. antiSMASH analyses can be incorporated at this stage. CORASON is used to find conserved genomic contexts surrounding enzyme genes identified after EvoMining or ARTS

been observed that duplicated copies in certain core families act as a mechanism of resistance. For these reasons, ARTS infers the existence of an antibiotic BGC when duplicated copies are present in the BGC vicinity.

Using ARTS, it is possible to analyze a previously computed antiSMASH job, a genome, or even a metagenome. In these later cases, an antiSMASH job will be run before proceding with ARTS workflow. ARTS will use the antiSMASH results to identify BGCs that could be used for antibiotic synthesis and it will cluster all BGC families that satisfy its screening criteria using BiG-SCAPE as previously described. ARTS is powerful tool for the targeted mining of antibiotics and is available at http://arts.ziemertlab.com/.

5.5 DeepBGC

As described throughout this chapter, numerous bioinformatics tools have been developed to facilitate natural product genome mining. From BLAST to antiSMASH, programmers have tried to implement rules to define BGCs based on reference genes similarity and protein domain composition. To take advantage of the increasing genomic data some approaches have implemented more generalizable machine learning approaches that provide a greater ability to discover new BGC genomic elements, such as ClusterFinder [57]. However, there are some intrinsic limitations in the HMM-based tools employed. The set of processes carried out by HMM is not capable of preserving the relationships between distant entities. This means that HMM-based tools cannot capture complex information between clearly related entities [58], limiting their ability to detect BGC.

Very recently, the genome mining tool DeepBGC, which employs a deep learning approach, was published. DeepBGC implements two types of processes together: recurrent neural networks (RNN) and vector representations of essential protein domains (Pfam). With these implementations, it can detect relationships between adjacent and distant genomic entities. DeepBGC uses a bidirectional short and long-term memory RNN (BiLSTM) and a grammar-jump neural network similar to word2vec (pfam2-vec). Compared to Clusterfinder, DeepBGC appears to improve the identification of BGC from known classes of bacterial genomes and endows it with great potential for detecting new classes of BGC. Finally, DeepBGC predictions were enhanced with generic random forest classifiers that allow classification of BGCs based on product class and molecular activity of compounds [59]. This tool could be used with bacterial genomes and microbiome metagenomic samples. The microbiome analysis could allow associations with disease phenotypes and have important clinical impacts for translating microbiome data to therapeutic interventions. DeepBGC can be considered as a step forward in genome mining, since it significantly improves the precision of classic BGC identification, as well as the identification of novel sequence signatures even without being present in the database training repositories. This is actually possible due to its efficient data/relationship extrapolation capabilities. DeepBGC, available at https://github.com/Merck/deepbgc, compliments current data mining practices and allows the scientist to optimize the search for new BGCs and their natural products.

6 Concluding Remarks

Evolutionary genome mining for the discovery and characterization of NP biosynthesis has developed into practical genomic tools and analysis pipelines, some of which are described above. While we have moved from speculative linkages of the past to robust evolutionary hypotheses of BGCs and the genes that comprise them, it is important to recognize that this subfield is still in its infancy. Methods to link the evolutionary histories of the genes and metabolites of NP-producing organisms will undoubtedly continue to develop and improve, just as evolutionary thinking continues to permeate all areas of NP research. As such, genome miners adopting an evolutionary mindset need to embrace evolution beyond anecdote, as this is what increases the likelihood and quality of the functional predictions done with current tools.

In other words, while evolution-driven tools ease the construction of unprecedented predictions, this depends on realizing that evolution is a dynamic process rather than a final point within a process. The natural forces that shape this evolutionary process are indeed embraced by evolutionary genome mining methods, but if they are not embraced by their users then an opportunity may be lost, and unsound predictions are likely to be posed. In each step therefore adoption of these methods will benefit from "thinking evolutionarily" while following the protocols described, in the sense of asking: what is behind this step that sustains an evolution outcome? What is the evolutionary mechanism underlying the selected method? During this process it is crucial to bear in mind the differences between species-level evolution and the evolution of BGCs, which can be driven by many of the forces sustaining evolutionary genome mining, and many others that remain to be discovered.

References

1. Bentley SD et al (2002) Complete genome sequence of the model actinomycete Streptomyces coelicolor A3(2). Nature 417:141–147
2. Chevrette MG, Currie CR (2019) Emerging evolutionary paradigms in antibiotic discovery. J Ind Microbiol Biotechnol 46:257–271
3. Chevrette MG et al (2020) Evolutionary dynamics of natural product biosynthesis in bacteria. Nat Prod Rep 37:566–599
4. Cruz-Morales P et al (2016) Phylogenomic analysis of natural products biosynthetic gene clusters allows discovery of Arseno-organic metabolites in model Streptomycetes. Genome Biol Evol 8:1906–1916
5. Chevrette MG et al (2019) The antimicrobial potential of Streptomyces from insect microbiomes. Nat Commun 10:516
6. Hurley A et al (2021) Tiny earth: a big idea for STEM education and antibiotic discovery. MBio 12:e03432-20
7. Montalbán-López M et al (2021) New developments in RiPP discovery, enzymology and engineering. Nat Prod Rep 38:130–239
8. Whitford CM, Cruz-Morales P, Keasling JD, Weber T (2021) The design-build-test-learn cycle for metabolic engineering of Streptomycetes. Essays Biochem 65(2):261–275. https://doi.org/10.1042/EBC20200132

9. Blin K et al (2019) antiSMASH 5.0: updates to the secondary metabolite genome mining pipeline. Nucleic Acids Res 47:W81–W87
10. Blin K et al (2017) antiSMASH 4.0-improvements in chemistry prediction and gene cluster boundary identification. Nucleic Acids Res 45:W36–W41
11. Narzisi G, Mishra B (2011) Comparing De novo genome assembly: the long and short of it. PLoS One 6:e19175
12. Liao Y-C, Lin S-H, Lin H-H (2015) Completing bacterial genome assemblies: strategy and performance comparisons. Sci Rep 5:1–8
13. Davis JJ et al (2020) The PATRIC bioinformatics resource center: expanding data and analysis capabilities. Nucleic Acids Res 48: D606–D612
14. Aziz RK et al (2008) The RAST server: rapid annotations using subsystems technology. BMC Genomics 9:75
15. Seemann T (2014) Prokka: rapid prokaryotic genome annotation. Bioinformatics 30:2068–2069
16. Delcher AL, Harmon D, Kasif S, White O, Salzberg SL (1999) Improved microbial gene identification with GLIMMER. Nucleic Acids Res 27:4636–4641
17. Hyatt D et al (2010) Prodigal: prokaryotic gene recognition and translation initiation site identification. BMC Bioinformatics 11:119
18. Devoid S et al (2013) Automated genome annotation and metabolic model reconstruction in the SEED and model SEED. Methods Mol Biol 985:17–45
19. Majoros WH, Pertea M, Salzberg SL (2004) TigrScan and GlimmerHMM: two open source ab initio eukaryotic gene-finders. Bioinformatics 20:2878–2879
20. van Santen JA, Kautsar SA, Medema MH, Linington RG (2021) Microbial natural product databases: moving forward in the multi-omics era. Nat Prod Rep 38:264–278
21. Sorokina M, Steinbeck C (2020) Review on natural products databases: where to find data in 2020. J Cheminform 12:20
22. Kautsar SA et al (2020) MIBiG 2.0: a repository for biosynthetic gene clusters of known function. Nucleic Acids Res 48:D454–D458
23. Blin K, Shaw S, Kautsar SA, Medema MH, Weber T (2021) The antiSMASH database version 3: increased taxonomic coverage and new query features for modular enzymes. Nucleic Acids Res 49:D639–D643
24. Medema MH et al (2011) antiSMASH: rapid identification, annotation and analysis of secondary metabolite biosynthesis gene clusters in bacterial and fungal genome sequences. Nucleic Acids Res 39:W339–W346
25. Wolf T, Shelest V, Nath N, Shelest E (2016) CASSIS and SMIPS: promoter-based prediction of secondary metabolite gene clusters in eukaryotic genomes. Bioinformatics 32:1138–1143
26. Kloosterman AM, Shelton KE, van Wezel GP, Medema MH, Mitchell DA (2020) RRE-Finder: a Genome-Mining Tool for Class-Independent RiPP Discovery. mSystems 5:e00267
27. Li W et al (2021) RefSeq: expanding the prokaryotic genome annotation pipeline reach with protein family model curation. Nucleic Acids Res 49:D1020–D1028
28. Kamra P, Gokhale RS, Mohanty D (2005) SEARCHGTr: a program for analysis of glycosyltransferases involved in glycosylation of secondary metabolites. Nucleic Acids Res 33: W220–W225
29. Caboche S, Leclère V, Pupin M, Kucherov G, Jacques P (2010) Diversity of monomers in nonribosomal peptides: towards the prediction of origin and biological activity. J Bacteriol 192:5143–5150
30. Stachelhaus T, Mootz HD, Marahiel MA (1999) The specificity-conferring code of adenylation domains in nonribosomal peptide synthetases. Chem Biol 6:493–505
31. Minowa Y, Araki M, Kanehisa M (2007) Comprehensive analysis of distinctive polyketide and nonribosomal peptide structural motifs encoded in microbial genomes. J Mol Biol 368:1500–1517
32. Khayatt BI, Overmars L, Siezen RJ, Francke C (2013) Classification of the adenylation and acyl-transferase activity of NRPS and PKS systems using ensembles of substrate specific hidden Markov models. PLoS One 8:e62136
33. Röttig M et al (2011) NRPSpredictor2--a web server for predicting NRPS adenylation domain specificity. Nucleic Acids Res 39: W362–W367
34. Chevrette MG, Aicheler F, Kohlbacher O, Currie CR, Medema MH (2017) SANDPUMA: ensemble predictions of nonribosomal peptide chemistry reveal biosynthetic diversity across Actinobacteria. Bioinformatics 33:3202–3210
35. Helfrich EJN et al (2021) Evolution of combinatorial diversity in trans-acyltransferase polyketide synthase assembly lines across bacteria. Nat Commun 12:1422
36. chevrm. chevrm/transPACT: transPACT v1.0.1. (2020). https://doi.org/10.5281/zenodo.4148258

37. Conway KR, Boddy CN (2012) ClusterMine360: a database of microbial PKS/NRPS biosynthesis. Nucleic Acids Res 41:D402–D407
38. Ichikawa N et al (2013) DoBISCUIT: a database of secondary metabolite biosynthetic gene clusters. Nucleic Acids Res 41:D408–D414
39. Sélem-Mojica N, Aguilar C, Gutiérrez-García K, Martínez-Guerrero CE, Barona-Gómez F (2019) EvoMining reveals the origin and fate of natural product biosynthetic enzymes. Microb Genom 5:e000260
40. Chevrette MG et al (2019) Taxonomic and metabolic incongruence in the ancient genus. Front Microbiol 10:2170
41. Cruz-Morales P et al (2013) The genome sequence of Streptomyces lividans 66 reveals a novel tRNA-dependent peptide biosynthetic system within a metal-related genomic island. Genome Biol Evol 5:1165–1175
42. Ausland C et al (2021) dbCAN-PUL: a database of experimentally characterized CAZyme gene clusters and their substrates. Nucleic Acids Res 49:D523–D528
43. Alcock BP et al (2020) CARD 2020: antibiotic resistome surveillance with the comprehensive antibiotic resistance database. Nucleic Acids Res 48:D517–D525
44. Palaniappan K et al (2019) IMG-ABC v.5.0: an update to the IMG/Atlas of Biosynthetic Gene Clusters Knowledgebase. Nucleic Acids Res 48:D422–D430
45. Bortolaia V et al (2020) ResFinder 4.0 for predictions of phenotypes from genotypes. J Antimicrob Chemother 75:3491–3500
46. van Santen JA et al (2019) The natural products atlas: an open access Knowledge Base for microbial natural products discovery. ACS Cent Sci 5:1824–1833
47. Medema MH, Takano E, Breitling R (2013) Detecting sequence homology at the gene cluster level with MultiGeneBlast. Mol Biol Evol 30:1218–1223
48. Navarro-Muñoz JC et al (2019) A computational framework to explore large-scale biosynthetic diversity. Nat Chem Biol 16:60–68
49. Kautsar SA, van der Hooft JJJ, de Ridder D, Medema MH (2021) BiG-SLiCE: A highly scalable tool maps the diversity of 1.2 million biosynthetic gene clusters. Gigascience 10: giaa154
50. Kautsar SA, Blin K, Shaw S, Weber T, Medema MH (2020) BiG-FAM: the biosynthetic gene cluster families database. Nucleic Acids Res 49: D490–D497
51. Alanjary M, Cano-Prieto C, Gross H, Medema MH (2019) Computer-aided re-engineering of nonribosomal peptide and polyketide biosynthetic assembly lines. Nat Prod Rep 36:1249–1261
52. Adamek M, Alanjary M, Ziemert N (2019) Applied evolution: phylogeny-based approaches in natural products research. Nat Prod Rep 36:1295–1312
53. Barona-Gómez F, Cruz-Morales P, Noda-García L (2012) What can genome-scale metabolic network reconstructions do for prokaryotic systematics? Antonie Van Leeuwenhoek 101: 35–43
54. Medema MH, Fischbach MA (2015) Computational approaches to natural product discovery. Nat Chem Biol 11:639–648
55. Mungan MD et al (2020) ARTS 2.0: feature updates and expansion of the antibiotic resistant target seeker for comparative genome mining. Nucleic Acids Res 48:W546–W552
56. Alanjary M et al (2017) The antibiotic resistant target seeker (ARTS), an exploration engine for antibiotic cluster prioritization and novel drug target discovery. Nucleic Acids Res 45:W42–W48
57. Cimermancic P et al (2014) Insights into secondary metabolism from a global analysis of prokaryotic biosynthetic gene clusters. Cell 158:412–421
58. Choo KH, Tong JC, Zhang L (2004) Recent applications of hidden Markov models in computational biology. Genomics Proteomics Bioinformatics 2:84–96
59. Hannigan GD et al (2019) A deep learning genome-mining strategy for biosynthetic gene cluster prediction. Nucleic Acids Res 47:e110

Chapter 9

Inducing Global Expression of Actinobacterial Biosynthetic Gene Clusters

Meghan A. Pepler, Xiafei Zhang, Hindra, and Marie A. Elliot

Abstract

Bacteria produce an impressive array of bioactive specialized metabolites, with *Streptomyces* (and the actinobacteria more generally) being unusually diverse and prolific producers. However, the biosynthetic potential of these organisms has yet to be fully explored, as many of the biosynthetic gene clusters that direct the synthesis of these natural products are transcriptionally silent under laboratory growth conditions. Here, we describe strategies that can be employed to broadly stimulate the expression of biosynthetic gene clusters in *Streptomyces* and their relatives, follow the transcription of these genes, and assess the antimicrobial activity of the resulting molecules.

Key words *Streptomyces*, Biosynthetic gene cluster, Transcription regulator, Gene expression, Bioassay, Antibiotic, Genetic engineering

1 Introduction

Bacteria are remarkable chemists and are capable of making complex natural products with diverse activities. These "specialized metabolites" are thought to promote competition and communication in their native environments [1, 2] and have been extensively co-opted for application in medicine (e.g., antibiotics, antifungals, anti-cancer agents) and agriculture (e.g., pesticides). With the onset of the genomic era came the realisation that some groups of bacteria, like the streptomycetes [3], myxobacteria [4, 5], and cyanobacteria [6], harbour even greater metabolic potential than had been initially realized. However, accessing this chemical diversity has proven to be challenging, as many of the biosynthetic clusters that direct the synthesis of these compounds are transcriptionally silent under standard laboratory conditions.

Meghan A. Pepler, Xiafei Zhang and Hindra contributed equally to this work.

Elizabeth Skellam (ed.), *Engineering Natural Product Biosynthesis: Methods and Protocols*,
Methods in Molecular Biology, vol. 2489, https://doi.org/10.1007/978-1-0716-2273-5_9,

Stimulating the expression of these poorly expressed biosynthetic gene clusters has been a major goal of the field. A multitude of strategies have been developed to circumvent or alleviate these expression issues [7], including genetic approaches using cluster-specific regulators (to stimulate specific biosynthetic clusters) [8] and global regulators (to stimulate multiple biosynthetic clusters) [9–11]; chemical genetic strategies using small molecule elicitors [12, 13]; microbial co-culture [14, 15]; and nutritional tactics [16] to more closely mimic the natural environment inhabited by these bacteria; and synthetic strategies like promoter refactoring, and cluster capture and heterologous expression [17, 18]. Here, we describe a genetic approach to stimulating the expression of biosynthetic gene clusters, outline strategies to monitor changes in gene expression, and describe methods to assess the antimicrobial (antibacterial or antifungal) properties of newly stimulated specialized metabolites in *Streptomyces* bacteria and their close relatives. While the described approaches focus on the global activation of biosynthetic gene cluster expression/natural product synthesis, the same strategies could be employed to overexpress a cluster-situated activator and promote expression of a specific biosynthetic cluster.

2 Materials

2.1 Overexpressing a Regulatory Gene of Interest

1. Integrating plasmid vector (e.g., pSET152) carrying a constitutively active promoter (e.g., *ermE**), driving the expression of a gene encoding a global regulator of specialized metabolism [e.g., *crp*; constitutively active allele of *afsQ1*; dominant negative mutant variant of *lsr2* [9–11]] (*see* **Note 1**).

2.2 Conjugation from Escherichia coli *into* Streptomyces

1. *Escherichia coli* ET12567/pUZ8002 [19] (*see* **Note 2**).
2. *Streptomyces* spore/actinobacterial stock of desired parent strain.
3. Antibiotics for *E. coli* ET12567 strain and pUZ8002 plasmid selection (25 μg/mL kanamycin; 25 μg/mL chloramphenicol), and integrating plasmid vector selection (depends on resistance genes carried by the integrating plasmid vector).
4. 25 mg/mL nalidixic acid.
5. 10% glycerol solution.

2.3 Media for E. coli *Growth and Growth of Desired* Streptomyces *Exconjugants*

1. Lysogeny broth (LB) [20] (for *E. coli* growth): Mix together 10 g tryptone, 5 g yeast extract, and 10 g NaCl, with distilled water (dH_2O); make the volume up to 1 L with dH_2O. Autoclave to sterilize.

2. Mannitol-soy flour (MS) agar [21] (Conjugation medium): Mix together 20 g mannitol, 20 g soya bean meal/flour, and 20 g agar with tap water, and make the volume up to 1 L. Sterilize by autoclaving.
3. Maltose-yeast extract-malt extract (MYM) broth/agar [22] (*Streptomyces* growth medium): Mix together 4 g maltose, 4 g yeast extract, and 10 g malt extract (add 20 g agar if wanting to use for plates) with dH_2O, and make the volume up to 1 L. Autoclave to sterilize.
4. 2× yeast extract-tryptone (YT) broth [23] (*Streptomyces* spore heat shock medium): Mix together 16 g tryptone, 10 g yeast extract, and 5 g NaCl with dH_2O, and make the volume up to 1 L. Autoclave to sterilize.

2.4 RNA Extraction (See Note 3)

1. 4 mm^2 glass beads. Autoclave the beads twice at 121 °C for 20 min.
2. $5\%_{w/v}$ sodium *N*-lauroyl sarcosinate: Dissolve 5 g of sodium *N*-lauroyl sarcosinate in 80 mL of sterile distilled deionized water ($ddiH_2O$) in a sterile 100 mL glass bottle. Store the solution at room temperature.
3. Lysis solution: 4 M guanidine thiocyanate, 25 mM trisodium citrate dihydrate, $0.5\%_{w/v}$ sodium *N*-lauroyl sarcosinate, $0.8\%_{v/v}$ β-mercaptoethanol. Per cell pellet: add 2.36 g guanidine thiocyanate, 0.037 g trisodium citrate dihydrate, and 0.5 mL sodium *N*-lauroyl sarcosinate ($5\%_{w/v}$) in a sterile 15 mL Falcon tube. Fill to 4 mL with sterile $ddiH_2O$ and dissolve the components with vigorous vortexing. Add 0.04 mL β-mercaptoethanol to the solution and bring the total volume to 5 mL with sterile $ddiH_2O$. Mix the solution by inverting the tube (*see* **Note 4**).
4. Phenol:chloroform:isoamyl alcohol (50:50:1). Combine 150 mL phenol, 150 mL chloroform, and 3 mL isoamylalcohol in a clean 400 mL brown glass bottle. Gently invert the bottle 10 times to mix the solution. Store the solution in the fridge and let it sit overnight before using.
5. 3 M CH_3COONa/CH_3COOH buffer: dissolve 24.6 g of CH_3COONa in 70 mL $ddiH_2O$ and adjust the pH of the solution to 6.0 using glacial CH_3COOH (17.4 M). Bring the total volume to 100 mL with $ddiH_2O$. Autoclave the solution twice at 121 °C for 20 min. Store the solution at room temperature.
6. 100% isopropanol.
7. $70\%_{v/v}$ ethanol. Add 70 mL of ethanol to a clean 250-mL graduated cylinder. Bring the total volume to 100 mL with $ddiH_2O$. Transfer the solution to a sterile 200-mL glass bottle. Store the solution at room temperature.

8. DNase buffer (10×) and DNase (5000 U) (we use TURBO™ DNase, Fisher Scientific).
9. 10 × TBE: 1 M Tris, 1 M boric acid, 0.02 M EDTA. Dissolve 121.1 g of Tris base, 61.8 g of boric acid, and 7.4 g EDTA in 800 mL dH_2O in a large beaker on a stir plate (with a stir bar). Bring the total volume to 1 L with dH_2O. Store the buffer at room temperature.
10. 1 × TBE. Mix 100 mL 10 × TBE solution with 900 mL dH_2O. Store the buffer at room temperature.
11. $2\%_{w/v}$ agarose gel. Dissolve 1.5 g of agarose in 75 mL 1 × TBE by microwaving the mixture for 1 min. Let the mixture cool for 15 min, then add 5 μL of 1% ethidium bromide and mix well. Pour the mixture into a gel casting tray and let cool for 20 min.
12. Wear gloves and a lab coat at all times when working with RNA, and use only sterile filtered tips and double-autoclaved microcentrifuge tubes.

2.5 Antimicrobial Bioassays

2.5.1 Media for Bioassays

1. Bennett's agar [23]: Mix together 10 g glucose, 1 g yeast extract, 1 g beef extract, 2 g N-Z amine A, and 15 g agar with $ddiH_2O$, before bringing the total volume to 1 L. Sterilize by autoclaving.
2. ISP4 supplemented agar [11]: Mix together 37 g Difco™ ISP Medium 4 (10 g soluble starch, 1 g K_2HPO_4, 1 g $MgSO_4$, 1 g NaCl, 2 g $(NH_4)_2SO_4$, 2 g $CaCO_3$, 1 mg $FeSO_4$, 1 mg $MnCl_2$, 1 mg $ZnSO_4$, 20 g agar) with 1 g maltose, 1 g mannitol, 1 g glycerol, and 1 g sucrose, in $ddiH_2O$, and then bring the total volume to 1 L. Sterilize by autoclaving.
3. Lysogeny Broth: *see* Subheading 2.3.
4. Difco™ nutrient agar: Mix 23 g Difco™ Nutrient agar (3 g beef extract, 5 g peptone, 15 g agar) with $ddiH_2O$, and then bring the total volume to 1 L. Sterilize by autoclaving.
5. YPD broth/agar: Mix 20 g peptone and 10 g yeast extract (add 20 g agar if wanting solid medium), and then make volume up to 900 mL with $ddiH_2O$ and autoclave. Add 100 mL of $20\%_{w/v}$ glucose or dextrose (prepared and autoclaved separately) into 900 mL broth mixture.

2.5.2 Possible Indicator Strains (See **Note 5**)

1. *Micrococcus luteus*: Gram-positive bacterium—highly susceptible to most antibiotics.
2. *Bacillus subtilis*: Gram-positive bacterium—reasonably susceptible to many antibiotics.
3. *Staphylococccus aureus*: Gram-positive bacterium—depending on the strain, may be resistant to many antibiotics.

4. *E. coli* MG1655 (or another lab strain): Gram-negative bacterium.
5. *E. coli* BW25113 Δ*bamB*Δ*tolC* [24]: strain with defective membrane that enables antibiotic access to the periplasmic space (and inner membrane/cytoplasm) by compounds that otherwise could not traverse the outer membrane, and that further prevents antibiotic export/efflux once inside the cell.
6. *Saccharomyces cerevisiae*: fungus.

3 Methods

3.1 Genetic Manipulation

3.1.1 Moving Construct from E. coli Cloning Host into E. coli ET12567/pUZ8002

1. Isolate and purify your regulator-overexpressing plasmid (and the equivalent empty plasmid vector as a control), according to the manufacturer's instructions for your plasmid isolation kit of choice.
2. Make electrocompetent cells: Grow *E. coli* ET12567/pUZ8002 overnight at 37 °C in 10 mL LB containing kanamycin (25 μg/mL) and chloramphenicol (25 μg/mL). Inoculate 100 μL of this overnight culture into 10 mL of fresh LB containing kanamycin and chloramphenicol at the same concentrations. Grow this subculture at 37 °C while shaking at 200 rpm for ~3–4 h, until the OD_{600} is ~0.4. Collect the cells by centrifugation at 6000 × *g* for 5 min at 4 °C. Discard the supernatant and resuspend the cell pellet by gently mixing in 10 mL ice-cold 10% glycerol. Repeat the centrifugation and glycerol washing steps three times. After the final wash, resuspend the cell pellet in the remaining ~100 μL of 10% glycerol.
3. Transform plasmids (*see* **Note 6**): Mix 50 μL of your electrocompetent *E. coli* cell suspension with ~50 ng (1–2 μL) of plasmid DNA. Transfer the DNA-cell mixture into the bottom of a 0.2 cm ice-cold electroporation cuvette (ideally with no bubbles) and subject to electroporation. Time constants can vary depending on the electroporator, but if a spark is seen, there was likely residual salt in either your DNA or your cell suspension, and the experiment will need to be re-done using re-washed DNA/cells. Following electroporation, immediately add 1 mL of ice-cold LB to the shocked cells and incubate, with shaking, for 1 h at 37 °C.
4. Selection of successful transformants: Prepare two LB agar plates containing kanamycin (25 μg/mL), chloramphenicol (25 μg/mL), and the appropriate antibiotic for plasmid selection. Pipette 50 μL of the cell suspension onto one plate. Spin the microcentrifuge tube for 1 min at 6000 × *g* to pellet the remaining *E. coli* cells. Pour off the supernatant and resuspend the cell pellet in the residual ~50 μL of medium. Pipette the

remaining cells onto the second plate. Spread the dilute cell suspension on the first plate, and then spread the more concentrated cell suspension on the second plate (if the electroporation is very efficient, you will have single colonies on the first plate; if it is less efficient, you will have single colonies on the second plate). Incubate these plates at 37 °C overnight.

3.1.2 Conjugation from E. coli *into* Streptomyces *and Creating a Spore Stock*

1. Inoculate an *E. coli* ET12567/pUZ8002 transformant carrying the regulatory gene overexpression construct into 10 mL of LB containing kanamycin (25 μg/mL), chloramphenicol (25 μg/mL), and the plasmid-specific selectable antibiotic. Incubate this culture with shaking (200 rpm) at 37 °C overnight, before then inoculating 100 μL of this overnight culture into 10 mL of fresh LB medium supplemented with the same three antibiotics and at the same concentration as for the overnight culture. Grow this subculture with shaking for 3–4 h at 37 °C, until an OD_{600} of ~0.4 is reached. Collect the cells by centrifugation at 6000 × *g* for 5 min at 4 °C, and then wash the cells three times in fresh LB medium (*see* **step 2** in Subheading 3.1.1) to remove any antibiotics that might affect *Streptomyces* growth.
2. Add 10 μL (~10^8) of *Streptomyces* spores to 1 mL 2 × YT broth. Heat shock these spores at 50 °C for 10 min, and then cool to room temperature (*see* **Note 7**).
3. Mix the washed *E. coli* cell suspension with 500 μL of heat shocked spores and spin briefly in a microcentrifuge. Discard the supernatant and spread the cell/spore mixture onto MS medium (no antibiotics). At the same time, spread the remaining 500 μL of heat shocked spores onto an independent MS medium plate to serve as a negative control (prior to plating, spores can be concentrated prior by centrifuging, as described above). Incubate both plates at 30 °C for 5–8 h (*see* **Note 8**).
4. Overlay each inoculated plate with 1 mL sterile water containing 0.5 mg nalidixic acid (20 μL of 25 mg/mL stock; selectively kills *E. coli*) and 25 μL (depending on stock concentration) of antibiotic selecting for the overexpression plasmid. Continue incubating at 30 °C until robust (sporulating) *Streptomyces* growth is observed (*see* **Note 9**). Ensure that no *Streptomyces* growth is observed on the negative control plate (without *E. coli*) before proceeding.
5. From your experimental plate, streak out putative exconjugants for single colonies on MS or MYM agar medium, supplemented with selective antibiotics. Once individual colonies have grown, inoculate one into 5 mL MYM broth for growth overnight at 30 °C. Antibiotic supplementation is not required at this stage, provided your plasmid can integrate into the

chromosome of its host *Streptomyces* sp. From your overnight culture, spread 200 μL of this overnight culture onto MYM plates (*see* **Note 10**), and use the remainder to isolate genome DNA using your preferred method (*see* **Note 11**) to confirm the presence of your plasmid of interest within the exconjugant (see below). For the inoculated plates, incubate at 30 °C until the culture sporulates (when the white aerial hyphae turn color due to production of the mature spore pigment; pigment color differs depending on the *Streptomyces* sp.), at which point a spore stock can be made, and frozen in 40% glycerol for future use.

3.2 PCR Check for Strain Integrity

1. Design primers to confirm the presence of the overexpression plasmid in your exconjugants from Subheading 3.1.2. Different primer combinations are possible here. For the overexpression plasmid-carrying strain, it is typical to position one primer within the regulatory gene being overexpressed, and the other upstream or downstream of the cloned regulatory gene (specific to the vector backbone), oriented such that the primer pair amplifies an easily visualized PCR product (usually >300 bp) when using genomic DNA from your overexpression strain (isolated in Subheading 3.1.2) as template, and a parental (plasmid-free) strain as a negative control. For the plasmid alone-carrying control, it is common to use primers flanking the multiple cloning site, again, using the plasmid-free strain as a negative control.

3.3 Expression Analyses

3.3.1 RNA Extraction

1. Determining which gene clusters are activated or repressed in response to overexpression of a global regulator of specialized metabolism can be achieved using a variety of techniques, including RNA-seq (providing a comprehensive view of gene expression at a given timepoint), and quantitative or semi-quantitative PCR (providing expression information for genes within specific clusters of interest), and comparing cluster transcript levels in the overexpression strain, relative to the plasmid-alone control strain.
2. To isolate RNA for analysis using any of these techniques, add ~2 mL of glass beads into a pre-cooled, sterile 15-mL Falcon tube (*see* **Note 12**), then transfer ~5 mm^2 of collected cells into the tube (collected using a flame-sterilized metal spatula).
3. Add 5 mL ice-cold lysis solution to the cells and vortex the mixture for 2 min at maximum speed. Immediately add 5 mL of pre-cooled phenol:chloroform:isoamyl alcohol (50:50:1) to the suspension. Vortex the mixture for 30 s at maximal speed and then immediately move the mixture to ice and let it cool for 30 s. Repeat three times. Centrifuge the mixture at 7800 × *g* for 5 min at 4 °C. Carefully transfer the top aqueous phase to a

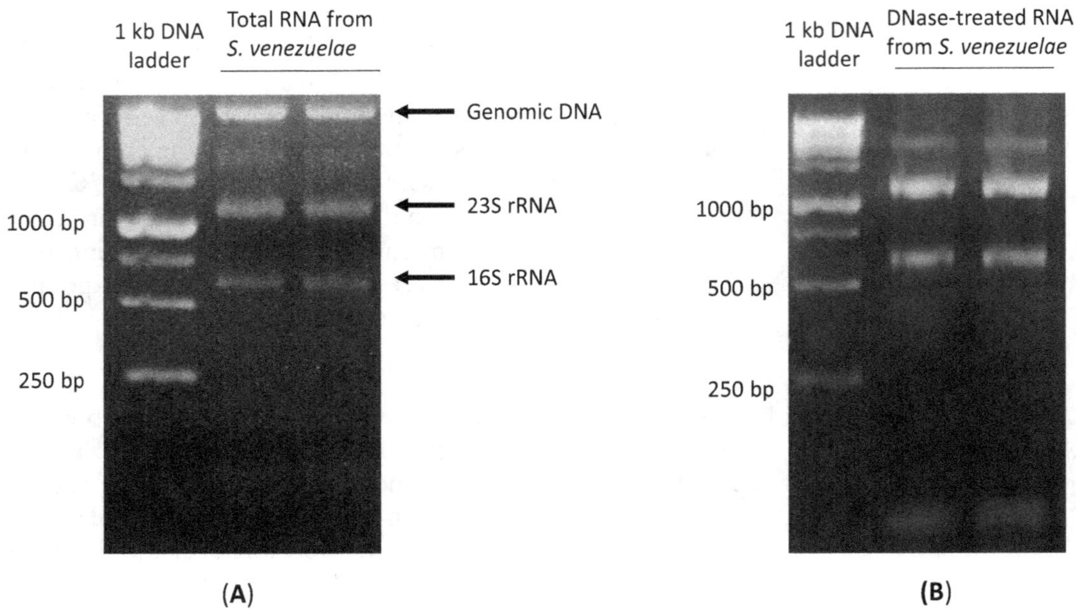

Fig. 1 RNA integrity check after total RNA extraction and DNase treatment. (**a**) Total RNA extracted from wildtype *S. venezuelae*, in duplicate. (**b**) Purified RNA after DNase treatment followed by phenol-chloroform extraction, in duplicate. Samples were run on 2% agarose gels at 100 V for 30 min

pre-cooled, sterile 15-mL Falcon tube, using a pipette with filtered tips (*see* **Note 13**).

4. Repeat **step 3** (phenol-chloroform extraction, excluding the addition of lysis solution) until little or no interface remains (this usually requires at least two extractions, and sometimes three).
5. Add 0.1 volume of ice-cold CH_3COONa/CH_3COOH buffer (3 M, pH 6) and an equal volume of ice-cold isopropanol (100%) to the recovered aqueous phase. Mix gently by inverting the Falcon tube 10 times. Incubate the mixture at −20 °C overnight (*see* **Note 14**).
6. Centrifuge the mixture at 7800 × *g* for 30 min at 4 °C, and then discard the supernatant. Wash the pellet with 500 μL of ethanol ($70\%_{v/v}$). Re-centrifuge at 7800 × *g* for 5 min at 4 °C, before again discarding the supernatant. Allow the pellet to air dry by placing the Falcon tube upside-down on a paper towel for 10 min.
7. Resuspend the nucleic acid pellet in nuclease-free water, and then transfer the nucleic acid suspension to a twice autoclaved 1.5-mL microcentrifuge tube. Run 1 μL of sample on a 2% agarose gel at 100 V for 30 min to check the integrity of the associated RNA (Fig. 1a). Store the samples at −80 °C.

8. To remove co-extracted DNA, mix the nucleic acid extract from **step 7** with 50 μL of DNase buffer (10×) and 10 μL of DNase in a twice autoclaved 1.5-mL microcentrifuge tube and bring the total volume to 500 μL with nuclease-free water. Incubate the reaction mixture at 37 °C for 1 h, and then add an additional 5 μL of DNase to the reaction and incubate for another hour (*see* **Note 15**).
9. Transfer the reaction from **step 8** to a twice autoclaved 1.5-mL microcentrifuge tube (*see* **Note 16**).
10. Immediately add one volume of phenol:chloroform:isoamyl alcohol (50:50:1) to the sample. Vortex the mixture for 30 s at maximum speed. Centrifuge the sample at 16,000 × *g* for 5 min at 4 °C. Transfer the upper, aqueous phase to a pre-cooled, twice autoclaved 1.5-mL microcentrifuge tube.
11. Repeat **steps 9** and **10** until little or no debris remains at the aqueous-organic interface.
12. Add 0.1 volume of ice-cold CH_3COONa/CH_3COOH buffer (3 M, pH 6) and an equal volume of ice-cold isopropanol (100%) to the recovered aqueous phase. Mix gently by inverting the tube 10 times. Incubate the mixture at −20 °C overnight.
13. Centrifuge the mixture at 16,000 × *g* for 30 min at 4 °C and discard the supernatant. Wash the pellet with 500 μL of ethanol ($70\%_{v/v}$). Centrifuge the mixture at 16,000 × *g* for 5 min at 4 °C. Discard the supernatant, and dry the pellet by placing the microcentrifuge tube up-side-down on a paper towel for 10 min.
14. Gently resuspend the pellet in nuclease-free water while sitting on ice (start with 50 μL, and if that appears to be too concentrated (RNA pellet does not seem to be moving into solution), add more water in 25 μL aliquots). Run 1 μL of sample on a 2% agarose gel at 100 V for 30 min to check RNA integrity (Fig. 1b) and assess the concentration using a NanoDrop 1000 spectrophotometer (note that this can sometimes overestimate RNA concentrations). Store RNA samples at −80 °C.
15. Check for DNA contamination of RNA samples by PCR using primers that amplify housekeeping genes (e.g., *hrdB* or *rpoB*), ideally designed to amplify a product smaller than 500 bp. It is also good practice to include a reaction in which genomic DNA is used as PCR template as a positive control for the PCR reaction (as you are hoping to see no product associated with your RNA samples).
16. Use 1 μg of RNA from step 14 as template in PCR along with dilutions of quantified genomic DNA as a positive control to confirm the absence of residual genomic DNA from the RNA

preparations (the genomic DNA dilutions allow you to determine the approximate quantity of any contaminating DNA in the RNA samples, and assess whether these samples require another round of DNase treatment).

17. Once the quality of the RNA samples has been confirmed, it can be used for quantitative/semi-quantitative (reverse transcription) PCR, or RNA-seq, to assess the relative transcript levels of biosynthetic clusters of interest [11, 25, 26].

3.4 Antimicrobial Bioassays (See Note 17)

3.4.1 Monitoring Antimicrobial Production Using the "Pancake" Bioassay Technique

1. Inoculate 1–5 μL droplets of *Streptomyces* spore suspensions on either Bennett's agar or ISP4 supplemented agar (or your "antimicrobial production" medium of choice) such that it will grow as a defined circular colony (~5–8 mm in diameter) (Fig. 2a). Incubate the plates at 30 °C for 3–7 days (it is useful to test the activity after different growth times, e.g., 3, 5, and 7 days). It is recommended to inoculate both your genetically modified strains and the corresponding empty plasmid control strain on the same plate to enable direct comparisons.
2. The day before beginning the bioassay (i.e., your *Streptomyces* cultures have been growing for 2–6 days), inoculate your indicator organism of choice into 5 mL of appropriate growth medium (LB for many bacteria; YPD broth for many fungi) and grow overnight at 30–37 °C, depending on the optimal growth temperature for your indicator organism of choice.
3. The next day, use the dense overnight culture to inoculate warm Difco™ nutrient agar (e.g., 1% inoculum into the warm—not hot—molten agar) and mix well. The nutrient agar can be sterilized on the same day or prepared in advance and melted in a microwave when needed. Pour the inoculated agar into a new petri plate (of the same size and shape as the *Streptomyces* inoculated plates) and let it set.
4. Once the inoculated agar has set, hold the plate upside down and pry up the agar using a sterile spatula (Fig. 2b). Transfer the inoculated agar ("indicator" layer) to the *Streptomyces* "production" plate and overlay it on top of the *Streptomyces* colony layer, to create a two layered "pancake" (Fig. 2c). If air bubbles are captured between the layers, remove them by pushing them out the side using the end of a sterile spatula.
5. Incubate the resulting layered plate for 12–48 h, with the indicator layer oriented on the top, at 30–37 °C (depending on the optimal growth temperature of the indicator strain). Bacterial indicators typically grow overnight, while *S. cerevisiae* may need up to 2 days.
6. Once indicator strains have grown (ideally as a lawn—if not, more inoculum is needed), assess whether growth is inhibited around the *Streptomyces* colonies. Clear zones of growth

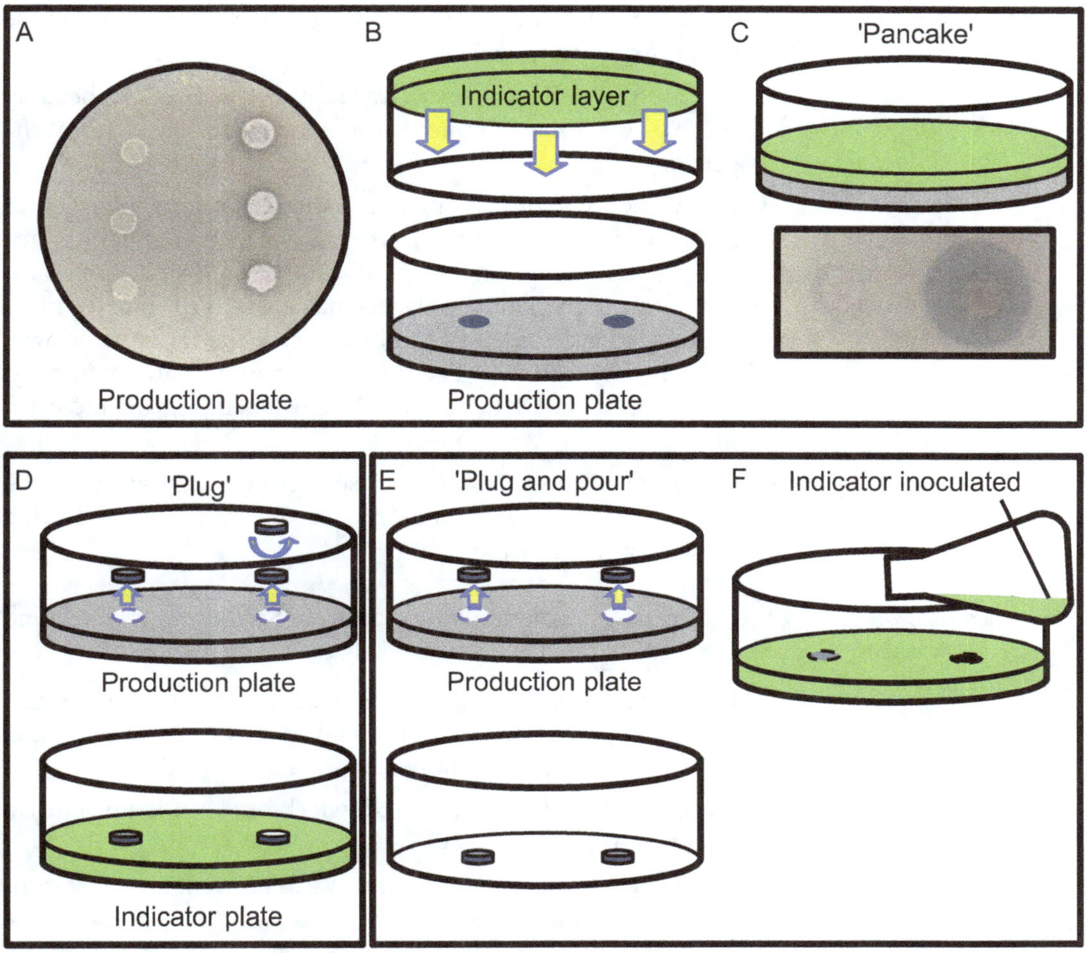

Fig. 2 Antimicrobial bioassays using three techniques. (**a**–**c**) Pancake method. (**a**) Grow circular *Streptomyces* colonies on an "antimicrobial production" medium. (**b**) Transfer the indicator-inoculated agar ("indicator layer") onto the *Streptomyces* plate. (**c**) This gives a two agar layer "pancake" (top), with typical visualization of the inhibition zone resulting from the production of diffusible antibiotics that kill/inhibit the growth of the indicator strain (bottom). (**d**) "Plug" technique: removing agar plugs (with *Streptomyces* on top) and placing these onto the indicator-inoculated agar. Left plug: agar-side down and right plug: biomass (vegetative or aerial hyphae)-side down. (**e**–**f**) "Plug and Pour" technique: similar to **d**, except the indicator-inoculated agar is poured after the plugs are placed onto an empty petri plate

inhibition typically indicate bacteriocidal/fungicidal activity, while fuzzy zones indicate bacteriostatic/fungistatic activity.

3.4.2 Monitoring Antimicrobial Production Using the "Plug" Bioassay Technique

(alternative to the Pancake method)

1. Inoculate *Streptomyces* cultures on production medium plates (*see* **step 1** in Subheading 3.4.1), or inoculate 5–10 μL spores into 5 mL MYM broth and grow at 30 °C overnight (or until dense), and then spread 50 μL of the dense liquid culture on production medium plates to ultimately generate a lawn on the

growth medium of choice. Grow for 3–7 days (or longer if the strain is slow to sporulate).

2. Prepare the indicator layer (*see* **steps 2** and **3** in Subheading 3.4.1) but do not remove the inoculated layer from the petri dish.
3. From the *Streptomyces* plates, remove an agar plug (with robustly growing *Streptomyces*), and place on the indicator strain-containing agar. The plugs can be removed using a variety of sterile implements (large end of 1000 μL pipette tip, a straw, spatulas, etc.), and can be placed either agar-side down, or biomass-side down (Fig. 2d). Incubate the plates with plug-side up for 12–48 h, at 30–37 °C (depending on indicator organism).
4. As for the Pancake method, assess growth inhibition of the indicator strain around the *Streptomyces* agar plugs following growth of the indicator.

3.4.3 Monitoring Antimicrobial Production Using the "Plug and Pour" Bioassay Technique

(alternative to the methods described in Subheadings 3.4.1 and 3.4.2)

1. Inoculate *Streptomyces* plates (*see* **step 1** in Subheading 3.4.2) and once desired growth is reached, remove a growth plug as described in **step 3** of Subheading 3.4.2.
2. Place the agar plug (agar-side down/biomass-side up) into an empty petri plate (Fig. 2e).
3. Prepare the indicator organism in warm nutrient agar, (*see* **steps 2** and **3** in Subheading 3.4.1) and once mixed, pour the inoculated agar into the petri plate containing the agar plugs. Stop pouring when the inoculated agar level reaches the top of the *Streptomyces* agar plug (Fig. 2f), and let the agar set.
4. Incubate the plates at 30–37 °C for 12–48 h (again, depending on the growth of the indicator bacterium or fungus).
5. Observe growth inhibition as above.

4 Notes

1. In order to overexpress your desired regulatory gene, your plasmid vector should contain an *ermE** promoter or another strong constitutive or inducible promoter. It should also be able to replicate in *E. coli*, have an *oriT* (to allow conjugation from *E. coli*), and carry an *attB* site alongside the corresponding phage integrase (to enable integration into the *Streptomyces* chromosome). Plasmid integration into the *Streptomyces* chromosome eliminates the need for continued antibiotic selection.

2. pUZ8002 carries the *tra* genes needed to promote conjugation, while the ET12567 strain is defective in its methylation capabilities, and the resulting unmethylated plasmid DNA is better able to circumvent the recipient *Streptomyces* restriction/modification defense systems; this is important for successful introduction of DNA into some *Streptomyces* spp.
3. RNA isolation can also be accomplished using a kit, although the yield is often much lower than for the method described here.
4. Make the lysis solution fresh before each use, and leave on ice.
5. There are many indicator strains you could employ; the ones suggested here represent a subset of the many possible options. If you wanted to test whether *any* antibacterial compound is being produced, *M. luteus* is an excellent indicator choice. If you are interested in assaying the production of new compounds with activity against specific antibiotic resistant organisms, resistant bacteria/fungi can be used instead.
6. In lieu of electroporating the plasmids into *E. coli*, classical heat shock-mediated transformation into chemically-competent *E. coli* cells is also effective.
7. Heat shocking spores can help to promote conjugation, but can also be lethal to certain *Streptomyces* species. It is worth doing an initial test to determine the heat-tolerance of the species you are working with, and if necessary, reducing the temperature or duration of the heat shock.
8. The non-selective incubation of *Streptomyces* and *E. coli* is to provide an opportunity for spore germination and initiation of germ tube outgrowth, and to allow for conjugation with *E. coli* to occur. For rapidly growing strains, overnight incubation may be too long before overlaying with selective antibiotics (as they would have initiated aerial development, which makes antibiotic overlays challenging, as aerial hyphae are hydrophobic); in these cases, you can incubate these cultures overnight at room temperature, or at 30 °C for 6–8 h. Conversely, if your strain grows slowly, or does not undergo efficient conjugation, plating your transconjugants on MS medium supplemented with 10 mM $MgCl_2$ can help. For these strains, it may take up to 24 h for a thin film of vegetative growth to appear (this is the ideal time to be applying your selective antibiotic overlay).
9. Different *Streptomyces* species have different conjugation capabilities. Using freshly made spore stocks (not frozen) can enhance conjugation efficiency. Many other strategies have been developed to assist with challenging conjugations; techniques for hard to conjugate systems have been described elsewhere [27].

10. For strains making agarase (where strains grow embedded in the agar medium), overlaying plates with cellophane discs before streaking for a lawn can help enormously with spore isolation.
11. The Norgen Biotek® Bacterial Genomic DNA Isolation Kit works well for isolating *Streptomyces* chromosomal DNA, as does the protocol described by Kieser et al. [23].
12. Wipe down workspace and instruments with $70\%_{v/v}$ ethanol and pre-cool the following materials on ice: 15-mL Falcon tubes, 1.5-mL microcentrifuge tubes, phenol:chloroform:isoamyl alcohol (50:50:1), CH_3COONa/CH_3COOH buffer (3 M, pH 6), isopropanol (100%), ethanol ($70\%_{v/v}$) and nuclease-free water.
13. Recover the glass beads and dry the beads in a fume hood. The beads can be washed and re-used.
14. If needed, samples can be kept at −20 °C for several days.
15. The amount of DNase required and the incubation time varies depending on the amount of genomic DNA in the RNA extracts, and on the efficacy of the DNase (enzymes from different suppliers can be more or less effective).
16. Transferring the DNA digestion reaction into a new tube can decrease subsequent DNA contamination in the final RNA sample.
17. Method 1 (Pancake technique) can provide information on how soluble/diffusible the produced metabolites are, whereas method 2 can distinguish whether the antimicrobial compounds are effectively secreted into the agar. Methods 2 and 3 can be used to efficiently test agar plugs from different "antibiotic production" plates, and can be readily tested with different indicator strains.

References

1. Romero D, Traxler MF, López D, Kolter R (2011) Antibiotics as signal molecules. Chem Rev 111:5492–5505. https://doi.org/10.1021/cr2000509
2. Davies J (2013) Specialized microbial metabolites: functions and origins. J Antibiot (Tokyo) 66:361–364. https://doi.org/10.1038/ja.2013.61
3. Hopwood DA (2007) *Streptomyces* in nature and medicine: the antibiotic makers. Oxford University Press, Oxford
4. Weissman KJ, Müller R (2010) Myxobacterial secondary metabolites: bioactivities and modes-of-action. Nat Prod Rep 27:1276–1295. https://doi.org/10.1039/c001260m
5. Schäberle TF, Lohr F, Schmitz A, König GM (2014) Antibiotics from myxobacteria. Nat Prod Rep 31:953–972. https://doi.org/10.1039/c4np00011k
6. Shah S, Akhter N, Auckloo B et al (2017) Structural diversity, biological properties and applications of natural products from cyanobacteria. A review. Mar Drugs 15:354. https://doi.org/10.3390/md15110354
7. Zhang X, Hindra, Elliot MA (2019) Unlocking the trove of metabolic treasures: activating silent biosynthetic gene clusters in bacteria and fungi. Curr Opin Microbiol 51:9–15. https://doi.org/10.1016/j.mib.2019.03.003

8. Hou B, Lin Y, Wu H et al (2018) The novel transcriptional regulator LmbU promotes lincomycin biosynthesis through regulating expression of its target genes in *Streptomyces lincolnensis*. J Bacteriol 200:e00447–e00417. https://doi.org/10.1128/JB.00447-17
9. Gao C, Hindra, Mulder D et al (2012) Crp is a global regulator of antibiotic production in *Streptomyces*. mBio 3:e00407-12. https://doi.org/10.1128/mBio.00407-12
10. Daniel-Ivad M, Hameed N, Tan S et al (2017) An engineered allele of *afsQ1* facilitates the discovery and investigation of cryptic natural products. ACS Chem Biol 12:628–634. https://doi.org/10.1021/acschembio.6b01002
11. Gehrke EJ, Zhang X, Pimentel-Elardo SM et al (2019) Silencing cryptic specialized metabolism in *Streptomyces* by the nucleoid-associated protein Lsr2. Elife 8:e47691. https://doi.org/10.7554/eLife.47691
12. Moon K, Xu F, Zhang C, Seyedsayamdost MR (2019) Bioactivity-HiTES unveils cryptic antibiotics encoded in actinomycete bacteria. ACS Chem Biol 14:767–774. https://doi.org/10.1021/acschembio.9b00049
13. Craney A, Ozimok C, Pimentel-Elardo SM et al (2012) Chemical perturbation of secondary metabolism demonstrates important links to primary metabolism. Chem Biol 19:1020–1027. https://doi.org/10.1016/j.chembiol.2012.06.013
14. Pishchany G, Mevers E, Ndousse-Fetter S et al (2018) Amycomicin is a potent and specific antibiotic discovered with a targeted interaction screen. Proc Natl Acad Sci 115:10124–10129. https://doi.org/10.1073/pnas.1807613115
15. Lee N, Kim W, Chung J et al (2020) Iron competition triggers antibiotic biosynthesis in *Streptomyces coelicolor* during coculture with *Myxococcus xanthus*. ISME J 14:1111–1124. https://doi.org/10.1038/s41396-020-0594-6
16. Rigali S, Titgemeyer F, Barends S et al (2008) Feast or famine: the global regulator DasR links nutrient stress to antibiotic production by *Streptomyces*. EMBO Rep 9:670–675. https://doi.org/10.1038/embor.2008.83
17. Lee N, Hwang S, Lee Y et al (2019) Synthetic biology tools for novel secondary metabolite discovery in *Streptomyces*. J Microbiol Biotechnol 29:667–686. https://doi.org/10.4014/jmb.1904.04015
18. Palazzotto E, Tong Y, Lee SY, Weber T (2019) Synthetic biology and metabolic engineering of actinomycetes for natural product discovery. Biotechnol Adv 37:107366. https://doi.org/10.1016/j.biotechadv.2019.03.005
19. MacNeil DJ, Gewain KM, Ruby CL et al (1992) Analysis of *Streptomyces avermitilis* genes required for avermectin biosynthesis utilizing a novel integration vector. Gene 111:61–68. https://doi.org/10.1016/0378-1119(92)90603-M
20. Miller J (1972) Experiments in molecular genetics. Cold Spring Harbor Laboratory, Cold Spring Harbor, New York.
21. Hobbs G, Frazer C, Gardner DJ et al (1989) Dispersed growth of *Streptomyces* in liquid culture. Appl Microbiol Biotechnol 31:272–277. https://doi.org/10.1007/BF00258408
22. Stuttard C (1982) Temperate phages of *Streptomyces venezuelae*: lysogeny and host specificity shown by phages SV1 and SV2. Microbiology 128:115–121. https://doi.org/10.1099/00221287-128-1-115
23. Kieser T, Bibb MJ, Buttner MJ et al (2000) Practical *Streptomyces* genetics. John Innes Foundation, Norwich, UK
24. King AM, Reid-Yu SA, Wang W et al (2014) Aspergillomarasmine A overcomes metallo-β-lactamase antibiotic resistance. Nature 510:503–506. https://doi.org/10.1038/nature13445
25. Tanaka Y, Kasahara K, Hirose Y et al (2013) Activation and products of the cryptic secondary metabolite biosynthetic gene clusters by rifampin resistance (*rpoB*) mutations in actinomycetes. J Bacteriol 195:2959–2970. https://doi.org/10.1128/JB.00147-13
26. Culp EJ, Yim G, Waglechner N et al (2019) Hidden antibiotics in actinomycetes can be identified by inactivation of gene clusters for common antibiotics. Nat Biotechnol 37:1149–1154. https://doi.org/10.1038/s41587-019-0241-9
27. Netzker T, Schroeckh V, Gregory MA et al (2016) An efficient method to generate gene deletion mutants of the rapamycin-producing bacterium *Streptomyces iranensis* HM 35. Appl Environ Microbiol 82:3481–3492. https://doi.org/10.1128/AEM.00371-16

Chapter 10

Engineering Modular Polyketide Biosynthesis in *Streptomyces* Using CRISPR/Cas: A Practical Guide

Jean-Malo Massicard, Li Su, Christophe Jacob, and Kira J. Weissman

Abstract

The CRISPR/Cas system, which has been widely applied to organisms ranging from microbes to animals, is currently being adapted for use in *Streptomyces* bacteria. In this case, it is notably applied to rationally modify the biosynthetic pathways giving rise to the polyketide natural products, which are heavily exploited in the medical and agricultural arenas. Our aim here is to provide the potential user with a practical guide to exploit this approach for manipulating polyketide biosynthesis, by treating key experimental aspects including vector choice, design of the basic engineering components, and trouble-shooting.

Key words CRISPR/Cas, Genome editing, *Streptomyces*, Polyketide biosynthesis

1 Introduction

The *Streptomyces* bacteria are widely recognized as among the most prolific producers of mega-enzyme derived polyketide natural products [1]. Since the discovery in the 1990s of the modular genetic architecture underlying their biosynthesis by polyketide synthases (PKSs), attempts have been made to redirect the pathways towards the generation of novel analogous by genetic engineering. However, the sheer size of the PKS genes (typically 5–50 kbp) and of the overall biosynthetic gene clusters (BGCs, upwards of 150 kbp) made such experiments both laborious and time-consuming [2, 3]. Fortunately, not long after its discovery, multiple laboratories began to adapt the clustered regularly interspaced short palindromic repeats (CRISPR)/CRISPR-associated protein (Cas) system to work in *Streptomyces*—inaugurating a revolution in our ability to "edit" PKS-based biosynthesis.

The most-commonly employed system, CRISPR-Cas9, works by inducing double-stranded breaks within target DNA at a user-defined position (the CRISPR portion acts like a "homing device"

Elizabeth Skellam (ed.), *Engineering Natural Product Biosynthesis: Methods and Protocols*,
Methods in Molecular Biology, vol. 2489, https://doi.org/10.1007/978-1-0716-2273-5_10,

guiding the Cas9 "molecular scissors" to an appropriate cut site). The cell's native homology-based DNA repair mechanisms are then exploited to introduce modifications into the DNA in the region of the break, based on user-supplied DNA "instructions." To date, this system has been utilized in *Streptomyces* to make a wide range of useful modifications to PKS BGCs, including removal of multi-gene fragments or entire pathways [4, 5], and on the single gene level, deletion/inactivation [4, 5], replacement [6], generation of gene fusions (e.g. to tags, reporter genes, etc.), and point mutations [2, 3, 7].

At the most basic level [3, 8, 9], CRISPR-Cas9 requires three functional elements: (1) Cas9, an endonuclease belonging to the bacterial adaptative immune system; (2) CRISPR RNA (crRNA), short (20 bp) RNA transcribed from the spacer sequence of the CRISPR array and complementary to the DNA target cleavage site (called the "protospacer sequence"); and (3) a *trans*-activating crRNA (tracrRNA), a short RNA which hybridizes with the crRNA to form an RNA duplex. In practice, the two RNA molecules are combined into a single-guide RNA (sgRNA), designed to target a specific DNA sequence [7]. The sgRNA binds to Cas9 to form a functional binary complex capable of cleaving DNA to generate double-stranded breaks (DSBs) (*see* **Note 1**).

Repair of the resulting double-strand break can occur via two distinct mechanisms: non-homologous end joining (NHEJ) [10] and homology directed repair (HDR) [11]. NHEJ introduces insertions and deletions at the target site via direct ligation of the two extremities, and thus tends to introduce errors. HDR, on the other hand, employs a homologous sequence to guide the repair– either one already present in the host or a customized sequence introduced by the experimenter (an "editing template"). It is the latter possibility which notably allows for engineering in and around the break region, to introduce desired genetic modifications.

In this context, our aim with this review is to provide the community, and in particular Ph.D. students and postdocs, with a practical guide to effectively using CRISPR/Cas9 and other CRISPR systems to modify PKS BGCs in *Streptomyces*. We will in particular discuss the advantages and disadvantages of this technique, difficulties that may be encountered during the experiments (particularly in the case of editing DNA regions with high sequence similarity, as characteristic of those encoding polyketide synthase multienzymes), and various trouble-shooting approaches. For complementary information concerning the detailed mechanism underlying CRISPR/Cas-based genome editing and its potential for driving natural products discovery ("genome mining"), we refer the reader to recent, expert reviews by Mougiakos et al. [8], Tong et al. [9], and Zhao et al [3].

To aid the reader, we present below a lexicon of key abbreviations:

Biosynthetic Gene Cluster (BGC): sets of operons that encode the biosynthetic enzymes constituting a natural product pathway.

Clustered Regularly Interspaced Short Palindromic Repeats (CRISPR): DNA array of small (30–45 nt) sequences found in bacteria and archaea, usually of foreign origin, which are separated by repeat sequences of similar size.

CRISPR-associated enzyme (Cas): Protein encoded from a *cas* gene generally found close to CRISPR array, which complexes with RNA to form a ribonucleoprotein.

dCas: Catalytically inactive Cas.

CRISPR locus: DNA sequence corresponding to the locus of a CRISPR array, the *cas* gene and in certain cases, ancillary modules.

CRISPR RNA (crRNA): RNA transcribed from the CRISPR locus, composed of the processed transcript of a spacer and a CRISPR array, which is a component of the CRISPR/Cas complex and guides the latter to the target.

Double-strand break (DSB): Simultaneous cleavage of the two DNA strands.

Single-guide RNA (sgRNA): Synthetic RNA sequence comprising fused crRNA and tracrRNA.

Protospacer: DNA sequence identical to a spacer sequence of the corresponding CRISPR array targeted for cleavage by a CRISPR/Cas system.

Protospacer adjacent motif (PAM): Short sequence of 2–8 bp (3 bp in our specific case), allowing the recognition and the targeting of the CRISPR-Cas systems (with the exception of type III).

Spacer: DNA sequence of 30–45 nucleotides derived from target genetic elements, which are located in the CRISPR array and flanked by repeat sequences.

Seed: RNA sequence consisting of the first 8–12 bp of a protospacer immediately adjacent to the PAM sequence; both are critical for a successful targeting.

Non-Homologous End Joining (NHEJ): DSB repair mechanism based on the activities of DNA-binding protein Ku and ligase LigD for the minimal NHEJ, and accessory NHEJ actors in the case of *Streptomyces*, that can introduce errors, including loss of nucleotides.

Homologous Directed Repair (HDR): DSB repair mechanism which relies on homologous recombination, and which is exploited in CRISPR/Cas-based genome editing.

Trans-activating CRISPR RNA (tracrRNA): RNA transcribed from the type II CRISPR loci which is a component of the CRISPR/Cas system.

1.1 Strategic Planning and Key Points

Our recommended general protocol involves the following distinct steps, which are further detailed in the following section.

1. Design the desired final construct in silico.
2. Choose the CRISPR/Cas "tool kit" (i.e., available vector set, Table 1) adapted to the target modification, generate the appropriate spacer (sgRNA) and editing templates, and insert them into the vector.
3. Transform the modified CRISPR/Cas vector into a non-methylating strain of *E. coli* to enable conjugative transfer to *Streptomyces* (importantly, the absence of DNA methylation prevents the *Streptomyces* restriction-modification system from acting on the introduced vector).
4. Conjugate the vector into the target *Streptomyces*, followed by careful screening for the desired genetic events.
5. Remove the vector from the *Streptomyces* via replicons present in the plasmid (e.g. temperature sensitive), which allows for subsequent rounds of modification.

The first step towards modifying the target *Streptomyces* chromosome is to design the desired final construct in silico. Having both the original sequence and the desired final construct as electronic files substantially facilitates the design of oligonucleotides to be used as primers for PCR amplification of the sgRNA and editing template(s), and for PCR screening of the obtained exconjugants. Several bioinformatics tools allow for modifying in silico the vector used for CRISPR/Cas genome editing (*see* below and Table 1 for guidelines as to vector choice). These include the free version of SnapGene (https://www.snapgene.come) and Primer-Blast (https://www.ncbi.nlm.nih.gov/tools/primer-plast) and other online commercial software, for primer design.

During this design process, the researcher must consider the cloning strategy that will be employed to insert the CRISPR/Cas components into the vector (i.e., restriction sites used, possibly including a 5′ arm of homology for Gibson assembly [12], compatible 5′ ends for Golden Gate assembly [13], or specific restriction sites [14]). In terms of the composition and conditions of the PCR reactions, these should be carried out according to the polymerase suppliers' recommendations, while online software is available to aid in identifying appropriate annealing temperatures.

1.2 General Considerations Prior to Adapting the Selected CRISPR/Cas Vector

Table 1 lists all of the vectors currently available for use in genome editing in *Streptomyces*, as well as their key attributes (these are discussed in more detail in the following sections). In choosing among the many options, an important factor is to select a vector which has already been demonstrated to function well in the target *Streptomyces* strain. Based on the experience in our laboratory (ease

Table 1
List of the clustered regularly interspaced short palindromic repeats (CRISPR)/CRISPR-associated (Cas) systems for genome editing in *Streptomyces*. The most recently-developed vectors appear at the end of the list

Editing plasmid	*Streptomyces* selectable marker	Cas protein and strain of origin	Promoter for Cas9 expression	Promoter for sgRNA expression	Plasmid clearance	Addgene reference or laboratory contact	Actinomycetes tested[a]
pCRISPomyces-1	$Apra^R$	Cas9 from *Streptococcus pyogenes*	Constitutive: *rpsL*p(XC)	Constitutive: *gapdh*p (EL)-crRNA and rpsLp(RP)-tracrRNA	Temp-sensitive replicon origin pSG5	61,736 [4]	*S. lividans*
pCRISPomyces-2	$Apra^R$	Cas9 from *Streptococcus pyogenes*	Constitutive: *rpsL*p(XC)	Constitutive: Gapdhp (EL)-sgRNA	Temp-sensitive replicon origin pSG5	61,737 [4]	*S. lividans* *S. viridochromogenes* *S. albus*
pKCas9dO	$Apra^R$	Cas9 from *Streptococcus pyogenes*	Inducible: *tipA*p	Constitutive: *j23199*p-sgRNA	Temp-sensitive replicon origin pSG5	62,552 [16]	*S. coelicolor* M145 *S. pristinaespiralis*
pCRISPR-Cas9	$Apra^R$	Cas9 from *Streptococcus pyogenes*	Inducible: *tipA*p	Constitutive: *ermE**p-sgRNA	Temp-sensitive replicon origin pSG5	125,686 [5]	*S. coelicolor* A3(2)
pCRISPR-Cas9-ScaligD	$Apra^R$	Cas9 from *Streptococcus pyogenes*	Inducible: *tipA*p	Constitutive: *ermE**p-sgRNA	Temp-sensitive replicon origin pSG5	125,688 [5]	*S. coelicolor* A3(2)
pCRISPR-dCas9	$Apra^R$	dCas9 from *Streptococcus pyogenes*	Inducible: *tipA*p	Constitutive: *ermE**p-sgRNA	Temp-sensitive replicon origin pSG5	125,687 [5]	*S. coelicolor A3(2)*
pWHU2653	$Apra^R$	Cas9 from *Streptococcus pyogenes*	Constitutive: *aac(3)IV*p	Constitutive: *ermE**p-sgRNA	Temp-sensitive replicon origin pSG5	Yuhui Sun group [16]	*S. coelicolor* M145

(continued)

Table 1
(continued)

Editing plasmid	*Streptomyces* selectable marker	Cas protein and strain of origin	Promoter for Cas9 expression	Promoter for sgRNA expression	Plasmid clearance	Addgene reference or laboratory contact	Actinomycetes tested[a]
pMWCas9	$Apra^R$	Cas9 from *Streptococcus pyogenes*	Inducible: *tipA*p	Constitutive: *ermE**p-sgRNA	pIJ101	Xudong Qu group [17]	*Sac. erythraea* *S. sp.* AL2110
pQS-gusA	$Apra^R$	Cas9 from *Streptococcus pyogenes*	Inducible: *tipA*p	Constitutive: *ermE**p-sgRNA	pIJ101	Chengzhang Fu group [18]	*S. coelicolor* M145 *Verrucosispora sp.* MS100137
pQS-idgS	$Apra^R$	Cas9 from *Streptococcus pyogenes*	Inducible: *tipA*p	Constitutive: *ermE**p-sgRNA	pIJ101	Chengzhang Fu group [18]	*S. coelicolor* M145 *Verrucosispora sp.* MS100137
pWHU2653-TRMA	$Apra^R$	Cas9 from *Streptococcus pyogenes*	Inducible: *tipA*p	Constitutive: *ermE**p-sgRNA	pIJ101	Xuming Mao group [19]	*S. coelicolor* M145
pKC1139-TRMA	$Apra^R$	Cas9 from *Streptococcus pyogenes*	Inducible: *tipA*p	Constitutive: *ermE**p-sgRNA	Temp-sensitive replicon origin pSG5	Xuming Mao group [19]	*S. coelicolor* M145
pKCCpf1	$Apra^R$	Cpf1 from *Francisella novicida*	Constitutive: *ermE**p	Constitutive: *kasO*p*- crRNA	Temp-sensitive replicon origin pSG5	Yinhua Lu group [20]	*S. coelicolor* M145 *S. hygroscopicus*
pKCCpf1-MsmE	$Apra^R$	Cpf1 from *Francisella novicida*	Constitutive: *ermE**p	Constitutive: *kasO*p*-crRNA	Temp-sensitive replicon origin pSG5	Yinhua Lu group [20]	*S. coelicolor* M145 *S. hygroscopicus*
pSETddCpf1	$Apra^R$	Cpf1 from *Francisella novicida*	Constitutive: *ermE**p	Constitutive:*kasO*p*-crRNA	/	Yinhua Lu group [20]	*S. coelicolor* M145 *S. hygroscopicus*

pCRISPomyces-Sth1Cas9	Apra[R]	Cas9 from *Streptococcus thermophilus*	Constitutive: *rpsL*p(XC)	Constitutive: Gapdhp (EL)- sgRNA	Temp-sensitive replicon origin pSG5	129,552 [21]	*S. albus* *S. sp* NRRL S-244 *S. roseosporus*
pCRISPomyces-SaCas9	Apra[R]	Cas9 from *Staphylococcus aureus*	Constitutive: *rpsL*p(XC)	Constitutive: Gapdhp (EL)- sgRNA	Temp-sensitive replicon origin pSG5	129,553 [21]	*S. albus* *S. sp* NRRL S-244 *S. roseosporus*
pCRISPomyces-FnCpf1	Apra[R]	Cpf1 from *Francisella novicida*	Constitutive: *rpsL*p(XC)	Constitutive: Gapdhp (EL)- sgRNA	Temp-sensitive replicon origin pSG5	129,554 [21]	*S. albus* *S. sp* NRRL S-244 *S. roseosporus*
pSET-dCas9	Apra[R]	dCas9 from *Streptococcus pyogenes*	Constitutive: *ermE**p	/	/	110,183 [22]	*S. coelicolor* M145
pSET-dCas9-actII-4-NT-S1	Apra[R]	dCas9 from *Streptococcus pyogenes*	Constitutive: *ermE**p	Constitutive: j23119-sgRNA	/	110,185 [22]	*S. coelicolor* M145
pCRISPR-cBEST	Apra[R]	Cas9n from *Streptococcus pyogenes*	Inducible: *tipA*p	Constitutive: ermEp*-sgRNA	Temp-sensitive replicon origin pSG5	125,689 [23]	*S. coelicolor* A3(2) *S. collinus* *S. griseofuscus*
pCRISPR-aBEST	Apra[R]	Cas9n from *Streptococcus pyogenes*	Inducible: *tipA*p	Constitutive: ermEp*-sgRNA	Temp-sensitive replicon origin pSG5	131,464 [23]	*S. coelicolor* A3(2) *S. collinus* *S. griseofuscus*
pKC-dCas9-CDA-ULstr	Apra[R]	dCas9 from *Streptococcus pyogenes*	Inducible: *tipA*p	Constitutive: j23119-sgRNA	Temp-sensitive replicon origin pSG5	Yinhua Lu group [24]	*S. coelicolor* M145 *S. rapamycinicus*
pCMU-4	Apra[R]	Cas9 from *Streptococcus pyogenes*	Inducible: *D4-TheoR*	Constitutive: Gapdhp (EL)-sgRNA	Temp-sensitive replicon origin pSG5	Takano group [25]	*S. lividans* *S. coelicolor* M145
pCMU4.4	Apra[R]	Cas9 from *Streptococcus pyogenes*	Constitutive: *ermE*p*	Constitutive: Gapdhp (EL)-sgRNA	Temp-sensitive replicon origin pSG5	Takano group [25]	*S. lividans* *S. coelicolor* M145

[a] Actinomycetes presented in the table are those which were tested in the original publication

of use, success rate, etc. [15]), we recommend pCRISPomyces-2 [4] and pCRISPR-Cas9 [5], but particular experimental constraints may direct you to an alternative, more recently-developed system.

1.3 Expression of the Cas9

In *Streptomyces*, pyogenes [25, 26], and therefore the majority of the discussion that follows will focus on Cas9. Although *Streptomyces* and *Streptococcus* are both genera of Gram-positive eubacteria, they are significantly divergent in terms of phylogeny. Indeed, *Streptomyces* has high genomic G+C content (often over 70%), whereas *Streptococcus* has a low average G+C content (ca. 41%). Moreover, analysis of *Streptococcus pyogenes* genes has revealed the presence of rare codons, including several *bldA* codons which are translation-level regulators of specialized metabolites in *Streptomyces* species. Therefore, the *cas9* gene used in CRISPR-Cas9 vectors has been optimized to take in account the *Streptomyces* codon bias, thereby ensuring the correct expression level of Cas9 in *Streptomyces* species. In addition to this *Streptomyces*-adapted Cas9, several other CRISPR-Cas9-derived technologies are available, enriching the toolbox for modifying *Streptomyces* genomes. For example, Cpf1 derived from *Francisella novicida* [20], which is a Class 2 type V Cas enzyme, has also been successfully used for genome editing of *Streptomyces* [21]. This alternative system notably provides the option of using a T-rich PAM (TTTN for Cpf1, as opposed to NGG for Cas9), in cases where Cas9 exhibits low efficacy.

The *cas9* gene (codon-optimized for *S. coelicolor*) is generally placed under the control of a strong constitutive promoter. Promoters used to date include *ermE*p* (as in the pCMU4.4 and pKCCpf1 editing plasmids, for example), which is a commonly used promoter for heterologous expression in *Streptomyces* [27], as well as *rpsL*p(XC) as in pCRISPomyces-1 and -2, which has been employed successfully to drive high-level heterologous expression of BGCs in various *Streptomyces* including *S. albus* J1074, *S. venezuelae* ISP5230, *S. coelicolor* M1146, and *S. avermitilis* SUKA16 [2]. Alternatively, gene expression can be placed under the control of an inducible promoter, such as *tipAp* [28] (as in the pCRISPR-Cas9 and pMWCas9 editing plasmids), which has also been widely used for heterologous expression of *Streptomyces* genes. Notably, inducible expression can increase the chances of obtaining exconjugants before Cas9 is rendered active [2]. Nevertheless, one important limitation with this system is the requirement for a copy of the thiostrepton responsive activator *tipA* in the target *Streptomyces* genome [29]. Another option is the theophylline riboswitch (TheoR) [30], which allows for inducing expression of the *cas9* gene using theophylline [25] as in the pCMU and pCMU-4 editing plasmids. The *theoR* gene encodes the elongation factor Tu (EF-Tu), which is an essential protein involved in protein translation [31].

Based on the cumulative results obtained with these vectors, a consensus has emerged that moderate expression of Cas9 is desirable, is achievable with the inducible promoters and varying concentrations of inducer, as Cas9 over-production can be cytotoxic [5]. As mentioned earlier, the inducible systems offer the additional advantage of allowing the transformation and editing steps to be separated (i.e., transformation takes place in the absence of inducer), which can be critical if the conjugation efficiency is low.

1.4 Design and Expression of sgRNA

Together, the Cas9/sgRNA complex targets a 20 nt sequence at the 3′ end of which is the PAM motif. The short length of the recognition sequence can be problematic, as if it is not unique in the genome, then Cas9 may cut at alternative chromosomal loci, leading to potentially detrimental off-target effects. This is particularly problematic in the case of *Streptomyces* with their high G+C content, which intrinsically increases the chances that two short sequence regions will be identical. Fortunately, several bioinformatics tools/websites have been developed to help with sgRNA design, including the free online software CRISPy-web (https://crispy.secondarymetabolites.org/#/input) [32, 33]. This tool notably allows for maximizing the editing efficiency at the selected ("on-target") site. On a practical level, the user uploads the genome sequence of the target *Streptomyces* (as a Genbank file or an antiSMASH job ID [34]), and specifies the target to be edited. The result is an output page containing a graphic interface which reveals the scanned region. Genes are shown and sgRNAs on the forward and reverse strands are displayed below the genes. A table provides further details on the identified sgRNAs, ranked by the least number of potential off-target matches. For more detail on the software, the reader is referred to the original description of the CRISPy-web software [32, 33]. Based on our experience, we recommend experimentally testing at least two sgRNAs for each target genetic modification within a *Streptomyces* genome (*see* **Note 2**).

Finally, it is noteworthy that sgRNAs can be employed in *Streptomyces* to achieve gene repression. This strategy, called "CRISPR interference (CRISPRi)" is based on an inactive form of Cas, either dCas9 [a nuclease-deficient Cas9 bearing two mutations (D10A and H840A)] [22] or ddCpf1 (a nuclease-deficient Cpf1 bearing an E1006A mutation) [20] (Table 1). By disabling the catalytic activity of Cas9, the binary sgRNA/Cas complex still binds to the target DNA, but the effect is to silence the gene rather than to cleave/modify it.

1.5 Design of Editing Templates for Deletion, Insertion and Mutation

In order to exploit the homologous repair mechanism for genome editing, a 2 kbp editing template must be included within the engineering vector. Critically, this template should contain DNA regions homologous to the target genome [homology "arms" (HR)] both upstream and downstream of the introduced double-

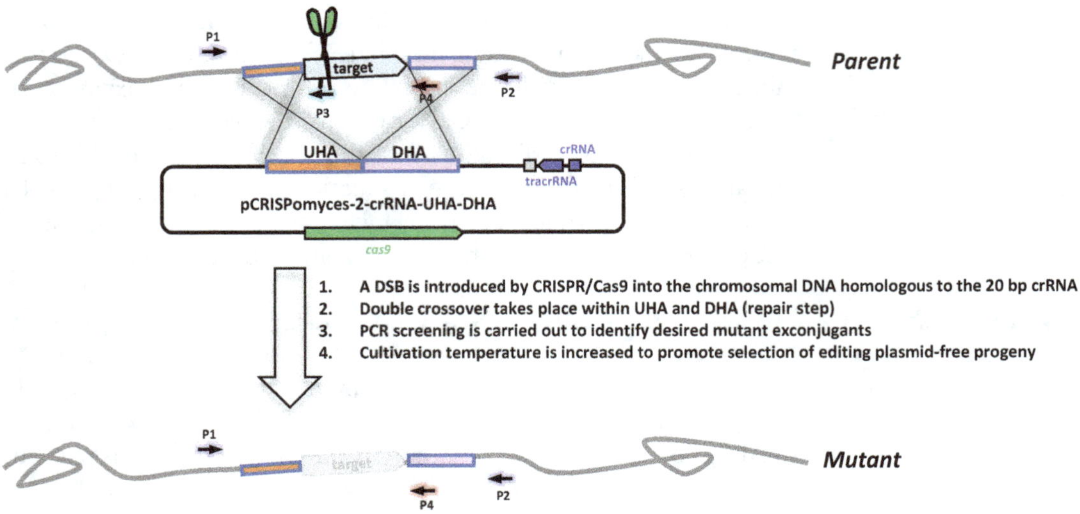

Fig. 1 Strategy for gene deletion using CRISPR/Cas9. In this model, a target-specific sgRNA cassette (crRNA +tracrRNA) is used to guide the endonuclease (Cas9, green scissors) to it target. The editing template is composed of the two homology arms that flank the target locus to be removed [UHA: upstream homology arm (orange) and DHA: downstream homology arm (pink)]. The suggested locations of primers (P1–P4) to confirm the deletion in the target genome by PCR-screening and sequencing are shown. Specifically, we use two sets of primers for PCR screening: P1 (that anneals outside of the 1 kb homology arms) and P4 (that anneals within the DHA) leading to a band smaller than 1.5 kbp in size from the desired strains, and P1 and P3 (that anneal within the deleted region), which result in a 1.1 kbp band, but only in the presence of parental genomic DNA. Primers P1 and P2, which are located at a distance of 150–500 nucleotides from the homology arms within the genomic DNA, can be used to sequence the edited chromosomal region

strand break (1 kbp on each side). The choice of which particular arms to use depends on the specific genome editing strategy to be carried out. In our laboratory, a central goal is to generate defined structural analogues of high-value polyketides, by targeting specific regions of the polyketide synthase-encoding genes (functional "domains" and "modules"). For this, we have successfully adapted the existing CRISPR-Cas9 vector systems to carry out not only gene deletions/inactivations, but fusions, exchanges, and point mutations [15].

Gene deletion is the strategy that has been most actively pursued to date in *Streptomyces* [3]. To achieve this type of modification, the homology arms must match regions upstream and downstream of the locus to be eliminated (1 kbp each) (Fig. 1). In this way, we were notably able to remove DNA fragments ranging from 0.44 kbp to 36 kbp in the chromosome of *S. ambofaciens* [15]. Although it has been reported that it is possible to introduce several modifications simultaneously (i.e., by using multiple sgRNAs, a method referred to as multiplex targeting [4, 24, 35, 36]), in our experience, the same result can be obtained rapidly by introducing multiple, consecutive changes.

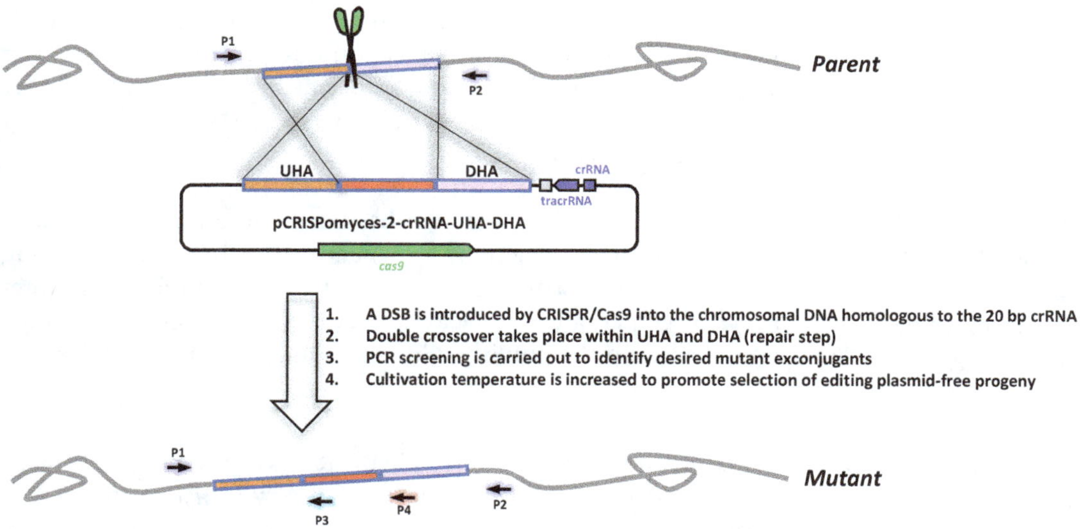

Fig. 2 Strategy for gene fusion using CRISPR/Cas9. In this model, a target-specific sgRNA cassette (crRNA +tracrRNA) is used to guide the endonuclease (Cas9) to it target (green scissors). The editing template is composed of the two homologous arms (UHA and DHA) flanking the target locus, fused to the gene of interest (Red box) to be inserted. The suggested locations of primers (P1–P4) to confirm the insertion in the target genome by PCR-screening and sequencing, are shown. Specifically, we used two sets of primers for PCR screening: P1 (that anneals outside of the 1 kb homology arms) and P3 (that anneals within the insertion region) leading to a band from the desired strains, and P1 and P4 (that anneal within the DHA) producing a band from both the parental and the desired strain's genomic DNA, but of different sizes. Primers P1 and P2, which are located 150–500 nucleotides away from the homology arms within the genomic DNA, can be used to sequence the edited chromosomal region

It is also possible to fuse a target sequence to another gene (for example, encoding a fluorescent marker to allow visualization *in cellulo*) (Fig. 2). In this case, the editing template contains the sequence to be inserted sandwiched between the two homology arms. Depending on whether the sequence is to be fused at the 5′ or 3′ end of the target gene, it will then constitute a part of the upstream or downstream homology arm. In terms of the Cas cut site, it needs to be just adjacent to either the 5′ or 3′ end of the gene to be modified, while respecting the constraints of the required PAM sequence. One difficulty encountered with this strategy is that the Cas9 target is also present within the editing template. To avoid unwanted cleavage, it is therefore necessary to modify the sequence in the editing template (which corresponds to the sgRNA) to alter its codon usage but without introducing changes into the encoded protein sequence, so that the edited chromosome is no longer a substrate for Cas9. This can be conveniently achieved by employing a small synthetic DNA fragment covering this region during assembly of the editing template (whereas the remainder of the template can be obtained by PCR).

In the case of site-directed mutagenesis, although it has been reported that this type of modification can be achieved in one step [37, 38], in our hands it is more efficient to first delete the target gene sequence, and then re-introduce a mutated version containing the desired alteration [15].

1.6 Transfer of CRISPR Plasmids into the Target Streptomyces by Intergeneric Conjugation

Once the CRISPR-Cas9 system has been chosen and the sgRNA (s) as well as the editing template assembled in the vector following the developers' instructions, the next step is to modify the target *Streptomyces* genome. Subheading 3 details our recommended protocol, based on the published literature and our own experience.

1.7 Potential Issues with the pSG5 Replicon and Revertants

Although the introduction of CRISPR/Cas systems has substantially improved genome engineering of *Streptomyces*, unexpected recombination events have been reported [17]. These have been attributed, at least in some cases, to the temperature-sensitive pSG5 replicon used for plasmid curing after double crossover [41], which is present in the majority of current CRISPR-based plasmids (Table 1). For example, Mo et al. [17] reported a case in which the pCRISPR-Cas9 plasmid has been integrated into the genome of *Streptomyces* indicating that both gene deletion in the targeted region and unpredicted recombination events occurred [17]. The origin of the problems was traced to the pSG5 replicon, and indeed this element was discovered serendipitously to provoke high levels of spontaneous recombination within the rapamycin and tylosin PKSs [6]. This problem can be addressed by replacing pSG5 with a segregationally unstable pIJ101 replicon, as in pMWCas9 [17]. Although the precise mechanism by which pSG5 induces recombination is currently unclear, it has been proposed that the elevated temperature (37 °C) required to prevent the replication of pSG5 and necessary for plasmid clearance may provoke a bacterial emergency response, and with it, induction of DNA repair systems, leading to homologous recombination events between genes with high sequence similarity.

While editing the stambomycin PKS in *S. ambofaciens*, we also obtained a "wild-type revertant-like" mutant, using the pCRISPomyces-2 toolkit (Fig. 3) [15]. Using a crRNA sequence targeted to the ketosynthase (KS) domain of module 22 of the PKS, a "mid" fragment of approximately 600 bp containing the designed mutation and two nearly 1 kbp arms homologous to the regions flanking "mid" (UHA and DHA), were sequentially inserted into the pCRISPomyces-2 vector to generate the functional pCRISPomyces-2-crRNA16-UHAmidDHA vector. Using the standard protocol described below, several colonies were isolated for PCR screening, yielding candidates which contained the desired point mutations (Fig. 3). Thereafter, the cultivation temperature was

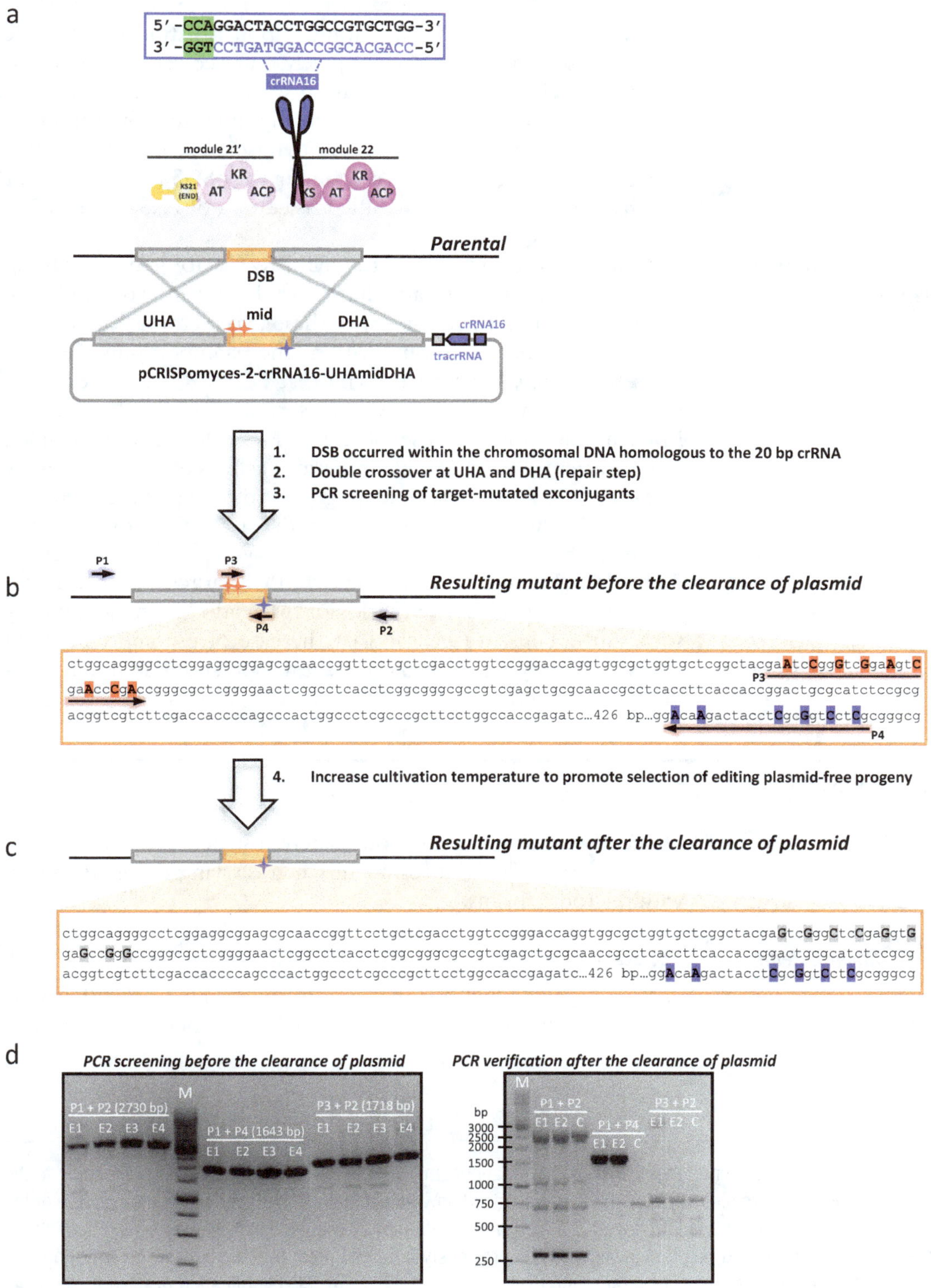

Fig. 3 Editing the stambomycin biosynthetic gene cluster in *S. ambofaciens* using a CRISPR-Cas9 system containing a pSG5 replicon. (**a**) In order to introduce point mutations (red stars) into the acyl carrier protein (ACP) domain of module 21 using the pCRISPomyces-2 toolkit, an sgRNA was designed against the KS domain of module 22. (**b**) PCR amplification using adapted primer pairs identified exconjugants (E1–E4) which

increased to 37 °C, coupled with screening for apramycin sensitivity following several rounds of plating. Upon isolation of genomic DNA and sequencing of the putatively mutated region, we found that the previously-verified mutation within the "mid" fragment had reverted to wild-type, presumably due to pSG5-induced recombination. Given that this is not an isolated observation, it is increasingly clear that vectors containing the pSG5 replicon are not optimal for carrying out CRISPR-Cas editing of PKS genes [17].

1.8 Comparison to PCR Targeting

A commonly used alternative to CRISPR/Cas for genome editing in *Streptomyces* is PCR targeting [44]. In recent work [15] we directly compared the efficiency of these two methods for generating comparable mutations within the stambomycin PKS in *S. ambofaciens*. Globally, we found that CRISPR/Cas-based engineering offered multiple advantages relative to PCR targeting. Most importantly, the use of CRISPR/Cas systems was much more efficient, with the target mutations achieved in approximately half the time. It can also be leveraged for direct genome modification, while PCR targeting requires the availability of a cosmid spanning the target and flanking regions. We also showed that the 33 bp "scar" sequence remaining after PCR targeting, when introduced on its own, reduced yields of the wild-type metabolites by some 30%. Thus, we believe that the ease of use and efficacy of CRISPR/Cas systems are likely to see it supplanting PCR targeting in the longer term.

2 Materials

Diligently follow all regulations/safety instructions of your institute/university/local government when carrying out experiments with microorganisms.

2.1 General Items

1. Agarose.
2. DNA gel loading dye (6×): 10 mM Tris–HCl pH 8.0, 1 mM EDTA, 25% glycerol (v/v), 0.12% Bromophenol blue (w/v).

Fig. 3 (continued) contained the desired point mutations. In addition, the DNA sequence corresponding to the selected crRNA16 was also mutated (blue star) to avoid unwanted cleavage by Cas9 within the genome of the newly-obtained mutants; the introduced changes were confirmed by sequencing (residues highlighted in blue and red). (**c**) Sequencing result showing that the previously-obtained point mutations have reverted to wild-type (highlighted in grey) after several round of cultivation at 37 °C. (**d**) PCR verification of the exconjugants before and after clearance of the CRISPR plasmid. Clearly, the primer pair P3/P2 does not work after the plasmid clearance, consistent with the sequencing result shown in (**b**). *DSB* double-strand break; *UHA/DHA* up-/downstream homologous arms; *P* primer; *E* exconjugant; *M* marker; *C* control strain (the parental strain)

3. DNA polymerases, e.g., DreamTaq with green buffer, Phusion High-Fidelity DNA polymerase with GC Buffer.
4. Solvents: ethanol (absolute), ethanol (75% (v/v)), hydrochloric acid (HCl), chloroform, isopropanol.
5. GeneRuler 1-kb DNA ladder (Thermo Fisher Scientific).
6. Glycerol ≥99.5% (w/v).
7. Liquid nitrogen.
8. 50 mg/mL Lysozyme.
9. 10 mg/mL RNase A.
10. Gel and PCR Clean-up Kit, e.g., NucleoSpin.
11. 20 mg/mL proteinase K.
12. 10% SDS.
13. 5 M NaCl.
14. 25 mg/mL chloramphenicol stock solution. The powder is dissolved in absolute ethanol (*see* **Note 3**).
15. 50 mg/mL kanamycin stock solution. The powder is dissolved in Milli-Q filtered water (*see* **Note 3**).
16. 25 mg/mL nalidixic acid stock solution. Nalidixic acid is poorly soluble in distilled water. We advise dissolving the powder in 0.1 M NaOH (NaOH is corrosive and thus must be manipulated with caution) (*see* **Note 3**).
17. Antibiotic to select for CRISPR plasmid (*see* Table 1).
18. Oligonucleotides (specific to the gene being mutated).

2.2 Media and Buffers

1. LB broth: 10 g/L of tryptone, 5 g/L of yeast extract, and 5 g/L of NaCl. Dissolve the powder in distilled water. Adjust the pH to 7.0 with 1 M NaOH before autoclaving (121 °C, 20 min). To make LB plates, add 17 g/L of agar before sterilization.
2. 2YT broth: 16 g/L of tryptone, 10 g/L of yeast extract, and 5 g/L of NaCl. Dissolve the powder in distilled water. Adjust the pH to 7.0 with 1 M NaOH before autoclaving (121 °C, 20 min).
3. SFM plates: 20 g/L fat-reduced soy flour, 20 g/L of D-mannitol, and 20 g/L of agar) dissolved in tap water and supplemented with 10 mM of $MgCl_2$ for *S. coelicolor* or *S. ambofaciens.*
4. ISP2 plates: 4 g/L of yeast extract, 10 g/L of malt extract, and 5 g/L of dextrose) for many *Streptomyces* strains, or TSA plates (17 g/L of tryptone, 3 g/L of soytone, 2.5 g/L of glucose, 5 g/L of NaCl, 2.5 g/L of K_2HPO_4) for *S. ambofaciens.*
5. ISP2 broth:4 g/L of yeast extract, 10 g/L of malt extract, and 5 g/L of dextrose. Dissolve the powder in distilled water. Adjust the pH to 7.0 with 1 M NaOH before autoclaving (121 °C, 20 min) (*see* **Note 4**).

6. 0.5 M EDTA: Dissolve the powder in an appropriate volume, mix and adjust the pH to 8.0 using NaOH (1 N). Top up the solution to a desired final volume and autoclave it at 121 °C for 20 min. Autoclaved EDTA solution can be stored at 4 °C for up to 3 months.
7. 50× TAE buffer: For 1 L, weigh 242 g of Tris base and dissolve it in ~800 mL of Milli-Q filtered water, mix and add 57.1 mL of 100% acetic acid and 100 mL of 0.5 M EDTA (pH 8.0). Adjust the solution to a final volume of 1 L. To make a 1× TAE buffer, mix 20 mL of 50× TAE buffer (v/v) with 980 mL of Milli-Q filtered water. Both 50× TAE and 1× TAE buffers can be stored at room temperature for up to 6 months.
8. SET buffer: 75 mM NaCl, 25 mM EDTA pH 8.0, 20 mM Tris–HCl, pH 7.5.

2.3 Equipment

1. 8-strip PCR tubes, cryotubes, falcon tubes, flat-cap well-sealing strips, microcentrifuge tubes, serological pipettes, syringe with non-absorbent cotton filter.
2. Aluminum foil, bacteriological petri dishes, parafilm, scissors, wooden toothpicks, tweezers.
3. Benchtop centrifuge, desktop microcentrifuge, minicentrifuge.
4. BlueCap bottles with blue GL45 lids.
5. Electroporator.
6. Thermal cycler, gel electrophoresis tank, and power source.
7. Heating block.
8. Incubator.
9. Laminar flow hoods.
10. Magnetic stirrer and magnetic stir bar.
11. Milli-Q water system.
12. Molecular imager (Gel Doc + Gel Documentation System with Image Lab software).
13. OD cuvettes and Optical density (OD) meter.
14. Pipetboy.
15. pH meter.
16. Shaker and shake flasks.
17. UV spectrophotometer.
18. Vortex.

2.4 Strains and Plasmids

1. *E. coli* ET12567 (pUZ8002).
2. *Streptomyces* strain of interest.
3. CRISPR plasmid of choice (*see* Table 1).

3 Methods

3.1 Preparation of E. coli *ET12567 (pUZ8002) Conjugation Donor Strain*

3.1.1 Preparation of Electroporation-Competent Cells

1. Inoculate the donor strain *E. coli* ET12567 (pUZ8002) from frozen glycerol stock or a single colony into 5 mL LB medium supplemented with 25 μg/mL chloramphenicol and 50 μg/ mL kanamycin and incubate at 37 °C overnight with 200 rpm shaking (*see* **Note 5**).
2. Add 0.5 mL of the overnight culture to 25 mL of LB medium supplemented with 25 μg/mL chloramphenicol and 50 μg/ mL kanamycin in a 250 mL Erlenmeyer flask.
3. Place the Erlenmeyer flask into a 37 °C shaking incubator and grow cells until the OD_{600nm} reaches ≃0.4 (generally 2–4 h are sufficient). Initially, we advise measuring the OD_{600nm} every hour until it reaches ≃0.3 and every 15 min thereafter until it reaches ≃0.4 (*see* **Note 6**).
4. Once an OD_{600nm} value of ≃0.4 is reached, transfer the culture to a chilled 50 mL centrifuge tube, and incubate on ice for 30 min with occasional tube inversion.
5. Centrifuge the cells for 7 min at 4000 × *g* at 4 °C and carefully remove all of the supernatant. You can use a micropipette equipped with a 1000 μL tip to remove as much liquid as possible (*see* **Note 7**).
6. Add 25 mL ice-cold 10% (v/v) glycerol to the cell pellet in the bottom of the centrifuge tube and gently resuspend cells with a micropipette equipped with a 1000 μL tip. Add an additional 25 mL 10% (v/v) glycerol to the centrifuge tube, seal and carefully invert to mix. Centrifuge tubes again for 7 min at 4000 × *g*, and carefully remove all of the supernatant as in **step 5**.
7. Add 1 mL ice-cold 10% (v/v) glycerol and then transfer cells to a 1.5 mL microcentrifuge tube (*see* **Note 8**).
8. Centrifuge tubes for 5 min at 4000 × *g* at 4 °C. Carefully remove the supernatant and resuspend the pellet in 250 μL ice-cold 10% (v/v) glycerol, which will provide 4–5 electroporations. Divide this suspension into 50 μL aliquots in ice-cold 1.5 mL microfuge tubes, and incubate on ice until the next step (*see* **Note 9**).

3.1.2 Electroporation of E. coli *ET12567 (pUZ8002) with the CRISPR Plasmid*

1. Chill on ice one electroporation cuvette per transformation (*see* **Note 10**).
2. In a 1.5 mL microcentrifuge tube on ice (from the previous step), mix 10–50 ng of salt-free CRISPR plasmid with 50 μL of electroporation-competent cells.
3. Introduce the plasmid DNA into the cells by electroporation (*see* **Note 11**).

4. Immediately after electroporation, add 1 mL LB medium to the cuvette using a micropipette equipped with a 1000 μL tip and incubate the tubes for at least 2 h with shaking at 37 °C.
5. Plate several 1:10 dilutions of the electroporation mix using LB solid medium supplemented with 25 μg/mL chloramphenicol, 50 μg/mL kanamycin and the appropriate antibiotic for the CRISPR plasmid.
6. Incubate plates at 37 °C for at least 16 h.

3.2 Preparation of the Streptomyces Conjugation Acceptor Strain

1. Plate the *Streptomyces* strain onto sporulation plates and incubate several days at 30 °C (*see* **Note 12**).
2. Add 10 mL of distilled water (*see* **Note 13**).
3. Scrape the surface of the culture gently with a sterilized cotton swab (*see* **Note 14**).
4. Aspirate the spore suspension using a sterile pipette.
5. Place the spore suspension in a 50 mL centrifuge tube containing a home-made filter system (*see* **Note 15**).
6. Pour the filtered spore suspension into a centrifuge tube and centrifuge at 2000 × *g* for 10 min at room temperature. Carefully and rapidly remove the supernatant and discard it (*see* **Note 16**).
7. Resuspend the spores from a well-sporulating plate in 1–2 mL of 2YT or sterile water in a 15 mL centrifuge tube (*see* **Note 17**).

3.3 Intergeneric Conjugation

1. Use an *E. coli* ET12567 (pUZ8002) sample from the previous step to inoculate 3–5 mL of LB medium supplemented with 25 μg/mL chloramphenicol, 50 μg/mL kanamycin and the appropriate antibiotic for the chosen CRISPR plasmid, and incubate overnight at 37 °C with 200 rpm shaking.
2. The following day, use 0.5 mL from the previous preculture to inoculate 25 mL of LB medium supplemented with 25 μg/mL chloramphenicol, 50 μg/mL kanamycin and the appropriate antibiotic for the CRISPR plasmid, and incubate at 37 °C with 200 rpm shaking until the culture reaches an OD_{600nm} of 0.3–0.5.
3. Transfer the *E. coli* ET12567 (pUZ8002) culture into a 50 mL centrifuge tube and centrifuge it at 4000 × *g* for 7 min at room temperature. Carefully decant the supernatant and resuspend the pellet in 25 mL of LB medium. Repeat this step twice and resuspend the final pellet in 2.5 mL LB medium (*see* **Note 18**).
4. Mix 500 μL *E. coli* ET12567 (pUZ8002) suspension from the previous step with 100 μL heat-shocked spores in a 1.5 mL microcentrifuge tube, by inversion (*see* **Note 19**).

5. Split the mixture onto several sporulation plates containing 10 mM $MgCl_2$ (for example plate 400 μL, 150 μL, and 50 μL and complement with 100 μL of LB medium to achieve proper spreading), then air-dry the plates in a laminar flow hood for 5–10 min.
6. Incubate the plates at 30 °C for 16–24 h.
7. The following day, overlay the conjugation plates with the appropriate antibiotic for the CRISPR plasmid and 50 μg/mL nalidixic acid (to kill the *E. coli*) and incubate for 3–5 days at 30 °C (*see* **Note 20**).
8. When the aerial mycelia of exconjugants appear, pick 100 with a sterilized wooden toothpick and transfer them to a fresh sporulation plate supplemented with the antibiotic for the CRISPR plasmid and 50 μg/mL of nalidixic acid. Incubate for 2–5 days at 30 °C.

3.4 Genotype Screening by Colony PCR

1. Scratch 0.1 cm^2 of mycelia from the previous plates using a sterile wooden toothpick, and transfer it to 10 μL pure DMSO in a PCR tube (*see* **Note 21**).
2. Incubate for 10 min at 100 °C using a heating block, and then shake vigorously (*see* **Note 22**).
3. Centrifuge at maximum speed for 10 s to pellet the mycelial debris (*see* **Note 23**).
4. Use 1 μL of the resulting solution as template DNA for colony PCR. The primers used depend on the engineering strategy (*see* recommendations in Figs. 1 and 2).
5. Perform the PCR as follows (*see* **Note 24**):

Components	Volume (μL)	Phase	Time	Temperature	Number of cycles
DNA template	1	Initial denaturation	2 min	95 °C	1
dNTP (1.25 mM each)	4	Denaturation	30 s	95 °C	30
Oligodeoxyribonucleotide Forward (10 mM)	0.5	Hybridization	30 s	X °C	
Oligodeoxyribonucleotide Reverse (10 mM)	0.5	Extension	XX min	72 °C	
DMSO (100%)	1.25	Final extension	10 min	72 °C	1
Dream Taq green buffer (10×)	2.5				
Dream Taq (1.25 U/μL)	0.5				
ddH_2O	13.75 μL				
Total volume	25 μL				

X: *The hybridization temperature depends on the primer pair used and is easily determined using the supplier's software*
XX: *The elongation time depend on the size of the amplified fragment (for example, DreamTaq requires 1 min/kbp)*

6. Analyze 10 μL of the above PCR reaction on an agarose gel (1% w/v) with 1× TAE buffer. Make sure to include a marker to compare the size of the fragments obtained. Run the agarose gel at a voltage of 1–5 V/cm (distance between the electrodes) for 45 min (*see* **Note 25**).
7. After the electrophoresis, visualize the bands using a Gel doc system and the adapted software.
8. If bands of the expected size are obtained, extract them from the gel using a Gel extraction kit following the manufacturer's protocol.
9. Measure the concentration and evaluate the quantity of each extracted DNA fragment using a spectrophotometer.
10. Check the extracted fragments by sequencing, using the same primers as for the colony PCR (*see* **Note 26**).

3.5 Curing CRISPR Plasmids Containing a Temperature-Sensitive Replicon

Most of the plasmids (Table 1) incorporate the heat-sensitive pSG5 origin of replication, and therefore can be eliminated from the host *Streptomyces* by increasing the growth temperature. The plasmid can be eliminated by growing the edited *Streptomyces* strains either on solid or in liquid medium, as we do not observe any significant difference in the efficiency of plasmid curing between these two strategies. As elimination in liquid culture has already been described several times in the literature [4, 5], the following is a protocol for curing onto solid medium.

1. Pick the mycelium of edited *Streptomyces* strains with a sterilized wooden toothpick, and transfer it onto a non-selective and a selective plate, so that single colonies are obtained, and incubate for 2–5 days at 37 °C (*see* **Note 27**).
2. Repeat this step until clones unable to grow on a selective medium are obtained (*see* **Note 28**).

3.6 Genotype Confirmation

3.6.1 Isolation of *Streptomyces* Genomic DNA

In order to confirm results obtained by colony PCR and to sequence the engineered region, it is necessary to isolate the *Streptomyces* genomic DNA. The following small-scale method has been proven to be effective for almost all *Streptomyces* strains in our laboratory, and the isolated genomic DNA is reliable for most genetic manipulations, including amplification of genes-of-interest, digestion, and large fragment cloning (ca. 10 kbp). However, the quality of this genomic DNA is insufficient to use it for sequencing purposes. In this case, the use of the Wizard genomic DNA extraction kit (Promega) is highly recommended.

1. Grow *Streptomyces* mycelium in 10 mL of an optimum culture medium for the *Streptomyces* strain used by inoculating 10 μL spore suspension and shaking at 200 rpm, 28 °C/30 °C for 2–3 days (*see* **Note 29**).

2. Remove 5 mL of the 10 mL culture and centrifuge at 10,000 × *g* for 2 min to harvest the mycelia (ca. 50–100 mg).
3. Wash the mycelia twice with 1 mL of distilled water, resuspend in 1 mL SET buffer (NaCl 75 mM, EDTA pH 8 25 mM, Tris–HCl 20 mM pH 7.5) and mix by vortexing for 20 s.
4. Add 20 μL of lysozyme (50 mg/mL) and incubate the mixture at 37 °C in a water bath for 0.5–1 h or until the cell suspension become slightly viscous due to cell wall degradation.
5. Add 10 μL RNase A (10 mg/mL) and mix by inversion. Incubate the mixture at 37 °C for 0.5 h.
6. Add 30 μL high quality Proteinase K (20 mg/mL) and 150 μL 10% SDS, and mix by inversion. Incubate the mixture in a 55 °C water bath for 40–60 min (until the mixture become transparent).
7. Add 400 μL 5 M NaCl and let the tube cool down at room temperature.
8. Add 1 mL chloroform to denature proteins in the lysis mixture, followed by mixing 30 min in a rotary shaker and centrifugation at 10,000 × *g* for 10 min.
9. Transfer the top layer to a new tube and add 0.6 volumes of isopropanol. Mix by inversion until the appearance of white filamentous precipitates, followed by centrifugation at 10,000 × *g* for 2 min. Decant the tube and rinse the DNA twice with 500 μL 75% ethanol.
10. Decant the tube again (using a pipette to remove residual ethanol solution if needed), dry the tube at room temperature on the bench for 5–15 min, and then finally add 200 μL distilled water.
11. Incubate the tube in a 55 °C water bath for 10 min, mix by inversion, and cool to room temperature.
12. Measure the concentration of the DNA using a UV spectrophotometer (NanoDrop, Thermo Fisher Scientific) and then store at −20 °C.

3.6.2 PCR Amplification of the Edited Genomic DNA

To complete the genetic modification process and to proceed to subsequent experiments with the obtained genetically modified strains (*notably due to the genetic variability from one generation to another of* Streptomyces *and even within the same generation* [39, 40]), it is necessary to again verify by PCR the genetic modifications are present. At this stage we employ a high-fidelity polymerase to limit errors during PCR amplification, and use the extracted genomic DNA as a template.

1. Perform the PCR as follows (*see* **Note 30**):

Components	Quantity	Phase	Time	Temperature	Number of cycles
DNA template	80–100 ng	Initial denaturation	2 min	98 °C	1
dNTP (1.25 mM each)	6	Denaturation	30 s	98 °C	30
Oligodeoxyribonucleotide Forward (10 mM)	0.5	Hybridization	30 s	X °C	
Oligodeoxyribonucleotide Reverse (10 mM)	0.5	Extension	XX min	72 °C	
DMSO (100%)	2.5	Final extension	10 min	72 °C	1
High-fidelity polymerase GC rich buffer (5×)	5				
High-fidelity polymerase (2 U/μL)	0.5				
ddH_2O	10 μL				
Total volume	25 μL				

X: The hybridization temperature depends on the primer pair used and is easily determined using the supplier's software.
XX: The elongation time depend on the size of the fragment.

2. Analyze the DNA fragments using the same protocol as for colony PCR, and confirm by sequencing.

4 Notes

1. An in-built safety mechanism ensures that Cas9 only cuts at DNA sequences adjacent to so-called protospacer adjacent motifs (PAMs), thereby avoiding cuts within its own CRISPR array. This feature must be accounted for during experimental design.
2. Each plasmid system has its specific requirements for the insertion of sgRNAs, and therefore there is no universal protocol. As a consequence, the corresponding relevant primary publication should be consulted in each case (Table 1).
3. For the antibiotics used, each powder should be dissolved in the appropriate solvent, mixed/vortexed and then filtered through a 0.2-μm filter in a laminar flow hood, and then aliquoted into sterilized 1.5 mL tubes. The antibiotic stock can be stored at −20 °C for up to 1 year.
4. To ensure selection pressure, the media are cooled to a temperature of 50–60 °C and are supplemented with suitable antibiotics.

5. Plasmid pUZ8002 is an RK2 derivative with a defective *oriT*, which is not self-transmissible but supplies mobilization functions for *oriT*-containing plasmids [42].
6. It is essential not to exceed an OD_{600nm} of 0.4.
7. We recommend starting to cool the centrifuge when the cells are put on ice. From this point on, the cells will be increasingly competent and thus fragile. Therefore, all subsequent mixing or suspensions of cells should be done gently, in the absence of vortexing, and as rapidly as possible.
8. This step can be repeated to ensure cell suspension is free of salts.
9. It is possible to store the electrocompetent cells at −80 °C so that they are "ready-to-use." If a larger stock of competent cell is needed, increase the initial culture volume, and correspondingly adjust the number of centrifuge tubes. For example, for 100 mL of culture, divide it into four 25 mL aliquots, and follow the protocol described above. At the end of the protocol, aliquot 50 μL of the final suspensions into ice-cold 1.5 mL microfuge tubes and snap-freeze them in liquid nitrogen. The electrocompetent cells can be stored at −80 °C for up to 6 months.
10. In order to choose the appropriate cuvette size and electroporation conditions, we refer the reader to the electroporator instruction manual.
11. The time constant should be beyond 5 ms for optimal results. Lower time constants indicate problems with the cells, the DNA, the electroporation cuvettes or even the equipment. If a click is heard at the time of electroporation, this is probably due to residual traces of salts either in the competent cells or in the volume of plasmid supplied, and means that the electroporation needs to be repeated.
12. *Streptomyces* are unusual soil bacterial in that they grow in the form of a substrate mycelium on the soil surface, and later, as an aerial mycelium, from which the spores are produced. As the spores are robust, it is common to store *Streptomyces* as spore suspensions in strain collections. Therefore, the ability to efficiently sporulate *Streptomyces* and make spore suspensions are crucial steps in working with these microbes. For example, the *S. coelicolor* M145 strain when grown on SFM plates sporulates after 5 days at 30 °C. Although SFM agar plates work for a large majority of *Streptomyces*, such as *S. coelicolor*, *S. lividans*, *S. albus*, and *S. ambofaciens*, it is nonetheless worth experimenting with different media to determine the best conditions for a particular strain. We recommend collecting the spore suspension the day of the conjugation to increase the efficiency of the process. In the case of non-sporulating *Streptomyces*,

such as *whi* and *bld* mutant strains, an alternative method is to obtain mycelia using a medium such as YEME (containing 34% sucrose) by growing for 36–48 h, centrifuge to collect the mycelia, wash three times with 10.3% sucrose solution, and finally store the mycelia in 10.3% sucrose solution at −80 °C.

13. *Streptomyces* spores are rather resistant to osmotic damage and therefore can be resuspended in distilled water. Alternatively, the user can resuspend the spores in 2YT.
14. This step allows for dispersing the spores in the suspension solution.
15. This step allows for elimination of residual agar and mycelial debris. A "home-made" filter system consists of a 15 mL syringe containing non-absorbent cotton. Alternatively, a cotton pad can be placed directly on the plate, and the spore suspension aspirated through it (in this case, subsequent filtration isn't necessary). To maximize the amount of spores collected, repeat this step once. In the case of *Streptomyces* strains which sporulate poorly, collect spores from more than one plate or use mycelia.
16. The spore pellet easily detaches from the tube after centrifugation, and therefore caution must be used during this step.
17. The spores can be used for up to 1 month if stored at 4 °C. For longer conservation at −80 °C, resuspend the spores in 2 mL of 20% glycerol and freeze.
18. This step allows for obtaining cells in antibiotic-free LB medium. We strongly advice carrying out all of the recommended washing steps.
19. Depending on the Streptomyces species, some spores need to be heat-shocked for 10 min at 50 °C before this step, depending on the *Streptomyces* species. For example, spores from *S. ambofaciens* don't require a heat-shock, whereas those from *S. coelicolor* do [43].
20. We recommend adding the antibiotics sequentially starting with that dictated by the CRISPR plasmid, and then nalidixic acid in 300 μL sterile water, and then air-drying the plates in a laminar flow hood for 5–10 min. This volume is recommended for the nalidixic acid because of its poor water solubility. The growth time to achieve exconjugants will depend on which *Streptomyces* is used. In the case of prolonged incubation of conjugation plates, wrap the plates in aluminum foil to limit dehydration of the medium. The plates must be incubated until aerial mycelia form.
21. We test approximately 50 clones per experiment. For colony PCR, we recommend scratching the mycelia before you can see the sporulation.

22. In our lab, this simple incubation is sufficient to lyse *Streptomyces* cells. It is possible to increase the efficiency of this step by an additional incubation at −20 °C in a freezer for 15 min. Both the incubations at 100 °C and −20 °C can be repeated if necessary.
23. We observe that it is better to homogenize by gently pipetting the remaining solution containing genomic DNA to obtain reproducible results. Centrifuge again if debris remains in the solution.
24. We use a Taq polymerase for routine colony PCR. We recommend amplifying fragments with a size less than 2.5 kb. Indeed, longer fragments may be difficult to amplify and thus false negatives can be obtained, leading to loss of correctly genetically modified exconjugants.
25. Because the conductivity decreases as a function of time, check the gel every 10 min. Stop the gel when the migration front reaches the bottom of the gel.
26. If a single band is obtained at the expected size, you can use the remaining PCR reaction to purify the band using a PCR extraction kit to increase the yield of the product for sequencing. At this stage, it is possible to observe a mix of wild-type and mutated phenotypes. Indeed, sgRNAs exhibit variable efficiencies and it is possible that <100% of the genomic DNA will be modified. In addition, the aerial hyphae used for colony PCR are not compartmentalized and therefore have multiple copies of genomic DNA. As a consequence, we prefer to carry out screening using a single genomic DNA, and therefore start from a spore. To do this, re-streak the exconjugants and after a sporulation step, scratch the surface of one clone with a wooden toothpick and transfer it onto a plate supplemented with the appropriate antibiotics. Analyze your exconjugants again to confirm the obtained genotype.
27. The incubation time depends on the generation time of the *Streptomyces* species.
28. Clones that are no longer able to grow on a selective medium no longer contain the CRISPR plasmid. Due to the high frequency of chromosomal recombination events in *Streptomyces*, we recommend limiting successive sub-culturing. This advice also applies when using liquid media.
29. As many *Streptomyces* strains, including *S. coelicolor* and *S. lividans*, tend to grow as rather compact masses, an Erlenmeyer flask containing a stainless-steel spring or sterile glass beads (3 mm diameter) favors dispersed growth.
30. We typically use the high-fidelity polymerase Phusion Hot Start II DNA Polymerase (Thermo Fisher).

Acknowledgments

We acknowledge financial support from the Agence Nationale de la Recherche (grant number ANR-16-CE92-0006-01, PKS STRUCTURE to K.J.W.), the Université de Lorraine, the Centre National de la Recherche Scientifique (CNRS), and the IMPACT Biomolecules project of the Lorraine Université d'Excellence (Investissements d'avenir - ANR 15-004).

References

1. van der Heul HU, Bilyk BL, McDowall KJ, Seipke RF, van Wezel GP (2018) Regulation of antibiotic production in Actinobacteria: new perspectives from the post-genomic era. Nat Prod Rep 35:575–604. https://doi.org/10.1039/C8NP00012C
2. Alberti F, Corre C (2019) Editing streptomycete genomes in the CRISPR/Cas9 age. Nat Prod Rep 36:1237–1248. https://doi.org/10.1039/c8np00081f
3. Zhao Y, Li G, Chen Y, Lu Y (2020) Challenges and advances in genome editing technologies in *Streptomyces*. Biomol Ther 10:734. https://doi.org/10.3390/biom10050734
4. Cobb RE, Wang Y, Zhao H (2015) High-efficiency multiplex genome editing of *Streptomyces* species using an engineered CRISPR/Cas system. ACS Synth Biol 4:723–728. https://doi.org/10.1021/sb500351f
5. Tong Y, Charusanti P, Zhang L, Weber T, Lee SY (2015) CRISPR-Cas9 based engineering of Actinomycetal genomes. ACS Synth Biol 4:1020–1029. https://doi.org/10.1021/acssynbio.5b00038
6. Wlodek A, Kendrew SG, Coates NJ, Hold A, Pogwizd J, Rudder S, Sheehan LS, Higginbotham SJ, Stanley-Smith AE, Warneck T, Nur-E-Alam M, Radzom M, Martin CJ, Overvoorde L, Samborskyy M, Alt S, Heine D, Carter GT, Graziani EI, Koehn FE, McDonald L, Alanine A, Rodríguez Sarmiento RM, Chao SK, Ratni H, Steward L, Norville IH, Sarkar-Tyson M, Moss SJ, Leadlay PF, Wilkinson B, Gregory MA (2017) Diversity oriented biosynthesis via accelerated evolution of modular gene clusters. Nat Commun 8:1206. https://doi.org/10.1038/s41467-017-01344-3
7. Jinek M, Chylinski K, Fonfara I, Hauer M, Doudna JA, Charpentier E (2012) A programmable dual-RNA-guided DNA endonuclease in adaptive bacterial immunity. Science 337:816–821. https://doi.org/10.1126/science.1225829
8. Mougiakos I, Bosma EF, de Vos WM, van Kranenburg R, van der Oost J (2016) Next generation prokaryotic engineering: the CRISPR-Cas toolkit. Trends Biotechnol 34:575–587. https://doi.org/10.1016/j.tibtech.2016.02.004
9. Tong Y, Weber T, Lee SY (2019) CRISPR/Cas-based genome engineering in natural product discovery. Nat Prod Rep 36:1262–1280. https://doi.org/10.1039/c8np00089a
10. Bertrand C, Thibessard A, Bruand C, Lecointe F, Leblond P (2019) Bacterial NHEJ: a never-ending story. Mol Microbiol 111:1139–1151. https://doi.org/10.1111/mmi.14218
11. Sander JD, Joung JK (2014) CRISPR-Cas systems for genome editing, regulation and targeting. Nat Biotechnol 32:347–355. https://doi.org/10.1038/nbt.2842
12. Gibson DG, Young L, Chuang RY, Venter JC, Hutchison CA, Smith HO (2009) Enzymatic assembly of DNA molecules up to several hundred kilobases. Nat Methods 6:343–345. https://doi.org/10.1038/nmeth.1318
13. Engler C, Kandzia R, Marillonnet S (2008) A one pot, one step, precision cloning method with high throughput capability. PLoS One 3:e3647. https://doi.org/10.1371/journal.pone.0003647
14. Tong Y, Whitford CM, Blin K, Jørgensen TS, Weber T, Lee SY (2020) CRISPR-Cas9, CRISPRi and CRISPR-BEST-mediated genetic manipulation in streptomycetes. Nat Protoc 15:2470–2502. https://doi.org/10.1038/s41596-020-0339-z
15. Su L, Hotel L, Paris C, Brachmann A, Piel J, Jacob C, Aigle B, Weissman KJ (2021) Successes, surprises and pitfalls in modular polyketide synthase engineering: generation of ring-contracted stambomycins. Nat Commun. In revision. https://doi.org/10.21203/rs.3.rs-222743/v1
16. Huang H, Zheng G, Jiang W, Hu H, Lu Y (2015) One-step high-efficiency CRISPR/

Cas9-mediated genome editing in *Streptomyces*. Acta Biochim Biophys Sin 47:231–243. https://doi.org/10.1093/abbs/gmv007

17. Mo J, Wang S, Zhang W, Li C, Deng Z, Zhang L, Qu X (2019) Efficient editing DNA regions with high sequence identity in actinomycetal genomes by a CRISPR-Cas9 system. Synth Syst Biotechnol 4:86–91. https://doi.org/10.1016/j.synbio.2019.02.004
18. Wang Q, Xie F, Tong Y, Habisch R, Yang B, Zhang L, Müller R, Fu C (2020) Dual-function chromogenic screening-based CRISPR/Cas9 genome editing system for actinomycetes. Appl Microbiol Biotechnol 104: 225–239. https://doi.org/10.1007/s00253-019-10223-4
19. Wang K, Zhao QW, Liu YF, Sun CF, Chen XA, Burchmore R, Burgess K, Li YQ, Mao XM (2019) Multi-layer controls of Cas9 activity coupled with ATP synthase over-expression for efficient genome editing in *Streptomyces*. Front Bioeng Biotechnol 7:304. https://doi.org/10.3389/fbioe.2019.00304
20. Li L, Wei K, Zheng G, Liu X, Chen S, Jiang W, Lu Y (2018) CRISPR-Cpf1-assisted multiplex genome editing and transcriptional repression in *Streptomyces*. Appl Environ Microbiol 84: e00827–e00818. https://doi.org/10.1128/AEM.00827-18
21. Yeo WL, Heng E, Tan LL, Lim YW, Lim YH, Hoon S, Zhao H, Zhang MM, Wong FT (2019) Characterization of Cas proteins for CRISPR-Cas editing in streptomycetes. Biotechnol Bioeng 116:2330–2338. https://doi.org/10.1002/bit.27021
22. Zhao Y, Li L, Zheng G, Jiang W, Deng Z, Wang Z, Lu Y (2018) CRISPR/dCas9-mediated multiplex gene repression in *Streptomyces*. Biotechnol J 13:e1800121. https://doi.org/10.1002/biot.201800121
23. Tong Y, Whitford CM, Robertsen HL, Blin K, Jørgensen TS, Klitgaard AK, Gren T, Jiang X, Weber T, Lee SY (2019) Highly efficient DSB-free base editing for streptomycetes with CRISPR-BEST. Proc Natl Acad Sci U S A 116: 20366–20375. https://doi.org/10.1073/pnas.1913493116
24. Zhao Y, Tian J, Zheng G, Chen J, Sun C, Yang Z, Zimin AA, Jiang W, Deng Z, Wang Z, Lu Y (2020) Multiplex genome editing using a dCas9-cytidine deaminase fusion in *Streptomyces*. Sci China Life Sci 63:1053–1062. https://doi.org/10.1007/s11427-019-1559-y
25. Su L, Hôtel L, Paris C, Chepkirui C, Brachmann AO, Piel J, Jacob C, Aigle B, Weissman KJ (2022) Engineering the stambomycin modular polyketide synthase yields 37-membered mini-stambomycins. Nat Commun 13:515. https://doi.org/10.1038/s41467-022-27955-z
26. Mali P, Esvelt KM, Church GM (2013) Cas9 as a versatile tool for engineering biology. Nat Methods 10:957–963. https://doi.org/10.1038/nmeth.2649
27. Wilkinson CJ, Hughes-Thomas ZA, Martin CJ, Böhm I, Mironenko T, Deacon M, Wheatcroft M, Wirtz G, Staunton J, Leadlay PF (2002) Increasing the efficiency of heterologous promoters in actinomycetes. J Mol Microbiol Biotechnol 4:417–426
28. Murakami T, Holt TG, Thompson CJ (1989) Thiostrepton-induced gene expression in *Streptomyces lividans*. J Bacteriol 171:1459–1466. https://doi.org/10.1128/jb.171.3.1459-1466.1989
29. Yun BS, Hidaka T, Kuzuyama T, Seto H (2001) Thiopeptide non-producing *Streptomyces* species carry the *tipA* gene: a clue to its function. J Antibiot 54:375–378. https://doi.org/10.7164/antibiotics.54.375
30. Rudolph MM, Vockenhuber MP, Suess B (2013) Synthetic riboswitches for the conditional control of gene expression in *Streptomyces coelicolor*. Microbiology 159:1416–1422. https://doi.org/10.1099/mic.0.067322-0
31. Sprinzl M (1994) Elongation factor Tu: a regulatory GTPase with an integrated effector. Trends Biochem Sci 19:245–250. https://doi.org/10.1016/0968-0004(94)90149-x
32. Blin K, Pedersen LE, Weber T, Lee SY (2016) CRISPy-web: an online resource to design sgRNAs for CRISPR applications. Synth Syst Biotechnol 1:118–121. https://doi.org/10.1016/j.synbio.2016.01.003
33. Blin K, Shaw S, Tong Y, Weber T (2020) Designing sgRNAs for CRISPR-BEST base editing applications with CRISPy-web 2.0. Synth Syst Biotechnol 5:99–102. https://doi.org/10.1016/j.synbio.2020.05.005
34. Blin K, Shaw S, Steinke K, Villebro R, Ziemert N, Lee SY, Medema MH, Weber T (2019) antiSMASH 5.0: updates to the secondary metabolite genome mining pipeline. Nucleic Acids Res 47:W81–W87. https://doi.org/10.1093/nar/gkz310
35. Zhang J, Zhang D, Zhu J, Liu H, Liang S, Luo Y (2020) Efficient multiplex genome editing in *Streptomyces* via engineered CRISPR-Cas12a systems. Front Bioeng Biotechnol 8:726. https://doi.org/10.3389/fbioe.2020.00726
36. Najah S, Saulnier C, Pernodet JL, Bury-Moné S (2019) Design of a generic CRISPR-Cas9 approach using the same sgRNA to perform gene editing at distinct loci. BMC Biotechnol

19:18. https://doi.org/10.1186/s12896-019-0509-7

37. Wendt KE, Ungerer J, Cobb RE, Zhao H, Pakrasi HB (2016) CRISPR/Cas9 mediated targeted mutagenesis of the fast-growing cyanobacterium *Synechococcus elongatus* UTEX 2973. Microb Cell Factories 15:115. https://doi.org/10.1186/s12934-016-0514-7
38. Zhang JJ, Moore BS (2020) Site-directed mutagenesis of large biosynthetic gene clusters via oligonucleotide recombineering and CRISPR/Cas9 targeting. ACS Synth Biol 9: 1917–1922. https://doi.org/10.1021/acssynbio.0c00265
39. Volff JN, Altenbuchner J (1998) Genetic instability of the *Streptomyces* chromosome. Mol Microbiol 27:239–246. https://doi.org/10.1046/j.1365-2958.1998.00652.x
40. Zhang Z, Du C, de Barsy F, Liem M, Liakopoulos A, van Wezel GP, Choi YH, Claessen D, Rozen DE (2020) Antibiotic production in *Streptomyces* is organized by a division of labor through terminal genomic differentiation. Sci Adv 6:eaay5781. https://doi.org/10.1126/sciadv.aay5781
41. Muth G (2018) The pSG5-based thermosensitive vector family for genome editing and gene expression in actinomycetes. Appl Microbiol Biotechnol 102:9067–9080. https://doi.org/10.1007/s00253-018-9334-5
42. Kieser T, Bibb M, Buttner M, Chater K, Hopwood D (2000) Practical *Streptomyces* genetics. John Innes Foundation, Norwich
43. Kim MK, Ha HS, Choi SU (2008) Conjugal transfer using the bacteriophage phiC31 *att/int* system and properties of the *attB* site in *Streptomyces ambofaciens*. Biotechnol Lett 30: 695–699. https://doi.org/10.1007/s10529-007-9586-0
44. Gust B, Challis GL, Fowler K, Kieser T, Chater KF (2003) PCR-targeted *Streptomyces* gene replacement identifies a protein domain needed for biosynthesis of the sesquiterpene soil odor geosmin. Proc Natl Acad Sci U S A 100:1541–1546. https://doi.org/10.1073/pnas.0337542100

Chapter 11

CRISPR/Cas9-Based Methods for Inactivating Actinobacterial Biosynthetic Genes and Elucidating Function

Audam Chhun and Fabrizio Alberti

Abstract

The CRISPR/Cas9 technology allows fast and marker-less genome engineering that can be employed to study secondary metabolism in actinobacteria. Here, we report a standard experimental protocol for the deletion of a biosynthetic gene in a *Streptomyces* species, using the vector pCRISPomyces-2 developed by Huimin Zhao and collaborators. We also describe how carrying out metabolite analysis can reveal the putative biosynthetic function of the inactivated gene.

Key words CRISPR/Cas9, pCRISPomyces-2, Actinobacteria, Biosynthetic gene, Natural products

1 Introduction

In the last decade, the advent of CRISPR/Cas9 has opened new avenues for genome engineering [1]. Toolkits have been made available that allow for the CRISPR/Cas9-based editing of organisms in all domains of life, including actinobacteria (also known as actinomycetes) [2–5]. Actinobacteria are a phylum of Gram-positive bacteria and are considered one of the most prolific sources of antibiotics known to humankind [6]. CRISPR/Cas9-based genome engineering is being used extensively in these microorganisms to engineer natural product biosynthetic gene clusters (BGCs) [7].

Here, we describe the standard workflow used by us to inactivate biosynthetic genes in actinobacteria via the CRISPR/Cas9-based plasmid pCRISPomyces-2 (pCm2), developed by Huimin Zhao and collaborators [2]. We provide an overview of how we then elucidate the function of such genes through metabolite analysis. As a working example, we are showing the knock-out of the non-ribosomal peptide synthetase gene *SclN* from the scleric acid BGC [8].

Elizabeth Skellam (ed.), *Engineering Natural Product Biosynthesis: Methods and Protocols*,
Methods in Molecular Biology, vol. 2489, https://doi.org/10.1007/978-1-0716-2273-5_11,

2 Materials

Prepare all solutions using deionized water and analytical grade reagents. Store all reagents at room temperature unless otherwise stated.

2.1 Golden Gate Assembly

1. 5 M NaOH.
2. 30 mM HEPES pH 7.8: weigh out 0.7149 g of HEPES and transfer to a beaker. Add water to a volume of 100 mL. Use a stir bar and stir plate to mix. Once completely dissolved, adjust pH to 7.8 with 5 M NaOH.
3. T4 DNA ligase and buffer (NEB).
4. *Bbs*I (NEB).
5. Nuclease-free water.
6. PCR tubes.
7. Sterile Eppendorf tubes.
8. Chemically competent *Escherichia coli* NEB5alpha.
9. Water bath.
10. Sterile syringe filters, 0.22 μm.
11. 100 mg/mL apramycin antibiotic stock: weigh out 1 g of apramycin and transfer to a 15 mL falcon tube. Add water to a volume of 10 mL and mix by inverting the tube until completely dissolved. Filter sterilize the antibiotic solution using a 0.22 μm syringe filter. Aliquot the antibiotic stock in ten 1.5 mL Eppendorf tubes and store at −20 °C.
12. LB agar plates supplemented with 50 μg/mL apramycin: weigh out 5 g of sodium chloride, 5 g of tryptone, 2.5 g of yeast extract, 7.5 g of agar, and add to a 1 L Duran bottle. Dissolve in approximately 400 mL of deionized water and adjust the pH to 7.0 with 5 M NaOH. Top up the solution to 500 mL with deionized water and autoclave at 121 °C for 20 min. Let the medium cool down to approximately 55 °C and add 250 μL of 100 mg/mL apramycin antibiotic stock. Mix well and pour the medium onto Petri dishes in a biological safety cabinet, then let the plates cool down and store them in the fridge.
13. LB broth: weigh out 5 g of sodium chloride, 5 g of tryptone, 2.5 g of yeast extract, and add to a 1 L Duran bottle. Dissolve in approximately 400 mL of deionized water and adjust the pH to 7.0 with 5 M NaOH. Top up the solution to 500 mL with deionized water and autoclave at 121 °C for 20 min.
14. Sterile plastic loops.
15. GeneJET Plasmid Miniprep Kit (Thermo Fisher Scientific).

2.2 Gibson Assembly

1. *Xba*I (NEB).
2. 10× CutSmart buffer (NEB).
3. Quick CIP (NEB).
4. Nuclease-free water.
5. PCR tubes.
6. GeneJET PCR Purification Kit (Thermo Fisher Scientific).
7. Q5 High-Fidelity 2× master mix (NEB).
8. GelRed (10,000×).
9. 5 M NaOH.
10. 1 L of 50× Tris Acetate-EDTA (TAE) buffer: first prepare 500 mL of 0.5 M (ethylenediamenetetraacetic acid) EDTA solution. For this, weigh out 93.06 g of EDTA disodium salt and dissolve in approximately 400 mL of water. Adjust pH to 8.0 with 5 M NaOH before topping up to 500 mL with water. Autoclave at 121 °C for 20 min. Second, weigh out 242 g of tris base and dissolve in 700 mL of water. Add to the dissolved tris base 57.1 mL of glacial acetic acid and 100 mL of 0.5 M EDTA solution. Adjust volume to 1 L.
11. 1% (w/v) agarose gels in 1× TAE buffer containing 1× GelRed: weigh out 1 g of agarose powder and add to 100 mL of 1× TAE buffer in a beaker. Dissolve agarose by heating the mixture in a microwave until homogeneously clear. Let it cool down to approximately 55 °C and add 10 μL of GelRed. Mix thoroughly. Pour the melted agarose into a gel casting tray and let it cool down at room temperature until solidified.
12. NEBuilder HiFi DNA Assembly 2× Master Mix (NEB).
13. Chemically competent *E. coli* NEB5alpha.
14. Water bath.
15. Sterile plastic loops.
16. GeneJET Plasmid Miniprep Kit (Thermo Fisher Scientific).

2.3 Introduction of Plasmid DNA into Actinomycetes

1. *E. coli* ET12567/*pUZ8002* chemically competent or electro-competent cells.
2. Sterile wedge-shaped spreaders.
3. Sterile syringe barrels fitted with sterile cotton wool, and sterile syringe plungers.
4. Sterile falcon tubes.
5. Water bath.
6. Sterile Eppendorf tubes.
7. 5 M NaOH.

8. 25 mg/mL kanamycin antibiotic stock: weigh out 250 mg of kanamycin and transfer to a 15 mL falcon tube. Add water to a volume of 10 mL and mix by inverting the tube until completely dissolved. Filter sterilize the antibiotic solution using a 0.22 μm syringe filter. Aliquot the antibiotic stock in ten 1.5 mL Eppendorf tubes and store at −20 °C.
9. 25 mg/mL chloramphenicol antibiotic stock: weigh out 250 mg of chloramphenicol and transfer to a 15 mL falcon tube. Add water to a volume of 10 mL and mix by inverting the tube until completely dissolved. Filter sterilize the antibiotic solution using a 0.22 μm syringe filter. Aliquot the antibiotic stock in ten 1.5 mL Eppendorf tubes and store at −20 °C.
10. 25 mg/mL nalidixic acid antibiotic stock: weigh out 250 mg of nalidixic acid and transfer to a 15 mL falcon tube. Add water to a volume of 10 mL and mix by inverting the tube until completely dissolved. Filter sterilize the antibiotic solution using a 0.22 μm syringe filter. Aliquot the antibiotic stock in ten 1.5 mL Eppendorf tubes and store at −20 °C.
11. LB broth with 25 μg/mL kanamycin, 25 μg/mL chloramphenicol, 50 μg/mL apramycin: weigh out 5 g of sodium chloride, 5 g of tryptone, 2.5 g of yeast extract, and add to a 1 L Duran bottle. Dissolve in approximately 400 mL of deionized water and adjust the pH to 7.0 with 5 M NaOH. Top up the solution to 500 mL with deionized water and autoclave at 121 °C for 20 min. Let the medium cool down to room temperature, then add 500 μL of 25 mg/mL kanamycin antibiotic stock, 500 μL of 25 mg/mL chloramphenicol antibiotic stock, and 250 μL of 100 mg/mL apramycin antibiotic stock.
12. LB agar plates supplemented with 50 μg/mL apramycin: weigh out 5 g of sodium chloride, 5 g of tryptone, 2.5 g of yeast extract, 7.5 g of agar, and add to a 1 L Duran bottle. Dissolve in approximately 400 mL of deionized water and adjust the pH to 7.0 with 5 M NaOH. Top up the solution to 500 mL with deionized water and autoclave at 121 °C for 20 min. Let the medium cool down to approximately 55 °C and add 250 μL of 100 mg/mL apramycin antibiotic stock. Mix well and pour the medium onto Petri dishes in a biological safety cabinet, then let the plates cool down and store them in the fridge.
13. 2× YT: weigh out 8 g of tryptone, 5 g of yeast extract, 2.5 g of sodium chloride and add to a 1 L Duran bottle. Dissolve in approximately 400 mL of deionized water and adjust the pH to 7.0 with 5 M NaOH. Top up the solution to 500 mL with deionized water and autoclave at 121 °C for 20 min.
14. SFM + 10 mM $MgCl_2$ agar plates: weigh out 10 g of soya flour, 10 g of mannitol, 10 g of agar, 0.476 g of magnesium

dichloride and add to a 1 L Duran bottle. Dissolve in 500 mL of tap water and autoclave at 121 °C for 20 min. Let the medium cool down to approximately 55 °C, mix well and pour the medium onto Petri dishes in a biological safety cabinet, then let the plates cool down and store them in the fridge.

15. SFM agar plates containing 50 μg/mL apramycin (and 25 μg/mL nalidixic acid): weigh out 10 g of soya flour, 10 g of mannitol, 10 g of agar and add to a 1 L Duran bottle. Dissolve in 500 mL of tap water and autoclave at 121 °C for 20 min. Let the medium cool down to approximately 55 °C and add 250 μL of 100 mg/mL apramycin antibiotic stock (and 500 μL of 25 mg/mL nalidixic acid antibiotic stock), mix well and pour the medium onto Petri dishes in a biological safety cabinet, then let the plates cool down and store them in the fridge.

2.4 Validation of Mutants

1. FastDNA Spin Kit for Soil (MP Biomedicals).
2. Sterile plastic loops.
3. PCR tubes.
4. Q5 High-Fidelity 2× Master Mix (NEB).
5. Nuclease-free water.

2.5 Plasmid Clearance

1. Sterile plastic loops.
2. YEME: weigh out 1.5 g of yeast extract, 1.5 g of malt extract, 2.5 g of peptone, 5 g of glucose and add to a 1 L Duran bottle. Dissolve in 500 mL of deionized water and autoclave at 121 °C for 20 min.
3. Sterile falcon tubes.
4. Sterile wedge-shaped spreaders.
5. SFM agar plates: weigh out 10 g of soya flour, 10 g of mannitol, 10 g of agar and add to a 1 L Duran bottle. Dissolve in 500 mL of tap water and autoclave at 121 °C for 20 min. Let the medium cool down to approximately 55 °C, mix well and pour the medium onto Petri dishes in a biological safety cabinet, then let the plates cool down and store them in the fridge.

2.6 Preparation of a Spore Stock

1. Sterile plastic loops.
2. Sterile wedge-shaped spreaders.
3. Sterile deionized water.
4. Sterile syringe barrels fitted with sterile cotton wool, and sterile syringe plungers.
5. Sterile falcon tubes.
6. Sterile serological pipettes.

7. 25% (w/v) glycerol: measure out 125 g of glycerol and top up with deionized water to 500 mL. Autoclave at 121 °C for 20 min.
8. Sterile cryopreservation vials.

2.7 Metabolite Purification

1. Ethyl acetate.
2. 37% HCl.
3. Separation funnel.
4. Anhydrous magnesium sulfate.
5. Filter paper discs.
6. HPLC-grade acetonitrile.
7. HPLC-grade water.
8. HPLC-grade formic acid.
9. Micro-centrifugal filter (nylon membrane, pore size 0.2 μm, volume 750 μL).
10. Snap-seal amber glass vial (2 mL).
11. Column Zorbax Eclipse Plus C18, with dimensions 2.1 × 100 mm and 1.8 μm particle size.

3 Methods

3.1 Design of sgRNAs and HR Arms for pCm2 Retargeting

Gene knock-out using pCm2 requires the design of (a) single guide RNAs (sgRNAs) targeting the Cas9 to the gene(s) to be inactivated, and (b) homologous recombination (HR) arms to allow for double crossover recombination events (Fig. 1).

1. Download the genome of your strain of interest as a GenBank file and upload it to the CRISPY-web tool (crispy.secondarymetabolites.org) [9].
2. Scan the genomic region to be deleted for possible protospacers, such that they would function with the Cas9 enzyme, with the 3′ protospacer adjacent motif (PAM) NGG and to be 20-bp for the full protospacer. To delete *SclN* (3249 bp), two protospacers located at the extremities of the gene were selected. If the genomic region to be deleted is smaller than 1 kb, the design of a single sgRNA located centrally within the target is sufficient for genomic deletion (*see* **Note 1**).
3. When dual-protospacers are needed, design a 499-bp synthetic construct based on the two selected protospacers following the structure

5'-**bbsI**-acgc-protospacer1-gRNAscaffold-T7term-Pgapdh-protospacer2-gttt-**bbsI**-3'

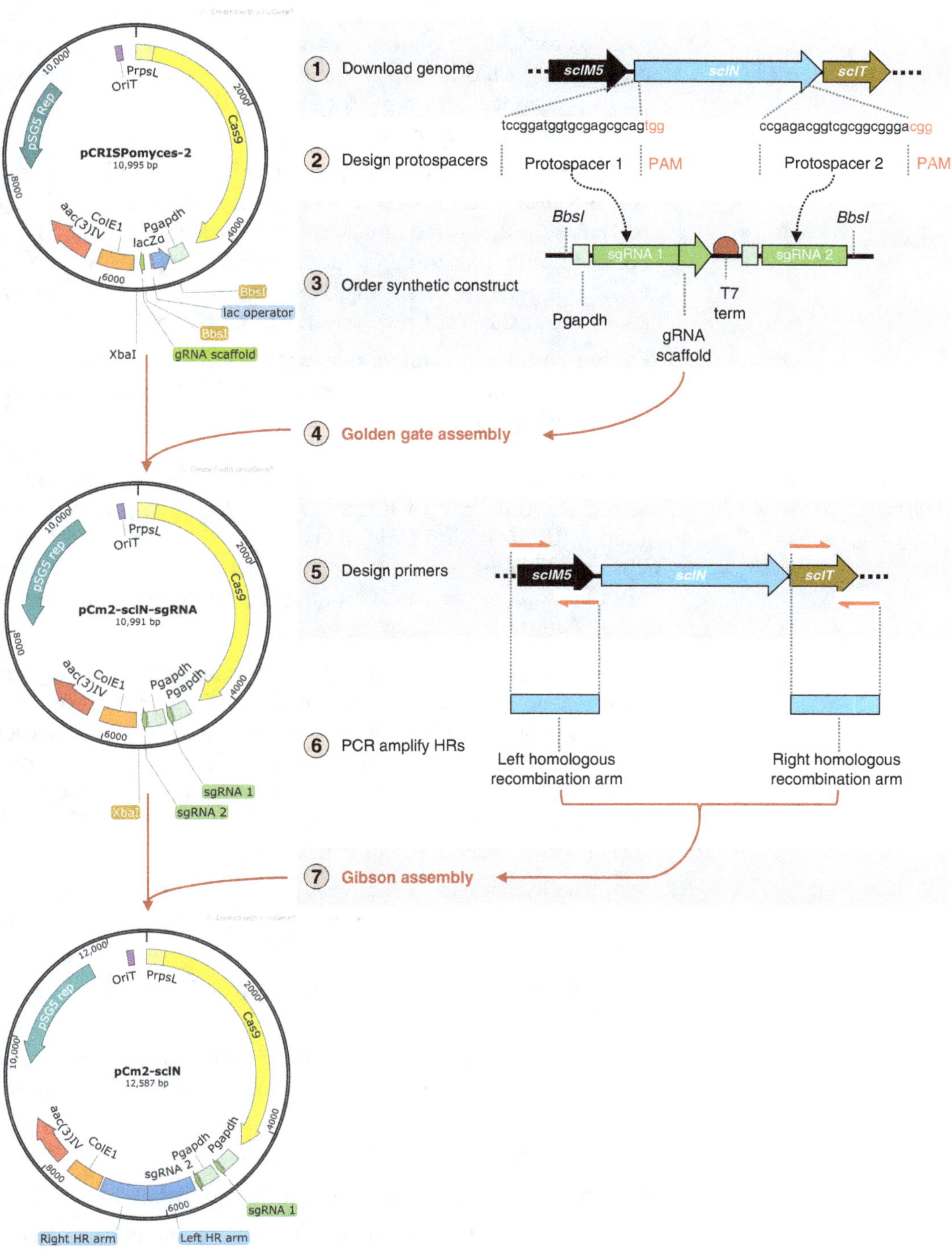

Fig. 1 Workflow of pCRISPomyces-2 retargeting for gene deletion

If a single sgRNA is required, it can be more simply synthesized as two 24-bp oligonucleotides that are the reverse complement of each other, such that the annealed product results in double-stranded 20-bp protospacers with 4-bp *BbsI*-compatible overhangs (*see* **Note 2**).

4. Design oligonucleotide pairs to amplify two ~1-kb HR arms corresponding to genomic sequences flanking the chosen sgRNA(s) but located outside the region to be deleted (*see* **Note 3**). Primers should include overhang sequences allowing for cloning of the resulting HR arms in pCm2 by Gibson Assembly at the *Xba*I restriction site (*see* **Note 4**).
5. Design additional primers that would allow you to sequence-verify the correct insertion of your sgRNA(s) and HR arms into pCm2, using Sanger sequencing (*see* **Note 5**).

3.2 Construction of Retargeted pCRISPomyces-2

After the design of the sgRNA(s) and the HR arms, pCm2 can be retargeted for the gene of interest to be deleted by cloning the designed sgRNA(s) using Golden Gate Assembly and the HR arms by Gibson Assembly.

All steps are to be carried out at room temperature unless otherwise stated.

1. If a single spacer was designed, anneal the corresponding oligonucleotides as follows: resuspend (or dilute, as appropriate) oligonucleotides to 100 μM in water; mix together 5 μL of each oligonucleotide with 90 μL 30 mM HEPES at pH 7.8; heat mixture in a thermal cycler to 95 °C for 5 min, followed by a linear decrease in temperature to 4 °C at 0.1 °C/s (if a dual-spacer was designed, directly proceed to **step 2**).
2. Set up a Golden Gate assembly reaction with the single spacer/ synthetic construct bearing the dual guide RNA cassettes, as for Fig. 2a.
3. Use a thermal cycler to incubate the reaction as illustrated in Fig. 2b.
4. Transform 3–5 μL of the Golden Gate reaction to *E. coli* NEB5alpha by heat shock, following manufacturer's guidelines.
5. Centrifuge transformed cells for 3 min at 6000 × *g*.
6. Discard supernatant to leave about 100 μL of media.
7. Resuspend cells into the remaining medium and plate them out onto LB agar plates, supplemented with 50 μg/mL apramycin (*see* **Note 6**).
8. Incubate plates overnight at 37 °C.
9. Pick a few individual transformant colonies with a sterile loop and inoculate each one into 10 mL of LB supplemented with

a

Component	Volume (μL)	Comments
pCRISPomyces-2 (backbone)	X[a]	For 100 ng
sgRNA (insert)	0.3	
T4 Ligase Buffer (NEB)	2	
T4 Ligase (NEB)	1	
BbsI (NEB)	1	Stored at -80 °C
Nuclease-free water	X[b]	For Vf = 20 μL

[a] *To be determined based on the pCm2 concentration of the miniprep used.*
[b] *To be determined based on the volume of pCm2 miniprep added.*

b

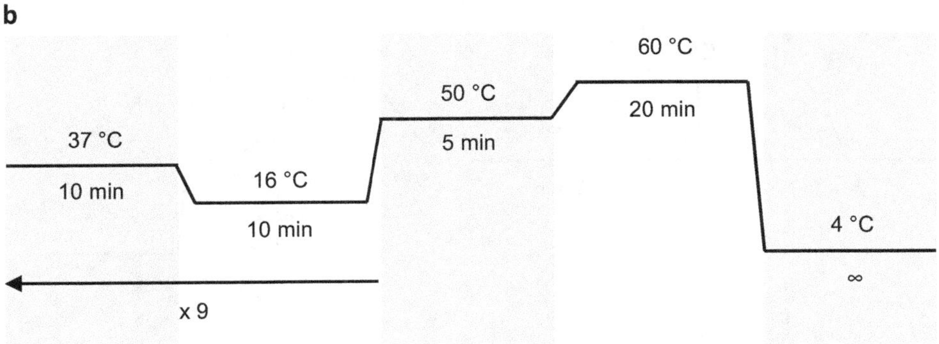

Fig. 2 Golden Gate reaction

50 μg/mL apramycin in a falcon tube and incubate cultures overnight at 37 °C, with shaking at 180 rpm.

10. Recover plasmid using a Miniprep kit, following manufacturer's protocol, and sequence-verify the obtained plasmids for correct sgRNA cassettes cloning, using the primers designed in Subheading 3.1, **step 5**.
11. To prepare the sgRNA-bearing pCRISPomyces-2 vector for Gibson assembly of the HR arms, set up a *Xba*I restriction digest, as for Fig. 3a.
12. Use a thermal cycler to incubate the reaction at 37 °C for 1 h.
13. Add 1 μL of Quick CIP (alkaline phosphatase) to the reaction mixture.
14. Incubate the reaction at 37 °C for another 1 h, followed by an inactivation step at 80 °C for 20 min, as for Fig. 3b.
15. Clean up the digested DNA using a PCR Purification Kit, following the manufacturer's protocol.

a

Component	Volume (μL)	Comments
XbaI (NEB)	1	
pCRISPomyces(sgRNA)	X[a]	For 1 μg
10X CutSmart buffer	5	
H_2O	X[b]	For Vf = 50 μL
Quick CIP (NEB)	1	To add after the first incubation step

[a] *To be determined based on the pCm2(sgRNA) concentration of the miniprep obtained.*
[b] *To be determined based on the volume of pCm2(sgRNA) miniprep added.*

b

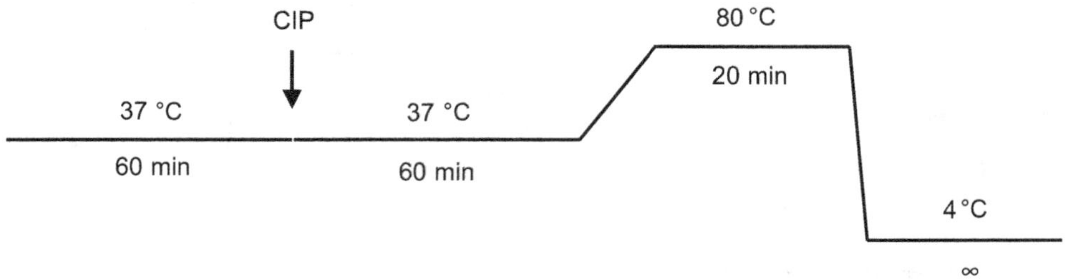

Fig. 3 *XbaI* restriction digest reaction

16. Amplify by PCR the two HR arms from your strain's genomic DNA using the high-fidelity Q5 polymerase and the primers designed in Subheading 3.1, **step 4**, by setting up a reaction as for Fig. 4a.
17. Use a thermal cycler to incubate the reactions as illustrated in Fig. 4b.
18. Confirm PCR amplification of the HR arms through gel electrophoresis by loading 5 μL of the PCR products in a 1% (w/v) agarose gel in 1× TAE buffer containing 1× GelRed; load an appropriate DNA ladder and run at 100 V for 60 min.
19. Visualize the gel in a UV transilluminator.
20. Clean up the remaining 45 μL of the PCR products using a PCR Purification Kit.
21. In order to clone the two PCR-amplified HR arms into the linearized and dephosphorylated pCRISPomyces-2 vector bearing the sgRNA(s), set up a Gibson Assembly reaction, as indicated in Fig. 5a (*see* **Note 7**).

a

Component	Volume (μL)	Comments
Q5® High-Fidelity 2X Master Mix	25	
10 μM Forward Primer	2.5	
10 μM Reverse Primer	2.5	
Purified genomic DNA	1	
Nuclease-free water	19	For Vf = 50 μL

b

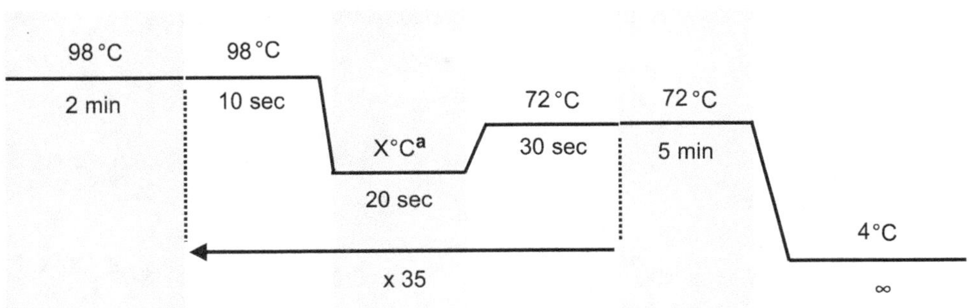

[a] *The optimal annealing temperature should be determined based on the primer sequences used (see **Note 3**).*

Fig. 4 Amplification of Homology Recombination Arms by PCR

22. Use a thermal cycler to incubate the reactions as illustrated in Fig. 5b.
23. Transform 2–5 μL of the Gibson Assembly reaction to *E. coli* NEB5alpha by heat shock, following the manufacturer's guidelines.
24. Centrifuge transformed cells for 3 min at 6000 × *g*.
25. Discard supernatant to leave about 100 μL of media.
26. Resuspend cells into the remaining medium and plate them out onto LB agar plates, supplemented with 50 μg/mL apramycin.
27. Incubate plates overnight at 37 °C.
28. Pick a few individual transformant colonies with a sterile loop and inoculate each one into 10 mL of LB supplemented with 50 μg/mL apramycin in a falcon tube and incubate cultures overnight at 37 °C, with shaking at 180 rpm.
29. Recover plasmid using a Miniprep kit, following the manufacturer's protocol, and sequence-verify the obtained plasmids for

a

Component	Volume (μL)	Comments
NEBuilder HiFi DNA Assembly 2X Master Mix	10	
HRs A	X[a]	For a 2:1 molar ratio insert:vector
HRs B	X[a]	
Digested vector	X[b]	For 100 ng
Nuclease-free water	X[c]	For Vf = 20 μL

[a] *To be determined based on the concentration of the amplicons obtained.*
[b] *To be determined based on the concentration of the digested pCm2(sgRNA).*
[c] *To be determined based on the volume of pCm2(sgRNA) and HRs added.*

b

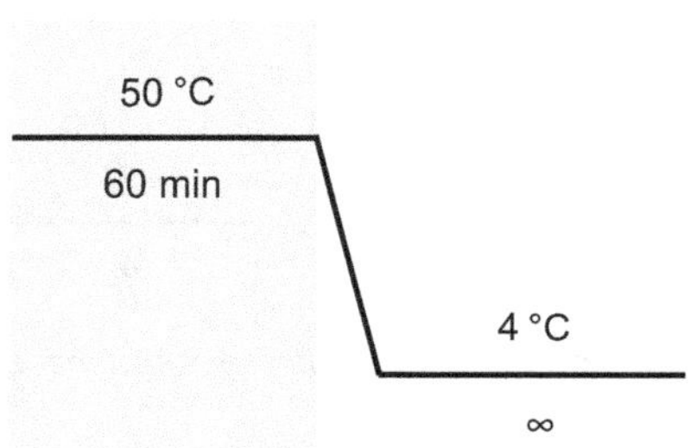

Fig. 5 Gibson Assembly of HRs into pCRISPomyces-2

correct HR arms cloning, using the primers designed in Subheading 3.1, **step 5**.

3.3 Introduction of Plasmid DNA into Actinomycetes

The most commonly used procedure for transfer of plasmid DNA into actinomycetes is intergeneric conjugation from *Escherichia coli*, which we are describing in detail here. Alternative procedures, such as protoplast transformation, may also be used and are described elsewhere [10].

All steps are to be carried out at room temperature unless otherwise stated.

1. Introduce the pCm2 plasmid bearing both the sgRNA and HR arms into *E. coli* ET12567/*pUZ8002* with your preferred transformation method, e.g., heat shock of chemically competent cells or electroporation of electrocompetent cells, and select transformant colonies by plating them onto LB agar plates containing 25 μg/mL kanamycin, 25 μg/mL chloramphenicol, 50 μg/mL apramycin and incubate at 37 °C overnight.

2. Inoculate a single colony of *E. coli* ET12567/*pUZ8002*/*pCm2* into 10 mL of LB with 25 μg/mL kanamycin, 25 μg/mL chloramphenicol, 50 μg/mL apramycin in a falcon tube and incubate at 37 °C overnight (16–20 h), with shaking at 180 rpm.
3. The morning after, inoculate 200 μL of the overnight *E. coli* ET12567/*pUZ8002*/*pCm2* culture into a falcon tube containing 10 mL of LB with 25 μg/mL kanamycin, 25 μg/mL chloramphenicol, 50 μg/mL apramycin (*see* **Note 8**) and incubate at 37 °C, with shaking at 180 rpm until an $OD_{600} = 0.4$ is reached (*see* **Note 9**).
4. While the *E. coli* culture is incubating, harvest *Streptomyces* spores by carefully pouring 10 mL of 2× YT onto the surface of the bacterial lawn.
5. Use a sterile wedge-shaped spreader to scrape off the spores; using a sterile 5 mL serological pipette, transfer the spore suspension into a sterile syringe barrel fitted with sterile cotton wool.
6. Fit the plunger into the syringe barrel, push the spore suspension through the cotton wool and decant into a sterile falcon tube.
7. Centrifuge the spore suspension at 6000 × *g* for 5 min, discard the supernatant, and resuspend the pelleted spores in 1 mL of 2× YT (*see* **Note 10**).
8. Store the spores at 4 °C until the *E. coli* strain is ready to be used (*see* **Note 11**).
9. When the *E. coli* culture has reached the desired $OD_{600} = 0.4$, harvest the cells by centrifugation at 6000 × *g* for 5 min, discard the supernatant, and wash the pellet in 10 mL of LB to remove the antibiotics by pipetting up and down.
10. While performing the washing steps on the *E. coli* cells, heat shock the actinomycete spores at 50 °C for 10 min, then leave them to cool at room temperature.
11. Repeat the centrifugation at 6000 × *g* for 5 min and wash the pelleted *E. coli* cells again in 10 mL of LB, as above.
12. Harvest the *E. coli* cells at 6000 × *g* for 5 min, resuspend the cell pellet in 1 mL of LB (enough for two conjugation reactions), and transfer them into a sterile Eppendorf tube.
13. In another sterile Eppendorf tube, mix together 500 μL of actinomycete spores and 500 μL of *E. coli* cells.
14. Centrifuge at 6000 × *g* for 2 min, remove 900 μL of supernatant, and resuspend the actinomycete/*E. coli* mixture in the remaining 100 μL by gently pipetting up and down.

15. Make a 1:10 dilution of the actinomycete/*E. coli* mixture in a final volume of 100 μL of 2× YT.
16. Plate out the neat mixture, as well as the 1:10 dilution, on SFM + 10 mM MgCl2 agar plates without any antibiotics (*see* **Note 12**).
17. Let the plates dry under a laminar flow cabinet for 10 min and incubate overnight (16–20 h) at 30 °C (*see* **Note 13**).
18. The morning after, overlay the conjugation plate with a 1 mL sterile solution of 0.5 mg nalidixic acid and 1.25 mg apramycin in sterile deionized water.
19. Spread the antibiotic solution with a sterile wedge-shaped spreader, taking care to scrape off the bacteria from the surface of the plate (*see* **Note 14**).
20. Let the plate dry under a laminar flow cabinet for 15–30 min, then incubate at 30 °C until exconjugants appear (2–5 days).
21. Pick individual actinomycete exconjugants with a sterile loop and streak each one for single colonies onto fresh SFM agar plates containing 50 μg/mL apramycin and 25 μg/mL nalidixic acid. Incubate the plates at 30 °C for 2–3 days.
22. Subculture individual colonies onto fresh SFM agar plates containing 50 μg/mL apramycin (*see* **Note 15**).

3.4 Validation of Mutants

In order to validate that the CRISPR/Cas9-based editing has occurred, a PCR-based screening is recommended, which may be followed by Illumina® whole-genome sequencing to rule out that any off-target effects have happened (*see* **Note 16**).

All steps are to be carried out at room temperature unless otherwise stated.

1. Purify the genomic DNA of the actinomycete exconjugants, as well as of the parental strain, using the MP Biomedicals FastDNA Spin Kit for Soil. Pick up the bacteria with a sterile loop from fresh plates and process approximately 100 mg of each exconjugant (*see* **Note 17**).
2. Use 1 μL of purified genomic DNA to set up a PCR amplification reaction in a 25 μL volume as for Fig. 6a (*see* **Note 18**).
3. Use a thermal cycler to incubate the reactions as illustrated in Fig. 6b (*see* **Note 19**).
4. Analyze the PCR amplification reaction through gel electrophoresis in a 1% (w/v) agarose gel in 1× TAE buffer containing 1× GelRed; load an appropriate DNA ladder and run at 100 V for 60 min.
5. Visualize the gel in a UV transilluminator (Fig. 7).

a

Component	Volume (μL)	Final concentration
Q5® High-Fidelity 2X Master Mix	12.5	1X
10 μM forward primer[a]	1.25	500 nM
10 μM reverse primer[a]	1.25	500 nM
Purified genomic DNA	1	10-100 ng
Nuclease-free water	9	For Vf = 25 μL

[a] see ***Note 18***.

b

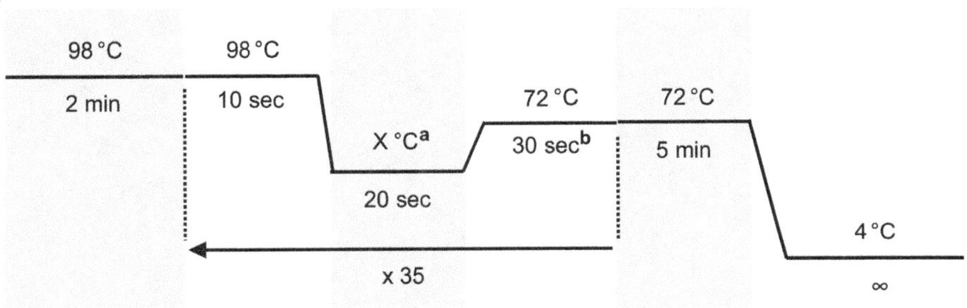

[a] *The optimal annealing temperature should be determined based on the primer sequences used (*see ***Note 3****).*
[b] see ***Note 19***.

Fig. 6 PCR amplification reaction for mutant validation

3.5 Plasmid Clearance

Once correct CRISPR/Cas9-mediated genome editing has been confirmed, we recommend clearing off the pCm2 plasmid. This may enable the user to perform a further round of CRISPR/Cas9-mediated genome editing.

1. Using a sterile loop, pick up a single colony of an exconjugant mutant strain and use it to inoculate 10 mL of YEME in a sterile falcon tube.
2. Incubate at 39–40 °C (*see* **Note 20**), with shaking at 180 rpm for 3–5 days.
3. Make serial dilutions (up to 1:10,000) of the bacterial growth in YEME using sterile Eppendorf tubes.
4. Using a sterile wedge-shaped spreader, plate out the serial dilutions (from 1:100 to 1:10,000) on non-selective SFM agar plates to obtain single colonies. Incubate the plates at 39–40 °C for 2–3 days.
5. Pick up single colonies from the above plate and streak each one onto a fresh SFM agar plate (*see* **Note 21**). Incubate the plate at 39–40 °C for 2–3 days.

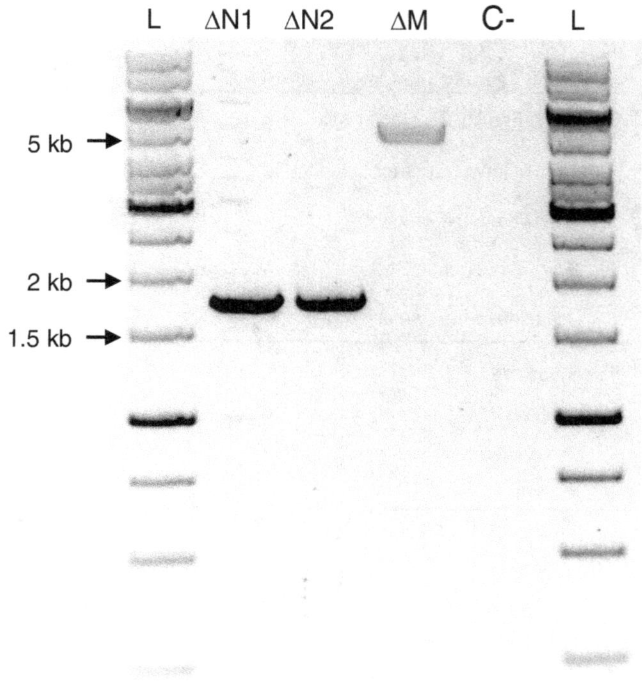

Fig. 7 CRISPR/Cas9-mediated deletion of *SclN*. Expected amplicon size from *S. albus/scl ΔsclM4 ΔsclN* (lanes ΔN1, ΔN2): 1747 bp; from *S. albus/scl ΔsclM4* (lane ΔM): 4995 bp. Lanes L: Thermo Scientific GeneRuler 1 kb DNA Ladder. (Reproduced from Ref. 8 with permission from the Royal Society of Chemistry)

6. Restreak colonies on a fresh SFM agar plate and replicate it onto another SFM agar plate containing 50 μg/mL apramycin (*see* **Note 22**). Incubate the plates at 30 °C for 2–3 days.
7. Colonies that are unable to grow on the SFM agar plate containing 50 μg/mL apramycin will have lost the pCm2 plasmid and can therefore be subjected to a second round of CRISPR/Cas9-based genome editing (*see* **Note 23**).

3.6 Preparation of a Spore Stock

1. Pick up a loopful of apramycin-sensitive exconjugant from the above SFM agar plate, resuspend the bacteria in 100 μL of YEME by pipetting up and down.
2. Using a sterile wedge-shaped spreader, plate out the bacteria on an SFM agar plate. Incubate at 30 °C for 2–3 days.
3. Pour 10 mL of sterile water onto the bacterial lawn. Use a sterile wedge-shaped spreader to scrape off the spores.
4. Using a sterile 5 mL serological pipette, transfer the spore suspension into a sterile syringe barrel fitted with sterile cotton wool.

5. Fit the plunger into the syringe barrel, push the spore suspension through the cotton wool and decant into a sterile falcon tube.
6. Centrifuge the spore suspension at 3000 × *g* for 5 min, discard the supernatant, and resuspend the pelleted cells in 1 mL 25% (w/v) glycerol.
7. Store the spore stock at −80 °C.

3.7 Metabolite Purification and LC-MS Analysis

In order to investigate the effect of the CRISPR-mediated deletion on metabolite production, analysis of the culture supernatant by liquid chromatography-mass spectrometry (LC-MS) can be carried out. In our example, we looked for a halt in the production of scleric acid resulting from the deletion of the *SclN* gene (*see* **Note 24**).

1. Grow both your mutant and wild-type strains in liquid media, as appropriate for the production of the compound of interest.
2. Add an equal volume of ethyl acetate to the volume of supernatant.
3. Adjust the pH of the medium/ethyl acetate mixture to pH 3 using 37% HCl.
4. Transfer the mixture to a separation funnel, shake several times to mix the two phases then let them sit to achieve separation.
5. Collect the ethyl acetate phase into a beaker.
6. Collect the aqueous phase in a separate beaker and add again an equal volume of ethyl acetate to it.
7. Transfer the mixture back to a separation funnel, shake several times to mix the two phases then let them sit to achieve separation.
8. Pool the ethyl acetate phase with the one collected in **step 5**.
9. Add a spoonful of anhydrous magnesium sulfate to the pooled ethyl acetate phase and stir to mix.
10. Filter out the magnesium sulfate from the ethyl acetate phase by using a disc of filter paper mounted on a funnel.
11. Evaporate the ethyl acetate under reduced pressure.
12. Dissolve the dried residue in 500 μL of 50:50 (v/v) HPLC-grade methanol/water.
13. Filter the resuspended organic fraction to remove any residual aggregates using a micro-centrifugal filter with nylon membrane and pore size 0.2 μm.
14. Transfer the sample to a snap-seal amber glass vial and keep at −20 °C until analysis.
15. Inject 5–20 μL of the prepared organic extract through a reverse phase column connected to a LC-MS.

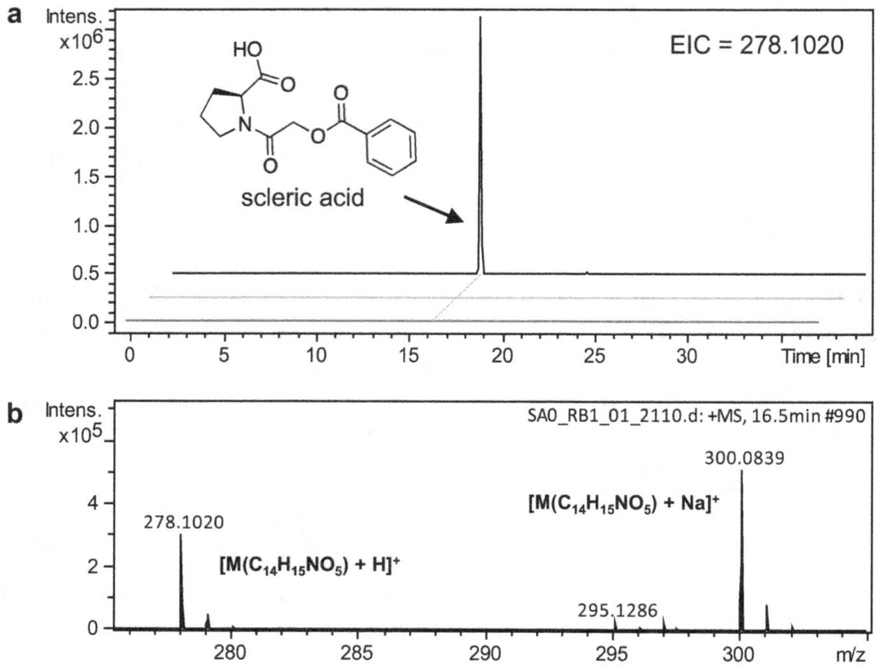

Fig. 8 Metabolite analysis on control strain and biosynthetic gene deletion mutants. (**a**) LC-MS detection of metabolites produced from *S. albus/scl ΔsclM4* (black), *S. albus/scl ΔsclM4 ΔsclN1* (green), and *S. albus/scl ΔsclM4 ΔsclN2* (magenta). Extracted ion chromatograms in positive mode for $m/z = 278.1020$ are shown, highlighting accumulation of scleric acid in *S. albus/scl ΔsclM4* (black trace). (**b**) High-resolution mass spectrum in positive mode for scleric acid. (Reproduced from Ref. 8 with permission from the Royal Society of Chemistry)

16. Separate metabolites using a linear gradient of elution from 95: 5 solvent A/solvent B to 0:100 solvent A/solvent B over 30 min, with a flow rate of 1 mL/min. Solvent A and B being water (with 0.1% formic acid) and acetonitrile (with 0.1% formic acid), respectively.
17. Operate the mass spectrometer in positive ion mode with a 50–1500 *m/z* scan range.
18. Compare the metabolic profiles of the mutant and wild-type strains using an appropriate analysis software by extracting the ion chromatograms for the expected mass of the studied compound (Fig. 8). Inspect the base-peak chromatogram to look for potential intermediates that may accumulate in the deletion mutant and could give clues on the role of the deleted gene.

4 Notes

1. Protospacers should be selected, when possible, to be on the non-coding strand, to have four purines at the 3′ of the protospacer and to be unique in the genome, even with 1-bp mismatch to avoid off-target effect of the Cas9.

2. Here is an example of a dual-sgRNA construct designed for *SclN* knock-out:

5'- **GAGACATCTTTGAAGACAA**acgcTCCGGATGGTGCGAGCGCAGg

ttttagagctagaaatagcaagttaaaataaggctagtccgttatcaacttgaaaaagtggcaccgagtcggtgc

ttttttagcataaccccttggggcctctaaacgggtcttgaggggttttttggctgctccttcggtcggacgtgcgt

ctacgggcaccttaccgcagccgtcggctgtgcgacacggacggatcgggcgaactggccgatgctggga

gaagcgcgctgctgtacggcgcgcaccgggtgcggagcccctcggcgagcggtgtgaaacttctgtgaatg

gcctgttcggttgcttttttatacggctgccagataaggcttgcagcatctgggcggctaccgctatgatcgggg

cgttcctgcaattcttagtgcgagtatctgaaaggggatacgcCCGAGACGGTCGCGGCGGG

Agttt**AAGTCTTCTTTCACGTGGC-3'**

In case of a single sgRNA, with the protospacer sequence TCCGGATGGTGCGAGCGCAG for example, order the oligonucleotides pair (a) 5′-ACGCTCCGGATGGTGCGAGCGCAG-3′ and (b) 5′-AAACCTGCGCTCGCACCATCCGGA-3′, with the underlined sequence representing the *Bbs*I--compatible overhang.

3. The annealing temperature for specific primer pairs can be calculated using the NEB Tm Calculator tool (http://tmcalculator.neb.com).

4. We recommend amplifying HR arms of a length between 0.8 and 1 kb. For design simplicity, we suggest designing the reverse and forward primers of the left and right HR arms, respectively, as the reverse complement of each other. We also suggest adding the 25 nucleotides upstream and downstream of the *Xba*I cut site as overhangs of the forward primer of the left HR arm and the reverse primer of the right HR arm, respectively. Here is an example of such oligonucleotides for the cloning of the HR arms flanking the *SclN* gene:

   pCm2-sclN LA FF.

   5′-tcggttgccgccgggcgtttttatCGTGCCACGCACGGACGT-3′.

   pCm2-sclN LA RR.

   5′-GGATGTGGTGGTGCTCATCCGGAGCGCAGGCTCGTG-3′.

   pCm2-sclN RA FF.

   5′-CACGAGCCTGCGCTCCGGATGAGCACCACCACATCC-3′.

   pCm2-sclN RA RR.

   5′-gcggccttttacggttcctggcctCTGGTCGGTCGAGGACCC-3′

5. To verify sgRNA sequence(s), we recommend designing a primer binding to the terminator downstream the gRNA scaffold. In our experience, the primer 5-′-GAGCGTTCTGAACAAATCC-3′ allows efficient sequencing of the sgRNA region of pCm2. To check HR arms, we suggest designing primers binding to the terminator downstream the gRNA scaffold and the plasmid origin of replication. A potential oligonucleotide pair is 5-′-TTGTTCAGAACGCTCGGT-3′ and 5-′-TTTCGCCACCTCTGACTTG-3′.
6. Alternatively, cells can be plated onto selective LB plates supplemented with IPTG and X-gal for blue-white screening. We have found, however, that the high efficiency of the Golden Gate assembly results in picking successful clones when simply choosing 3–5 transformants at random.
7. We advise the reader to refer to the manufacturer's guidelines on NEBuilder HiFi DNA Assembly (international.neb.com) for further details on optimal reaction setup and molar calculations. We also recommend preparing an additional reaction where the inserts (i.e., HR arms) are replaced by nuclease-free water. Obtention of transformants issued from this reaction allows the user to assess for residual undigested vector and will guide the number of transformants to screen.
8. In parallel, we recommend setting up another LB culture where the antibiotic amounts are halved, because in our experience the full antibiotic amounts may prevent the cells from reaching the desired OD before the end of the working day.
9. This is expected to take between 3 and 5 h depending on the amount of initial inoculum and the concentration of antibiotics used.
10. This is enough for two conjugation reactions; if conjugation with more than two plasmids is being carried out during the experiment, then we recommend preparing spore stocks from more than one initial plate.
11. Some protocols recommend storing the spores at 4 °C overnight, in order to allow for pre-germination and improve the efficiency of the conjugation.
12. Alternative media to SFM may be more appropriate for the growth of specific actinomycete strains or may be trialed in parallel. We recommend the reader to refer to reference 9 for a comprehensive list of potential media to use.
13. Some actinomycete strains tend to overgrow if incubated overnight at 30 °C, therefore a lower incubation temperature, e.g., 25 °C, may be preferable.

14. Some protocols recommend to simply tilt the plate to allow the antibiotic solution to reach the entirety of the bacterial lawn and to not use a spreader to scrape off the bacteria from the surface of the agar medium; we have found that this can lead to a high background of untransformed colonies growing on the conjugation plate, especially when working with actinomycetes that form a highly hydrophobic lawn.
15. We recommend that plates are divided into 4 × 4 sectors so that up to 16 exconjugants can be streaked onto each plate. At this stage, the nalidixic acid selection can be dropped because the *E. coli* would have not survived the two previous rounds of selection.
16. We advise the reader to refer to the manufacturer's protocols (http://illumina.com) for instructions on library preparation, if performing Illumina whole-genome sequencing.
17. We recommend screening at least 5–6 exconjugants, to increase chances of having positive clones, as well as to have biological replicates for the subsequent metabolite analysis.
18. The primers for PCR amplification screening should be designed so that they can only bind to the genomic DNA and not to the HR arms included in the pCm2 plasmid. We recommend designing the forward primer so that it will bind approximately 50 bp upstream of the left HR arm, and the reverse primer so that it will bind approximately 50 bp downstream of the right HR arm. This will enable you to selectively amplify from the target genomic DNA locus.
19. Use an extension time that will allow you to amplify the longest expected amplicon; i.e., when using CRISPR/Cas9 to perform a gene deletion, aim to amplify the wild-type sequence.
20. Plasmid pCRISPomyces-2 contains a temperature sensitive origin of replication pSG5 [11], which means that the plasmid is only replicated when the bacteria are grown at temperatures below 34 °C; in our experience, effective plasmid clearance can be achieved when actinomycetes are grown at a temperature of 39 °C or above. If the specific actinomycete strain under study cannot grow at such high temperatures, we advise to attempt plasmid clearance at lower temperatures but always above 34 °C.
21. We recommend that a plate is divided into 4 × 4 sectors so that up to 16 colonies can be streaked onto it.
22. Instead of doing replica plating, it might be easier to streak sequentially the same colony first on the SFM agar plate and then on the SFM agar plate containing 50 μg/mL apramycin, without sterilizing/changing the loop in between streaks of the same colony.

23. After plasmid clearance, we recommend to validate once more that the desired genome editing is still in place, following the procedure described in Subheading 3.4. Actinomycetes form multinucleated mycelia, therefore a mixed population of transformed/untransformed nuclei may arise and, in non-selective conditions, may result in propagation of the untransformed nuclei.

24. The following steps are optimized for the isolation of organic compounds but should be adapted based on the nature of the molecule of interest.

References

1. Jinek M, Chylinski K, Fonfara I et al (2012) Programmable dual-RNA—guided DNA endonuclease in adaptive bacterial immunity. Science 337(6096):816–821. https://doi.org/10.1126/science.1225829
2. Cobb RE, Wang Y, Zhao H (2015) High-efficiency multiplex genome editing of Streptomyces species using an engineered CRISPR/Cas system. ACS Synth Biol 4(6):723–728. https://doi.org/10.1021/sb500351f
3. Huang H, Zheng G, Jiang W et al (2015) One-step high-efficiency CRISPR/Cas9-mediated genome editing in Streptomyces. Acta Biochim Biophys Sin Shanghai 47(4): 231–243. https://doi.org/10.1093/abbs/gmv007
4. Tong Y, Charusanti P, Zhang L et al (2015) CRISPR-Cas9 based engineering of Actinomycetal genomes. ACS Synth Biol 4(9): 1020–1029. https://doi.org/10.1021/acssynbio.5b00038
5. Zeng H, Wen S, Xu W et al (2015) Highly efficient editing of the Actinorhodin polyketide chain length factor gene in Streptomyces coelicolor M145 using CRISPR/Cas9-CodA(sm) combined system. Appl Microbiol Biotechnol 99(24):10575–10585. https://doi.org/10.1007/s00253-015-6931-4
6. Genilloud O (2017) Actinomycetes: still a source of novel antibiotics. Nat Prod Rep 34(10):1203–1232. https://doi.org/10.1039/c7np00026j
7. Alberti F, Corre C (2019) Editing Streptomycete genomes in the CRISPR/Cas9 age. Nat Prod Rep 36(9):1237–1248. https://doi.org/10.1039/C8NP00081F
8. Alberti F, Leng DJ, Wilkening I et al (2019) Triggering the expression of a silent gene cluster from genetically intractable bacteria results in Scleric acid discovery. Chem Sci 10(2): 453–463. https://doi.org/10.1039/c8sc03814g
9. Blin K, Pedersen LE, Weber T et al (2016) CRISPy-web: an online resource to design sgRNAs for CRISPR applications. Synth Syst Biotechnol 1(2):118–121. https://doi.org/10.1016/j.synbio.2016.01.003
10. Kieser T, Bibb MJ, Buttner MJ et al (2000) Practical Streptomyces genetics: a laboratory manual. John Innes Foundation, Norwich
11. Muth G, Nußbaumer B, Wohlleben W et al (1989) A vector system with temperature-sensitive replication for gene disruption and mutational cloning in Streptomycetes. Mol Gen Genet 219(3):341–348. https://doi.org/10.1007/BF00259605

Chapter 12

Understanding and Manipulating Assembly Line Biosynthesis by Heterologous Expression in *Streptomyces*

Lihan Zhang, Takayoshi Awakawa, and Ikuro Abe

Abstract

Assembly line enzymes, including polyketide synthases and nonribosomal peptide synthetases, play central roles in the construction of complex natural products. Due to the sequential biochemistry processed in each domain, the domain architecture of the assembly line enzymes strictly correlates with the product molecule. This colinearity makes assembly line enzymes an ideal target for rational reprogramming. Although many of the past engineering attempts suffered from decreased product yield, recent advancements in the bioinformatic analysis and engineering design now provide new opportunity to work on these modular mega-enzymes. This chapter describes the methods for analyzing and engineering the assembly line enzymes, including module and domain analysis needed for designing the engineering of assembly line biosynthesis, and the expression vector construction with an example of two-vector heterologous expression system in *Streptomyces*.

Key words Assembly line engineering, Heterologous expression, Bioinformatic analysis, *Streptomyces*, Biosynthetic pathway

1 Introduction

Nature has developed a plethora of complex natural products through the evolution of biosynthetic enzymes. One of the most sophisticated biosynthesis is the use of templated, modular enzymes comprising multiple domains within a single polypeptide [1]. Polyketide synthase (PKS) [2] and nonribosomal peptide synthetase (NRPS) [3] are the two main family of modular, assembly line-type megaenzymes most frequently observed in filamentous actinobacteria such as *Streptomyces* [4]. Because of the sequential biochemistry processed in each domain, the domain architecture of the assembly line enzymes strictly correlates with the product compound. This colinearity makes assembly line enzymes an ideal target for rational reprogramming. Since the discovery of these assembly lines in the 1990s, numerous attempts have been made to rationally engineer the assembly line enzymes but resulted in limited success

Elizabeth Skellam (ed.), *Engineering Natural Product Biosynthesis: Methods and Protocols*,
Methods in Molecular Biology, vol. 2489, https://doi.org/10.1007/978-1-0716-2273-5_12,

hampered by a severe decrease in product yield in many cases [5]. However, recent advancements in bioinformatics, molecular cloning, and evolution-based engineering now provide new opportunities to work on these modular megaenzymes [6, 7].

The biosynthetic logic of PKS and NRPS shares commonality in that each module catalyzes condensation of a monomer—an acetate/propionate unit by a ketosynthase (KS) domain in PKS and an amino acid by a condensation (C) domain in NRPS—that are tethered on an acyl/peptidyl carrier protein (ACP/PCP). The elongated intermediate is then modified by processing domains, to complete one elongation cycle. Since their discovery, a module is defined by a set of domains starting from KS/C and ending with ACP/PCP [1]. Recent engineering studies and evolutionary analyses, however, suggest that both KS and C domains indeed play the gatekeeping role and have a strong coevolutionary tie with the upstream carrier protein [8, 9]. This led to the proposal of a new module definition in PKS [10] and an exchange unit concept in NRPS [9, 11], both of which put a carrier protein in the middle of the module and ending with a KS or C domain. As the number of sequenced biosynthetic gene clusters increase drastically, we envision that better understanding of these assembly line enzymes can be gained from bioinformatic and evolutionary analysis, which ultimately may lead to a breakthrough in modular enzyme engineering.

The assembly line enzymes are encoded in large, repetitive genes colocalized as a gene cluster spanning several dozens of kilobase pairs to even more than 200 kbp in length [12]. This nature makes them challenging for precise enzyme manipulation in native producer strains; hence many engineering attempts employed with small, one or a few modules constructs, with synthetic intermediates as substrates [13–15]. To engineer the final product of assembly line enzymes, heterologous expression is a key technique in PKS and NRPS studies [16–18]. In addition, by taking advantage of modern molecular cloning methods, a systematic gene deletion series of individual biosynthetic genes can be generated faster in a heterologous expression system than knocking out each gene in a native producer strain [19]. The biotechnology tools including seamless cloning [20], Red/ET recombineering [21], TAR cloning [22], RecE direct cloning [23], and CRISPR-Cas system [24] have offered solid foundations for heterologous expression of assembly line gene clusters [25].

In this chapter we first discuss the bioinformatics protocols for analyzing assembly line enzymes that will aid better understanding and engineering of assembly line enzymes. We then describe the protocol of the two-vector expression system in *Streptomyces*. The two-vector system enables flexible investigation of a biosynthetic pathway by gene deletion and complementation [19, 26, 27]. We use bioengineering of PKS-NRPS biosynthetic pathway, which in our experience can take the best advantage of heterologous expression, as an example to describe each step needed for heterologous expression of a gene cluster. General methods for genetic manipulation and heterologous expression of *Streptomyces* are summarized in several resources [28–31]. Finally, we note that the strategy discussed here is not limited to biosynthetic studies of assembly line enzymes but widely applicable for biosynthetic pathway investigation.

2 Materials

2.1 Bioinformatic Analysis of Actinobacterial Biosynthetic Gene Clusters

We list several programs that we find useful according to each purpose. Other equivalent programs can also be employed.

1. Bioinformatics platform: Geneious Prime (*see* **Note 1**).
2. Molecular cloning: SnapGene software.
3. ORF finder: Frameplot4.0 [32] (http://nocardia.nih.go.jp/fp4/).
4. Conserved domain search: CD-search [33], HMMER [34].
5. Homolog search: BLAST [35].
6. Phylogenetic analysis platform: MEGA X [36].
7. Multiple sequence alignment: MAFFT [37].
8. Tree construction: FastTree2.0 [38].
9. Tree visualization: iTOL [39].
10. Biosynthetic gene cluster analysis: antiSMASH [40].
11. Structural modeling: SWISS-MODEL [41].

2.2 Construction of Heterologous Expression Vector for *Streptomyces*

1. Antibiotic stock solutions: ampicillin (50 mg/ml in dH_2O); apramycin (50 mg/ml in dH_2O); kanamycin (50 mg/ml in dH_2O); chloramphenicol (25 mg/ml in ethanol); thiostrepton (50 mg/ml in DMSO).
2. L-Arabinose: 1 M in dH_2O.
3. LB-Miller broth: 10 g Bacto-tryptone, 5 g yeast extract, 10 g NaCl per liter dH_2O.
4. LB-agar: 10 g Bacto-tryptone, 5 g yeast extract, 10 g NaCl, 15 g agar per liter dH_2O.
5. *E. coli* strains: DH5α; BW25113/pKD76.
6. Plasmids: pUC19; pRT801; pTYM19ep [42].
7. PCR thermocycler.
8. Primestar Max DNA polymerase (Takara, Japan).
9. Restriction enzymes.
10. In-Fusion HD cloning kit (Takara, Japan).

11. Miniprep kit.
12. DNA purification kit.
13. Forward and reverse primers.

2.3 Transformation of Streptomyces

1. *Streptomyces lividans* TK21, *S. albus* J1024, or *S. avermitilis* SUKA [43].
2. Antibiotic stock solutions: ampicillin (50 mg/ml in dH_2O); apramycin (50 mg/ml in dH_2O); kanamycin (50 mg/ml in dH_2O); chloramphenicol (25 mg/ml in ethanol); thiostrepton (50 mg/ml in DMSO).
3. TSB broth: 30 g Tryptone Soya Broth powder per liter dH_2O.
4. YEME broth: 340 g sucrose, 10 g glucose, 3 g yeast extract, 3 g malt extract, 5 g Bacto peptone per liter dH_2O. Add sterile 2 ml 2.5 M $MgCl_2{\cdot}6H_2O$ and 50 ml 20% glycine after autoclave.
5. MS agar: 20 g D-mannitol, 20 g soy flour (local supermarket), 20 g agar per liter tap water. Add sterile 2 ml 2.5 M $MgCl_2{\cdot}6H_2O$ after autoclave.
6. R5 agar: 103 g sucrose, 10 g glucose, 0.25 g K_2SO_4, 10.12 g $MgCl_2{\cdot}6H_2O$, 5 g yeast extract, 0.1 g casamino acid, 5.73 g TES, 3 g L-proline, 2 ml trace element (see below), 20 g agar per liter dH_2O. Add sterile 10 ml 0.5% KH_2PO_4, 4 ml 5 M $CaCl_2{\cdot}2H_2O$, 7 ml 1 M NaOH after autoclave and mix thoroughly.
7. Trace element: 40 mg $ZnCl_2$, 20 mg $FeCl_3{\cdot}2H_2O$, 10 mg $CuCl_2{\cdot}2H_2O$, 10 mg $MgCl_2{\cdot}4H_2O$, 10 mg $Na_2B_4O_7{\cdot}10H_2O$, 10 mg $(NH_4)_6Mo_7O_{24}{\cdot}4H_2O$ per liter dH_2O.
8. Conc. P buffer: 103 g sucrose, 0.25 g K_2SO_4, 2.02 g $MgCl_2{\cdot}6H_2O$, 2 ml trace element per 800 ml dH_2O.
9. P buffer: 80 ml Conc. P buffer, 10 ml 0.25 M TES pH 7.2, 10 ml 3.68% $CaCl_2{\cdot}2H_2O$, 1 ml 0.5% KH_2PO_4, prepare sterile individual solutions and mix them with the above order (*see* **Note 2**).
10. Sucrose solution: 103 g sucrose per liter dH_2O.
11. PEG solution: 1 g PEG 1000 3.85 ml sterile P buffer, autoclave PEG first and then add P buffer, mix thoroughly.
12. Lysozyme buffer: 20 mg lysozyme (Nacalai, Japan) in 10 ml P buffer.
13. Cotton filter (*see* **Note 3**).

3 Methods

3.1 Bioinformatics Analysis of Assembly Line Enzymes

The first step for the successful engineering of the assembly line enzymes is to correctly understand the domains and modules in the enzyme. In public genome database, these genes are often misannotated because of their large size, and their highly repetitive sequences tend to cause errors in ORF annotation or in next-generation sequencing itself. This section covers ORF prediction, module prediction, domain and motif detection, phylogenetic analysis, and sequence-based structural analysis (Fig. 1).

1. Find open reading frames (ORFs) that (i) span 2500 bp or longer and (ii) have >85% or higher GC content at the third letter of each codon by a ORF analysis program like Frameplot4.0 (Fig. 2) (*see* **Note 4**).
2. Determine the start codon. The start codon is usually associated with a ribosome binding site (RBS) of AG-rich region at 6–10 bp upstream (*see* **Note 5**). For modules that accept an intermediate from previous modules, the N-terminal contains a docking domain of ~40 AAs for PKS and ~20 AAs for NRPS [44].
3. Perform domain detection. Input amino acid sequence into a domain search program such as CD-search or HMMER. BLAST can also be used in case comparison to other enzymes is important.
4. Extract module sequences. Extract the amino acid sequences of each module from the ORFs by cutting at the ACP-KS boundary, or PCP-C boundary (Fig. 3) (*see* **Note 6**).
5. Multiple sequence alignment (MSA) of modules. Construct MSA by MAFFT with E-INS-i method (*see* **Note 7**).
6. Active site analysis. Whether the domain is active or not can be predicted by the active site residues of each domain (e.g., active site Cys of KS shown in Fig. 1). The active site residues of PKS and NRPS are reviewed in [3, 45].)
7. Specificity prediction. antiSMASH gives good prediction for substrate specificity of AT (for PKS) and A (for NRPS) domains. Submit the gene cluster information in Genbank format to antiSMASH. In modular PKSs, sequence motifs that reflect the substrate specificity of AT and KR domains can also be easily identified from MSA (Fig. 4). The C domain in NRPS is known to form different clades in phylogenetic tree according to its function (*see* **step 9**).
8. Extract MSA of the domain of interest from the entire module MSA.

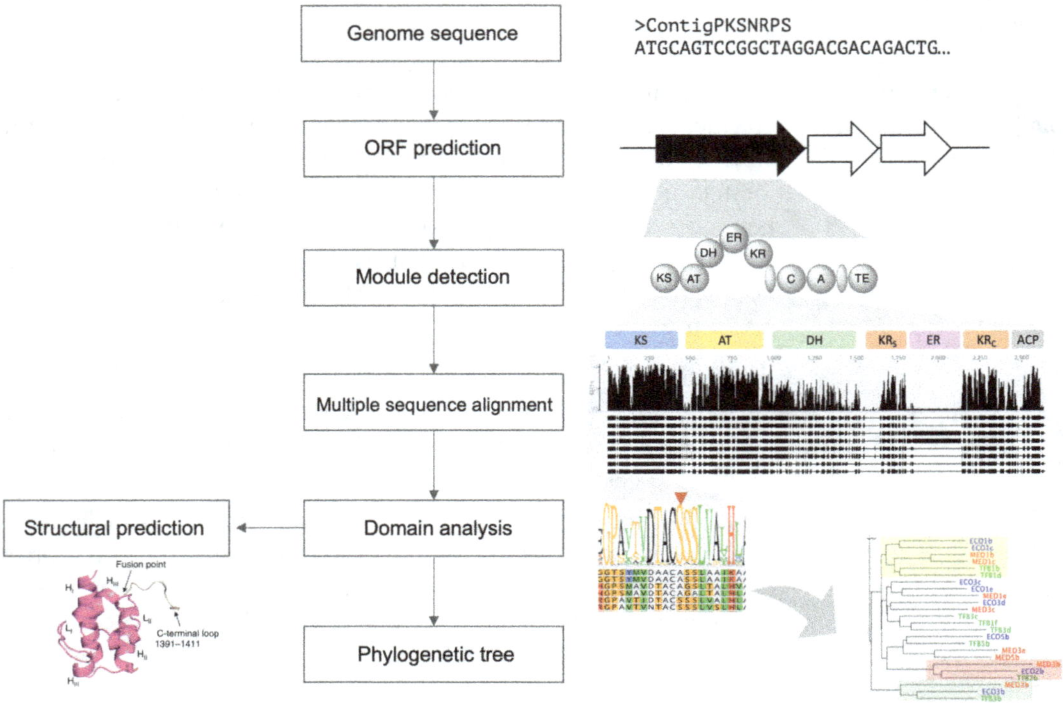

Fig. 1 Overall scheme of bioinformatic analysis for assembly line enzymes

9. Generate a phylogenetic tree of each domain. (Fig. 5) Using MSA of a domain of interest to construct trees by FastTree with WAG model (*see* **Note 8**).
10. Structural prediction of a domain of interest. Input the AA sequence to SWISS-MODEL and search for template. From the resulting templates, select the best scoring template (*see* **Note 9**), then build a model.
11. Based on the secondary structure predicted in **step 10**, find a conformationally flexible loop without secondary structure that does not interrupt domain–domain interaction as a candidate site for enzyme engineering site.
12. Alternatively, find a linker region that connects domains but with high sequence homology, to use as a candidate site for homologous recombination site.

3.2 Construction of Heterologous Expression Vector for *Streptomyces*

The expression vector is critical for successful heterologous expression. *E. coli-Streptomyces* shuttle vectors are most frequently used due to its convenience in gene manipulation. The components needed for an expression shuttle vector includes: (1) replication site; (2) resistance marker for *E. coli*; (3) replication site or phage attachment site, and integrase for *Streptomyces*; (4) resistance marker for *Streptomyces*; and (5) a promoter region. When

Fig. 2 Result of Frameplot analysis. (**a**) Six reading frames illustrated by three colors. Arrows represent possible start codons. (**b**) Third-letter GC content chart. From 3122 to 4720 bp shows an ORF (the ORF arrow added by authors for clarity) with the green line on top, indicating that the green reading frame has the highest third-letter GC content. The broken line shows 70% GC content line. (**c**) Amino acid sequence of the selected ORF. The ORF can be selected by clicking the above arrows. (**d**) Nucleotide sequence including −120 region and ribosome binding site underlined

Domain search result (4160 aa)

Model	No.	Description	Position (aa)	Score	E-value
AMP-binding	2	AMP-binding enzyme	484-903, 1568-1957	657.6	5.2e-201
Condensation	2	Condensation domain	3-296, 1090-1387	387.3	3.5e-119
ketoacyl-synt	1	Beta-ketoacyl synthase, N-terminal domain	2149-2397	208.9	8.8e-65
KR	1	KR domain	3549-3729	195.2	7.8e-61
Acyl_transf_1	1	Acyl transferase domain	2674-2982	187.9	2.9e-58
PP-binding	3	Phosphopantetheine attachment site	987-1050, 2046-2109, 3793-3855	150.2	1.4e-47
Ketoacyl-synt_C	1	Beta-ketoacyl synthase, C-terminal domain	2405-2528	140.1	2.9e-44
Thioesterase	1	Thioesterase domain	3888-4148	109.1	3.8e-34
DH	1	DH domain	3062-3234	39.2	7.4e-13
Epimerase	1	NAD dependent epimerase/dehydratase family	3552-3721	23.1	4.3e-08

Show details Hide details -- | -- Erase data

Fig. 3 Domain analysis by PKS/NRPS-domain Finder (http:/avermitilis.ls.kitasato-u.ac.jp/pfam-pks/), which also uses hidden Markov model (HMM)-based prediction. Each domain is colored with description and likelihood score. This protein contains three modules (C-A-PCP)-(C-A-PCP)-(KS-AT-DH-KR-ACP-TE) in traditional definition

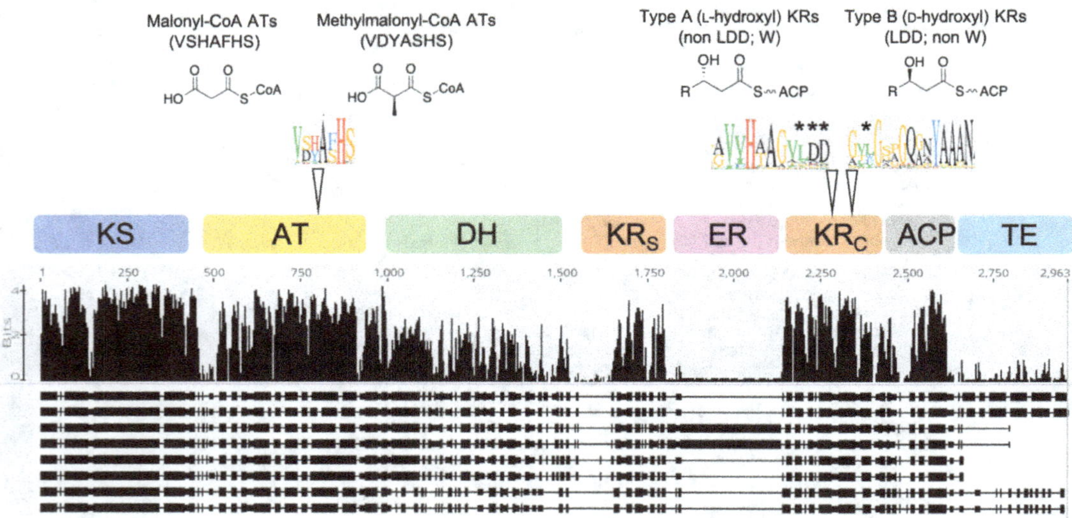

Fig. 4 Multiple sequence alignment (MSA) of PKS domains with specificity motif of AT and KR highlighted by sequence logo

conjugation method is used for transformation, an *oriT* region is also needed.

The engineering of assembly line enzymes can be divided into two strategies: point mutation and domain/module recombination. Point mutation has advantage in its simple design and is particularly useful in altering substrate specificity of a target domain, but often results in significantly decreased yield [46]. Domain/module recombination is a promising approach, with a number of recent engineering results showing equivalent

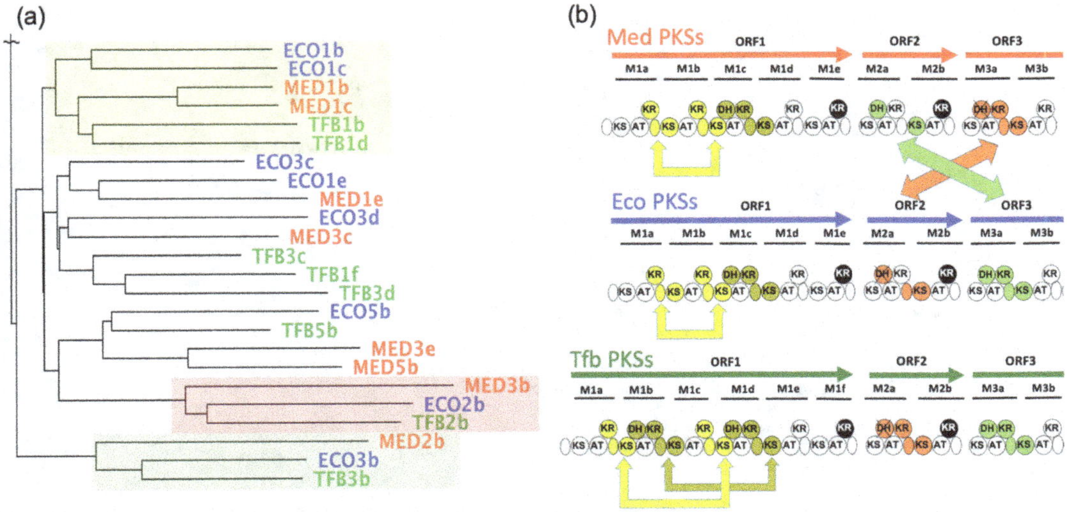

Fig. 5 Phylogenetic analysis of KS domains from three aminopolyol PKSs, modified from ref. [8]. (**a**) Maximum likelihood tree of KS domains from Med, ECO, and TFB gene clusters. Colored branches show the closest distance in phylogeny and suggest gene duplication events. (**b**) Integrated results of phylogenetic analysis of each domain showing gene recombination events

yield to the wild-type pathway [11, 47, 48]. Here we use an AT domain exchange in a PKS gene as an example to describe the engineering protocol. The two-plasmid expression system can express relatively large gene clusters and allows convenient manipulation of a target gene. The overall scheme is shown in the following (Fig. 6).

1. Using primers designed for In-Fusion cloning (*see* **Note 10**), PCR amplify the biosynthetic genes and put adjacent to each other so that they can form a new operon (*see* **Note 11**). Purify the PCR product by agarose gel electrophoresis.
2. Perform In-Fusion reaction (*see* **Note 10**) with *Eco*RI and *Hind*III digested pUC19 at 50 °C for 15 min, transform into *E. coli* DH5a to obtain pUC19-1.
3. Select resistant colonies; confirm that the target genes are correctly cloned by PCR or restriction enzyme digestion. Extract the correct vector.
4. PCR amplify the ϕC31, *tsr*^{R}, and P*ermE** region from pTY-M19ep (*see* **Note 12**) with overhanging primer targeting at the *Hind*III site of pUC19-1.
5. Perform In-Fusion reaction with *Hind*III digested pUC19-1 at 50 °C for 15 min and transform into *E. coli* DH5a.
6. Check the heterologous expression elements are correctly inserted to obtain pExpress-1.

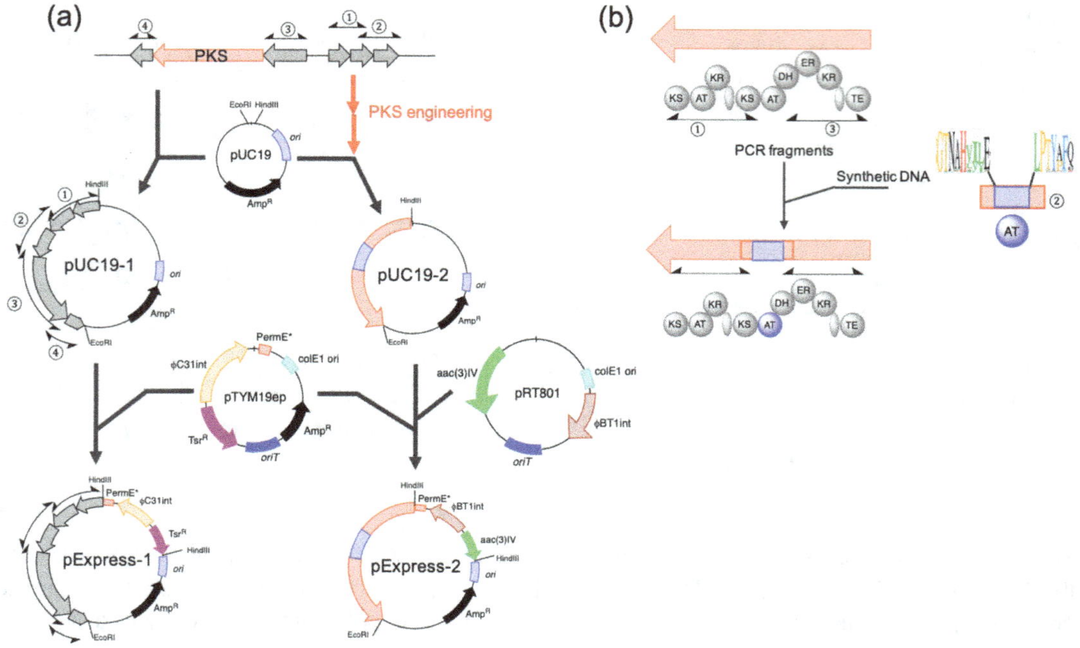

Fig. 6 Overall scheme of two-plasmid system for PKS engineering by heterologous expression. (**a**) Plasmid construction scheme. (**b**) Scheme of AT swapping by In-Fusion cloning. Blue AT domain is replaced with original AT domain at the GTNAH motif at the end of KS and the LPTYxF motif at the end of post-AT linker [50]. Synthetic DNA fragment is prepared with longer margin corresponding to the original sequence. The obtained PKS with AT swap will then be cloned into a heterologous expression vector

7. PCR amplify KS-AT-KR-ACP-KS and DH-ER-KR-ACP-TE fragments with hangover for In-Fusion cloning (Fig. 6b) (*see* **Note 13**).
8. Prepare synthetic DNA that has the target AT sequence and original sequence at both N- and C-terminals, adjusted with the above obtained PCR fragment for In-Fusion ligation (Fig. 6b).
9. Perform In-Fusion ligation using the PCR fragments, synthetic DNA with target AT, and pUC19 vector. Check the constructed vector to obtain pUC19-2.
10. Introduce the heterologous expression elements of ϕBT1 (*see* **Note 14**) and *aac(3)IV*, and P*ermE** to the pUC19-2.

If a cosmid/bacterial artificial chromosome (BAC) library is available, an expression vector can be easily made by simple gene targeting to introduce heterologous expression elements using Red/ET technology.

11. Transform the cosmid/BAC vector of interest into *E. coli* BW25113/pKD78 (*see* **Note 15**). Culture at 30 °C with appropriate antibiotics for the vector selection and 25 μg/ml chloramphenicol.

12. Prepare electrocompetent cells of the *E. coli* BW25113/ pKD78 with the target vector. Culture 100 ml of LB with appropriate antibiotics (for the target vector selection), 25 μg/ml chloramphenicol, and 10 mM L-arabinose (final conc.) in 500 ml Erlenmeyer flask at 30 °C at 200 rpm for ~4 h to reach $OD_{600} = 0.2–0.4$. Wash twice with 10% glycerol, resuspend in 100 μl of 10% glycerol. Flash freeze and store at −80 °C if needed.
13. PCR amplify ϕBT1 and *aac(3)IV* region from pRT801 with overhanging primer targeting at flanking region of the target biosynthetic gene cluster for λ RED mediated recombination. Purify the PCR product by agarose gel electrophoresis.
14. Transform the PCR fragment of ϕBT1 and *aac(3)IV* into the electrocompetent cell obtained from **step 2** by standard electroporation. Culture at 37 °C for overnight with the appropriate antibiotics and 50 μg/ml apramycin.
15. Check the insert by PCR or restriction enzyme digestion and extract the cosmid/BAC vector (*see* **Note 16**).

3.3 Transformation of Streptomyces

The *Streptomyces lividans* strain has been widely used for heterologous expression host due to the absence of strong DNA restriction system. Other hosts, such as *S. albus*, *S. avermitilis*, and *S. coelicolor*, are also frequently used for secondary metabolite production (*see* **Note 17**). For the protocol of another major transformation method conjugation, several references can be found elsewhere [28, 31].

1. Grow *S. lividans* spore suspension or glycerol stock on ISP2 agar for 5 days.
2. Inoculate 2 × 2 cm agar piece on 100 ml of YEME medium in 500 ml baffled Erlenmeyer flask and incubate at 28 °C for 36–40 h at 200 rpm.
3. Harvest the culture by centrifugation at 4000 × *g* for 10 min at room temperature, discard supernatant, and resuspend the cell pellet in 20 ml Sucrose solution.
4. Centrifuge again at the same condition, discard the supernatant, resuspend the cells in 10 ml P buffer.
5. Add 10 ml Lysis Buffer, incubate at 30 °C for 30–60 min with gentle mixing (*see* **Note 18**).
6. Filtrate by sterilized cotton.
7. Centrifuge at 4000 × *g* for 10 min at room temperature, discard supernatant, resuspend in 20 ml of P buffer.
8. Repeat **step 6**, then centrifuge at the same condition, discard supernatant, resuspend in 2 ml of P buffer.

9. Dispense 50 μl. (For storage, freeze by liquid nitrogen and store at −80 °C *see* **Note 19**).

The following describes the transformation of protoplast cells.

10. (If using preserved protoplast stock, thaw protoplast on ice.) Prepare ×1, ×5, and ×25 dilution using P buffer to test appropriate concentration for obtaining single colony transformant.
11. Add 1–5 μl plasmid/cosmid solution in 50 μl to the protoplasts, then immediately add 200 μl PEG solution, mix by gentle pipetting, then gently spread on R5 plate; wait for plate to be completely dry.
12. Incubate at 28 °C for 12–16 h, then overlay appropriate antibiotics; wait for plate to be completely dry.
13. Colonies can be seen after 48 h of incubation at 28 °C. Pick a colony by a toothpick, streak on a antibiotics-containing MS agar plate. Visible growth should be seen after 48 h incubation at 28 °C.
14. Pick a colony and soak in 5 μl of DMSO for 30 min. Use 1 μl for colony PCR to check the insert if needed.
15. Select multiple colonies for metabolic profiling (*see* **Note 20**).

4 Notes

1. Geneious Prime can handle most bioinformatic analysis tasks including molecular cloning, multiple sequence alignment, phylogenetic analysis, and next-generation sequencing assembly.
2. 500 μl of P buffer is needed for one protoplast-PEG transformation. 200 ml of P buffer is needed for *Streptomyces* protoplast preparation.
3. Put cotton in a syringe and sterilize.
4. In general, PKS or NRPS enzymes have more than 1000 amino acids, but fragmented enzymes also rarely exist. *Streptomyces* are known to have codon bias of the third letter GC content being exceptionally high. Frameplot4.0 is also useful for strains that are AT-rich and have codon letter biases.
5. The start codon is not always associated with a predictable RBS. The start codon of GTG is also common in *Streptomyces.*
6. Due to the repetitive nature of assembly line enzymes, analyses based on modules give more information and are preferred in many cases. The new module definition ending with KS/C domain often spans for two proteins, because the PKS/NRPS genes tend to end with a carrier protein and start with a KS/C

domain. Therefore, for module extraction, applying traditional module definition is more convenient.

7. Only sequences with homology can give meaningful MSA, and modules containing rare functional domains like methyltransferase domains may need to be treated separately if they disrupt the alignment. Do not put both NRPS modules and PKS modules in one MSA. If dealing with >1000 sequences, FFT-NS-i method can accelerate the calculation time but is less accurate.

8. When comparing multiple homologous assembly line enzymes, a phylogenetic tree can tell the possible evolutionary scenario or at least in a rough resolution can help distinguish different groups. When phylogenetic information is not needed, pairwise sequence identity is a convenient alternative to assess domain similarity.

9. The best template can be judged by sequence identity and the global model quality estimation (GMQE) value (sequence identity 0–100%, GMQE value 0–1, both the higher the better).

10. Other cloning methods such as Gibson assembly are also applicable. We noticed that ligation of more than five fragments has significantly less efficiency by In-Fusion cloning, though we have succeeded in seven fragments ligation for the maximum. The accurate molar ratio of each fragment is important when dealing with multiple fragments.

11. *Streptomyces* genome is GC-rich and addition of 3–5% DMSO or 1 M betaine to the PCR reaction mix is beneficial for successful PCR.

12. ϕC31 region consists of ϕC31 integrase gene and attP region. These elements can also be obtained from DNA synthesis or other plasmids such as: ϕC31, pSET152; *tsr*^{R}, pGM190; P*ermE** pUWL201.

13. Due to the repetitive nature of PKS gene sequences, we recommend selecting the less conserved region to amplify PCR fragment and adjust the length of inserted AT domain by DNA synthesis.

14. Other integration systems such as R4 or TG1 [49] can also be used.

15. Larger vectors, typically over 100 kbp, are prone to physical sheering. Avoid vortexing. Cutting the pipette tip and gently pipetting is recommended.

16. Commercial cosmid/BAC extraction kits can be used. General DNA precipitation by ethanol/isopropanol also works fine. Avoid vortexing and pipette gently to prevent physical sheering.

17. *S. lividans* lacks DNA restriction system and can accept methylated DNA. Strains that have strong DNA restriction systems like *S. coelicolor* need non-methylated DNA template for transformation. *E. coli* strains such as JM109 and ET12567 can be used for this purpose.
18. Depending on the amount and type of cells used, incubation time can be prolonged. Addition of achromopeptidase 1 mg/ml can facilitate protoplast formation for difficult strains such as *S. coelicolor*.
19. Protoplasts are generally stable for storage for more than a year at deep freezer.
20. To avoid the effect of accidental gene recombination of the introduced genes and other undesired events in the expression host, screening of multiple colonies is highly recommended.

Acknowledgments

This work was supported by National Natural Science Foundation for Young Scientists of China (82003631) and National Natural Science Foundation of China General Program (22177092) to L. Z., Grant-in-Aid for Scientific Research from the Ministry of Education, Culture, Sports, Science and Technology, Japan (JSPS KAKENHI Grant Number JP16H06443, JP 20KK0173, and JP20H00490), Japan Science and Technology Agency (JST SICORP Grant No. JPMJSC1701), the New Energy and Industrial Technology Development Organization (NEDO, Grant Number JPNP20011), and AMED (Grant Number JP21ak0101164) to I. A., and JP17H04763, JP19H04641, JP21H02636, UTEC-UTokyo FSI Research Grant Program, Kato Memorial Bioscience Foundation, and The Asahi Glass Foundation, to T.A.

References

1. Fischbach MA, Walsh CT (2006) Assembly-line enzymology for polyketide and nonribosomal peptide antibiotics: logic, machinery, and mechanisms. Chem Rev 106:3468–3496
2. Hertweck C (2009) The biosynthetic logic of polyketide diversity. Angew Chem Int Ed Engl 48:4688–4716
3. Sussmuth RD, Mainz A (2017) Nonribosomal peptide synthesis-principles and prospects. Angew Chem Int Ed Engl 56:3770–3821
4. Moumbock AFA, Gao M, Qaseem A et al (2020) StreptomeDB 3.0: an updated compendium of streptomycetes natural products. Nucleic Acids Res 49:D600–D604
5. Weissman KJ (2016) Genetic engineering of modular PKSs: from combinatorial biosynthesis to synthetic biology. Nat Prod Rep 33:203–230
6. Alanjary M, Cano-Prieto C, Gross H et al (2019) Computer-aided re-engineering of nonribosomal peptide and polyketide biosynthetic assembly lines. Nat Prod Rep 36:1249–1261
7. Nivina A, Yuet KP, Hsu J et al (2019) Evolution and diversity of assembly-line polyketide synthases. Chem Rev 119:12524–12547
8. Zhang L, Hashimoto T, Qin B et al (2017) Characterization of giant modular PKSs provides insight into genetic mechanism for

structural diversification of aminopolyol polyketides. Angew Chem Int Ed Engl 56:1740–1745

9. Bozhuyuk KAJ, Fleischhacker F, Linck A et al (2018) De novo design and engineering of non-ribosomal peptide synthetases. Nat Chem 10:275–281
10. Keatinge-Clay AT (2017) Polyketide synthase modules redefined. Angew Chem Int Ed Engl 56:4658–4660
11. Bozhuyuk KAJ, Linck A, Tietze A et al (2019) Modification and de novo design of non-ribosomal peptide synthetases using specific assembly points within condensation domains. Nat Chem 11:653–661
12. Hashimoto T, Hashimoto J, Kozone I et al (2018) Biosynthesis of quinolidomicin, the largest known macrolide of terrestrial origin: identification and heterologous expression of a biosynthetic gene cluster over 200 kb. Org Lett 20:7996–7999
13. Menzella HG, Reid R, Carney JR et al (2005) Combinatorial polyketide biosynthesis by de novo design and rearrangement of modular polyketide synthase genes. Nat Biotechnol 23: 1171–1176
14. Walker MC, Thuronyi BW, Charkoudian LK et al (2013) Expanding the fluorine chemistry of living systems using engineered polyketide synthase pathways. Science 341:1089–1094
15. Marsden AF, Wilkinson B, Cortes J et al (1998) Engineering broader specificity into an antibiotic-producing polyketide synthase. Science 279:199–202
16. Komatsu M, Komatsu K, Koiwai H et al (2013) Engineered Streptomyces avermitilis host for heterologous expression of biosynthetic gene cluster for secondary metabolites. ACS Synth Biol 2:384–396
17. Huo L, Hug JJ, Fu C et al (2019) Heterologous expression of bacterial natural product biosynthetic pathways. Nat Prod Rep 36: 1412–1436
18. Zhang JJ, Tang X, Moore BS (2019) Genetic platforms for heterologous expression of microbial natural products. Nat Prod Rep 36: 1313–1332
19. Hu Z, Awakawa T, Ma Z et al (2019) Aminoacyl sulfonamide assembly in SB-203208 biosynthesis. Nat Commun 10:184
20. Gibson DG, Young L, Chuang R-Y et al (2009) Enzymatic assembly of DNA molecules up to several hundred kilobases. Nat Methods 6: 343–345
21. Zhang Y, Buchholz F, Muyrers JP et al (1998) A new logic for DNA engineering using recombination in Escherichia coli. Nat Genet 20:123–128
22. Zhang JJ, Yamanaka K, Tang X et al (2019) Direct cloning and heterologous expression of natural product biosynthetic gene clusters by transformation-associated recombination. Methods Enzymol 621:87–110
23. Fu J, Bian X, Hu S et al (2012) Full-length RecE enhances linear-linear homologous recombination and facilitates direct cloning for bioprospecting. Nat Biotechnol 30:440–446
24. Tong Y, Whitford CM, Blin K et al (2020) CRISPR-Cas9, CRISPRi and CRISPR-BEST-mediated genetic manipulation in streptomycetes. Nat Protoc 15:2470–2502
25. Kudo K, Hashimoto T, Hashimoto J et al (2020) In vitro Cas9-assisted editing of modular polyketide synthase genes to produce desired natural product derivatives. Nat Commun 11:4022
26. Awakawa T, Zhang L, Wakimoto T et al (2014) A methyltransferase initiates terpene cyclization in teleocidin B biosynthesis. J Am Chem Soc 136:9910–9913
27. Awakawa T, Fujioka T, Zhang L et al (2018) Reprogramming of the antimycin NRPS-PKS assembly lines inspired by gene evolution. Nat Commun 9:3534
28. Kieser T, Bibb MJ, Buttner MJ et al (2000) Practical streptomyces genetics, vol 291. John Innes Foundation, Norwich
29. Liu R, Deng Z, Liu T (2018) Streptomyces species: ideal chassis for natural product discovery and overproduction. Metab Eng 50:74–84
30. Nybo SE, Shepherd MD, Bosserman MA et al (2010) Genetic manipulation of Streptomyces species. Curr Protoc Microbiol. Chapter 10: Unit 10E 13
31. Rebets Y, Kormanec J, Luzhetskyy A et al (2017) Cloning and expression of metagenomic DNA in Streptomyces lividans and subsequent fermentation for optimized production. Methods Mol Biol 1539:99–144
32. Ishikawa J, Hotta K (1999) FramePlot: a new implementation of the frame analysis for predicting protein-coding regions in bacterial DNA with a high G + C content. FEMS Microbiol Lett 174:251–253
33. Lu S, Wang J, Chitsaz F et al (2020) CDD/SPARCLE: the conserved domain database in 2020. Nucleic Acids Res 48:D265–D268
34. Potter SC, Luciani A, Eddy SR et al (2018) HMMER web server: 2018 update. Nucleic Acids Res 46:W200–W204

35. Altschul SF, Gish W, Miller W et al (1990) Basic local alignment search tool. J Mol Biol 215:403–410
36. Kumar S, Stecher G, Li M et al (2018) MEGA X: molecular evolutionary genetics analysis across computing platforms. Mol Biol Evol 35:1547–1549
37. Katoh K, Rozewicki J, Yamada KD (2019) MAFFT online service: multiple sequence alignment, interactive sequence choice and visualization. Brief Bioinform 20:1160–1166
38. Price MN, Dehal PS, Arkin AP (2010) FastTree 2--approximately maximum-likelihood trees for large alignments. PLoS One 5:e9490
39. Letunic I, Bork P (2019) Interactive tree of life (iTOL) v4: recent updates and new developments. Nucleic Acids Res 47:W256–W259
40. Blin K, Shaw S, Steinke K et al (2019) antiSMASH 5.0: updates to the secondary metabolite genome mining pipeline. Nucleic Acids Res 47:W81–W87
41. Waterhouse A, Bertoni M, Bienert S et al (2018) SWISS-MODEL: homology modelling of protein structures and complexes. Nucleic Acids Res 46:W296–W303
42. Onaka H, Taniguchi S, Ikeda H et al (2003) pTOYAMAcos, pTYM18, and pTYM19, actinomycete-Escherichia coli integrating vectors for heterologous gene expression. J Antibiot (Tokyo) 56:950–956
43. Komatsu M, Uchiyama T, Ōmura S et al (2010) Genome-minimized Streptomyces host for the heterologous expression of secondary metabolism. Proc Natl Acad Sci U S A 107:2646–2651
44. Richter CD, Nietlispach D, Broadhurst RW et al (2008) Multienzyme docking in hybrid megasynthetases. Nat Chem Biol 4:75–81
45. Keatinge-Clay AT (2012) The structures of type I polyketide synthases. Nat Prod Rep 29: 1050–1073
46. Kushnir S, Sundermann U, Yahiaoui S et al (2012) Minimally invasive mutagenesis gives rise to a biosynthetic polyketide library. Angew Chem Int Ed Engl 51:10664–10669
47. Miyazawa T, Hirsch M, Zhang Z et al (2020) An in vitro platform for engineering and harnessing modular polyketide synthases. Nat Commun 11:80
48. Sugimoto Y, Ishida K, Traitcheva N et al (2015) Freedom and constraint in engineered noncolinear polyketide assembly lines. Chem Biol 22:229–240
49. Baltz RH (2012) Streptomyces temperate bacteriophage integration systems for stable genetic engineering of actinomycetes (and other organisms). J Ind Microbiol Biotechnol 39:661–672
50. Yuzawa S, Deng K, Wang G et al (2017) Comprehensive in vitro analysis of acyltransferase domain exchanges in modular polyketide synthases and its application for short-chain ketone production. ACS Synth Biol 6:139–147

Chapter 13

Heterologous Expression, Purification, and Characterization of Type II Polyketide Synthase Acyl Carrier Proteins

Grayson S. Hamrick, Casey H. Londergan, and Louise K. Charkoudian

Abstract

The enzymes that comprise type II polyketide synthases (PKSs) are powerful biocatalysts that, once well-understood and strategically applied, could enable cost-effective and sustainable access to a range of pharmaceutically relevant molecules. Progress toward this goal hinges on gaining ample access to materials for in vitro characterizations and structural analysis of the components of these synthases. A central component of PKSs is the acyl carrier protein (ACP), which serves as a hub during the biosynthesis of type II polyketides. Herein, we share methods for accessing type II PKS ACPs via heterologous expression in *E. coli*. We also share how the installation of reactive and site-specific spectroscopic probes can be leveraged to study the conformational dynamics and interactions of type II PKS ACPs.

Key words Heterologous expression, Acyl carrier protein, Type II polyketide, Polyketide synthase, Polyaromatic polyketide

1 Introduction

Recent technical advances in bioinformatics and computational biology have led to the discovery of intriguing biosynthetic gene clusters (BGCs) in a large swath of organisms [1, 2]. BGCs encode suites of enzymes that cooperate to manufacture molecules thought to mediate an organism's ecological interactions. These molecules, known as specialized metabolites (and commonly referred to as secondary metabolites or, more broadly, natural products), play a central role in medicine both as therapeutics and as inspiration for new pharmaceutically relevant compounds [3]. Because biosynthesis mediated by these enzyme assemblies requires only earth-abundant, natural starting materials and uses water as a solvent, mixing-and-matching enzymes encoded by different BGCs could provide cost-effective and sustainable access to a host of pharmaceutically relevant "unnatural natural products." Further biochemical and structural analyses of enzymes encoded by BGCs are

Elizabeth Skellam (ed.), *Engineering Natural Product Biosynthesis: Methods and Protocols*, Methods in Molecular Biology, vol. 2489, https://doi.org/10.1007/978-1-0716-2273-5_13,

necessary to make these goals a reality. However, the chemical innovation coordinated by the enzymes encoded in these BGCs is often difficult to access, as the native organisms harboring BGCs of interest are often resistant to culture under laboratory conditions. Heterologous expression of the proteins encoded by BGCs can bypass these issues, opening the door for the investigation of previously out-of-reach products [4]. This approach can also centralize the production of target protein products from disparate sources into one strain, thereby saving both time and resources. Therefore, identifying a heterologous expression system to produce the enzymes encoded within BGCs is an important step toward harnessing biosynthetic protein assembly lines for access to new chemical diversity. Several genetically engineered strains of *Escherichia coli* have emerged as popular heterologous host species for such applications due to their straightforward culture conditions, rapid growth, well-studied genetics, and flexibility for protein expression [5–7].

Heterologous expression in *E. coli* is an effective strategy for the study of polyketide synthases (PKSs), multienzyme assembly lines encoded by BGCs that manufacture an array of pharmaceutically relevant chemicals. The products of type II PKSs are particularly lucrative targets for heterologous expression in *E. coli* because of the chemical diversity of this class of molecules and their strong track record for serving as widely used therapeutic agents [3]. Molecules manufactured by type II PKSs include important antibiotics such as tetracycline and doxorubicin and anticancer agents such as doxycycline [8–10].

A central component of type II PKSs is the acyl carrier protein (ACP), which serves as a hub of activity during the manufacturing of polyketides. The ACP is a small (~9 kDa), helical, labile protein that is posttranslationally modified at a conserved serine residue at the *N*-terminus of helix II with the addition of a 4′-phosphopantetheine (Ppant) "arm." The terminal thiol of the 18 Å Ppant arm is the point of attachment for malonyl-CoA-based molecular building blocks and β-keto intermediates throughout the assembly of the polyketide. The *malonyl*-ACP interacts iteratively with the acylated ketosynthase-chain length factor heterodimer (KS-CLF; sometimes referred to as the KSα and KSβ) through repeated decarboxylative Claisen-like condensation reactions to extend the polyketide chain by two carbon atoms (Fig. 1). Together, the ACP and KS-CLF are known as the "minimal type II PKS" and perform the essential task of building the backbone of the polyketide chain. This chain is modified by tailoring enzymes to create the final polyaromatic product.

Understanding ACP–enzyme interactions may give researchers the ability to engineer hybrid PKSs. These PKSs would consist of enzymes encoded by a variety of evolutionarily diverged BGCs and could provide sustainable access to "unnatural natural products"

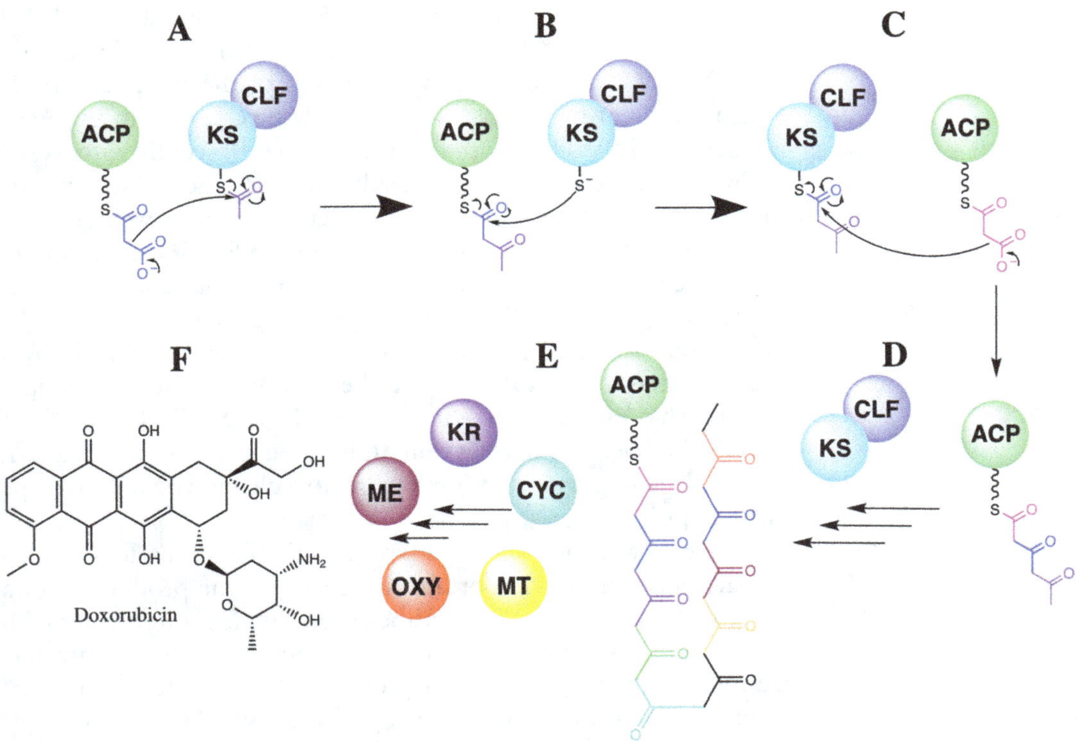

Fig. 1 An overview of the type II polyketide biosynthesis of doxorubicin, a well-known type II polyketide, highlighting the crucial role that the acyl carrier protein (ACP) plays in the biosynthetic process. The ACP is primed by malonyl transferase to add a reactive malonyl moiety to the free thiol of the Ppant arm, which is denoted by a squiggly line. (**a**) The primed ACP interacts with an acylated ketosynthase-chain length factor (KS-CLF) in a Claisen-like decarboxylative condensation reaction, resulting in a diketide tethered to the ACP as a thioester. (**b**) The thiolate of the active site serine residue of the KS-CLF acts as the nucleophile in a nucleophilic acyl substitution reaction, shifting the growing acyl chain back to the KS-CLF. (**c** and **d**) These steps repeat as the ACP and KS-CLF interact to iteratively produce the polyketide backbone of the final product. (**e**) A cohort of tailoring enzymes, including cyclase (CYC), methyltransferase (MT), oxygenase (OXY), methyl esterase (ME), and ketoreductase (KR) modify the completed polyketide backbone to finish the synthesis of the secondary metabolite (**f**)

with novel structures and biologically relevant activities. Type II PKSs share a common ancestor with fatty acid synthases (FASs), which suggests the possible compatibility between the components of type II PKSs and FASs for the future construction of combinatorial synthases and the production of novel, hybrid products with promising applications [11–13].

Type II PKSs are found across a variety of species [11]. The most well-studied type II PKSs are those found in Actinomycetes, a phylum of bacteria that are notoriously onerous to work with in the lab [14, 15]. Heterologous expression of type II PKSs could improve the efficiency of characterizing these systems; however, challenges in expressing soluble Actinobacterial KS-CLF heterodimers in heterologous hosts–including *E. coli*–has slowed the application of these techniques. In 2015, a bioinformatics study led by Hillenmeyer and Charkoudian suggested that the BGCs responsible for the transcription and translation of the minimal type II PKS are present in many strains anciently diverged from the Actinobacteria [11]. The great majority of these recently discovered pathways remain uncharacterized; neither their biosynthetic machinery nor their molecular products have been well studied. Fortunately, many of these putative type II PKS BGCs are found in phyla such as the Firmicutes and Proteobacteria, and hence are attractive candidates for heterologous expression in tractable, non-Actinobacterial heterologous hosts such as the well-known proteobacterium *E. coli*. In just the past year, researchers have shown that genes encoding type II PKS KS-CLFs from non-Actinobacterial species can indeed be expressed in *E. coli*, thereby opening a new era for the in vitro characterization of type II PKSs [14–17].

With newfound access to type II PKS KS-CLFs for in vitro characterization, it is important to streamline the production of a range of ACPs and develop methods to study their interactions with KS-CLFs. These advances will enable the molecular-level characterization of protein-protein and protein-substrate interactions that facilitate the production of type II polyketides and inform bioengineering studies to produce hybrid synthases capable of manufacturing "unnatural natural products." Herein we present our optimized methods for the cloning, *E. coli* heterologous expression, and purification of type II PKS ACPs. We also share methods to determine the molecular weight (MW) of ACPs using a single quadrupole liquid chromatography–mass spectrometry (LC-MS), assess the solvation environment of the ACP Ppant arm and its molecular cargo using site-specific vibrational spectrometry, and visualize ACP–KS interactions using a tandem mechanistic cross-linking/colorimetric assay.

2 Materials

2.1 General Equipment

1. Thermal cycler.
2. Pipettes and tips.
3. Agarose gel electrophoresis materials.
4. 0.2 mL microcentrifuge tubes.
5. 1.5 mL microcentrifuge tubes.
6. Nuclease-free H_2O.
7. Thermo Scientific™ NanoDrop 2000 or other UV-visible (UV-vis) spectrophotometer.
8. LB broth supplemented with 50 μg/mL kanamycin.

9. Orbital shaker with heating/cooling capabilities.
10. LB agar plates supplemented with 50 μg/mL kanamycin.
11. Inoculating loop/sterile toothpick.
12. Sterile glass beads.
13. Incubator.
14. Benchtop Centrifuge.
15. Hot water bath.

2.2 Amplification of ACP Genes

1. DNA polymerase (such as Q5® High-Fidelity DNA Polymerase).
2. Zymo DNA Clean & Concentrator™ Kit.
3. (Optional) Oligonucleotide PCR primers, designed in Subheading 3.1.

2.3 Gibson Assembly®

1. pET-28a (+) vector.
2. NdeI and EcoRI restriction endonucleases.
3. Zymoclean™ Gel DNA Recovery Kit.
4. Gibson Assembly® Master Mix.

2.4 Preparation and Transformation of Plasmids

1. Chemically competent *E. coli* cells optimized for molecular cloning (e.g., DH5α or XL1 Blue).
2. QIAprep® Spin Miniprep Kit.
3. Microcentrifuge.
4. Chemically competent *E. coli* cells optimized for protein production (e.g., BL21 or BAP1 [18]).
5. Parafilm.

2.5 Expression of ACPs in E. coli

1. 2 L Baffled culture flask with sponge top or other loose-fitting top to allow for aeration.
2. Autoclave.
3. 1 M Isopropyl β-D-1-thiogalactopyranoside (IPTG).
4. High-speed floor centrifuge.
5. Large (~500 mL) centrifuge tubes compatible with floor centrifuge.
6. 50 mL conical tubes.
7. Spatula.

2.6 Purification and On-Column Phosphopantetheinylation of ACPs

1. Lysis Buffer: 50 mM sodium phosphate, pH 7.6, 300 mM NaCl, 10 mM imidazole, and 10% (v/v) glycerol.
2. Ultrasonic homogenizer with tip probe.
3. Ni-NTA agarose affinity resin.

4. Fritted gravity column with stopcock and cap.
5. Wash Buffer: 50 mM sodium phosphate, pH 7.6, 300 mM NaCl, and 30 mM imidazole.
6. 50 mM sodium phosphate buffer, pH 7.6.
7. 50 mM dithiothreitol (DTT).
8. 250 mM $MgCl_2$.
9. 50 mM Coenzyme A or relevant Coenzyme A derivative.
10. 150 μM R4-4 Sfp enzyme [19, 20].
11. Elution Buffer: 50 mM sodium phosphate, pH 7.6, 100 mM NaCl, and 300 mM imidazole).
12. Dialysis materials. We use Thermo Scientific™ Slide-A-Lyzer™ Dialysis Cassette (2 K MWCO) with corresponding cassette buoys and syringes.
13. Bucket, 2 L or larger.
14. Sodium dodecyl sulfate–polyacrylamide gel electrophoresis (SDS-PAGE) materials.

2.7 Cyanylation of Holo-ACP Ppant Arm

1. Protein quantification assay materials. We use Pierce™ Coomassie (Bradford) Protein Assay Kit.
2. Ellman's Reagent (5,5′-dithiobis-(2-nitrobenzoic acid) or DTNB).
3. PD-10 desalting column.
4. Sodium cyanide (NaCN), reagent grade.
5. Double distilled water (ddH_2O).
6. Centrifugal concentration tubes, 3 kDa MWCO.
7. Liquid nitrogen.
8. Cryogenic dewar.
9. Cryogenic tongs.
10. Cryogenic gloves.

2.8 Collecting and Analyzing a Vibrational Spectrum

1. Bruker Optics Vertex 70 FT-IR with photovoltaic MCT detector.
2. Constant N_2 purge source: either a dry N_2 tank or, more ideally and sustainably, the output from a Parker-Balston IR purge gas generator.
3. Harrick 13 mm diameter demountable liquid cell.
4. General purpose wipes (i.e., Kimwipes™).
5. OPUS software package.
6. Origin software package.

2.9 Colorimetric ACP–KS Mechanistic Cross-Linking Assay

1. Aliquot of desired KS (or KSCLF).
2. Ellman's Reagent (5,5′-dithiobis-(2-nitrobenzoic acid) or DTNB).

2.10 Tracking Ppant Sequestration Activity Using Raman Spectroscopy

1. 275 μM ACP in 50 mM sodium phosphate buffer, pH 7.6.
2. 2.5 mM DTT.
3. 23 mM $MgCl_2$.
4. 18 mM ATP, pH adjusted to 7.6.
5. 0.8 μM Acyl-acyl carrier protein synthetase (AasS) in Tris buffer, pH 7.6.
6. 4.6 mM carboxylic acid probe (4-pentynoic acid (C5), 7-octynoic acid (C8), or 12-tridecynoic acid (C13)), stock in isopropanol.
7. CW-Raman spectrometer.
8. β-mercaptoethanol (BME).

2.11 Liquid Chromatography–Mass Spectrometry (LC-MS) Analysis of ACPs

1. Liquid chromatography and mass spectrometer (LC-MS) equipped for electrospray ionization mass spectrometry.
2. Waters XBridge Protein BEH C4 column.
3. Solvent A: water + 0.1% formic acid.
4. Solvent B: acetonitrile + 0.1% formic acid.

3 Methods

3.1 Amplification of ACP Genes for Assembly into Expression Vectors

There are two popular options for the acquisition of the gene encoding ACPs of interest for Gibson Assembly® into an expression vector. The target gene can be ordered from a synthetic biology company, or the gene can be amplified via polymerase chain reaction (PCR) from genomic DNA (either purchased or extracted from the wild-type host). This latter option will be described here.

1. The algorithm that we use for primer design (written in Python), can be found in the Supporting Information, titled Gibson_assembly_primer_design.py. Simply execute the file (on a machine with Python3 installed) in a command line window (such as Terminal on Mac or Linux, or Command Prompt on Windows) and enter the coding sequence (5′ to 3′) of the gene of interest when prompted. The results in the command line represent primer sequences designed to insert the gene of interest into the NdeI/EcoRI endonuclease restriction sites of the pET-28a (+) expression vector, either via traditional restriction digestion/ligation assembly or Gibson Assembly® (*see* **Notes 1–4**) [21]. Other software programs, such as SnapGene and Serial Cloner, offer robust primer design algorithms as well.

2. Using an online calculator (such as this Thermo Fisher Scientific version: http://www.thermoscientificbio.com/webtools/tmc/), determine the melting temperatures of the primer pair. The difference between the melting temperatures of the primers should be roughly less than or equal to 5 °C (*see* **Note 5**). Synthesize the primers or order them from a synthetic biology company.
3. Following the protocol provided by the polymerase manufacturer, perform PCR using the designed primers and target genomic DNA (*see* **Notes 6** and **7**). Use an annealing temperature 3 °C warmer than the lower of the two melting temperatures. Purify the amplified gene block using a Zymo DNA Clean & Concentrator™ Kit (*see* **Note 8**).

3.2 Gibson Assembly® and Transformation of Plasmid into Competent Cells

1. Prepare pET-28a (+) vector by digesting it with both NdeI and EcoRI restriction enzymes, as directed by the supplier of the enzymes (*see* **Note 9**). This will linearize the circular plasmid by cutting it at the NdeI and EcoRI restriction sites, preparing the vector for compatibility with the previously designed Gibson Assembly® primers (*see* **Note 10**). After the restriction digestion, purify the linearized pET-28a (+) vector. We use a Zymoclean™ Gel DNA Recovery Kit for this step.
2. Once the vector has been prepared and the primers have been obtained, set up the reaction by combining the reagents described in Table 1 in a 0.2 mL microcentrifuge tube.

Table 1
For typical assemblies, Gibson Assembly® has the highest rate of success when the molar ratio of the digested vector-to-DNA insert is 1:3 (*see* Notes 11 and 12). Determine the concentration of the two DNA segments (the DNA insert and the digested plasmid) in ng/μL. Then use the following formula to calculate the molar concentration of each component: $\frac{pmol}{\mu L} = \frac{(ng/\mu L) \cdot 1000}{(\#bp) \cdot 650\ Da}$, where ng/μL is the measured concentration of DNA, # bp is the length of the DNA segment, 650 Da is an approximation of the average mass of a nucleotide base, and 1000 is a conversion factor. Add enough of each DNA segment to the mixture to achieve the desired concentration. The total volume of DNA ($x + y$) should not exceed 5 μL. If the volume $x + y$ is less than 5 μL, add $z = 5-(x + y)$ μL of nuclease-free water to make up the difference

Reagent	Volume (μL)
DNA insert	x
Linearized pET-28a (+) vector	y
Nuclease-free H_2O	z
Gibson assembly® master mix	15
Total	20

3. Place the 0.2 mL microcentrifuge tube with prepared Gibson Assembly® mixture in a thermal cycler at 50 °C for 1 h.
4. After 1 h, transfer the entire 20 μL reaction mixture to 100 μL of chemically competent *E. coli* cells in a 1.5 mL microcentrifuge tube. Incubate on ice for 30 min (*see* **Note 13**).
5. Transfer competent cells to a 42 °C water bath for exactly 90 s.
6. Remove cells from the water bath and incubate on ice for 2 min.
7. Add 200 μL of lysogeny broth (LB) without antibiotic to the microcentrifuge tube and shake for 1 h at 200 RPM and 37 °C.
8. Remove the culture from the shaker and transfer the complete mixture onto a LB plate prepared with the appropriate antibiotic. We typically use pET-28a (+) for ACP expression, which encodes kanamycin resistance (50 μg/mL) (*see* **Note 14**). Spread the prepared cells around the plate using sterile glass beads or other sterile glass surface.
9. Leave the plate open in a sterile environment (either a sterile hood or near an open flame) for 10–15 min to allow the plate to dry.
10. Incubate the dry plate, with lid on and agar-side up, at 37 °C. Colonies harboring the assembled plasmid should begin to appear within 24–48 h (*see* **Note 15**).
11. Once colonies have appeared, continue by preparing the amplified plasmid or store the plate, sealed with parafilm at 4 °C, until ready to proceed.

3.3 Preparation of Amplified Plasmid and Transformation into Expression Strain

1. Pick a single colony from the selective plate prepared previously and inoculate a ~5 mL culture in LB liquid medium containing the appropriate selective antibiotic (e.g., 50 μg/mL kanamycin). Incubate overnight (for 12–16 h) at 37 °C, shaking at 200 RPM (*see* **Notes 16** and **17**).
2. Harvest the cells by centrifugation (5400 × *g*) for 10 min at 4 °C. Remove all traces of supernatant by inverting the open centrifuge tube until all of the LB has been drained. If the cells have been pelleted correctly, they will remain securely attached to the bottom of the tube upon inversion.
3. Purify the plasmid from the cell pellets using a kit such as the QIAprep Spin Miniprep Kit and a microcentrifuge (*see* **Note 18**).
4. After purification, measure the concentration of DNA and assess purity using NanoDrop UV-vis. An A_{260}/A_{280} ratio of ~1.7–2.0 indicates pure DNA. Additionally, confirm the sequence fidelity of the putative plasmid by sequencing (*see* **Note 19**).

5. Store purified plasmids at −20 °C (or −80 °C for long-term storage) and thaw before use. Once the plasmid sequence has been confirmed, transform into a production strain of *E. coli* (either BL21 or BAP1) as follows in **steps 6–11** (*see* **Note 20**).
6. Add 1 μL of plasmid to 100 μL of chemically competent cells.
7. Incubate on ice for 30 min.
8. Heat shock in a water bath at 42 °C for 90 s.
9. Remove from the water bath and incubate on ice for 2 min.
10. Add 200 μL of LB and incubate at 37 °C, shaking at 200 RPM, for 1 h.
11. Plate 100 μL of culture onto LB semisolid agar supplemented with kanamycin and incubate overnight at 37 °C.

3.4 Expression of ACPs in E. coli

1. Inoculate a seed culture of transformed cells containing the expression plasmid of choice in 10 mL of liquid LB treated with kanamycin (*see* **Note 21**).
2. Incubate overnight at 37 °C, shaking at 200 RPM.
3. Prepare 1 L of LB in a 2 L baffled culture flask and sterilize by autoclave.
4. Cool to room temperature and add 50 μg/mL kanamycin (for pET-28a (+)) (*see* **Note 22**).
5. In a sterile hood (or using aseptic technique), inoculate 1 L of prepared LB with the seed culture (which should now be cloudy due to overnight growth of cells). Incubate at 37 °C, shaking at 200 RPM.
6. Monitor the OD_{600} of the production culture by measuring the absorbance at 600 nm. Continue incubating until the OD_{600} reaches 0.4–0.6 (*see* **Notes 23** and **24**).
7. At appropriate OD_{600}, collect a 1 mL aliquot in a microcentrifuge tube and freeze. This sample will be used for subsequent SDS-PAGE analysis as a "preinduction control." Then add 250 μL of 1 M isopropyl β-D-1-thiogalactopyranoside (IPTG) to each culture in a sterile hood and shake the cultures at 18 °C, 125 RPM overnight for 14–18 h to induce the expression of the target ACP.
8. After shaking overnight, collect a 1 mL aliquot in a microcentrifuge tube and freeze for subsequent SDS-PAGE analysis as a "postinduction control."
9. Pour into large (~500 mL) centrifuge tubes and balance. Balancing is very important when using a high-speed centrifuge.
10. Collect cells by centrifugation. (4500 × *g*, 15 min, 4 °C) (*see* **Note 25**).

11. While spinning down, label a 50 mL conical tube for each protein and incubate on ice.
12. After centrifugation is complete, pour off supernatant. Scrape cells into the prepared conical tube using a clean spatula.
13. Either freeze the collected cells at −80 °C to be lysed at a later date or continue with the purification procedure as follows.

3.5 Purification and On-Column Phosphopantetheinylation of ACPs

Unless otherwise indicated, all materials should be kept at 4 °C (on ice at all times and in a cold room if possible).

1. With the sample on ice, gently pipette the cell pellet from ~1 L of ACP culture in 20 mL Lysis Buffer until it is evenly suspended. Do not vortex.
2. Sonicate using an ultrasonic probe at 12 W in ten pulses (30 s on, 30 s off) to lyse cells (*see* **Note 26**).
3. Centrifuge at 17,000 × *g* for 45 min to remove cell debris; the His_6-tagged target protein should be soluble and remain suspended in the supernatant (*see* **Note 27**).
4. Meanwhile, prepare 1.5 mL Ni-NTA agarose affinity resin. Resuspend resin slurry in the bottle by shaking gently and transfer ~3 mL to a conical tube. Centrifuge at 500 × *g* for 5 min (using a benchtop centrifuge) to settle the resin; remove the supernatant. Add 5 column volumes (CVs) of Lysis Buffer and invert gently to equilibrate resin. Centrifuge at 500 × *g* for 5 min to settle the resin; remove the supernatant while being careful to not discard the resin. Repeat equilibration with 5 CVs of Lysis Buffer, centrifuge, and remove the supernatant.
5. After the centrifugation in **step 3** has reached completion, collect 50 μL of the supernatant for future SDS-PAGE analysis. Transfer the remaining supernatant (containing the target protein) to the tube containing the equilibrated, settled resin.
6. Combine resin and protein by shaking on an orbital shaker at 200 RPM for 15 min.
7. Add mixed protein/resin to a fritted gravity column and wash with 5–10 CVs of Lysis Buffer.
8. Measure the A_{280} of the Wash Buffer using a spectrophotometer to establish a baseline absorbance (likely between 0.02 and 0.04). Wash the resin with Wash Buffer until the A_{280} returns to the baseline (i.e., until the absorbance of the eluent is similar to that of the Wash Buffer).
9. To the column, add 1–2 mL of 50 mM sodium phosphate buffer (pH 7.6) and allow the buffer to run through. Keep the resin wet, leaving about 100–200 μL of the buffer above the resin. Collect 50 μL of the flow through for future SDS-PAGE analysis.

10. Prepare 4 mL of the 'Sfp reaction mixture' by combining the following (*see* **Note 28**):
    (a) 2.4 mL of 50 mM phosphate buffer, pH 7.6.
    (b) 1.2 mL of 50 mM dithiothreitol (DTT) (*see* **Note 29**).
    (c) 160 μL of 250 mM $MgCl_2$.
    (d) 200 μL of 50 mM CoA or CoA derivative.
    (e) 80 μL of 150 μM R4–4 Sfp enzyme.
11. Add the resulting mixture to the column, cap/seal, and leave overnight at room temperature. Gentle rocking overnight using a nutating rocker is advised, but not required.
12. The next day, transfer the column to a 4 °C cold room. Wash the column with Wash Buffer to remove excess CoA until the absorbance of the eluent returns to the baseline (CoA absorbs strongly at 260 nm) (*see* **Note 30**). Collect 50 μL of the wash for future SDS-PAGE analysis.
13. Elute the now *holo*-ACP using Elution Buffer. Collect 10–12 × 1 mL fractions and measure the absorbance at 280 nm. Evaluate the 260/280 ratio to determine protein quantity and purity within the sample. Collect 50 μL of the eluent for future SDS-PAGE analysis.
14. Pool together all the fractions that have protein and measure the 280 nm absorbance of the pooled fraction again. Depending on the ACP, a reasonable titer is 15–25 mg of protein per liter of culture.
15. Dilute a small sample to 2–10 μg/mL in 50 mM sodium phosphate, pH 7.6 and determine the *holo*/*apo* ratio by LC-MS (*see* **Note 31**). *See* Subheading 3.10 for details.
16. Verify the overexpression of the target ACP by SDS-PAGE, including samples of the following: preinduction and postinduction controls from Subheading 3.4, supernatant from **step 5**, flow through from **step 9**, wash from **step 12**, and eluent from **step 13**.
17. Dialyze the pooled protein overnight at 4 °C against 2–3 L of 50 mM sodium phosphate, pH 7.6. The next day, concentrate and store for future use (*see* **Note 32**).

3.6 Cyanylation of Holo-ACP Ppant Arm

1. Determine the desired amount of ACP to cyanylate. We typically aim to make 100 μL of 1.5 mM ACP-SCN (~1.5×10^{-7} mol) to ensure sufficient protein for analysis by IR and titration assays (*see* **Note 33**).
2. If necessary, thaw frozen *holo*-ACP samples on ice. If the *holo*-ACP is stored in multiple aliquots, combine and concentrate to a volume of ~1 mL or less. Reactions will occur in 1.5 mL microcentrifuge tubes.

3. Determine the concentration of the *holo*-ACP via a protein quantification assay and/or transmission UV-vis.
4. Some carrier proteins (such as DEBS *holo*-ACP2(2) and Yersiniabactin *holo*-ArCP) form disulfide bonds between solvent exposed Ppant arms, creating ACP-homodimers. This prevents reaction with Ellman's reagent (5,5′-dithiobis-(2-nitrobenzoic acid) or DTNB). If your carrier protein does not form homodimers, skip **steps 5–7** and move to **step 8**.
5. Reduce disulfide bonds between Ppant arms by adding 100 times molar excess dithiothreitol (DTT). Mix by gentle pipetting up and down and incubate for 30 min at room temperature.
6. Remove excess DTT using a PD-10 column, which uses Sephadex G-25 Medium as a size-exclusion chromatographyChromatography matrix to separate high molecular weight materials (proteinsProteins) from low molecular weight materials (salts, small-molecule reagents, and solvents). Collect 0.25 mL fractions to maximize protein concentration of eluent.
7. Use a protein quantification assay or transmission UV-vis to identify fractions with protein. Pool these fractions and remeasure concentration. Move on to **step 8**.
8. Add eight times molar excess of predissolved DTNB (which is more soluble in a high ionic strength buffer) to the ACP sample. Mix gently by pipetting up and down. The solution should turn pale yellow due to the release of TNB^{2-}. Incubate for at least 30 min at room temperature, or on ice for more sensitive ACPs.
9. Follow the reaction using UV-vis at A_{325} for DTNB and A_{412} for TNB^{2-}. Refer to the Ellman's Reagent Protocol by Thermo Fisher Scientific for details on how to calculate the free sulfhydryl concentration of the reaction over time: https://www.thermofisher.com/order/catalog/product/22582?SID=srch-hj-22582#/22582?SID=srch-hj-22582. Also note that the exact extinction coefficient for the side product has been shown to exhibit thermochromism and that this can affect the calculated yield. The reaction is considered complete once the moles of TNB^{2-} reach that of the ACP in solution (*see* **Note 34**).
10. Add 55× molar excess of dissolved NaCN to the mixed ACP/DTNB/TNB^{2-} solution. Mix by pipetting up and down. The solution will turn dark yellow/orange. Incubate the reaction at room temperature for 30 min. The activated ACP-TNB will react with CN^{-} ions to produce a cyanylated ACP, as illustrated in Fig. 2 (*see* **Note 35**). Note that NaCN and aqueous cyanide ions are toxic chemicals. Be sure to refer to the material safety data sheet (MSDS) for information on how to properly store,

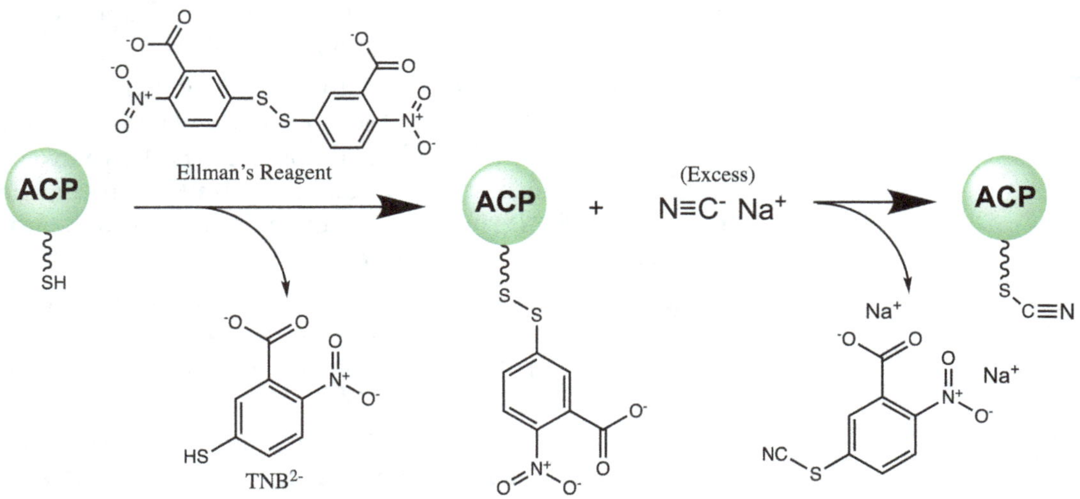

Fig. 2 Reaction scheme of the conversion of *holo*-ACP to the cyanylated ACP (ACP-SCN). The free thiol of the Ppant arm acts as a nucleophile to break up the disulfide bond of Ellman's Reagent, activating the ACP for reaction with aqueous cyanide ions. Cyanide attacks the sulfur of the Ppant arm thiol, releasing TNB^{2-}.

handle, and dispose of NaCN solutions, which should always be made on the smallest scale possible and immediately disposed of after use.

11. While the reaction incubates, prepare a PD-10 desalting column for separation of excess cyanylation reagents. Details of this step should be obtained from the manufacturer's protocol. Equilibrate column with 20 mL double distilled water (ddH_2O), followed by 20 mL of the same buffer used for the cyanylation reaction solution. Discard the flow-through in a waste beaker.
12. Apply the ACP-SCN reaction solution to the prepared PD-10 column. It is preferable to perform the following steps at 4 °C, although not necessary for most ACPs. Apply 0.5 mL of the reaction sample to the PD-10 column (*see* **Notes 36** and **37**). Allow the sample to enter into the packed bed and discard any flow-through.
13. Elute protein from the column. Prepare a rack of 15 microcentrifuge tubes labeled 1–15. Add 500 μL of equilibration buffer (50 mM phosphate buffer) and collect in the first microcentrifuge tube. Repeat the addition of buffer to the column 14 times, collecting 500 μL eluate in separate microcentrifuge tubes each time. The protein will elute faster than the excess reagents and side products (in the yellow solution) because it is larger.

14. Determine the ACP-containing factions by measuring the concentration of protein via transmission UV-vis and discard fractions with substantial absorptions above 300 nm. These absorptions are the result of reaction side products.
15. Combine fractions that contain protein and concentrate using centrifugal concentration tubes with a 3 kDa molecular weight cut off (MWCO). Remember from **step 1** that a good target concentration for future assays is 1.5 mM.
16. Analyze the sample by liquid chromatography–mass spectrometry (LC-MS; *see* Subheading 3.10) and/or obtain a spectrum of the ACP-SCN via IR (*see* Subheading 3.7).
17. Flash-freeze cyanylated proteins and store at −80 °C (*see* **Note 38**).
18. Run buffer through the used PD-10 column until all cyanide is eluted (~25 mL), then rinse the column with ddH_2O for storage. Carefully dispose of the eluted cyanide waste as directed by the MSDS.

3.7 Collecting and Analyzing a Vibrational Spectrum of ACP-SCN [22]

1. Prepare the FT-IR instrument for data collection by cooling the photovoltaic mercuric cadmium telluride (MCT) detector with liquid nitrogen. Refill completely every 5–6 h (*see* **Note 39** for general instructions associated with other FTIR spectrometers).
2. Collect the IR background of the buffer as follows in **steps 3–13**.
3. Rinse the CaF_2 windows of a Harrick 13 mm diameter demountable liquid cell (with 2 CaF_2 windows and a 56 μm Teflon spacer) with water and dry with a Kimwipe.
4. Place the Teflon spacer on one of the CaF_2 windows and place the two into the liquid cell.
5. Pipette 6–10 μL of the appropriate background buffer onto the CaF_2 window. Be very careful to avoid making bubbles as they will disrupt the IR signal.
6. Carefully drop the second window onto the sample inside the cell, again avoiding the formation of bubbles.
7. Quickly place the retaining ring and screw cap on top and seal the cell tight. Do not over-tighten, as this will crack the brittle CaF_2 windows.
8. Place the entire cell into the Vertex 70 IR.
9. Close the IR lid and wait about 10–15 min for the chamber to automatically purge itself of CO_2.
10. Set up OPUS data acquisition. Open OPUS and click the "Measurement" tab to open the run parameters. Include scanning from 1000 to 4000 cm^{-1} for 512 scans at 10 kHz scanner

velocity and 2 cm^{-1} resolution (*see* **Note 40**). Make sure that the signal is adjusted so that interferogram saturation is avoided when the blank (solvent+buffer) sample is present.

11. When all parameters are set, return to the basic tab and click "Background Single Channel" to collect the solvent reference spectrum.
12. Click "Sample Single Channel" to also collect a spectrum of the solvent blank (*see* **Notes 41** and **42**).
13. The (blank) spectrum file will appear on the left-hand column once it has finished processing. Choose the "ssc" sample block, Go to File → Save File As, and save your file as a .dpt text file in a desired location for further processing outside of OPUS.
14. Remove the cell from the IR chamber and rinse with deionized water.
15. Carefully add ~7 μL of ≥1 mM ACP-SCN in the same buffer solution onto the windows as done in **step 6.** When adding the buffer, avoid bubble formation and use the exact same windows and spacer as in **step 6**.
16. Place the liquid cell into the chamber and purge the instrument's internal atmosphere.
17. Return to the basic tab and click "Sample Single Channel."
18. Look again to the left column for your file. Select the absorbance block and right click to save the absorbance file ("abs" block) in .dpt format. If you plan to attempt manual averaging, you should also save the "ssc" block in the same way. The "rsc" block contains the reference spectrum collected above in **step 11**.
19. To correct the baseline of the absorbance spectrum in the CN stretching region and highlight the CN peak for further analysis (*see* **Note 43**), follow **steps 20–25**.
20. Open Origin and go to File → Import → Single ASCII and select the absorbance file saved in **step 13** to open the .dpt file of the scan data.
21. Delete all data points far away from the nitrile stretching region. Do not delete 2100–2210 cm^{-1}.
22. Duplicate the data twice by right clicking the tab at the bottom and clicking "Duplicate." The first data set is for record keeping. The second data set is for making a "spectral hole" window. The third data set is for subtracting the reference background.
23. The window in which any ACP-SCN will appear is from 2145 to 2180 cm^{-1}. Highlight these data points in the window. Right click and "Clear" to make a hole in the data where the CN signal should be. Graph the data with the hole by selecting

the forward slash at the bottom left-hand corner. At the top of the window select Analysis → Fitting → Polynomial Fit → Open Dialogue. Fit the remaining data (without the CN peak region) to a high-order polynomial (at least seventh degree).

24. Go to the third duplicated window and highlight both columns [A(X) and B(Y)]. Select Analysis → Data Manipulation → Subtract Reference Data → Open Dialogue. A Data Manipulation subtract_ref window will pop up. The Input 1 set will already be filled with the data you highlighted from the third window. Fill the empty Reference Data by clicking on the red button. On the bottom tabs, click FitPolynomialCurve1 (which is the fitted polynomial to the data with the hole). Select the first two columns. Return to the Data Manipulation Window and hit OK. Return to the third duplicated window. The B (Y) column should now contain Subtracted values. Plot the two columns.
25. Save the resulting absorbance spectrum in the CN region and export the graph, if desired, using the file dropdown menu.
26. If interested in further improving signal-to-noise ratio beyond what the instrument can provide in single scans, manually average multiple spectra of 512 scans as follows in **steps 27–31**.
27. Collect five (or more) solvent spectra and the same number of sample spectra, as outlined in **steps 2–18**. These can be from new refills of the same cell or from different illumination spots or orientations of the same sample.
28. Import the "ssc" data blocks, exported to .dpt files, from Steps 13 and 18 into Origin using the "Import Multiple ASCII" dialogue. The x data (frequencies in cm^{-1}) should all be the same; place all of the y data into one worksheet.
29. Create two new columns. In the first new column, use "Set Column Values" to create a sum of the blank single channel data blocks, and in the second new column, do the same for the sample (ACP-SCN) single channel data blocks.
30. Add one more new column, and use "Set Column Values" to create a new absorbance spectrum by manually calculating the absorbance spectrum, which is equal to

$$-\log_{10}\left(\frac{\text{Sum of Samples Column}}{\text{Sum of Blanks Column}}\right)$$

31. Plot this last column, which is the signal-averaged absorbance spectrum from repeated samples and repeated blanks. Follow the baseline procedure in **steps 26–31** to further reveal and isolate the CN stretching peak.

3.8 Colorimetric ACP–KS Mechanistic Cross-Linking Assay Utilizing Ellman's Reagent (DTNB) (See Note 44) [23]

3.8.1 Preparation of ACP-TNB-

1. Add 8 molar equivalents of 25-mM Ellman's reagent (5,5-′-dithiobis-(2-nitrobenzoic acid) or DTNB) in 50 mM phosphate buffer, pH 7.6, to *holo*-ACP (0.8–1.1 mM in 50 mM sodium phosphate buffer, pH 7.6) for conversion (*see* **Note 45**). Incubate at room temperature for 60–90 min. The reaction mixture will turn bright yellow, as shown in Fig. 3.
2. While the reaction mixture incubates, wash a PD-10 column with 5–10 CVs of 50 mM phosphate, pH 7.6 buffer to remove any ethanol or antimicrobial solution remaining from storage of the column.
3. Label 15 1.5 mL microcentrifuge tubes from 1 to 15.
4. After the DTNB reaction mixture has incubated for 90 min, apply 1–2 mL of the mixture to the column (*see* **Note 46**).
5. Collect 400–500 μL fractions, adding reaction mixture to the column as necessary to prevent the column from running dry (*see* **Note 47**).
6. Measure the OD_{280} of each fraction. Two products should be isolated: the target product, ACP-TNB$^-$ (which absorbs at 280 nm), and the byproduct, free DTNB/TNB^{2-} (which absorbs at 315 and 412 nm) (*see* **Note 48**).
7. Pool fractions containing pure ACP-TNB$^-$ and measure the absorbance. Save 10 μL for analysis by liquid chromatography–mass spectrometry (LC-MS). Concentrate the rest of the purified protein using a centrifugal concentration column with a 3 kDa MWCO.

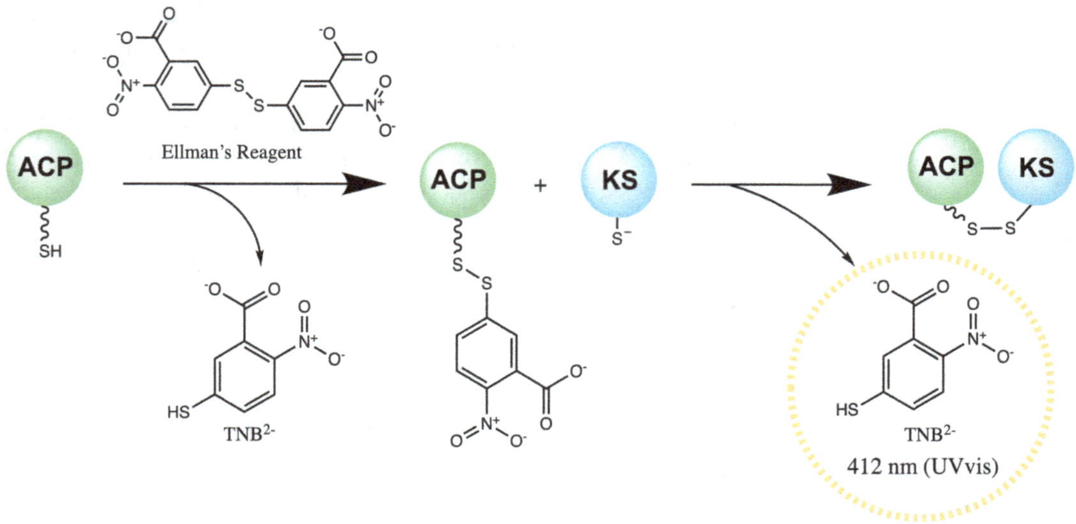

Fig. 3 Reaction scheme of activating *holo*-ACP to obtain/visualize ACP–KS interactions. The ACP Ppant arm is converted into a mix disulfide using DTNB. ACP-TNB$^-$ is then activated to form a covalent cross-link with the KS active site thiolate upon mechanistically relevant binding of the ACP and KS. The cross-linking reaction can be observed by SDS PAGE or by the release of aqueous TNB^{2-}, which absorbs at 412 nm and turns the solution yellow

8. Measure the concentration of ACP-TNB$^-$ in the sample using a protein quantification assay (following the manufacturer's specifications) or by UV-vis spectroscopy.

3.8.2 Quantification of Cross-Linking Activity

1. Ensure that the DTNB reaction successfully attached the TNB$^-$ probe to the *holo*-ACP by LC-MS using the sample saved in **step 7** above. *See* Subheading 3.10 for details.
2. Combine purified ACP-TNB$^-$ with the desired ketosynthase (either a ketosynthase homodimer, KS, or a ketosynthase-chain length factor heterodimer, KS-CLF) to make a 20 μL solution in 50 mM phosphate, pH 7.6 buffer with target concentrations of 25 μM KS and 50 μM ACP-TNB$^-$.
3. Incubate the reaction mixture at room temperature for exactly 20 min. A yellow color-change is an indication of reactivity. Upon cross-linking of ACP-TNB$^-$ with the KS, TNB^{2-} is released, which absorbs at 412 nm. After 20 min, measure the absorbance of the reaction mixture at 412 nm. Use the TNB^{2-} extinction coefficient (14,150 M^{-1} cm^{-1} at 25 °C at 412 nm) to calculate the number of moles of TNB^{2-} released for each reaction.
4. Cross-linking can additionally be observed via SDS-PAGE under nonreducing conditions as a band with a molecular weight of the cross-linked ACP–KS complex. Under reducing conditions (5% BME), the cross-linking disulfide bond is reduced and the complex separates, appearing as two separate bands (one for the ACP and one for the KS, or one each for the KS and CLF, if applicable).

3.9 Tracking Ppant Sequestration Activity Using Raman Spectroscopy

3.9.1 Loading of Alkyne Probe-Labeled Acyl Chain [24]

1. In a glass vial, combine the following, with a total volume on a 1 mL scale and final concentrations as given (*see* **Note 49**):
   (a) 275 μM *holo*-ACP in 50 mM sodium phosphate buffer, pH 7.6.
   (b) 2.5 mM DTT.
   (c) 23 mM $MgCl_2$.
   (d) 18 mM ATP, pH adjusted to 7.6 (*see* **Note 50**).
   (e) 0.8 μM Acyl-acyl carrier protein synthetase (AasS) in Tris buffer, pH 7.6 (*see* **Note 51**).
   (f) 4.6 mM carboxylic acid probe (4-pentynoic acid (C5), 7-octynoic acid (C8), or 12-tridecynoic acid (C13)), stock in isopropanol.
2. Shake the reaction mixture at 100 RPM for 16 h at 37 °C.
3. Centrifuge at 17,000 × *g* for 5 min to precipitate a pellet of insoluble species.

4. Decant the supernatant onto a PD-10 desalting column to separate the probe-loaded ACP from salts and excess substrate.
5. Identify fractions containing protein using transmission UV-vis. Unloaded carboxylic acid can be identified by a peak at 260 nm and should elute after the probe-loaded protein. Combine any fractions containing protein and lacking unloaded carboxylic acid.
6. Concentrate pooled fractions to 1–3 mM using a centrifugal filter with a 3 kDa MWCO.

3.9.2 Data Collection

The methods described here are for a home-built CW-Raman spectrometer [25]. However, similar methods can be applied to other similar dispersive Raman instruments, substituting instrument-specific instructions as required by the manufacturer and exact hardware and optical setup, which can be highly variable (*see* **Note 52**).

1. Focus a 532 nm DPSS CW laser attenuated to 80 mW incident power vertically through a 1 mM diameter glass capillary filled with 1–5 μL of sample.
2. Collect scattered light at 90° to the incoming excitation using an f/1.2 camera lens, collimate the collected light, and reject the Rayleigh scattering using an appropriate filter.
3. Focus the remaining collected light into the slit of a 500 mm single monochromator with a 600 grooves/mm diffraction grating blazed at 500 nm and oriented at an angle such that the OH stretching band of water appears at the long-wavelength edge of the spectrum.
4. Collect the spectrum on a back-illuminated, liquid-nitrogen cooled CCD camera. Bin the camera output to produce a single spectrum (scattered light counts vs wavelength).
5. Collect spectra in exposures of 1 min (exposure time should be kept below detector thresholds for saturation) for up to 2 h of total accumulation time.
6. Convert the wavelength to Raman shift for each column of the camera output.
7. Analyze spectra as described in Subheading 3.7, **steps 19–25**, importing the raw data from the scans directly into Origin. The same assumptions apply to using this analysis for Raman data as for IR (*see* **Note 43**).

3.10 Liquid Chromatography–Mass Spectrometry (LC-MS) Analysis of ACPs

1. ACPs can be analyzed using a single quadrupole LC-MS designed for routine mass-selective detection. While ACPs are typically ~8 to 10 kDa, they can be detected within the <1500 m/z range by deconvoluting sets of peaks observed in the mass spectrum representing multiple changed states. First,

calculate the expected molecular weights of the ACPs of interest (in the *apo* form) using the ExPASyProtParam Tool, found here: https://web.expasy.org/protparam/. Note that the addition of the Ppant arm leads to a Δ + 340 Da, so add 340 Da to the weight calculated for the *apo* form to find the predicted molecular weight of the *holo* form. Conversion of *holo*-ACP to ACP-SCN (Δ + 26 Da) or ACP-TNB- (Δ + 196 Da) can also be readily observed by LC-MS analysis.

2. Inject 20 μL of a 0.1 mg/mL ACP solution (in 50 mM sodium phosphate buffer, pH 7.6) to an LC-MS equipped to perform electrospray ionization mass spectrometry (ESI-MS) with a Waters XBridge Protein BEH C4 Column (or similar 300 Å, 3.5 μm, 2.1 mm × 50 mm column) heated to 45 °C (*see* **Note 53**). Use the following solvent gradient (where solvent A = water + 0.1% formic acid and solvent B = acetonitrile + 0.1% formic acid): 0–1 min 95% A, 3.1 min 5% A, 4.52 min 5% A, 4.92–9 min 95% A.
3. ACPs typically elute at ~4–4.5 min, which can be observed by peak in the LC trace that absorbs at 280 nm. The corresponding mass spectrum will likely look like a single set (for ACPs in a single form) or multiple sets (for ACPs in multiple forms) of 8–10 peaks corresponding to the +7 to +14 charge states.
4. Deconvolute the mass peaks produced by the run using ESIprot, a free online software found at https://www.bioprocess.org/esiprot/esiprot_form.php [26]. The resulting value can be compared to molecular weight predictions for the proteins of interest to determine if the ACP has been converted to the *holo* form (*see* **Notes 54** and **55**).

4 Notes

1. Two particularly helpful resources for understanding and applying Gibson Assembly® are Gibson, et al., *Nature Methods* (2009) and https://www.addgene.org/protocols/gibson-assembly/. It is advisable to understand the method before applying the primer design algorithm above.
2. We typically clone our genes of interest into pET-28a (+) at the NdeI/EcoRI cut sites to obtain the *N*-terminal His_6-tagged ACP. The His_6-tag can then be cleaved using thrombin.
3. Note that the program will run prematurely if carriage returns (new paragraphs) are included within the copied-and-pasted sequence. This will result in “Command not found” errors and can be solved by removing carriage returns before copying the sequence.

4. The primers produced by this algorithm are 49 base pairs in length. Typically, this length is within the less expensive range of oligonucleotides offered by manufacturers such as Eurofins Genomics.
5. While the melting temperatures of the primers should ideally be within ~5 °C of each other, we have observed success with wider gaps (up to 10 °C). Nonetheless, there are several ways to alter the melting temperature of a primer. Generally, adding more base pairs increases the temperature, while decreasing the length of the primer decreases the temperature. An A-T pair forms only two hydrogen bonds and thus has a lower melting temperature than a G-C pair, which forms three hydrogen bonds. Through trial-and-error, these guiding principles can be combined to produce primers that accurately reflect the gene sequence while also having an acceptable melting temperature.
6. PCR can be performed with a variety of polymerases. Information and protocols for the use of one such polymerase, the Q5® High-Fidelity DNA Polymerase, can be found here: https://www.neb.com/protocols/2013/12/13/pcr-using-q5-high-fidelity-dna-polymerase-m0491.
7. Actinomycete genomes (and other type II PKS BGCs) are often rich in GC content (>60%). The increased bond strength between this pair of nucleic acids (as opposed to AT pairs) can lead to difficulties in PCR. If our sequence is particularly high in GC, we add up to 5% DMSO (v/v) or a "GC enhancer" (volume according to the manufacturer's guidelines) to the PCR reaction mixture. Several companies produce GC enhancers. We use the Q5 High GC Enhancer by New England BioLabs.
8. Information, including protocols, for the Zymo DNA Clean & Concentrator™ Kit can be found here: https://www.zymoresearch.com/collections/dna-clean-concentrator-kits-dcc.
9. Additional useful information about restriction digestion can be found on the New England BioLabs website: https://www.neb.com/protocols/2012/12/07/optimizing-restriction-endonuclease-reactions.
10. To ensure linearization of the vector, we recommend transforming chemically competent cells with the doubly digested (linearized) vector and plating on LB agar plates supplemented with kanamycin. Do the same with intact pET-28a (+). The sample transformed with the linearized vector should not exhibit any growth on the antibiotic-treated plate, while the sample transformed with the intact plasmid should grow many colonies. DNA gel electrophoresis can also be used to verify the

successful digestion of the vector, as circularized DNA runs differently than linear DNA.

11. An easy way to assure the correct ratio between the DNA segments is to dilute the more concentrated component to match the concentration of the lesser concentrated component and simply add 1 μL of plasmid, 3 μL of insert, and 1 μL of nuclease-free water.
12. New England BioLabs, the producer of Gibson Assembly® materials, suggests that a total of 0.02 to 0.5 pmol of DNA be used when assembling 1 or 2 inserts into a vector.
13. Any host strain for routine cloning applications using plasmid vectors can work for this step. We typically use XL1 Blue or DH5α cells.
14. Target concentrations (in μg/mL) for several widely used antibiotics: Ampicillin—100; Kanamycin—50; Carbenicillin—100; Chloramphenicol—25; Spectinomycin—50; Tetracycline—10. Note: Tetracycline is a type II polyketide!
15. The replication of a large plasmid (i.e., a plasmid with a gene insert larger than a single ACP) is a resource-intensive process for cells, so expect the colonies of strains containing larger plasmids to take longer to appear. However, plasmids appearing after 72 h of incubation are typically "satellite colonies" and very likely do not contain the desired plasmid.
16. Growth for greater than 16 h is not recommended. Cells will begin to lyse and plasmid yields will be reduced.
17. Be sure to use a tube or flask with a volume of at least four times the volume of the culture and that the cap is loose enough to allow appropriate aeration (although not so loose as to allow contamination).
18. Information, including protocols, for the QIAprep Spin Miniprep Kit can be found here: https://www.qiagen.com/us/products/discovery-and-translational-research/dna-rna-purification/dna-purification/plasmid-dna/qiaprep-spin-miniprep-kit/#orderinginformation
19. We use the tube sequencing services provided by Eurofins Genomics: https://www.eurofinsgenomics.com/en/products/dna-sequencing/all-sequencing-options/
20. This step can be carried out before sequencing results are received, just be sure to keep track of which strains have been transformed with confirmed plasmids and which have not. Sequencing results are often returned the day immediately following shipment, so it is unlikely that strains with a faulty plasmid will get too far down the pipeline before the results come back. Restriction digestion can also be utilized (using NdeI and EcoRI) to cut the putative plasmid at the NdeI and

EcoRI cut sites. These fragments can then be run on an agarose gel to confirm the size of the insert.

21. Production of the target protein can easily be scaled up by doubling (tripling, etc.) the number of production cultures grown on Day 2. In this case, inoculate at least one seed culture for every planned production culture.
22. We keep 1 mL aliquots of a number of different antibiotic stock solutions at 1000× the target concentration so we can simply add 1 aliquot of antibiotic to 1 L of LB to reach the correct concentration. For example, we keep 1 mL aliquots of kanamycin at 50 mg/mL, which results in a 50 μg/mL final concentration upon addition to 1 L of LB.
23. Typically, cultures supplemented with a single antibiotic will reach an OD of 0.4–0.6 in about 2–3 h. Note that *E. coli* has a doubling time of ~20 min.
24. You can blank with water, as LB does not absorb at 600 nm.
25. We use a Beckman-Coulter Avanti floor centrifuge with a JA-10 rotor at 4,424 × *g* to harvest the cells.
26. There are a number of other lysing methods, such as freeze–thaw cycling, French press, and commercially available lysis reagents. We recommend sonication due to its speed and ease of use.
27. We use a Beckman-Coulter Avanti floor centrifuge with a JA-18 rotor at 24,985 × *g* to remove cell debris.
28. If low on materials, this reaction can be run with as little as 2 mL of reaction mixture. Simply scale ingredients accordingly.
29. DTT is a strong reducing agent and must be made fresh every time. It is normal for DTT to turn the Ni resin brown.
30. It usually takes about 30–50 mL of Wash Buffer to bring the absorbance at 260 nm close to zero.
31. The Sfp reaction is quite robust, so the goal is to reach 100% *holo*-ACP. If the target ACP is not efficiently converted from *apo*- to *holo*- form by R4-4 Sfp, the efficacy of AcpS-type phosphopantetheinyl transferases (PPTase) and/or use of the native PPTase can be explored [19].
32. We generally store our ACPs in 500 μL aliquots at ~1–2 mM.
33. The Harrick 13 mm IR liquid cell that we use requires 7 μL of (at least) 1 mM ACP per run. Be aware that conversion of *holo*-ACP to ACP-SCN does not always go to completion, according to LC-MS results. The protocol outlined here cyanylates 1–2 mg of ACP.
34. A_{412} measures TNB^{2-}, the side product of the reaction between DTNB and the free thiol of the Ppant arm. In general, for every mole of TNB^{2-} in solution, a mole of ACP has been

converted to the target mixed disulfide. If the target ACP has solvent-exposed thiols from surface cysteines, it is possible for these to react with the DTNB, although we tend to observe preferential modification of the Ppant arm due to its steric accessibility. If off-target modification becomes a problem, any solvent-exposed cysteine residues can be mutated to serine residues.

35. CN^- ions also react with remaining DTNB, which is largely what leads to the orange-colored solution. Keep in mind that CN^- is a strong base, so sufficient buffer capacity is required to keep the pH from rising and interfering with the successful formation of the covalently linked SCN functional group.
36. Excess cyanide and other species will not be effectively separated from the protein if more than 0.5 mL of the reaction solution is applied to the column at a time.
37. Repeated for emphasis: NaCN and aqueous cyanide ions are toxic chemicals. Be sure to refer to the MSDS for information on how to properly store, handle, and dispose of NaCN solutions.
38. Some less-stable ACPs require flash freezing with a cryoprotectant. If precipitation, typically white and wispy, is observed upon freeze-thawing, we add 5–30% glycerol (v/v) before freezing.
39. Detecting a single nitrile probe signal at millimolar concentration (or even below) can be accomplished with a number of different commercial FTIR spectrometers, but what is absolutely necessary is a stable and high signal-to-noise detector. We use a photovoltaic HgCdTe (MCT) detector produced by Kolmar, Inc. because it provides an excellent balance between sensitivity and robustness in that the signal does not easily saturate but very small differences in transmitted light can be detected. Other groups use InSb detectors, which are very sensitive in this particular spectral region, are NOT generally useful at lower frequencies, and saturate very easily (thus the source intensity must be dialed way back to work with InSb in a way that is less necessary with MCT). The most important hardware to make this kind of FTIR experiment work successfully is an excellent, liquid N_2-cooled detector and a stable N_2 purge source.
40. In H_2O samples, the region from approximately 1900–2800 cm^{-1} will contain usable data for probe detection, but on either side of this region there will be nonsensical data due to complete absorption of the IR source by H_2O. While it can be useful to collect data outside of this narrower window to view other signals from the protein, best signal-to-noise is obtained (and faster scanning can be accomplished) by limiting

each scan to 1900–2700 cm^{-1}. The nitrile stretching band is broad enough that scanning at high resolution (i.e., lower than 1 cm^{-1}) is not productive, but 2 cm^{-1} provides an upper bound on being able to collect a high-fidelity lineshape that is not affected by instrument resolution.

41. Single scans may be used to visualize the purge and the disappearance of water and CO_2 from the spectrum. CO_2 will show up on the spectrum as a strong doublet centered at about 2345 cm^{-1} overshadowing the area of interest corresponding to the thiocyanate probe, and water vapor appears as a series of sharp peaks spread across two broad spectral regions. No light will escape the sample from 3300 to 2900 or 1500 to 1650 cm^{-1}, so those regions should be ignored when H_2O is the solvent. Removal of these signals via a continuous dry N_2 purge provides much cleaner signals, especially from small peaks like the nitrile stretch of ACP-SCN at low-mM or lower concentrations.

42. If you plan to manually average multiple spectra in the following ACP scans, collect the blank sample scan five times and give each a different name (i.e., Buffer 1, Buffer 2, etc.). This is particularly advisable when working with a small concentration of ACP with poor signal-to-noise ratio. Flip the orientation of the cell at least once and remember to allow the chamber to repurge after moving or replacing the cell.

43. This baseline subtraction scheme assumes that there are no signals outside the "hole" region, which is not technically the case; the CN stretching band appears on top of the very broad, solute-sensitive combination band of liquid H_2O that is expected to shift between the solvent reference spectrum and the protein-containing sample spectrum. The baseline-subtraction scheme outlined here (which fits the regions immediately outside the peak region to an arbitrary but very flexible polynomial function, then subtracts the fit from the actual data) can introduce artifacts in the CN stretching lineshape. These artifacts can appear as "wiggles" or negative absorption features at the edges of the postprocessed lineshape. The presence of such artifacts suggests that the "hole" might be incorrect or that otherwise the assumptions made in this rudimentary baseline correction procedure are breaking down.

44. This colorimetric assay was developed and rigorously tested using the ketosynthase FabF from the *E. coli* fatty acid synthase. Whether it can or cannot be applied to other KSs and KS-CLFs is still under investigation, although preliminary results indicate that it likely can be.

45. "Molar equivalent" refers to the ratio of the number of molecules of one species relative to the number of molecules of

another. In this case, we want there to be 8 DTNB molecules for every ACP molecule.

46. The column should be run almost dry of buffer before the sample is added; adding the sample with more than 50–100 μL of buffer on top of the column will dilute the sample.
47. Collecting larger fractions will make it more difficult to isolate a pure sample of ACP-TNB^-.
48. In our experiments, ACP-TNB^- typically elutes in or around fractions 4, 5, and 6, while free DTNB/TNB^{2-} elutes after fraction 7.
49. Prepare in a glass vial, as plastic tubes may contain stray carboxylic acids that compete with the desired probe for loading onto the ACP.
50. The addition of more ATP may help if the reaction does not run to completion.
51. Acyl-acyl carrier protein synthetase (AasS) is derived from *Vibrio harvei* and is a promiscuous ligase for the addition of fatty acids (such as C5, C8, and C13) to the terminal thiol of the ACP's Ppant arm.
52. Successful detection of single-probe signals from aliphatic alkynes in low-mM (or lower) concentration samples via CW Raman depends mainly on two factors: a relatively intense scattering light source (visible is best as UV could lead to sample damage for long exposure) and a back-illuminated camera (usually CCD or EM-CCD). FT-Raman instrumentation will typically not be able to collect these signals with high fidelity: a dispersive instrument with an excellent detector is required. Time-resolved or pulsed higher-order Raman techniques should be able to acquire these probe signals. Dispersion gratings should be selected to provide a balance between spectral resolution (which should be on the order of 2–3 cm^{-1} in the alkyne CC stretching region, at maximum) and overspreading the few scattered photons across the available camera pixels.
53. We use an Agilent G6125BW LC-MS for our analysis of proteins.
54. Several posttranslational modifications to the ACP are often observed via LC-MS. We most commonly observe the loss of methionine (−Met) and the gluconoylation of the His_6-tag (+gluconoylation). A spreadsheet containing these modifications is available upon request.
55. A large error in the ESIprot result indicates the presence of multiple species in the sample, such as both *apo*- and *holo*-ACP, and the sets of peaks corresponding to each species have been mixed when selecting peaks for deconvolution.

Acknowledgments

The authors would like to thank Dr. Bashkim Kokona (Haverford College (HC)) Chemistry Department) for his technical guidance. We are also grateful to Matt Johnson (HC '14), Emily Winesett (HC '16), Sam Epstein (HC '18), Kofi Acheampong (HC '19), Gabriel Braun (HC '19), Danielle Jacobsen (HC '19), Dr. Karl Johnson (HC Biology Department), Adam Huff (formerly HC Chemistry Department), and members of the Charkoudian Lab for their contributions to the development of these methods. Our research related to acyl carrier proteins is currently supported by NIH 2R15GM120704 (L.K.C.), NSF CAREER Award CHE1652424 (L.K.C.), Cottrell Scholars Award 24350 (L.K.C.), a Henry Dreyfus Teacher Scholar Award (L.K.C.), and the Beckman Scholarship (G.S.H.).

References

1. Kautsar SA, Blin K, Shaw S, Navarro-Muñoz JC, Terlouw BR, van der Hooft JJJ, van Santen JA, Tracanna V, Suarez Duran HG, Pascal Andreu V, Selem-Mojica N, Alanjary M, Robinson SL, Lund G, Epstein SC, Sisto AC, Charkoudian LK, Collemare J, Linington RG, Weber T, Medema MH (2020) MIBiG 2.0: a repository for biosynthetic gene clusters of known function. Nucleic Acids Res 48(D1): D454–D458. https://doi.org/10.1093/nar/gkz882
2. Medema MH, Blin K, Cimermancic P, de Jager V, Zakrzewski P, Fischbach MA, Weber T, Takano E, Breitling R (2011) antiSMASH: rapid identification, annotation and analysis of secondary metabolite biosynthesis gene clusters in bacterial and fungal genome sequences. Nucleic Acids Res 39: W339–W346. https://doi.org/10.1093/nar/gkr466
3. Newman DJ, Cragg GM (2020) Natural products as sources of new drugs over the nearly four decades from 01/1981 to 09/2019. J Nat Prod 83(3):770–803. https://doi.org/10.1021/acs.jnatprod.9b01285
4. Charov K, Burkart MD (2020) Chapter Sixteen—Quantifying protein-protein interactions of the acyl carrier protein with solvatochromic probes. In: Chenoweth DM (ed) Methods in Enzymology, vol 638. Chemical Tools for Imaging, Manipulating, and Tracking Biological Systems: Diverse Methods for Prokaryotic and Eukaryotic Systems. Academic Press, Cambridge, pp 321–340
5. Gopal GJ, Kumar A (2013) Strategies for the production of recombinant protein in Escherichia coli. Protein J 32(6):419–425. https://doi.org/10.1007/s10930-013-9502-5
6. Kaur J, Kumar A, Kaur J (2018) Strategies for optimization of heterologous protein expression in *E. coli*: roadblocks and reinforcements. Int J Biol Macromol 106:803–822. https://doi.org/10.1016/j.ijbiomac.2017.08.080
7. Zerbs S, Giuliani S, Collart F (2014) Small-scale expression of proteins in E. coli. Meth Enzymol 536:117–131. https://doi.org/10.1016/B978-0-12-420070-8.00011-8
8. Vázquez-Laslop N, Mankin AS (2018) How macrolide antibiotics work. Trends Biochem Sci 43(9):668–684. https://doi.org/10.1016/j.tibs.2018.06.011
9. Pickens LB, Tang Y (2009) Decoding and engineering tetracycline biosynthesis. Metab Eng 11(2):69–75. https://doi.org/10.1016/j.ymben.2008.10.001
10. Pickens LB, Tang Y (2010) Oxytetracycline biosynthesis. J Biol Chem 285(36): 27509–27515. https://doi.org/10.1074/jbc.R110.130419
11. Hillenmeyer ME, Vandova GA, Berlew EE, Charkoudian LK (2015) Evolution of chemical diversity by coordinated gene swaps in type II polyketide gene clusters. PNAS 112(45): 13952–13957. https://doi.org/10.1073/pnas.1511688112
12. Keatinge-Clay AT, Maltby DA, Medzihradszky KF, Khosla C, Stroud RM (2004) An antibiotic factory caught in action. Nat Struct Mol Biol 11(9):888–893. https://doi.org/10.1038/nsmb808

13. Cai W, Zhang W (2018) Engineering modular polyketide synthases for production of biofuels and industrial chemicals. Curr Opin Biotechnol 50:32–38. https://doi.org/10.1016/j.copbio.2017.08.017
14. Cummings M, Peters AD, Whitehead GFS, Menon BRK, Micklefield J, Webb SJ, Takano E (2019) Assembling a plug-and-play production line for combinatorial biosynthesis of aromatic polyketides in *Escherichia coli*. PLoS Biol 17(7):e3000347. https://doi.org/10.1371/journal.pbio.3000347
15. Klein JG, Wu Y, Kokona B, Charkoudian LK (2020) Widening the bottleneck: heterologous expression, purification, and characterization of the Ktedonobacter racemifer minimal type II polyketide synthase in *Escherichia coli*. Bioorg Med Chem 28(20):115686. https://doi.org/10.1016/j.bmc.2020.115686
16. Du D, Katsuyama Y, Horiuchi M, Fushinobu S, Chen A, Davis TD, Burkart MD, Ohnishi Y (2020) Structural basis for selectivity in a highly reducing type II polyketide synthase. Nat Chem Biol 16(7):776–782. https://doi.org/10.1038/s41589-020-0530-0
17. Liu X, Hua K, Liu D, Wu Z-L, Wang Y, Zhang H, Deng Z, Pfeifer BA, Jiang M (2020) Heterologous biosynthesis of type II polyketide products Using *E. coli*. ACS Chem Biol 15(5):1177–1183. https://doi.org/10.1021/acschembio.9b00827
18. Zhang H, Skalina K, Jiang M, Pfeifer BA (2012) Improved E. coli erythromycin a production through the application of metabolic and bioprocess engineering. Biotechnol Prog 28(1):292–296. https://doi.org/10.1002/btpr.702
19. Mofid MR, Finking R, Marahiel MA (2002) Recognition of hybrid peptidyl carrier proteins/acyl carrier proteins in nonribosomal peptide Synthetase modules by the 4-′-Phophopantetheinyl transferases AcpS and Sfp. J Biol Chem 277(19):17023–17031. https://doi.org/10.1074/jbc.M200120200
20. Sunbul M, Marshall NJ, Zou Y, Zhang K, Yin J (2009) Catalytic turnover-based phage selection for engineering the substrate specificity of Sfp phosphopantetheinyl transferase. J Mol Biol 387(4):883–898. https://doi.org/10.1016/j.jmb.2009.02.010
21. Gibson DG, Young L, Chuang R-Y, Venter JC, Hutchison CA, Smith HO (2009) Enzymatic assembly of DNA molecules up to several hundred kilobases. Nat Methods 6(5):343–345. https://doi.org/10.1038/nmeth.1318
22. Johnson MNR, Londergan CH, Charkoudian LK (2014) Probing the Phosphopantetheine arm conformations of acyl carrier proteins using vibrational spectroscopy. J Am Chem Soc 136(32):11240–11243. https://doi.org/10.1021/ja505442h
23. Acheampong KK, Kokona B, Braun GA, Jacobsen DR, Johnson KA, Charkoudian LK (2019) Colorimetric assay reports on acyl carrier protein interactions. Sci Rep 9(1):15589. https://doi.org/10.1038/s41598-019-51554-6
24. Beld J, Finzel K, Burkart Michael D (2014) Versatility of acyl-acyl carrier protein Synthetases. Chem Biol 21(10):1293–1299. https://doi.org/10.1016/j.chembiol.2014.08.015
25. Epstein SC, Huff AR, Winesett ES, Londergan CH, Charkoudian LK (2019) Tracking carrier protein motions with Raman spectroscopy. Nat Commun 10(1):2227. https://doi.org/10.1038/s41467-019-10184-2
26. Winkler R (2010) ESIprot: a universal tool for charge state determination and molecular weight calculation of proteins from electrospray ionization mass spectrometry data. Rapid Commun Mass Spectrom 24(3):285–294. https://doi.org/10.1002/rcm.4384

Chapter 14

Cyanobacterial Genome Sequencing, Annotation, and Bioinformatics

Jonna Teikari, Martin Baunach, and Elke Dittmann

Abstract

Cyanobacteria are collectively a globally important monophyletic phylum of bacteria. They have attracted a lot of attention, not only because they are rich sources of natural bioactive products, including toxic substances, but also because they play an important role in global nitrogen and carbon cycles, and are capable of maintaining versatile environmental niche adaptations. A vast number of cyanobacterial genomes have become available due to fast development of sequencing technologies, but effort is still needed to comprehensively understand the molecular basis of their diversity. Here, we introduce a basic pipeline for the cyanobacterial genome sequencing project that can be employed to complete the whole cyanobacterial genome. The pipeline includes DNA extraction from the cyanobacterial culture of interest, hybrid genome sequencing, and genome assembly and annotation. At the end of the chapter, we briefly introduce genome mining tools and one successful genome mining example from our laboratory. This chapter provides general guidance regarding the sequencing project and thus includes several references for alternative methods and tools so that the reader can easily modify the pipeline according to the needs of the laboratory.

Key words Cyanobacteria, De novo genome sequencing, PacBio, Illumina, Genome mining, Natural products

1 Introduction

De novo genome sequencing, the sequencing of a novel genome without available reference genome, has revolutionized our understanding of cyanobacterial diversity and their niche adaptation strategies to inhabit a great variety of habitats. Study of cyanobacterial genomes is still a robust way to delineate novel and unknown metabolic pathways, and rapid decreases in sequencing costs has made the genomes available even to smaller research groups. Nowadays, many laboratories eagerly unravel the heredity of their own organisms to understand the ecological roles or capabilities that may be employed in biotechnological applications [1]. The size range of approximately 3000 sequenced cyanobacterial genomes (as of 31.12.2020) spans from the smallest genome of the marine

Elizabeth Skellam (ed.), *Engineering Natural Product Biosynthesis: Methods and Protocols*,
Methods in Molecular Biology, vol. 2489, https://doi.org/10.1007/978-1-0716-2273-5_14,

Candidatus Atelocyanobacterium thalassa UCYN-A (1.44 Mbp) [2] to the largest genome of *Scytonema hofmanni* PCC 7110 (12.07 Mbp) [3]. Within this broad genomic size range, only 621 genes are present in all species (core genome) [4]. The core genome consists mostly of well-known pathways in primary metabolism, transcription, translation, and DNA replication, whereas much more uncertainty exists outside of the core genome [4], making cyanobacteria an interesting field of study.

There are dozens of ways to navigate a de novo genome sequencing project from a fresh cyanobacterial culture to human-readable file formats. Even though there are several options to combine methods and tools to obtain the whole genome, the main workflow always covers the same steps: (i) DNA extraction; (ii) library preparation and library sequencing; (iii) genome assembly; and (iv) genome annotation. Many cyanobacterial genome projects still rely on short-read Illumina (or Ion Torrent) sequencing due to their cost efficiency and capability to provide a large amount of data cheaply (60–80× genome coverage recommended). Short reads usually result in fragmented genome assembly, because the cyanobacterial genome consists of many genomic repeats, elements that occur multiple times in the genomes [5]. Thus, we highly recommend a hybrid genome-sequencing approach that combines long- and short-read sequencing and tackles the limitations of a single platform in order to gain an accurate whole cyanobacterial genome sequence and highly accurate annotation.

The sequenced cyanobacterial genomes are a tremendously valuable resource, since they encode blueprints for innumerable novel chemical entities in the form of secondary metabolite biosynthesis gene clusters (SMGC). Endowed with a plethora of bioactivities, like cytotoxic, antibacterial, or antiviral properties, secondary metabolites are crucial molecular mediators in microbial ecology and provide indispensable drugs. Nonetheless, with the emergence of antibiotic resistance, growing threat of viral pandemics, and increased incidence of cancer, novel chemical entities are urgently needed. Cyanobacteria are prolific producers of natural products and an emerging source for drug discovery [6, 7]. The identification of promising SMGCs and discovery of their precious products via genome mining is key to the successful discovery of novel bioactive compounds.

The workflow described in this chapter (Fig. 1) combines long-read sequences and a paired-end short-read Illumina platform, which is currently a common approach to uncover a full cyanobacterial genome [8–11]. As cyanobacterial genera, and even species, vary a lot in their geno- and phenotypes, some steps in the protocol might need further species-specific optimization. This chapter will end with a short summary introducing popular genome mining tools and databases, together with a recent successful genome-

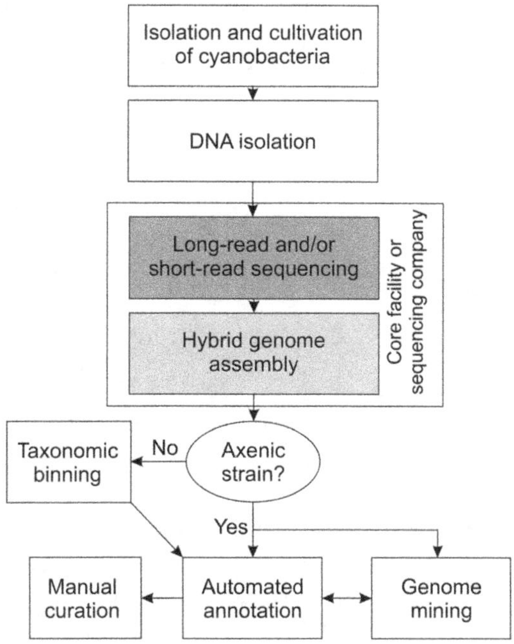

Fig. 1 Flow-chart of the cyanobacterial genome sequencing project

mining workflow from our laboratory. For more detailed information on tools and mining strategies, we recommend a recent excellent protocol dedicated exclusively to the mining of bacterial genomes for secondary metabolite gene clusters [12].

2 Materials

All buffers are prepared in autoclaved ultrapure water and analytical-grade reagents. Nitrile gloves should be worn for all steps involving phenol or chloroform, which are performed in the chemical fume hood. Phenol–chloroform waste must be collected into properly labeled containers and phenol–chloroform-contaminated consumables, such as tubes, pipette tips, and gloves, discarded into separate sealable containers. Nitrile gloves should always be used while touching ethidium bromide–contaminated working spaces and all waste collected into separate containers.

2.1 DNA Extraction

1. Lysis buffer: 50 mM Tris–HCl, pH 8.0, 100 mM EDTA, 0.1 M NaCl.
2. Lysozyme.
3. 10 mg/mL Proteinase K.
4. Phenol buffered in pH 7.5.
5. Chloroform–isoamyl alcohol 24:1.

6. TE buffer: 10 mM Tris–HCl, pH = 8.0, 1 mM EDTA.
7. 20 mg/mL RNase.
8. 10% cetyltrimethylammonium bromide (CTAB).
9. 5 M NaCl.
10. 100% ethanol.
11. Sucrose.

2.2 Quality Check

1. Agarose.
2. Ethidium bromide.
3. Tris–acetate–EDTA (TAE) running buffer.
4. DNA gel loading dye.
5. DNA ladder.

2.3 Bioinformatics

A powerful computer is needed to handle the sequencing data effectively. Assembly programs require at least 16 GB of RAM and 8 logical cores to finish a small bacterial genome, whereas to map one million Illumina reads with average length of 100 nucleotides, roughly 1 GB of RAM memory is required. The better the computer is, however, the less computing time is needed, as individual tasks are executed more rapidly. Most of the assembly and annotation programs are freely available, developed by the scientific community, and operate on a UNIX system such as Linux or MacOS. More easily accessible genome-analysis tools are continuously developed that are available online to address the increased demand to understand the biology of the novel genomes. All bioinformatics tools have their own tutorials providing detailed information of the installation, parameters, and troubleshooting. Only the most important parameters in terms of cyanobacterial genome sequencing are described in this chapter.

3 Methods

3.1 Extraction of High–Molecular-Weight Genomic DNA

High–molecular-weight, unshared, and uncontaminated genomic DNA is necessary to generate a good-quality DNA library for sequencing. One challenge in cyanobacterial genome sequencing is the maintenance of axenic cyanobacterial cultures due to the long evolutionary dependency of cyanobacteria to live in association with heterotrophic bacterial communities. Sequencing heavily contaminated DNA often results in unfinished genome assemblies and in the worst cases, causes incorrect annotations [13]. However, sometimes it is justified to sequence an entire bacterial community. In such cases, cyanobacterium is usually the dominating organism in the culture and, despite contamination, genome assemblers are capable of finishing long scaffolds and full natural product gene

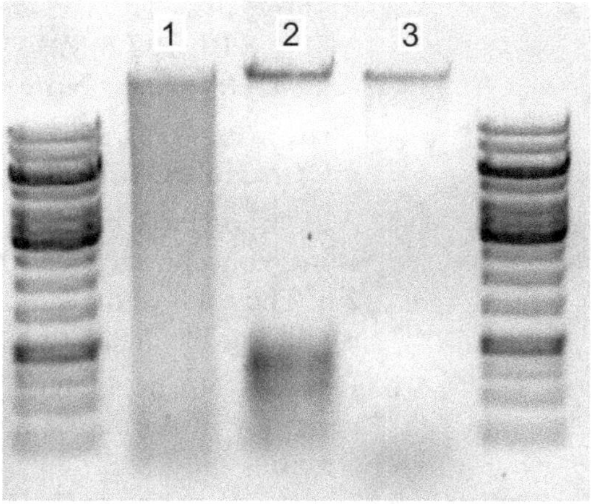

Fig. 2 DNA isolated from axenic *Nostoc punctiforme* PCC73102 using commercial column and phenol–chloroform extraction (described here). (1) Sheared DNA extracted using commercial kit, (2) phenol–chloroform extraction before RNA removal, (3) sample 2 after RNA degradation ready for sequencing. Ladder = GeneRuler™ 1 kb DNA ladder, Thermo Scientific

clusters. If planning to start from a complex ecosystem, more data is often required to achieve longer sequence stretches and to gain an answer for the research questions. Axenic cyanobacterial cultures are highly recommended as the starting material for cyanobacterial genome sequencing and thus this protocol follows the pipeline for axenic cyanobacterial sequencing.

Cyanobacteria DNA extraction itself can be challenging due to rigid cell walls covered by extracellular polysaccharides [14]. Gentle enzymatic cell breaking and addition of detergents are widely used in DNA extraction for sequencing, whereas mechanical disruption and commercial columns usually shear the DNA and should be avoided (Fig. 2). The quantity of DNA required depends on the chosen sequencing platform. For long-read sequencing, 2 μg is a sufficient amount of DNA, but tens of micrograms are recommended to enable the selection of the longest possible DNA fragments. For Illumina sequencing, only 50 ng of DNA is required for library amplification whereas Illumina sequencing is usually not an issue in terms of DNA quantity. It is always better to extract more DNA than needed because if more data is required later, the same DNA will be available for further usage.

1. Harvest 50 mL of axenic cyanobacterial culture in mid-logarithmic growth phase by centrifuging the culture at 5000 × *g* for 5 min (*see* **Note 1**).
2. Remove liquid carefully and wash cells by resuspending the pellet in 30 mL of lysis buffer. Collect cells at 5000 × *g* for 5 min.

3. Resuspend cells in 3 mL of buffer (50 mM Tris–HCl (pH = 8.0), 100 mM EDTA, 100 mM NaCl) and freeze the cells at −80 °C (*see* **Note 2**).
4. Thaw frozen cells. Add lysozyme 2 mg/mL and RNase 20 μg/mL (*see* **Note 3**). Incubate 1–2 h at 37 °C.
5. Add sarkosyl to a final concentration of 1% and proteinase K to a final concentration of 200 μg/mL. Incubate at 50 °C for 1–2 h. The color of the sample usually changes to dark or brownish during the incubation.
6. Extract DNA by adding 1:1 volume of phenol. Mix the tube gently by inverting for 5 min. Centrifuge at 12,000 × *g* for 5 min.
7. Collect upper water phase (*see* **Note 4**) and repeat phenol extraction until no interphase is visible.
8. Extract DNA by adding 1:1 volume of chloroform–isoamyl alcohol (24:1), mix, centrifuge, and collect the water phase as at **steps 6** and 7.
9. Precipitate DNA by adding 2:1 volume of 100% ethanol. Centrifuge at 12,000 × *g* for 20 min at 4 °C and remove supernatant.
10. Wash the pellet with 1 mL of 70% ethanol and centrifuge at 12,000 × *g* for 5 min at +4 °C.
11. Remove all liquid carefully using fine pipet and let pellet dry completely.
12. Dissolve DNA in 1 mL of TE buffer and store at −20 °C. Avoid freeze–thaw cycling of the sample. Use wide tips while pipetting DNA, avoid vigorous vortexing, and favor gentle mixing.
13. Check the quality of the DNA on 0.5% agarose gel, 1 h, 40 V and determine the concentration using Qubit fluorometer or NanoDrop spectrophotometer (*see* **Note 5**).

3.2 Hybrid Genome Sequencing

Outsourcing genome library preparation and genome sequencing is usually preferred, since library preparation and sequencing devices are expensive and their usage remains low, even in the bigger laboratories. However, despite outsourcing this part of the project, knowing the possibilities and limitations of the currently available sequencing platforms is crucial for everyone working on genome-sequencing projects. In general, suitable sequencing platforms for whole genomic approaches are divided into short-read (100–500 bp) and long-read (>10 kb) sequencing, both having their own strengths and weaknesses [15]. Pacific Biosciences' (PacBio) single-molecule, real-time (SMRT) sequencing and Oxford Nanopore Technologies' (ONT) long-read sequencing are currently the dominating long-read sequencing platforms on the market. These techniques commonly produce 10–50 kb long reads, which are beneficial in the assembly of genomic entities with repetitive regions. However, despite recent developments, both of the available long-read sequencing techniques are still notorious in their high error rates, which lowers the quality of the genome sequence remarkably [16, 17]. To overcome this challenge, an increased amount of sequence data can be used to improve consensus accuracy, which, in turn, increases the sequencing cost per base. Instead of sequencing a huge number of long reads, hybrid genome sequencing has become the preferred sequencing workflow, also among cyanobacterial researchers, to obtain high-quality reference genomes with moderate costs. Using hybrid approach, long reads are produced with relatively low coverage to provide a correct and finished genome structure, whereas sequencing errors are corrected using cheaper short-read data. There are several short-read sequencing platforms available, of which Illumina systems hold a vast part of the market share. For cyanobacterial hybrid sequencing, 2 × 300 bp paired-end MiSeq is one suitable platform, but as Illumina systems mainly differ in the sample throughput and read-length, all platforms are equally suitable to correct the long-read assembly. Roughly estimated, >10× coverage of long-read data and >1 Gbp of Illumina data is enough to provide a high-quality reference genome for axenic cyanobacteria. For more complex genomes, with a lot of repetitive elements and contaminations of the heterotrophic bacterial DNA community, deeper sequencing is needed.

3.3 Assembly of the Cyanobacterial Reference Genome

Typically, biologists perceive genome assembly as the most challenging step in genome-sequencing projects because i) assembly requires great computational power that may be lacking in the laboratory, and ii) tasks are often executed using the command line rather than graphical user interface (GUI). Sequencing centers and core facilities have responded to assembly issues by offering the assembly services with a low additional cost. Sometimes outsourcing genome assembly is justified, but even so, it is good to understand the basics of how to complete the genomic jigsaw puzzle. One alternative is to use the Galaxy platform that offers a nice GUI based platform for many bioinformatics tasks (https://usegalaxy.org/). Several step-by-step Galaxy tutorials are freely available for small bacterial genome assembly starting from different types of datasets, and these tutorials can be applied for cyanobacterial genome-sequencing projects. The assembly and annotation tasks described in this chapter have been designed so that they can also be executed on the Galaxy server.

Before proceeding to genome assembly, the quality of the sequencing data needs to be verified, and based on the quality report, contaminants need to be filtered and poor sequences trimmed. FastQC (*see* **Note 6**) is a simple tool to obtain an impression of the raw sequence data [18]. Two of the most important plots in the output file are "Per base sequence quality" and "Overrepresented sequences" first of which tells the accuracy of the sequence data and the second about the adapter leftovers (*see* **Note 7**). To ensure that low quality bases, contaminants, and leftover adapter sequences do not cause inaccuracies in the alignment, poor regions need to be trimmed from the short-read dataset. Cutadapt is the recommended tool to trim all unwanted base information and ensure that data will be aligned properly [19].

Check the data quality and adapter contamination using FastQC.

```
$ fastqc reads_R1.fq.gz
```

Remove leftover adapters, short reads, and low-quality bases.

```
$ cutadapt -a ADAPTER_F -A ADAPTER_R -m 100 -q 25 -o out_R1.
fastq -p out_R2.fastq reads_R1.fastq reads_R2.fastq
a = forward adapter
A = reverse adapter
q = minimum quality
m = minimum sequence length
```

In de novo genome assembly, sequenced DNA species are merged back to genomic elements, chromosomes, and plasmids, based on the overlapping read regions of the sequencing data. In hybrid assembly, genome structure is achieved by using the relatively low coverage of the long-read data and ambiguities are corrected by more accurate short-read Illumina data. Errors can also be eliminated by increased long-read sequencing depth, but due to the robustness and cost-efficiency, the Illumina platform is usually preferred to correct the final errors. Several bioinformatics tools exist for de novo hybrid genome assembly that either use long reads to connect short-read scaffolds (short-read-first) or where long reads are corrected by short reads before or after the assembly (long-read-first).

Unicycler is one approach that combines several tools into one package making genome assembly relatively straightforward [20] (Fig. 3). Unicycler first assembles short-read Illumina data using SPAdes [21] with read error corrections, a range of k-mer sizes, and several cleaning procedures to find the simplest possible assembly structure (*see* **Note 8**). Next, semiglobal alignment of long reads is applied to bridge the gaps in short-read alignment to make final consensus sequences. Although Unicycler uses short-read accuracy as far as possible, sometimes sequence string is only based on the long-read data. Thus, at the very final step, Unicycler applies Bowtie2 [22] and Pilon [23] to improve the draft assembly by removing mainly SNPs and indels (*see* **Note 9**). One additional advantage in

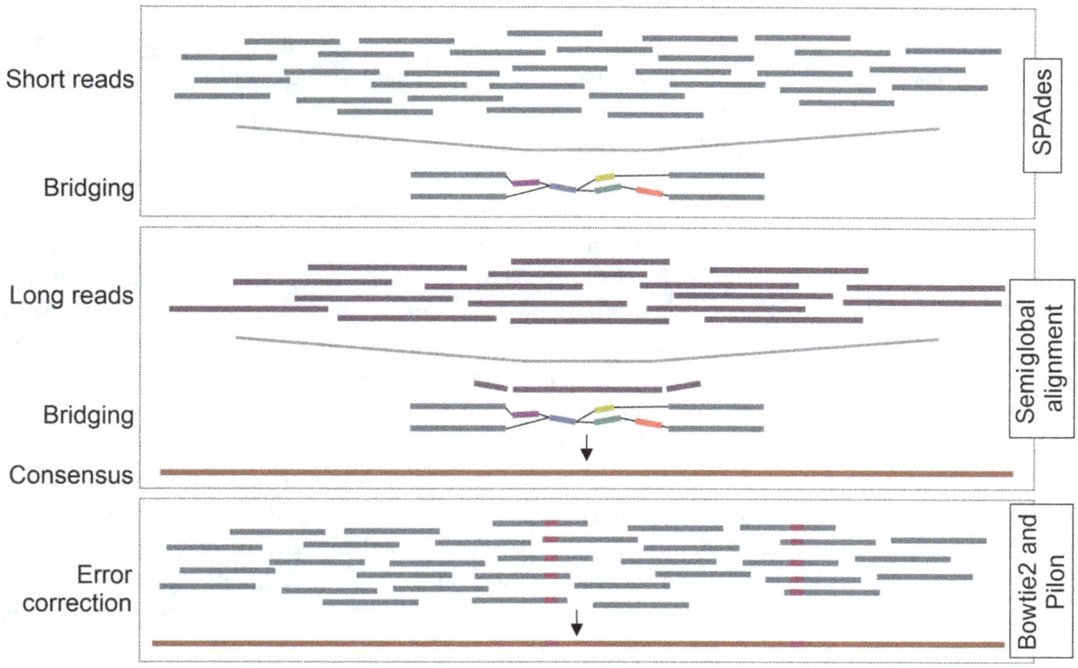

Fig. 3 Simplified graph of the Unicycler assembly pipeline described in this chapter. Hybrid assembly of bacterial genome tackles the ambiguities of the existing sequencing technologies, combining short and long-read sequences to generate complete and accurate final assembly. Details of the genome assembly algorithms are beyond this book chapter and complete description of the multistep assembly pipeline is described in the original publication [20]

Unicycler is its capability to consider genome elements circular and an additional circularization step and shifting of the genome assembly to start from the right position (usually *dnaA*) can be omitted from the further analysis (*see* **Note 10**). Final assembly can be evaluated using the Quast tool [24], which also provides statistics for the assembly.

Run Unicycler using paired-end short-read sequences and long-read data as an input (*see* **Note 7**).

```
$ unicycler -1 reads_R1.fastq -2 reads_R2.fastq -l long.fastq
--out out_dir --mode normal
mode = Bridging mode (note 11)
1 = FASTQ file of first short-reads in each pair
2 = FASTQ file of second short-reads in each pair
l = FASTQ or FASTA file of long-reads
```

Evaluate the quality of the assembly using Quast [24].

```
$ quast.py genome.fasta -o output_dir
```

3.4 Functional Annotation of the Cyanobacterial Genome

To understand the metabolic repertoire of the organism, the genome must be annotated. Six possible open reading frames (ORF) are identified and program decides which of the ORFs encode known, unknown, or hypothetical proteins. The task is relatively easy for cyanobacteria having high protein-coding density (>52%) [25] that renders automated annotation algorithms effective. Cyanobacterial genome annotation covers mostly the protein coding regions and well-known noncoding regions, such as 16S and 23S rRNAs, and tRNAs, despite the importance of the intergenic regions in gene regulation that have attracted increasing attention in recent years [26]. Promising pipelines to identify intergenic regions are available, but these methods are beyond the scope of this chapter [27].

None of the annotation algorithm is perfect and even the best tools yield errors resulting in the quick propagation of incorrect annotations in the reference database [28]. Inaccuracies in the databases have developed due to the usage of contaminated and low-quality reference genomes. Thus, critical curation of the genome annotation, for example by PCR or mutation studies, is essential especially if the main finding of the study relies strongly on the gene found by the annotation algorithm. In cyanobacterial research, Prokka [29], RAST online server ([30]), and NCBI Prokaryotic Genome Annotation Pipeline PGAP [31] are commonly used tools and a combination of the outputs from these three annotation programs usually provides a comprehensive overview of the known metabolic pathways (*see* **Note 12**). Annotated .gbk or .gff files can be visualized and manually corrected on the genome browsers, such as Artemis [32], and known metabolic pathways can be searched using BlastKOALA [33].

Annotate genome by Prokka.

```
$ prokka pilon.fasta --outdir prokka_annotation --prefix
mycyano
```

Visualize your annotation on Artemis genome browser.

```
$ art prokka_annotation/mycyano.gff
```

3.5 Genome Mining for the Discovery of Novel Cyanobacterial Natural Products

3.5.1 Overview of Computational Tools and Databases

The most widely used platform for the detection and analysis of secondary metabolite biosynthesis gene clusters (SMGCs) is currently antiSMASH (**anti**biotics and **S**econdary **M**etabolite **A**nalysis **SH**ell), which is available as a web-based application or as a standalone tool [34]. This platform allows the rapid genome-wide identification, annotation, and analysis of SMGC in the genomes of microorganisms. By using profile hidden Markov models [35] of genes that are specific for certain types of gene clusters, antiSMASH identifies the SMGCs of all known broad chemical classes. Besides detecting gene clusters, antiSMASH offers detailed sequence analysis by integrating and cross-linking various in silico secondary metabolite analysis tools and databases. For example, biosynthetic genes, regulators, and transporters are color coded and for modular enzymes like type I polyketide synthases (PKS) and nonribosomal peptide synthetases (NRPS) the different domains of a module are depicted and linked to predicted substrate specificity. Moreover, links to uncharacterized and known homologous gene clusters, as well as to defined subclusters, are provided for any predicted cluster. AntiSMASH also offers a rough prediction of core structures of the final products of type I PKS, NRPS, and lantipeptide biosynthesis gene clusters based on substrate specificity predictions of modules, collinearity rule, and the sequence of core peptides. However, cyclization and posttranslational modifications from tailoring enzymes are not considered and results should be interpreted with care.

An alternative to antiSMASH is PRISM (**PR**ediction **I**nformatics for **S**econdary **M**etabolomes). Started as a Java application and web server for the chemical structure prediction of genetically encoded nonribosomal peptides (NRPs) and type I and II polyketides (PKs) the latest published version (PRISM4) can detect 22 distinct natural product cluster types and predict the chemical structures of 16 different classes of secondary metabolites [36].

Besides general mining tools like antiSMASH or PRISM, various mining tools exist that are specialized to specific compound classes like SBSPKS (**S**tructure **B**ased **S**equence analysis of **PKS** and NRPS) [37] for polyketide synthase (PKS) and nonribosomal peptide synthetase (NRPS) gene clusters, RODEO (**R**apid **O**RF **D**escription and **E**valuation **O**nline) [38] and RiPPMiner [39] for ribosomal natural product (RiPP) biosynthesis gene clusters or BAGEL for bacteriocin biosynthesis gene clusters [40]. All of these tools represent high-confidence/low-novelty methods, since they utilize well-defined queries to identify signature genes or domains that are highly specific for known classes of biosynthesis pathways. Whilst this strategy quickly and reliably yields an overview of the SMGC repertoire of a single strain from its genome sequence it will inherently miss unknown types of gene clusters, which are highly likely to encode novel chemistry [41].

To address this shortcoming low-confidence/high-novelty algorithms are emerging. For example, the ClusterFinder algorithm seeks Pfam domains that occur frequently in biosynthetic gene clusters, such as from oxidoreductases, methyltransferases, coenzyme A (CoA) ligases, and cytochrome P450 enzymes, and with this detects patterns of gene functions that look SMGC-like [41, 42].

Recent community efforts to collect and catalogue data on secondary metabolite biosynthesis should be highlighted (Table 1), which can be of great value for data mining projects,

Table 1
Overview of computational tools and databases for the analysis of secondary metabolites and SMGCs

Tool/database	Web server URL	References
antiSMASH	https://antismash.secondarymetabolites.org/	[34]
PRISM	https://prism.adapsyn.com/	[36]
SBSPKS	http://www.nii.ac.in/sbspks2.html	[37]
RODEO	http://rodeo.scs.illinois.edu/	[38]
RiPPMiner	http://www.nii.ac.in/rippminer.html	[39]
BAGEL	http://bagel4.molgenrug.nl/	[40]
ClusterFinder	https://github.com/petercim/ClusterFinder	[42]
MIBiG	https://mibig.secondarymetabolites.org/	[43]
NP Atlas	https://www.npatlas.org/	[44]
Paired Omics	https://pairedomicsdata.bioinformatics.nl/	[45]
SMBP	http://www.secondarymetabolites.org	[47]

like the MIBiG consortium (**M**inimum **I**nformation about a **Bio**synthetic **G**ene cluster) focusing on biosynthesis gene clusters [43], NP Atlas focusing on chemical structures of natural products [44] or the Paired Omics Data Platform focusing on genotype/chemotype links ([44]. CyanoMetDB is a very recently introduced comprehensive database exclusively for cyanobacterial metabolites [46].

For more information on tools and databases that are available to mine, identify, and characterize natural product biosynthesis pathways and their producers based on 'omics data, we recommend a comprehensive review, which is accompanied by a comprehensive web portal called Secondary Metabolite Bioinformatics Portal (SMBP) [47].

3.5.2 Exemplary Workflow for the Mining of Cyanobacterial Genomes

Sequencing of microbial genomes often reveals a high number of SMGCs within a genome. However, this pronounced potential for the biosynthesis of secondary metabolites frequently is in stark contrast to the actual metabolic phenotype observed under standard laboratory conditions, owing to the fact that under these conditions, the majority of SMGCs are silent or poorly expressed. A promising strategy to bring these hidden metabolic treasures to light is the combination of *in silico* gene clusters and compound prediction, expression monitoring of target SMGCs and the variation of cultivation conditions. Cultivation under varying biotic and abiotic conditions has been shown to be an efficient way to trigger the expression of silent or poorly expressed SMGCs, a concept also known as the OSMAC (One Strain–Many Compounds) approach [48]. Expression monitoring is key to find optimal production

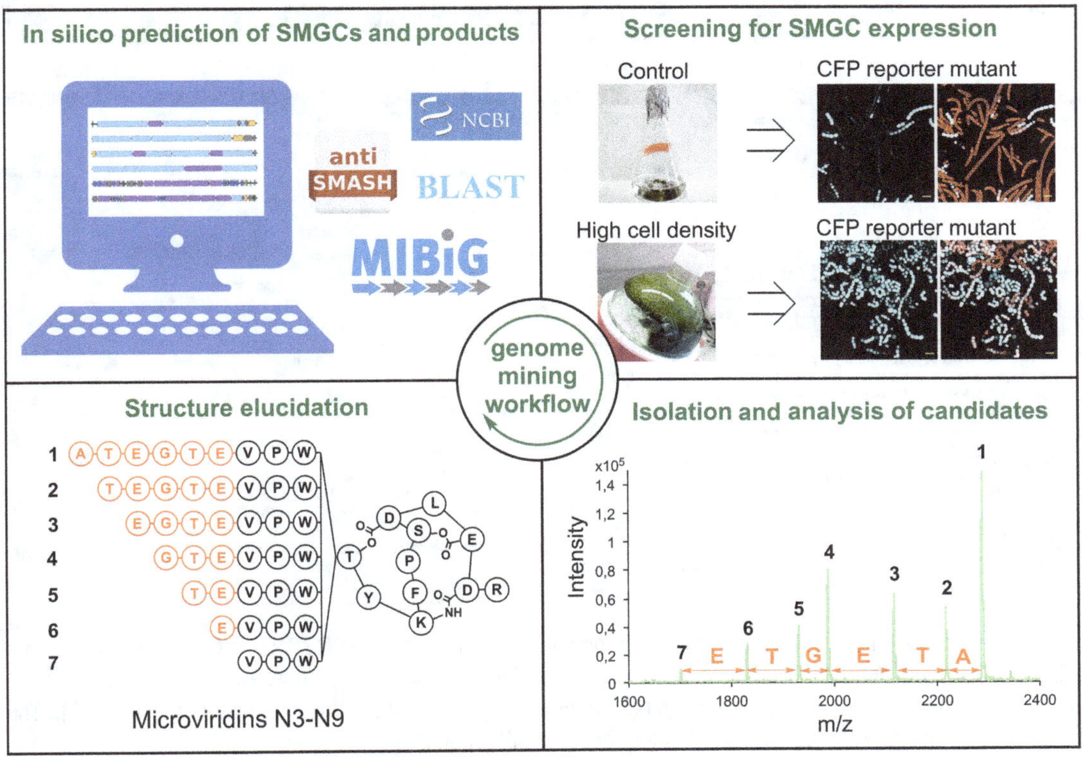

Fig. 4 Postsequencing workflow for mining of cyanobacterial genomes to identify novel compounds. As an example the discovery of unusual microviridins from *Nostoc punctiforme* PCC 73102 is chosen [49]

conditions. Finally, analytical concepts are mandatory to detect and confirm candidate compounds and might comprise screening for predicted features like mass range, chromo/fluorophores, polarity, bioactivity, or comparative metabolomics of producing and non-producing conditions.

A recent successful case of this strategy from our lab is the discovery of unusual microviridins from *Nostoc punctiforme* PCC 73102 (Fig. 4), a filamentous cyanobacterium capable of facultatively forming symbiotic relationships with diverse plant and fungal hosts [49].

In silico prediction of SMGCs on the chromosome of *N. punctiforme* PCC 73102 (NCBI accession number NC_010628) with antiSMASH version 5.2 reveals a high number of putative SMGCs (Fig. 5). This observation seems to be in line with the complex lifestyle of this bacterium. Notably, although being a model strain for cyanobacterial symbioses that has been investigated in great detail, only 4 of the 21 identified gene cluster regions have assigned products, namely, geosmin [50], heterocyst glycolipids [51], anabaenopeptins, and nostopeptolides [52], whereas the rest of the predicted SMGC regions are orphan. Intriguingly, region 4, which comprises the nostopeptolide

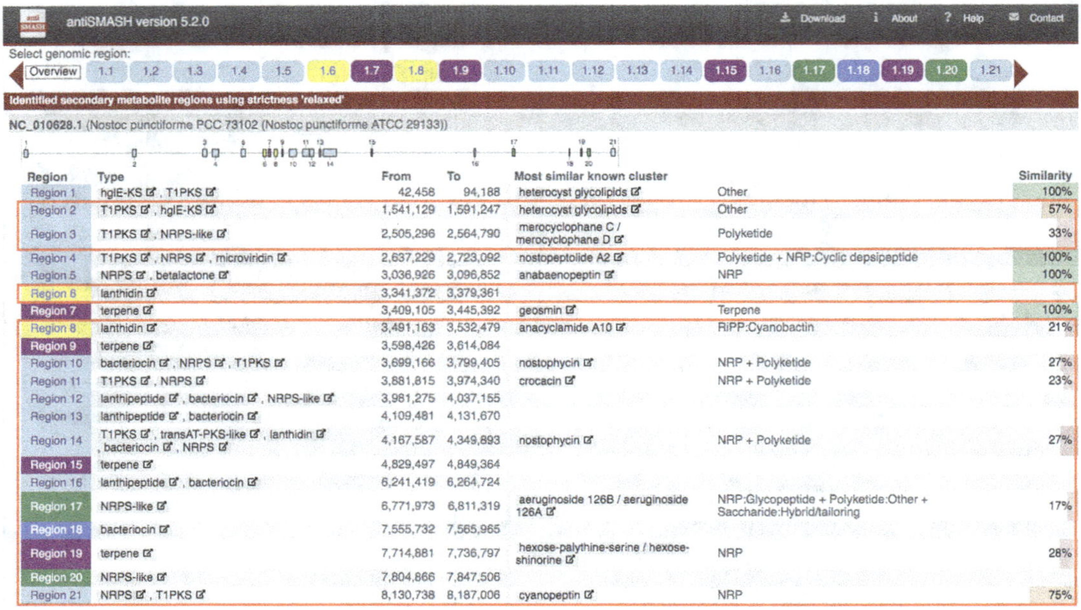

Fig. 5 In silico prediction of SMGCs on the chromosome of *N. punctiforme* PCC 73102 (NCBI accession number NC_010628) with antiSMASH version 5.221 identifies 21 gene cluster regions, only four of which have assigned products (geosmin [50], heterocyst glycolipids [51], anabaenopeptins and nostopeptolides [52]). The other SMGC regions, marked with red frames, are currently orphan

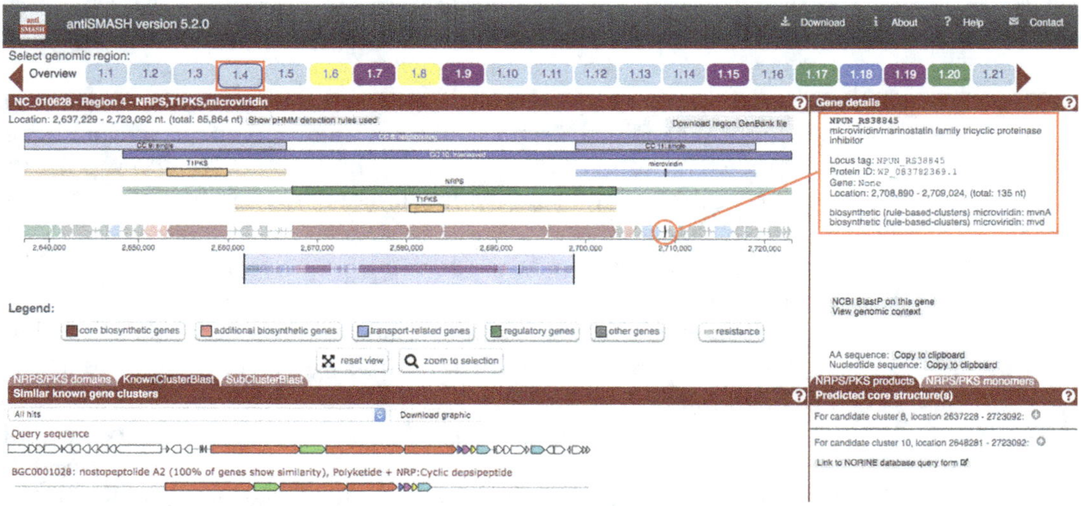

Fig. 6 Region 4, which comprises the nostopeptolide biosynthesis gene cluster also encodes a microviridin precursor that is recognized by antiSMASH. Following *in silico* prediction, seven novel microviridin congeners could be purified and structurally elucidated by a strategy comprising high cell density cultivation, CFP reporter strains and various analytical techniques [49]

biosynthesis gene cluster, also encodes a microviridin precursor that is recognized by antiSMASH (Fig. 6). Our group has a long-standing interest in these potent protease inhibitors, which belong to the class of ribosomal natural products [53]. Finally, by employing high cell density cultivation in comparison to conventional cultivation as control, a CFP reporter mutant to monitor expression of the microviridin (*mvd*) biosynthesis gene cluster and various techniques to isolate and analyze candidate compounds, seven novel microviridin congeners could be purified and structurally elucidated (Fig. 4) [49]. This example highlights the power of combining in silico methods with innovative cultivation techniques and expression monitoring of SMGCs for the discovery of novel cyanobacterial natural products.

4 Notes

1. Some cyanobacteria form gas vacuoles and due to this, collecting cells by centrifugation is impossible. Cells can also be filtered using 0.2 μM polycarbonate filters and washed on the filter with buffer (50 mM Tris–HCl (pH = 8.0), 100 mM EDTA, 100 mM NaCl). After washing the cells, they can be resuspended from the filter in 5 mL of the same buffer.
2. Freezing helps to lyse the cells but this step can also be omitted.
3. Especially some filamentous cyanobacteria such as *Aphanizomenon*, *Nodularia*, and *Oscillatoria* can be difficult to break, but increased osmolarity usually helps. Sucrose may be added to a final concentration of 20% (w/v) and lysozyme to the final concentration of 10 mg/mL. If sugar is added, the water phase usually sinks to the bottom of the tube after phenol and chloroform addition.
4. Some cyanobacterial cells are full of polysaccharides that may coprecipitate with the DNA and interfere (e.g., in PCR reactions). To remove the polysaccharides, NaCl may be added to a final concentration of 0.7 M and CTAB to 1%, then tubes incubated at 65 °C for 20 min. Chloroform extraction may then be performed as described for **step 8**, until no interphase or cloudy buffer phase is visible. DNA may then be precipitated and dissolved as described for **steps 9–12**.
5. NanoDrop also provides a rough estimate of the purity: for pure DNA A260/280 is close to 1.8 whereas higher values might indicate RNA contamination. In case of RNA contamination, RNase treatment is repeated and the sample purified with chloroform extraction as described in **Note 4**. NaCl is then added to a final concentration of 0.7 M.
6. Tools that are used in this chapter take FastQ file format as an input data, therefore data must be in the proper file format before starting the analysis. All sequencing platforms yield their own raw data types, including additional information beside

the sequencing data, however, based on our experience, sequencing centers provide converted fastq files together with original files for their customers.

7. FastQC quality score is low for long-read data and some of the latest technologies even lack phred score values on which FastQC relies, and FastQC is not optimized for long-read data quality control. Other tools, such as recently developed LongQC [54], are more suitable to view the quality of the long-read data. Again, based on our experience, sequencing companies only release quality controlled long-read data, which is already ready to use.
8. SPAdes uses the read-pair information for constructing simplified graphs and thus paired-end Illumina data is recommended.
9. Two main error types in cyanobacterial long-read sequencing are missed base(s) when same base occurs >3 times in a row (e.g., long-read ATCCCG, correct ATCCCCG) or SNPs (e.g., long-read ATCCGC, correct ATGCGC), both of which can comprise assembly errors and further affect the downstream analysis.
10. Most of the currently available genome assemblers consider all genomic elements linear and thus after the assembly, circular genomic elements contain the same sequences at the both ends. As cyanobacterial genomes and plasmids are circular, special programs are needed to find the similar sequences and based on these regions programs, circulate the genomic elements.
11. Unicycler contains three analysis bridging modes: conservative, normal, and bold. Conservative mode may produce more accuracy than scaffolded assembly, whereas bold assembly is contiguous but may contain sequencing errors.
12. PGAP can be run at NCBI after genome submission, or externally prior to submission. Whichever option is preferred, PGAP annotation makes NCBI submission very simple because annotation files are directly compatible with portal requirements. Other annotation programs provide all files required for the genome submission but additional converts or merging steps may be needed.

References

1. Micallef ML, D'Agostino PM, Al-Sinawi B et al (2015) Exploring cyanobacterial genomes for natural product biosynthesis pathways. Mar Genomics 21:1–12. https://doi.org/10.1016/j.margen.2014.11.009
2. Tripp HJ, Bench SR, Turk KA et al (2010) Metabolic streamlining in an open-ocean nitrogen-fixing cyanobacterium. Nature 464: 90–94. https://doi.org/10.1038/nature08786
3. Dagan T, Roettger M, Stucken K et al (2012) Genomes of Stigonematalean cyanobacteria (subsection V) and the evolution of oxygenic photosynthesis from prokaryotes to plastids.

Genome Biol Evol 5:31–44. https://doi.org/10.1093/gbe/evs117

4. Beck C, Knoop H, Steuer R (2018) Modules of co-occurrence in the cyanobacterial pan-genome reveal functional associations between groups of ortholog genes. PLoS Genet 14:1–20. https://doi.org/10.1371/journal.pgen.1007239
5. Shih PM, Wu D, Latifi A et al (2013) Improving the coverage of the cyanobacterial phylum using diversity-driven genome sequencing. Proc Natl Acad Sci U S A 110:1053–1058. https://doi.org/10.1073/pnas.1217107110
6. Singh RK, Tiwari SP, Rai AK, Mohapatra TM (2011) Cyanobacteria: an emerging source for drug discovery. J Antibiot (Tokyo) 64:401–412. https://doi.org/10.1038/ja.2011.21
7. Kehr JC, Picchi DG, Dittmann E (2011) Natural product biosyntheses in cyanobacteria: a treasure trove of unique enzymes. Beilstein J Org Chem 7:1622–1635. https://doi.org/10.3762/bjoc.7.191
8. Teikari JE, Hou S, Wahlsten M et al (2018) Comparative genomics of the Baltic Sea toxic cyanobacteria *Nodularia spumigena* UHCC 0039 and its response to varying salinity. Front Microbiol 9:356. https://doi.org/10.3389/fmicb.2018.00356
9. Yang Y, Lam V, Adomako M et al (2018) Phototaxis in a wild isolate of the cyanobacterium *Synechococcus elongatus*. Proc Natl Acad Sci 115:E12378–E12387. https://doi.org/10.1073/pnas.1812871115
10. Mobberley JM, Romine MF, Cole JK et al (2018) Draft genome sequence of cyanobacterium sp. strain HL-69, isolated from a benthic microbial mat from a magnesium sulfate-dominated hypersaline lake. Microbiol Resour Announc 6:e01583–e01517. https://doi.org/10.1128/genomeA.01583-17
11. Marston MF, Polson SW (2020) Whole-genome sequence of the cyanobacterium *Synechococcus sp.* strain WH 8101. Microbiol Resour Announc 9:e01593–e01519. https://doi.org/10.1128/MRA.01593-19
12. Adamek M, Spohn M, Stegmann E, Ziemert N (2017) Mining bacterial genomes for secondary metabolite gene clusters. In: Sass P (ed) Antibiotics: methods and protocols. Springer New York, New York, NY, pp 23–47
13. Cornet L, Meunier L, Van Vlierberghe M et al (2018) Consensus assessment of the contamination level of publicly available cyanobacterial genomes. PLoS One 13:1–26. https://doi.org/10.1371/journal.pone.0200323
14. Hoiczyk E, Hansel A (2000) Cyanobacterial cell walls: news from an unusual prokaryotic envelope. J Bacteriol 182:1191–1199. https://doi.org/10.1128/JB.182.5.1191-1199.2000
15. Besser J, Carleton HA, Gerner-Smidt P et al (2018) Next-generation sequencing technologies and their application to the study and control of bacterial infections. Clin Microbiol Infect 24:335–341. https://doi.org/10.1016/j.cmi.2017.10.013
16. Jain M, Koren S, Miga KH et al (2018) Nanopore sequencing and assembly of a human genome with ultra-long reads. Nat Biotechnol 36:338–345. https://doi.org/10.1038/nbt.4060
17. Wenger AM, Peluso P, Rowell WJ et al (2019) Accurate circular consensus long-read sequencing improves variant detection and assembly of a human genome. Nat Biotechnol 37:1155–1162. https://doi.org/10.1038/s41587-019-0217-9
18. Andrews S (2010) FastQC: a quality control tool for high throughput sequence data. Available online at http://www.bioinformatics.babraham.ac.uk/projects/fastqc/
19. Martin M (2011) Cutadapt removes adapter sequences from high-throughput sequencing reads. EMBnet. J 17:1. https://doi.org/10.14806/ej.17.1.200
20. Wick RR, Judd LM, Gorrie CL, Holt KE (2017) Unicycler: resolving bacterial genome assemblies from short and long sequencing reads. PLoS Comput Biol 13:1–22. https://doi.org/10.1371/journal.pcbi.1005595
21. Bankevich A, Nurk S, Antipov D et al (2012) SPAdes: a new genome assembly algorithm and its applications to single-cell sequencing. J Comput Biol 19:455–477. https://doi.org/10.1089/cmb.2012.0021
22. Langmead B, Salzberg SL (2012) Fast gapped-read alignment with bowtie 2. Nat Methods 9:357–359. https://doi.org/10.1038/nmeth.1923
23. Walker BJ, Abeel T, Shea T et al (2014) Pilon: An integrated tool for comprehensive microbial variant detection and genome assembly improvement. PLoS One 9:e112963
24. Gurevich A, Saveliev V, Vyahhi N, Tesler G (2013) QUAST: quality assessment tool for genome assemblies. Bioinformatics 29:1072–1075. https://doi.org/10.1093/bioinformatics/btt086
25. Ran L, Larsson J, Vigil-Stenman T et al (2010) Genome erosion in a nitrogen-fixing vertically transmitted endosymbiotic multicellular cyanobacterium. PLoS One 5:1–11. https://doi.org/10.1371/journal.pone.0011486

26. Muro-Pastor AM, Hess WR (2020) Regulatory RNA at the crossroads of carbon and nitrogen metabolism in photosynthetic cyanobacteria. Biochim Biophys acta Gene Regul Mech 1863:194477. https://doi.org/10.1016/j.bbagrm.2019.194477
27. Axmann IM, Kensche P, Vogel J et al (2005) Identification of cyanobacterial non-coding RNAs by comparative genome analysis. Genome Biol 6:R73. https://doi.org/10.1186/gb-2005-6-9-r73
28. Salzberg SL (2019) Next-generation genome annotation: we still struggle to get it right. Genome Biol 20:92. https://doi.org/10.1186/s13059-019-1715-2
29. Seemann T (2014) Prokka: rapid prokaryotic genome annotation. Bioinformatics 30:2068–2069. https://doi.org/10.1093/bioinformatics/btu153
30. Aziz RK, Bartels D, Best AA et al (2008) The RAST server: rapid annotations using subsystems technology. BMC Genomics 9:75. https://doi.org/10.1186/1471-2164-9-75
31. Tatusova T, DiCuccio M, Badretdin A et al (2016) NCBI prokaryotic genome annotation pipeline. Nucleic Acids Res 44:6614–6624. https://doi.org/10.1093/nar/gkw569
32. Carver T, Harris SR, Berriman M et al (2011) Artemis: an integrated platform for visualization and analysis of high-throughput sequence-based experimental data. Bioinformatics 28:464–469. https://doi.org/10.1093/bioinformatics/btr703
33. Kanehisa M, Sato Y, Morishima K (2016) BlastKOALA and GhostKOALA: KEGG tools for functional characterization of genome and metagenome sequences. J Mol Biol 428:726–731. https://doi.org/10.1016/j.jmb.2015.11.006
34. Medema MH, Blin K, Cimermancic P et al (2011) antiSMASH: rapid identification, annotation and analysis of secondary metabolite biosynthesis gene clusters in bacterial and fungal genome sequences. Nucleic Acids Res 39:W339–W346. https://doi.org/10.1093/nar/gkr466
35. Eddy SR (1998) Profile hidden Markov models. Bioinformatics 14:755–763. https://doi.org/10.1093/bioinformatics/14.9.755
36. Skinnider MA, Johnston CW, Gunabalasingam M et al (2020) Comprehensive prediction of secondary metabolite structure and biological activity from microbial genome sequences. Nat Commun 11:6058. https://doi.org/10.1038/s41467-020-19986-1
37. Khater S, Gupta M, Agrawal P et al (2017) SBSPKSv2: structure-based sequence analysis of polyketide synthases and non-ribosomal peptide synthetases. Nucleic Acids Res 45:W72–W79. https://doi.org/10.1093/nar/gkx344
38. Tietz JI, Schwalen CJ, Patel PS et al (2017) A new genome-mining tool redefines the lasso peptide biosynthetic landscape. Nat Chem Biol 13:470–478. https://doi.org/10.1038/nchembio.2319
39. Agrawal P, Khater S, Gupta M et al (2017) RiPPMiner: a bioinformatics resource for deciphering chemical structures of RiPPs based on prediction of cleavage and cross-links. Nucleic Acids Res 45:W80–W88. https://doi.org/10.1093/nar/gkx408
40. van Heel AJ, de Jong A, Song C et al (2018) BAGEL4: a user-friendly web server to thoroughly mine RiPPs and bacteriocins. Nucleic Acids Res 46:W278–W281. https://doi.org/10.1093/nar/gky383
41. Medema MH, Fischbach MA (2015) Computational approaches to natural product discovery. Nat Chem Biol 11:639–648. https://doi.org/10.1038/nchembio.1884
42. Cimermancic P, Medema MH, Claesen J et al (2014) Insights into secondary metabolism from a global analysis of prokaryotic biosynthetic gene clusters. Cell 158:412–421. https://doi.org/10.1016/j.cell.2014.06.034
43. Kautsar SA, Blin K, Shaw S et al (2020) MIBiG 2.0: a repository for biosynthetic gene clusters of known function. Nucleic Acids Res 48:D454–D458. https://doi.org/10.1093/nar/gkz882
44. van Santen JA, Jacob G, Singh AL et al (2019) The natural products atlas: an open access knowledge base for microbial natural products discovery. ACS Cent Sci 5:1824–1833. https://doi.org/10.1021/acscentsci.9b00806
45. Schorn M, Verhoeven S, Ridder L et al (2021) A community resource for paired genomic and metabolomic data mining. Nat Chem Biol 17(4):363–368
46. Jones MR, Pinto E, Torres MA, et al (2021) CyanoMetDB, a comprehensive public database of secondary metabolites from cyanobacteria. Water Res 196:117017. https://doi.org/10.1016/j.watres.2021.117017
47. Weber T, Kim HU (2016) The secondary metabolite bioinformatics portal: computational tools to facilitate synthetic biology of secondary metabolite production. Synth Syst Biotechnol 1:69–79
48. Bode HB, Bethe B, Höfs R, Zeeck A (2002) Big effects from small changes: possible ways to

explore nature's chemical diversity. Chem Biochem 3:619–627

49. Dehm D, Krumbholz J, Baunach M et al (2019) Unlocking the spatial control of secondary metabolism uncovers hidden natural product diversity in *Nostoc punctiforme*. ACS Chem Biol 14:1271–1279. https://doi.org/10.1021/acschembio.9b00240
50. Giglio S, Jiang J, Saint CP et al (2008) Isolation and characterization of the gene associated with geosmin production in cyanobacteria. Environ Sci Technol 42:8027–8032. https://doi.org/10.1021/es801465w
51. Campbell EL, Cohen MF, Meeks JC (1997) A polyketide-synthase-like gene is involved in the synthesis of heterocyst glycolipids in *Nostoc punctiforme* strain ATCC 29133. Arch Microbiol 167:251–258. https://doi.org/10.1007/s002030050440
52. Guljamow A, Kreische M, Ishida K et al (2017) High-density cultivation of terrestrial Nostoc strains leads to reprogramming of secondary metabolome. Appl Environ Microbiol 83:e01510-17. https://doi.org/10.1128/AEM.01510-17
53. Baunach M, Dittmann E (2020) Microviridins. In: Liu H-W, Begley TP (eds) Comprehensive Natural Products III. Elsevier, Oxford, pp 193–205
54. Fukasawa Y, Ermini L, Wang H et al (2020) LongQC: a quality control tool for third generation sequencing long read. G3 (Bethesda) 10:1193–1196. https://doi.org/10.1534/g3.119.400864

Chapter 15

Single Crossover to Inactivate Target Gene in Cyanobacteria

Jaimie Gibbons, Liping Gu, Yeyan Qiu, and Ruanbao Zhou

Abstract

Anabaena sp. PCC 7120 (hereafter *Anabaena* 7120) is a model cyanobacterium for studying pathways such as photosynthesis and nitrogen fixation along with many other metabolic pathways common to plants. In addition, since *Anabaena* 7120 forms specialized N_2-fixing cells, called heterocysts, to perform uniquely solar-powered, oxic nitrogen fixation under fixed-nitrogen depleted conditions, this cyanobacterium provides the unique opportunity to study cellular differentiation in bacteria. Since more than 155,810 sequenced prokaryotic genomes are currently available (Zhang et al., Microbiome 8(1):134, 2020), target gene inactivation, combined with analyses of the corresponding mutant's phenotype, has become a powerful tool to assess gene function through detecting a loss-of-function in the knockout mutant. In the method described here, a single crossover approach is used to knockout a target gene in *Anabaena* 7120. The method requires inserting an internal fragment of the target gene into the cyanobacterial integration vector pZR606 to create a knockout plasmid, and then is introduced to *Anabaena* 7120 via conjugative transformation. A single crossover, occurring via homologous recombination, disrupts the target gene, creating 3′- and 5′-deleted fragments (Fig. 1). The mutant containing the inactivated gene can then be studied to determine any loss of function, thereby defining the gene's function. This gene inactivation approach is based on an integrative vector pZR606 (Chen et al., Appl Microbiol Biotechnol 99:1779–1793, 2015), which may be broadly applied to gene inactivation in other cyanobacterial species as well as other prokaryotic organisms.

Key words Gene inactivation, Single crossover, Homologous recombination, Conjugal transformation

1 Introduction

With sequenced genomes becoming more readily available, specific gene inactivation becomes a viable approach to target individual genes for identification of function. *Anabaena* 7120 is a metabolically diverse cyanobacterium, with two of its core metabolic processes being photosynthesis and nitrogen fixation. Gene inactivation provides the opportunity to identify key genes and their functions within complex metabolic pathways such as these. In the method described here, a single crossover method is routinely used to disrupt the target gene in our laboratory. This

Elizabeth Skellam (ed.), *Engineering Natural Product Biosynthesis: Methods and Protocols*,
Methods in Molecular Biology, vol. 2489, https://doi.org/10.1007/978-1-0716-2273-5_15,

approach involves three key steps: construction of the cargo plasmid used for the gene knockout, conjugative transformation of the cargo plasmid to *Anabaena* 7120, and complementation of the knockout mutant.

In the first step, a knockout plasmid is created by cloning an internal fragment of the target gene into the integrative vector pZR606 [2]. Once the knockout plasmid is created, it is introduced into *Anabaena* 7120 via conjugative transformation. Following transformation of the knockout plasmid to wild-type *Anabaena* 7120, a single crossover event occurs, where the internal fragment contained within the knockout plasmid homologously recombines with the identical portion of the gene fragment present in the wild-type's chromosome. Through the single crossover, the entire plasmid is inserted into the wild-type chromosome, disrupting the target gene (Fig. 1). Finally, a complement plasmid, containing the intact target gene along with its native promoter, is constructed. The complement plasmid is introduced into the mutant via conjugative transformation, creating the complement strain. Phenotypic analysis comparing the wild type, mutant, and complement should reveal a loss-of-function in the mutant, which is recovered in the complement, thereby indicating the gene's role in the organism. If the loss of function in the mutant is not recovered in the complement, this indicates that the inactivated gene was not responsible for the loss-of-function in the mutant, possibly due to some polar effect on its downstream gene(s), or spontaneous mutation(s) somewhere else.

2 Materials

1. Petri dishes.
2. Glass flasks.
3. Microcentrifuge and 1.5 mL centrifuge tubes (sterile).
4. 15 mL conical centrifuge tubes (sterile).
5. 50 mL conical centrifuge tubes (sterile).
6. *Anabaena* sp. strain PCC 7120.
7. *Escherichia coli* NEB10β.
8. *E. coli* HB101.
9. Conjugal plasmid pRL443 and helper plasmid pRL623 [3].
10. TOPO™ TA Cloning™ Kit with pCR™2.1-TOPO vector and TOP 10 *E. coli* cells (Invitrogen).
11. pZR606, an integrative vector for *Anabaena* 7120 [2].
12. pZR670, a replication vector for *Anabaena* 7120 [2].
13. LB broth.

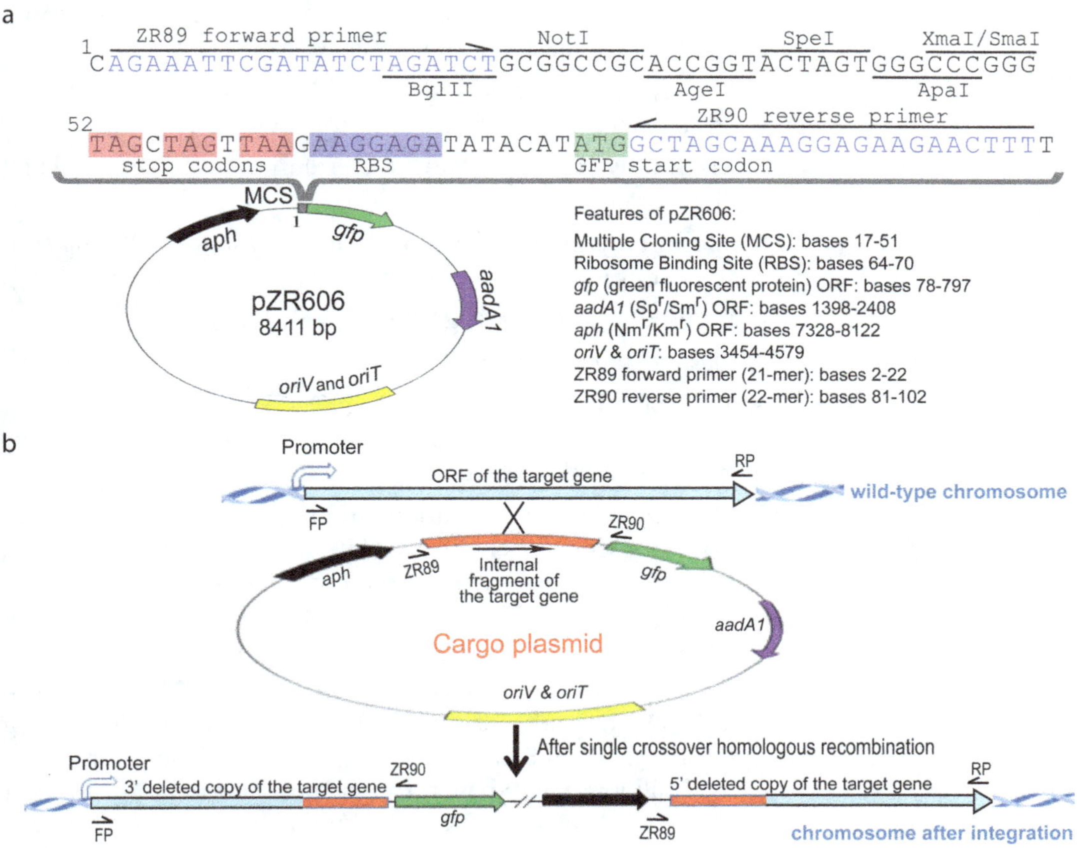

Fig. 1 Schematic representation of the strategy used to inactivate genes in *Anabaena* chromosomes and monitor promoter activity by single crossover recombination. (**a**) Pictorial representation of the pZR606 vector and its MCS. Primers ZR89 and ZR90 are for PCR verification of the insert in cargo plasmid. (**b**) Schematic depiction of single crossover homologous recombination. Cargo plasmid (plasmid 2) containing the internal fragment of a target gene inserted into the pZR606 at the MCS is integrated into *Anabaena* chromosome by single crossover recombination, which generates 3′-deleted and 5′-deleted copies of the target gene. The promoter-less *gfp* transcription driven by the target gene promoter and translation via the ribosome binding site upstream of the *gfp* ORF serves as a transcriptional marker. Primer pair FP + RP (forward and reverse primers of the target gene) was used to verify the wild-type gene; primer pair FP + ZR90 was used to distinguish the positive exconjugants (single crossover recombinants) from wild-type [2]

14. LB agar, used for growing *E. coli*. *Note: This brand of agar does not work for most cyanobacterial strains.*
15. Agar, required for growing *Anabaena* 7120 and other cyanobacteria (Fisher Scientific, catalog number: A360-500).
16. Ampicillin sodium salt (100 mg/mL stock solution in ddH_2O and filter sterilized and store at 4 °C).
17. Kanamycin sulfate (50 mg/mL stock solution in ddH_2O and filter sterilized and store at 4 °C).

18. Chloramphenicol (25 mg/mL stock solution in 95% ethanol and filter sterilized and store at 4 °C).
19. Neomycin trisulfate salt hydrate (50 mg/mL stock solution in ddH_2O and filter sterilized and store at 4 °C).
20. Erythromycin (10 mg/mL stock solution in 95% ethanol and filter sterilized and store at 4 °C).
21. Spectinomycin dihydrochloride pentahydrate (100 mg/mL stock solution in ddH_2O and filter sterilized and store at 4 °C).
22. Restriction enzymes (e.g., BglII, SpeI, XmaI [2]).
23. Plasmid Miniprep Kit.
24. Solution A: 500 mL 4% $MgSO_4{\cdot}7H_2O$, 500 mL 1.2% $CaCl_2{\cdot}2H_2O$, 500 mL 3.8% NaCl, 500 mL microelements stock solution (see below).
25. Microelements stock solution: 160 mL A & A Fe-EDTA solution (see below), 360 mg $MnCl_2{\cdot}4H_2O$, 61.1 mg $Na_2MoO_42H_2O$, 44 mg $ZnSO_2{\cdot}7H_2O$, 15.8 mg $CuSO_2{\cdot}5H_2O$, 572 mg H_3BO_3, 4.6 mg NH_4VO_3, 8 mg $CoCl_2{\cdot}6H_2O$, 1090 mL ddH_2O.
26. A & A Fe-EDTA solution: Dissolve 5.2 g KOH in 186 mL ddH_2O, then add 20.4 g $Na_2EDTA{\cdot}2H_2O$; dissolve 13.7 g $FeSO_4{\cdot}7H_2O$ in 364 mL ddH_2O; mix the two solutions; bubble Millipore filtered air through the solution until the color changes to brownish-orange (may take a few minutes to several hours). The final pH of FeEDTA solution is approximately 7.5.
27. Solution B: 28 g K_2HPO_4 in 500 mL ddH_2O.
28. KNO_3 + $NaNO_3$ solution: 8.48 g $NaNO_3$, 10.12 g KNO_3, 200 mL ddH_2O.
29. Allen and Arnon medium [4] plus nitrate [AA/8(N)] [5]: To prepare 1 L of medium combine 3.125 mL solution A, 0.781 mL solution B, and 5 mL KNO_3 + $NaNO_3$ solution to 991 mL ddH_2O.
30. Allen and Arnon agar medium plus nitrate (AA/N agar): To prepare 1 L of medium combine 25 mL solution A, 6.25 mL solution B, 5 mL KNO_3 + $NaNO_3$ solution, 10 g agar, 969 mL ddH_2O.
31. Immobilon-NC transfer membrane.
32. PCR reagents: Phusion™ high-fidelity DNA polymerase, Phusion™ buffer, dNTPs, and the required primers. The details for primers (ZR89, ZR90, FP, RP, ZR2679) can be found in Chen et al. [2].
33. Gel electrophoresis apparatus and Gel Extraction Kit.
34. Ligation reagents: T4 DNA ligase, 10× T4 DNA ligase buffer.
35. Water bath sonicator.

36. Platform shaker equipped with continuous fluorescent light illumination (ca. 50–100 μE/m^2 s).
37. Cyanobacteria Culture Room (constant at 30 °C) equipped with a digital temperature controlling system and fluorescent light bulbs illumination (ca. 50–150 μE/m^2 s).

3 Methods

3.1 Construction of Plasmid for Knocking out Target Gene

1. Amplify an internal fragment of the target gene by PCR. In the forward and reverse primers used to amplify the internal fragment, include recognition sequences for restriction enzymes (e.g., in the forward primer include a restriction enzyme cutting sequence for Restriction Enzyme 1; in the reverse primer include a restriction enzyme site for Restriction Enzyme 2; *see* **Note 1**).
2. Clone the PCR-amplified internal fragment into pTOPO2.1 (Km^R/Ap^R), creating plasmid 1. Transform Top10 *E. coli* cells with plasmid 1. Extract plasmid 1 from transformed *E. coli* cells, using a plasmid extraction kit.
3. Perform a restriction enzyme digestion of plasmid 1 with Restriction Enzyme 1 and Restriction Enzyme 2 to isolate the internal fragment of the target gene. Simultaneously, perform a restriction enzyme digestion of the integrative plasmid pZR606 (Fig. 1a) (Km^R/Nm^RSp^R) with Restriction Enzyme 1 and Restriction Enzyme 2. Perform DNA gel electrophoresis with the digestions. Cut the bands corresponding to the internal fragment size and pZR606 vector portion. Combine the cut bands in a single eppie tube. Use a gel extraction kit to extract the DNA from the gel.
4. Perform a ligation by combining 6 μL of the gel extraction solution of 2 DNA, 0.1 μL T4 DNA ligase (20 U/μL), 1 μL 10× T4 DNA ligase buffer, and 3 μL ddH_2O. Let the ligation sit at room temperature for 30 min. This step creates plasmid 2, the cargo/knockout plasmid (Km^R/Nm^RSp^R) (Fig. 1b). When this plasmid is introduced into *Anabaena* 7120, a single crossover event will occur, effectively knocking out the target gene (Fig. 1b).
5. Transform *E. coli* NEB 10β with plasmid 2.

3.2 Preparation of E. coli Strains

1. Grow *E. coli* HB101 containing pRL443 (Ap^R) and pRL623 (Cm^R) in 5 mL LB broth containing 100 μg/mL ampicillin and 25 μg/mL chloramphenicol overnight at 37 °C, 200 rpm. Also grow *E. coli* NEB 10β bearing the cargo plasmid (Km^R/Nm^RSp^R) in 5 mL LB broth containing 50 μg/mL kanamycin and 100 μg/mL spectinomycin overnight at 37 °C, 200 rpm.

For next step conjugal transformation of *Anabaena* 7120, all plasmids must be transformed into one of four *E. coli* strains (*see* **Note 2**).

2. Transfer 100 μL of the overnight *E. coli* cultures into 5 mL LB containing the appropriate antibiotics; grow for 3–4 h at 37 °C, 200 rpm until the cultures reach $OD_{600} \approx 0.5$.
3. Harvest the 5 mL *E. coli* cultures by centrifuging at 4000 × *g* at room temperature (22 °C) for 10 min.
4. Remove the supernatant and wash the pellet in 1 mL LB, transferring culture to a 1.5 mL eppie tube. Centrifuge at 13,000 × *g* at room temperature (22 °C) for 1 min. Repeat wash with 1 mL LB twice to completely remove antibiotic.
5. After final wash, resuspend pellets in 100 μL LB.
6. For mating, combine 100 μL of NEB 10β bearing the cargo plasmid with 100 μL HB101 harboring pRL443 and pRL623 (experimental group). For a negative control, add 100 μL LB to 100 μL HB101 harboring pRL443 and pRL623. Incubate at room temperature for 30 min to allow the cultures to mate.

3.3 Preparation of Anabaena 7120 Culture

1. Grow *Anabaena* 7120 in 50 mL AA/8(N) medium for approximately 7 days under continuous light (~50 μE/m^2 s) at 30 °C and shaken at 120 rpm until OD_{720} reaching ≈0.3.
2. Break the long filaments in the culture into single cells or average 2–3 cell filaments by sonicating the culture for 60–120 s using a water bath sonicator. Microscopic verification of fragmentation is required in this step.
3. Harvest the above 50 mL culture at 5000 × *g* at 22 °C for 10 min.
4. Remove the supernatant and resuspend the pellet in 1 mL AA/8(N).

3.4 Conjugal Transformation of a Cargo Plasmid into Anabaena 7120

1. Mix 150 μL of *Anabaena* 7120 from the resuspension (Subheading 3.3) with 150 μL of the mated *E. coli* mixture (experimental group, Subheading 3.2) or the control mixture (negative control group, Subheading 3.2), respectively, and incubate at room temperature for 1 h.
2. Evenly spread the conjugal mixtures onto autoclaved nitrocellulose membrane placed on top of AA(N) agar plate containing 5% LB (v/v) without antibiotic and incubate at 30 °C under continuous illumination (≈50 μE/m^2 s) for 48 h.
3. Transfer the membrane onto an AA(N) agar plate with 100 μg/mL neomycin (Nm) and 10 μg/mL spectinomycin (Sp); incubate under continuous light at 30 °C until neomycin/spectinomycin-resistant colonies appear (approximately 10–15 days) (Fig. 2a). *See* **Note 3**.

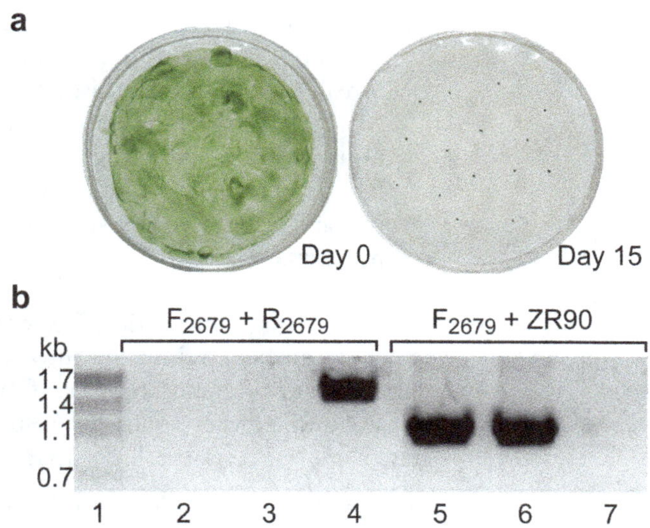

Fig. 2 Characterization of ava_2679 mutant in *Anabaena variabilis*. (**a**) The mating mixture containing the cargo plasmid pZR732 carrying ava_2679 internal fragment were spread onto Immobilon-nitrocellulose membrane atop AA/8(N) agar containing 50 μg/mL Nm for 0 (left panel) and 15 days (right panel). (**b**) Colony PCR verification of the ava_2679 mutant. The PCR product sizes amplified with ava_2679 forward (F2679) + reverse (R2679) (lanes 2–4) and F2679 + ZR90 (lanes 5–7) primer pairs were expected to be 1523 (lane 4) and 1068 bp (lanes 5 and 6), respectively. Lanes 2 and 5 and lanes 3 and 6 were two independent exconjugants/knockout mutants (designated SR2679), respectively. The 1068-bp PCR products amplified with F2679 + ZR90 were further sequenced to confirm ava_2679 gene disruption. The wild-type strain served as a negative control (lanes 4 and 7). The sizes of molecular size markers (lane 1) are indicated

3.5 Verification of Single Crossover Knockout Mutants

1. Restreak 10 or more potential knockout mutant colonies on AA/8(N) agar plate containing Nm (100 μg/mL) and Sp (10 μg/mL).
2. Screen for knockout mutants using colony PCR. To verify single crossovers, use a forward primer matching the chromosomal sequence upstream of the internal fragment and use reverse primer ZR90 matching the GFP gene downstream of the MCS in pZR606 (Fig. 2b, lanes 5–7). The wild-type target gene is screened for using a forward primer and reverse primer flanking the target gene's open reading frame (Fig. 2b, lanes 2–4). (Note that *Anabaena* 7120 has multiple chromosome copies; therefore a screened single crossover mutant colony may still contain the wild-type gene. If the mutant colonies have copies of both the mutated and wild-type genes present, *see* **Note 4** for steps to obtain homozygous single crossover mutants).

3. Suspend verified single crossover colonies in 100 μL AA/8N medium, and create serial dilutions of 1×, 10×, and 100×. Streak each dilution for isolation on an AA(N) agar plate with Nm100 (100 μg/mL neomycin) and Sp10 (10 μg/mL spectinomycin). This step is to isolate single cyanobacterial colonies free of *E. coli*. After obtaining isolated cyanobacterial colonies, repeat the preceding step to reverify the single crossover mutants.

3.6 Complementation Experiment

For the complementation experiment, the functional gene sequence must be reinserted into the knockout mutant to determine if the lost function returns. If the loss-of-function is reinstated in the complement strain, the results indicate that the knockout gene is indeed responsible for that function.

1. For the complement plasmid, a shuttle cargo plasmid is constructed (Fig. 3). The DNA sequence, containing the native promoter region along with the entire gene coding region of the target gene, is PCR amplified from *Anabaena* 7120, and then inserted into the MCS of the shuttle vector pZR670 (Cm^REm^R) to create the complement plasmid using similar approach for making above plasmid 2 (Subheading 3.1).
2. Transform NEB 10β with the complement plasmid.
3. A similar approach to Subheadings 3.2 and 3.4 above is used to transfer the complementing plasmid to the knockout mutant. Here, ensure that the knockout mutant rather than wild-type *Anabaena* 7120 is used for the conjugation.
4. Transfer the membrane to an AA (N) agar plate containing 100 μg/mL neomycin, 10 μg/mL spectinomycin, and 10 μg/mL erythromycin (Em10). Incubate under continuous light illumination (ca. 50 μE/m^2 s) at 30 °C. Colonies should appear in approximately 10–15 days.
5. Once colonies appear on the membrane, streak the colonies on an AA (N) Nm100Sp10Em10 agar plate. Incubate under continuous light illumination (ca. 50 μE/m^2 s) at 30 °C for 7–10 days. Restreak the colonies if necessary onto another AA (N) Nm100Sp10Em10 agar plate until the colonies appear to be free of *E. coli*.
6. Perform colony PCR with the conjugation colonies to verify the presence of complement plasmid, using one primer matching the inserted sequence and one primer matching the shuttle vector sequence of pZR670.
7. Examine the phenotype of at least three independent colonies that bearing the complement plasmid to verify that the function lost in the single crossover mutant is regained.

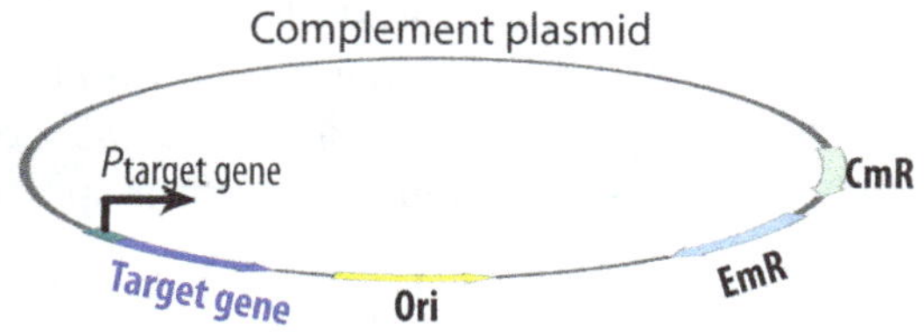

Fig. 3 A complement plasmid for complementation of the knockout mutant. The complement plasmid contains the target gene coding region along with its native promoter is inserted into a shuttle vector pZR670 ($Cm^R Em^R$). The complement plasmid must have a unique antibiotic resistance compared to the knockout mutant so that antibiotics may be used to select for knockout mutant colonies containing the complement plasmid

4 Notes

1. For Restriction Enzyme 1 and Restriction Enzyme 2, choose restriction enzymes that are also found in the MCS of the vector plasmid, so that the internal fragment can be cloned into the vector plasmid cut by the same or compatible restriction enzymes.
2. For conjugation experiments, all plasmids must be transformed into one of four *E. coli* strains. They are *E. coli* HB101, *E. coli* DH10B, *E. coli* NEB 10β, DH5α MCR.
3. Continue transferring the membrane to a new AA (N) antibiotic plate every 7 days until individual colonies appear. The lawn of cyanobacteria will die over time, leaving single colonies containing the single crossover.
4. For mutants that contain both the single crossover and the wild-type copy of the target gene, inoculate the colony into 10 mL AA/8(N) Nm100Sp10. Incubate the culture for 7 days under continuous light (~50 μE/m^2 s) at 30 °C and shaken at 120 rpm. Take the culture and sonicate it using a water bath sonicator for 90 s, or until all of the filaments in the culture are broken up into single cells or 2–3 cell short filaments. Prepare serial dilutions of the culture (1×, 10×, 100×) and plate 100 μL of the dilutions on AA(N) Nm100Sp10 agar plates. Incubate the plates under light (~50 μE/m^2 s) at 30 °C for 10 days, or until colonies form. Perform PCR with the colonies following Subheading 3.5, **step 2**. Repeat this step (growth, sonication, plating, and PCR) as needed until the PCR shows that the mutant is free of the wild-type gene.
5. For most cyanobacteria, Allen and Arnon medium (AA) can be replaced by BG11 medium (http://www-cyanosite.bio.purdue.edu/media/table/BG11.html) [6].

6. The threshold length for efficient homologous crossover/recombination was 330 bp [7], if the coding region of a target gene is shorter than 330 bp, we suggest that you should use our double crossover approach (*see* Chap. 10) to inactivate the target gene.

References

1. Zhang Z, Wang J, Wang J, Wang J, Li Y (2020) Estimate of the sequenced proportion of the global prokaryotic genome. Microbiome 8(1): 134. https://doi.org/10.1186/s40168-020-00903-z
2. Chen K, Xu X, Gu L, Hildreth M, Zhou R (2015) Simultaneous gene inactivation and promoter reporting in cyanobacteria. Appl Microbiol Biotechnol 99(4):1779–1793. https://doi.org/10.1007/s00253-014-6209-2
3. Elhai J, Vepritskiy A, Muro-Pastor AM, Flores E, Wolk CP (1997) Reduction of conjugal transfer efficiency by three restriction activities ofAnabaena sp. strain PCC 7120. J Bacteriol 179(6): 1998–2005
4. Allen MB, Arnon DI (1955) Studies on nitrogen-fixing blue-green algae. I. Growth and nitrogen fixation by anabaena Cylindrica Lemm. Plant Physiol 30(4):366–372
5. Hu NT, Thiel T, Giddings TH Jr, Wolk CP (1981) New Anabaena and Nostoc cyanophages from sewage settling ponds. Virology 114(1): 236–246
6. Rippka R, Deruelles J, Waterbury J, Herdman M, Stanier R (1979) Generic assignments, strain histories and properties of pure cultures of cyanobacteria. J Gen Microbiol 111(1):1–61
7. Biswas I, Gruss A, Ehrlich SD, Maguin E (1993) High-efficiency gene inactivation and replacement system for gram-positive bacteria. J Bacteriol 175(11):3628–3635. https://doi.org/10.1128/jb.175.11.3628-3635.1993

Chapter 16

Double Crossover Approach to Inactivate Target Gene in Cyanobacteria

Jaimie Gibbons, Liping Gu, and Ruanbao Zhou

Abstract

Anabaena sp. PCC7120 (hereafter *Anabaena* 7120) is a nitrogen-fixing, filamentous cyanobacterium. Given its diverse metabolism, it serves as an excellent model organism, particularly for studying cell differentiation, nitrogen fixation, photosynthesis, production of high-value chemicals, and synthetic biology. Gene knockout is a common approach to assess the function of gene products through assessing phenotypic loss of function. In the method described here, a double crossover approach is used to inactivate a target gene or target genes in *Anabaena* 7120. This method involves replicating the gene(s) from the wild-type genomic DNA and inserting them into an integrative plasmid vector. An internal portion of the genes may be removed and replaced with a GFP-Spectinomycin (*gfp-sp*) cassette. The plasmid is then introduced into *Anabaena* 7120 where a double crossover event occurs between the wild-type chromosome and the cargo plasmid, effectively replacing the wild-type gene with the disrupted gene from the plasmid. The *gfp-sp* cassette combined with the *sacB* gene serve as positive selection to identify double crossover mutants (Cai and Wolk (1990), 172(6):3138–3145, J. Bacteriol). Finally, the functional genes are cloned into another replicating plasmid vector to produce a cargo plasmid, which is conjugatively introduced into the mutant for a complementation test. By comparing the phenotypes among the wild-type, mutant, and complement, one should see a loss of function in the mutant which is recovered in the complement, thereby defining the function of the target gene. The double crossover approach described here for *Anabaena* PCC 7120 may be broadly applicable to the study of gene function in cyanobacteria and other prokaryotic organisms.

Key words Genetic engineering, Cyanobacteria, Double crossover, Homologous recombination, Conjugal transformation

1 Introduction

Double crossover gene inactivation is an effective approach to knockout a gene and determine its loss of function. In the method described here, one or more genes can be knocked out through a double crossover approach by deleting the internal portion of the gene(s) and replacing it with a GFP-Spectinomycin (*gfp-sp*) cassette. You do not have to delete the internal portion of a target gene, but an antibiotic cassette must be inserted within the coding

Elizabeth Skellam (ed.), *Engineering Natural Product Biosynthesis: Methods and Protocols*,
Methods in Molecular Biology, vol. 2489, https://doi.org/10.1007/978-1-0716-2273-5_16,

region of a target gene. The *gfp-sp* cassette here serves two purposes: first, a promoterless-*gfp* inserted to disrupt the gene(s) acts as a visible transcriptional marker to report the transcription of the target gene; second, spectinomycin cassette serves as a selection marker combined with *sacB* to identify and isolate double crossover knockout mutants [1]. This method of gene knockout involves three main steps: constructing a knockout plasmid, conjugatively transforming the knockout plasmid (cargo plasmid) into *Anabaena* 7120, and complementing the knockout mutant to verify the function of the knocked-out gene(s).

First, a plasmid is constructed, which contains the flanking regions of knocked out gene. The plasmid can be prepared using one of two methods. The first method involves amplifying the gene from the wild-type genomic DNA, inserting it into an integrative plasmid vector, deleting an internal portion of the gene, and replacing the deleted segment with a *gfp-sp* cassette. The second option uses overlap extension PCR [2] to selectively amplify the target gene while removing an internal fragment and inserting a *gfp-sp* cassette in the middle of the disrupted gene. Following overlap extension PCR, the PCR product is cloned into an integrative plasmid vector. The integrative plasmid vector used in both approaches contains the *sacB* gene, which is later used as a positive selection to identify double crossover mutants [1].

Following construction of the knockout plasmid, the plasmid is transformed to *Anabaena* 7120 by conjugation [3, 4]. In this step, *Escherichia coli* containing the knockout plasmid is mated with an *E. coli* strain containing the conjugal plasmid pRL443 and helper plasmid pRL623 [3]. Following *E. coli* mating, the *E. coli* cells are mated with wild-type *Anabaena* 7120 to transfer the knockout plasmid from the *E. coli* to *Anabaena* 7120. After the second mating, the mixture is plated on a nitrocellulose membrane on an AA(N) agar plate supplemented with 5% LB. After 2 days incubation, the membrane is transferred to an AA(N) agar plate containing antibiotics to select for the cyanobacterial cells that have been integrated with the knockout plasmid. Colonies from this plate are grown and undergo a second selection for double crossover mutants by plating the colonies on an AA(N) agar plate with the selective antibiotic and 5% sucrose. This step allows for selection against single crossover mutants as these mutants contain the *sacB* gene from the knockout plasmid while double crossover mutants do not contain *sacB*; *sacB*-containing single crossover mutants are lethal when sucrose is present [1, 5, 6]. Thus, this step ensures that only double crossover knockouts are selected.

After obtaining and verifying double crossover mutants by colony PCR, a complementation experiment is performed. First, a complement plasmid is constructed by inserting the intact wild-type gene along with its native promoter into a replicating plasmid vector, then the resulting plasmid is transformed into the knockout mutant by conjugation [7]. By comparing the phenotypes of the wild-type, knockout mutant, and complement, the knockout mutant should reveal a loss of function compared to the wild-type, and this loss of function is recovered in the complement.

2 Materials

1. Petri dishes.
2. Glass flasks.
3. 1.5 mL centrifuge tubes (sterile).
4. 15 mL conical centrifuge tubes (sterile).
5. 50 mL conical centrifuge tubes (sterile).
6. *Anabaena* sp. strain PCC 7120.
7. *Escherichia coli* NEB10β.
8. *E. coli* HB101.
9. Conjugal plasmid pRL443 and helper plasmid pRL623 [3].
10. TOPO™ TA Cloning™ Kit with pCR™2.1-TOPO vector and TOP 10 *E. coli* cells (Invitrogen).
11. pZR824, an integrative vector for *Anabaena* 7120 [8].
12. pZR666, an integrative vector for *Anabaena 7120* containing a *gfp-sp* cassette and *sacB* gene for counter-selection against single crossover mutants [8].
13. pZR2223, a replicating vector for *Anabaena* 7120 [8].
14. LB broth.
15. LB agar, used for growing *E. coli*. *Note: This brand of agar does not work for most cyanobacterial strains.*
16. Agar, required for growing *Anabaena* 7120 and other cyanobacteria (Fisher Scientific, catalog number: A360-500).
17. Ampicillin sodium salt (Ap).
18. Kanamycin sulfate (Km).
19. Chloramphenicol (Cm).
20. Erythromycin (Em).
21. Spectinomycin dihydrochloride pentahydrate (Sp). All the antibiotics are made as stock solution and stored at 4°C. How to make stock solutions are described in the accompanying Chap. 15.
22. Restriction enzymes: *Avr*II, *Bam*HI, *Bgl*II, *Nhe*I, *Not*I, *Sal*I, *Spe*I, *Xba*I, *Xma*I.
23. Plasmid Miniprep Kit.

24. Solution A: 500 mL 4% $MgSO_4 \cdot 7H_2O$, 500 mL 1.2% $CaCl_2 \cdot 2H_2O$, 500 mL 3.8% NaCl, 500 mL microelements stock solution (see below).
25. Microelements stock solution: 160 mL A&A Fe-EDTA solution (see below), 360 mg $MnCl_2 \cdot 4H_2O$, 61.1 mg $Na_2MoO_4 2H_2O$, 44 mg $ZnSO_2 \cdot 7H_2O$, 15.8 mg $CuSO_2 \cdot 5H_2O$, 572 mg H_3BO_3, 4.6 mg NH_4VO_3, 8 mg $CoCl_2 \cdot 6H_2O$, 1090 mL ddH_2O.
26. A&A FeEDTA solution: Dissolve 5.2 g KOH in 186 mL ddH_2O, then add 20.4 $Na_2EDTA \cdot 2H_2O$; dissolve 13.7 g $FeSO_4 \cdot 7H_2O$ in 364 mL ddH_2O; mix the two solutions; bubble Millipore filtered air through the solution until the color changes to brownish-orange (may take a few minutes to several hours). *Note: Fe-EDTA (Ethylenediaminetetraacetic acid, ferric sodium salt trihydrate) may be purchased from Fisher Scientific (Cat#:AC304680010).*
27. Solution B: 28 g K_2HPO_4 in 500 mL ddH_2O.
28. KNO_3 + $NaNO_3$ solution: 8.48 g $NaNO_3$, 10.12 g KNO_3, 200 mL ddH_2O.
29. Allen and Arnon medium plus nitrate [AA/8(N)]: To prepare 1 L of medium combine 3.125 mL solution A, 0.781 mL solution B, and 5 mL KNO_3 + $NaNO_3$ solution to 991 mL ddH_2O. Allen and Arnon medium (AA) can be replaced by BG11 medium (*see* **Note 5**).
30. Allen and Arnon agar medium plus nitrate (AA/N agar): To prepare 1 L of medium combine 25 mL solution A, 6.25 mL solution B, 5 mL KNO_3 + $NaNO_3$ solution, 10 g agar, 969 mL ddH_2O. When preparing the plates with 5% sucrose, filter-sterilized 50% sucrose was added to autoclaved LB or AA/agar medium to a final concentration of 5%.
31. Immobilon-NC transfer membrane.
32. Sucrose.
33. PCR reagents: high-fidelity DNA polymerase, buffer, dNTPs and primers. The details for required primers can be found in Gibbons et al. [8].
34. Gel electrophoresis apparatus and Gel Extraction Kit.
35. Water bath sonicator.
36. Platform shaker equipped with continuous fluorescent light illumination (ca. 50–100 $\mu E/m^2$ s).
37. Cyanobacteria Culture Room (constant at 30 °C) equipped with a digital temperature controlling system and Sylvania fluorescent light bulbs illumination (ca. 50–150 $\mu E/m^2$ s).
38. Microscope.

3 Methods

3.1 Construction of Plasmid for Knocking Out Target Gene

Here, alkane genes (*ado/aar*) are used as an example.

1. Amplify the target gene from *Anabaena* 7120 chromosomal DNA, using primers ZR241 and ZR242, homologous to the 5′ and 3′ ends of the target gene [8], respectively. If the coding region of a target gene is shorter than 660 bp, it is impossible to only use the coding region for double crossover knockout (*see* **Note 3**). In the design for primer ZR241, include a restriction enzyme recognition sequence for BamHI; in primer ZR242, include a restriction enzyme recognition sequence for AvrII. Clone PCR amplified *ado/aar* into pTOPO2.1 (Km^R/Ap^R), creating plasmid pZR932 (Fig. 1, Step 1).
2. Transform Topo10 *E. coli* cells with pZR932. Extract pZR932 from transformed *E. coli* cells, using a plasmid extraction kit.
3. Two restriction enzyme cutting sites within the target gene are needed to excise an internal fragment of the target gene and subsequently replace it with a GFP-antibiotic cassette. First, search the target gene to identify if any unique restriction enzyme cutting sites is already present in pZR932. If only one restriction enzyme cutting site is present or if no restriction enzyme sites are present, perform site directed mutagenesis (SDM) to introduce one or two restriction enzyme cutting sites into an appropriate location within the target gene sequence, respectively. In the alkane gene knockout, a restriction enzyme site for NheI was already present in *aar*. A second restriction enzyme site for NotI was introduced into *ado* via SDM to produce pZR933 (Fig. 1, Step 2).
4. Transform NEB 10β *E. coli* with pZR933; then, extract plasmid from transformed NEB 10β cells.
5. Enzymatic digestion of pZR933 with BamHI and AvrII, which flank the target genes. Clone the target gene into the multiple cloning sites of an integration vector pZR824 (Km^R/ Nm^R), cut with BglII and SpeI, to produce pZR934 (Fig. 1, Step 3). Plasmid pZR824 is a suicide vector that contains the gene *sacB*, which is lethal when sucrose is present [1, 5, 6, 8]. The suicide vector is required later for positive selection of double crossover mutants. Transform *E. coli* NEB 10β with pZR934; extract pZR934 from transformed NEB 10β cells.
6. Enzymatic digestion of pZR934 using NotI and NheI to remove an internal fragment of target gene. Insert the promoter-less gfp-sp cassette from pZR666 (cut with NotI and XbaI) into the vector portion of NotI & NheI digested pZR934 to produce the cargo plasmid (knockout plasmid) pZR935 ($Km^R/Nm^R/Sp^R$) (Fig. 1, Step 4).

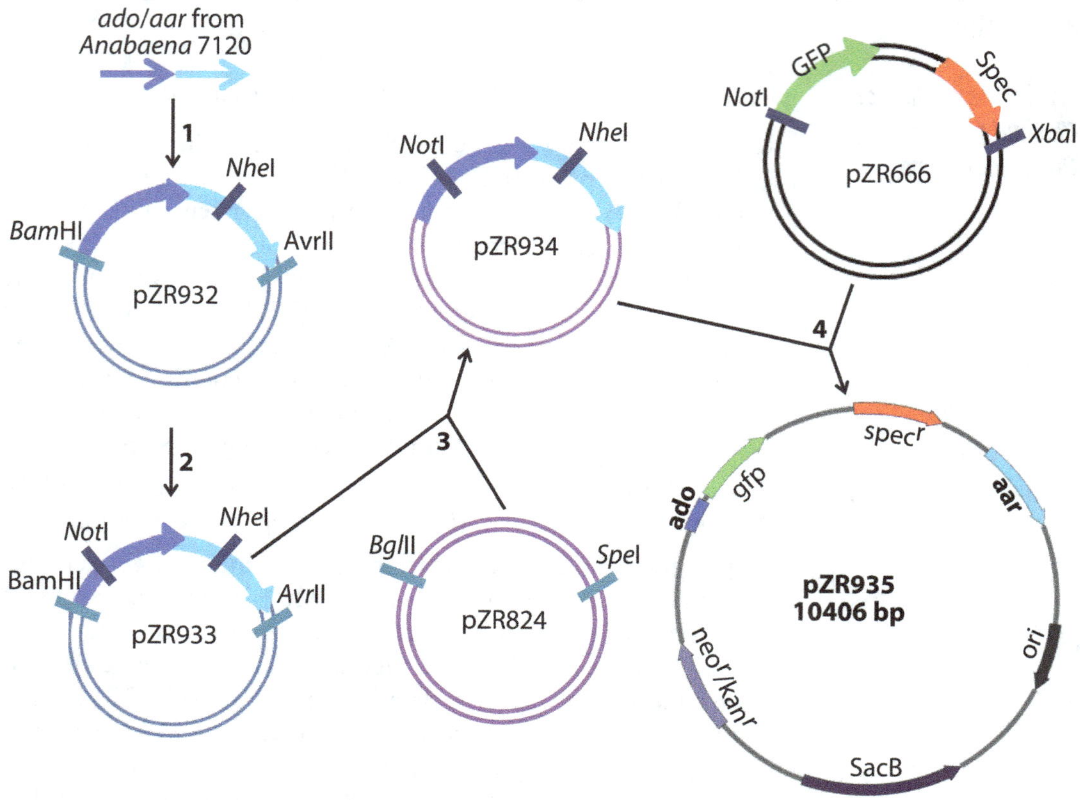

Fig. 1 Schematic illustration of pZR935 construction for knocking out *alr5283-alr5284*. Step 1: Amplified 2.7 kb fragment from *Anabaena* 7120 containing *alr5283-84* using primers ZR241, 242. The PCR product ligated to pCR2.1-TOPO vector to produce pZR932. Step 2: Site directed mutagenesis using primers ZR243, 244 to introduce NotI site within *alr5283* in pZR932, creating pZR933. Step 3: Digestion of pZR933 with BamHI and AvrII to obtain 2.7 kb *alr5283-84* mutated sequence. Ligated this sequence to BglII and SpeI cut pZR824 vector, creating pZR934. Step 4: NotI and XbaI cut out *gfp-sp* cassette from pZR666 ligated into NotI and NheI digested pZR934 to produce pZR935 [8]

7. If choosing this option to construct the knockout plasmid/cargo plasmid, follow with Subheading 3.3.

3.2 Construction of Plasmid for Knocking out Target Gene Via Overlap Extension PCR

This method (an alternative to Subheading 3.1) is highly recommend as it selectively amplifies the target gene while removing internal fragment

1. Design two primers (P1 and P2) to amplify the 5′ end of the target gene and two primers (P3 and P4) to amplify the 3′ end of the target gene, leaving the middle portion of the gene unamplified (Fig. 2, Overlap Extension PCR). Design primers P5 and P6 to amplify the *gfpsp* cassette (*gfp-sp*) from pZR666 [8]. The reverse primer for amplifying the 5′ region (P2) of the target gene must have a homologous sequence to the forward primer for amplifying the *gfp-sp* cassette (P5). Similarly, the

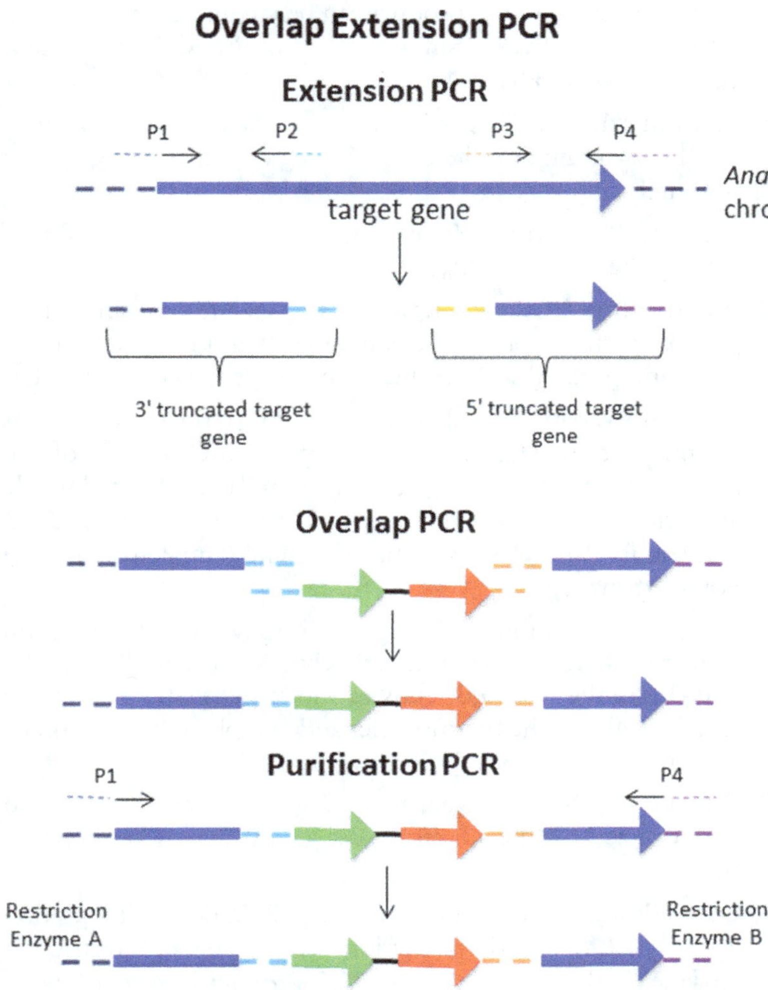

Fig. 2 Overlap extension PCR. Extension PCR: Primers for the 5′ end of the target gene (P1 and P2) and primers for the 3′ end of the target gene (P3 and P4) are designed and used to perform PCR to amplify the truncated 5′ and 3′ ends of the gene. Primers P5 and P6 are designed and used to amplify the *gfp-sp* cassette from pZR666 [8]. Primers P1 and P4 contain restriction enzyme sites (indicated by the dashed region). The dashed regions of primers P2 and P3 are homologous to the dashed regions of primers P5 and P6, respectively. Overlap PCR: the truncated 5′ and 3′ ends of the target gene are overlapped with the *gfp-sp* cassette via the homologous regions. Purification PCR: primers P1 and P4 are added to the overlap PCR reaction. The end product contains restriction enzyme sites that will be used to construct the knockout plasmid. Restriction enzyme A cutting sequence was encoded in primer P1, and restriction enzyme B was encoded in primer P4

forward primer for amplifying the 3′ region (P3) should have a homologous sequence to the reverse primer for amplifying the *gfp-sp* cassette (P6). The homologous regions are required so that in a following step, the 5′ and 3′ amplified regions can be overlapped with the *gfp-sp* cassette, thereby deleting an internal fragment of the gene while inserting the cassette. Primers P1 and P4 must also contain restriction enzyme cutting sites so the overlap extension PCR product can be cut and then is inserted into an integrative plasmid vector cut by same restriction enzymes (*see* **step 8**).

2. Perform extension PCR with wild-type *Anabaena* 7120 genomic DNA to amplify the 5′ and 3′ regions of the target gene separately.
3. Perform PCR with pZR666 using primers P5 and P6 to amplify the *gfp-sp* cassette.
4. Run the three PCR products from **steps 2** and **3** on an agarose gel, excise three correct DNA fragments and put them in one tube, and purify the DNA fragments using a gel extraction kit.
5. Perform overlap PCR (Fig. 2). Use the purified three DNA fragments from **step 4** as the template; use ½ to ¾ of the purified three DNA fragments as the templates. Do not include any primers in this overlapping PCR reaction. Run the PCR reaction for 15 cycles, using the annealing temperature for the homologous regions.
6. Perform purification PCR (Fig. 2). Remove the PCR reaction from the thermocycler. Immediately add primers P1 and P4 that flanks the 5′ and 3′ ends of the target gene. Return the reaction tube to the thermocycler and complete the reaction by running the PCR for 30 more cycles.
7. Run PCR product on an agarose gel, extract the expected correct size DNA band, and use a gel extraction kit to purify the DNA.
8. Digest both the PCR product (**step 7**) and the integration vector pZR824 with restriction enzyme A and restriction enzyme B, then ligate the cut PCR product into pZR824 to produce the knockout plasmid (Fig. 3).
9. Transform above knockout plasmid into *E. coli* NEB 10β cells.

3.3 Preparation of E. coli Strains

1. Grow *E. coli* HB101 containing pRL443 (Ap^R) and pRL623 (Cm^R) in 5 mL LB broth containing 100 μg/mL ampicillin and 25 μg/mL chloramphenicol overnight at 37 °C, 200 rpm. Also grow *E. coli* NEB 10β bearing the cargo plasmid ($Km^RNm^RSp^R$) in 5 mL LB broth containing 50 μg/mL kanamycin and 100 μg/mL spectinomycin overnight at 37 °C, 200 rpm. For next conjugal transformation of cyanobacteria, all plasmids must be transformed into one of four *E. coli* strains (*see* **Note 4**).
2. Transfer 100 μL of the overnight *E. coli* cultures into 5 mL LB containing the appropriate antibiotics; grow for 3 h at 37 °C, 200 rpm until the cultures reach OD_{600} ~ 0.5.
3. Harvest the 5 mL *E. coli* cultures by centrifuging at 4000 × *g* at room temperature (22 °C) for 10 min.

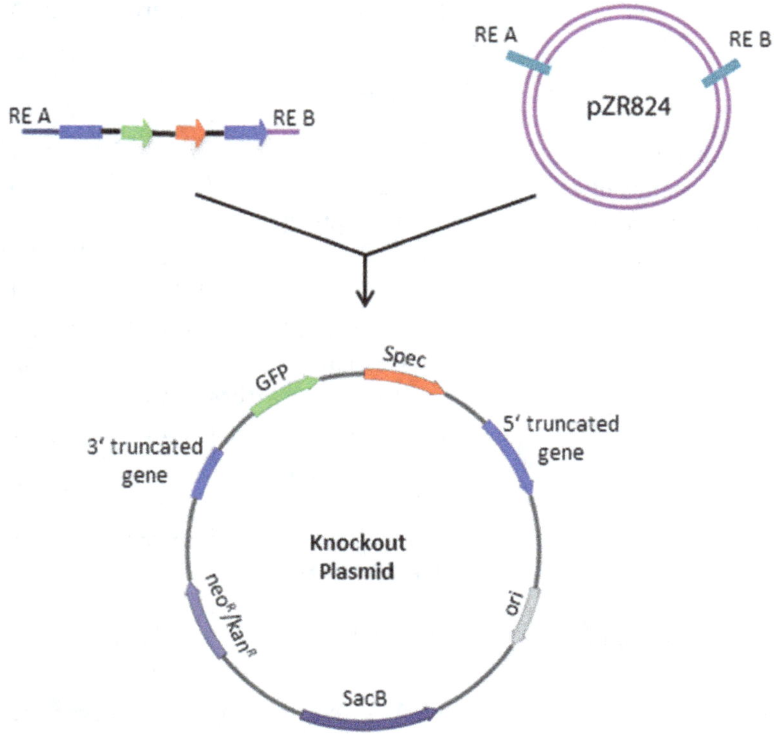

Fig. 3 Construction of knockout plasmid using overlap extension PCR product. Restriction enzymes cutting sequences A and B from the overlap extension PCR product (Fig. 2) are labeled RE A and RE B. The overlap extension PCR product cut by RE A & RE B ligated to an integrative vector pZR824 cut by RE A and RE B to produce the knockout plasmid

4. Remove the supernatant and wash the pellet in 1 mL LB, transferring culture to a 1.5 mL eppie tube. Centrifuge at 13,000 × *g* for 1 min. Repeat wash with 1 mL LB twice to completely remove the antibiotic.
5. After final wash, resuspend pellets in 100 μL LB.
6. For mating, combine 100 μL of NEB 10β bearing the cargo plasmid with 100 μL HB101 harboring pRL443 and pRL623 (experimental group). For a negative control, add 100 μL LB to 100 μL HB101 harboring pRL443 and pRL623. Incubate at room temperature for 30 min to allow the cultures to mate.

3.4 Preparation of *Anabaena 7120* Culture

1. Grow *Anabaena* 7120 in 50 mL AA/8(N) medium for approximately 7 days under continuous light (~50 μE/m^2 s) at 30 °C and shaken at 120 rpm until OD_{720} reaches ~0.3.
2. Break the long filaments in the culture into single cells or average 2–3 cell filaments by sonicating the culture for 60–120 s using a water bath sonicator. Microscopic verification of fragmentation is required in this step.

3. Harvest the 50 mL culture at 5000 × *g* at 22 °C for 10 min.
4. Remove the supernatant and resuspend the pellet in 1 mL AA/8(N).

3.5 Conjugal Transformation of a Cargo Plasmid into Anabaena 7120

1. Mix 150 μL of *Anabaena* 7120 from the resuspension (Subheading 3.4) with 150 μL of the mated *E. coli* mixture (experimental group, Subheading 3.3) or the control mixture (negative control group, Subheading 3.3), respectively, and incubate at room temperature for 1 h.
2. Evenly spread the conjugal mixtures onto autoclaved nitrocellulose membrane placed on top of AA(N) agar plate containing 5% LB (v/v) without antibiotic and incubate at 30 °C under continuous illumination (~50 μE/m^2 s) for 48 h.
3. Transfer the membrane onto an AA(N) agar plate with 10 μg/mL spectinomycin; incubate under continuous light at 30 °C until spectinomycin-resistant colonies appear (approximately 10–15 days).

3.6 Screening for Double Crossover Mutants

1. Restreak 10 or more potential knockout mutant colonies on AA/8(N) agar plate containing 10 μg/mL spectinomycin.
2. First, screen for single crossover mutants using colony PCR. Two primer sets can be used to verify a crossover has occurred. All primers are shown in Fig. 4. The first primer set contains a primer (ZR461) matching 49 bp upstream of primer ZR241 and a primer (ZR90) matching the *gfp* in the *gfp-sp* cassette. Using ZR461 and ZR90 to colony PCR looking for the expected 1121 bp band. The second primer set contains a primer (ZR462) matching 51 bp downstream of primer ZR242 and a primer (ZR1912) matching the *Spec^r* in the *gfp-sp* cassette. Using ZR462 and ZR1912 to colony PCR looking for the expected 2239 bp band. If only either of the primer sets is able to amplify expected band, this indicates only a single crossover occurred. If both primer sets are able to amplify the expected bands, this indicates that a double crossover occurred.
3. For colonies verified to contain single crossover mutants, grow the single crossover recombinants in 10 mL AA/8(N) 10 μg/mL spectinomycin for 7 days until culture reaches approximately OD_{700} 0.3.
4. After 7 days growth, sonicate the culture for 90 s using a water bath sonicator.
5. After sonication, remove 1 mL of the culture and centrifuge at 13,000 × *g* for 1 min. Resuspend harvested cell pellet in 100 μL AA/8(N).
6. Spread the cells on an AA/N agar plate containing 10 μg/mL spectinomycin and 5% sucrose (sucrose will provide negative

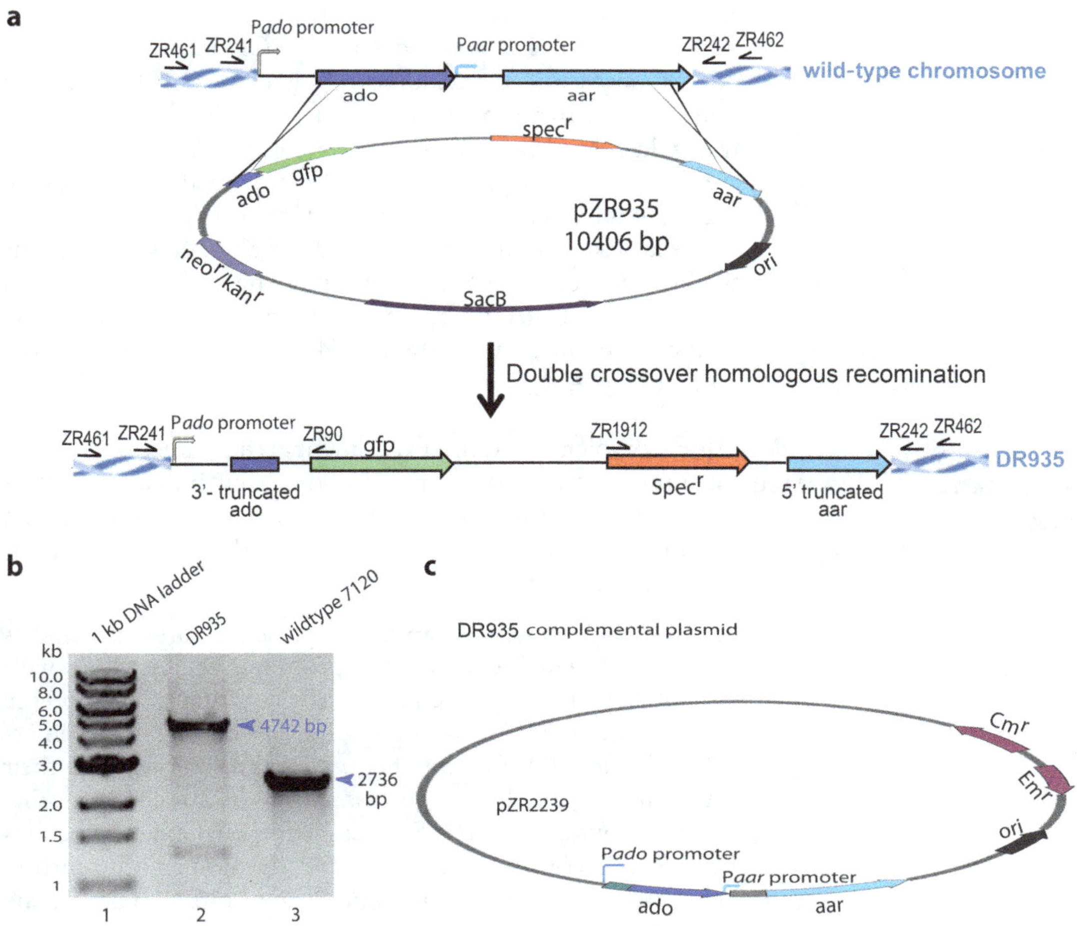

Fig. 4 Construction of *alr5283-alr5284* knockout mutant (DR935) and its complementing plasmid pZR2239. (**a**) 3′ deletion of *ado* (*alr5283*) and 5′ deletion of *aar* (*alr5284*) created by inserting a *gfp-sp* cassette between *alr5283-84* in *Anabaena* 7120 chromosome via double crossover/recombination with knockout plasmid pZR935. (**b**) PCR verification of DR935. Wild-type *Anabaena* 7120 had the intact *alr5283-84* gene sequence, which has a length of 2736 bp when amplified by ZR241, 242 (lane 3). DR935 contained the *gfp-sp* cassette inserted in the *alr5283-84* gene sequence, making the amplified gene sequence 4742 bp (lane 2). (**c**) Complement plasmid pZR2239 for DR935 complementation experiment. Plasmid pZR2239 contains *ado* and *aar* both under control of their native promoters [8]

selection for single crossovers [5, 6], which contain the *sacB* gene; therefore, only double crossover mutants should grow on the plate). Incubate the plate at 30 °C under light until colonies form (ca. 7–10 days).

3.7 Verification of Double Crossover Knockout Mutants

1. Restreak 10 or more potential double crossover mutant colonies, which were grown on an AA/N agar plate containing 10 μg/mL spectinomycin and 5% sucrose, onto AA/N 10 μg/mL spectinomycin agar plate.

2. Screen for double crossover knockout mutants using colony PCR (*see* **Note 1**). For the colony PCR reaction, use one primer upstream of the knockout gene (ZR241) and one primer downstream of the knockout gene (ZR242) (Fig. 4a). Use wild-type *Anabaena* 7120 as a control. In the wild type, the native gene sequence should be amplified; in the alkane gene example, 2736 bp was amplified in the wild type (Fig. 4b). In the knockout mutant, a larger DNA fragment is amplified as the insertion of the *gfp-sp* cassette into the middle of the gene increases the DNA bp region between the primers; in the alkane gene knockout mutant, 4742 bp was amplified in the mutant (Fig. 4b).

3.8 Complementation Experiment

For the complementation experiment, the functional gene sequence must be reinserted into the knockout mutant to determine if the lost function returns. If the loss of function is reinstated in the complement strain, the results indicate that the knocked out gene is responsible for that function.

1. For the complement plasmid, a replicating cargo plasmid is constructed, which contains the promoter region along with the entire gene coding sequence of the target gene. For the alkane complement plasmid (pZR2239 in Fig. 4), primers ZR1602 and ZR1603 amplified *ado* and *aar* along with their native promoters from the *Anabaena* 7120 chromosome. ZR1602 contained AatII/SalI restriction enzyme cutting sites and ZR1603 contained KpnI/XmaI restriction enzyme cutting sites, such that AatII/SalI-P-*ado*-P-*aar*-KpnI/XmaI was the gene product amplified by PCR.
2. The PCR product is first cloned into pTOPO2.1 TA cloning vector (Km^R/Ap^R) (Invitrogen), creating plasmid pZR2238. Transform Top10 *E. coli* cells with pZR2238, and extract pZR2238 from transformed *E. coli* cells.
3. Enzymatic digestion of plasmid pZR2238 using SalI and XmaI that flank the target genes (*ado* and *aar*). Clone the target genes into the multiple cloning sites of pZR2223 (Cm^R/ Em^R), cut by SalI and XmaI, to produce plasmid pZR2239 (Cm^R/ Em^R). Transform *E. coli* NEB 10β with plasmid pZR2239.
4. A similar approach to Subheadings 3.3–3.5 is used to transfer the complementing plasmid pZR2239 to the verified knockout mutant. Here, the knockout mutant rather than wild-type *Anabaena* 7120 is used for the conjugation.
5. Transfer the membrane to an AA(N) agar plate containing 10 μg/mL spectinomycin (Sp) and 10 μg/mL erythromycin (Em). Incubate under continuous light illumination (ca. 50 $\mu E/m^2$ s) at 30 °C. Colonies should appear in approximately 10–15 days.

6. Once colonies appear on the membrane, streak the colonies on an AA(N) agar plate containing Sp at 10 μg/mL and Em at 10 μg/mL. Incubate under continuous light illumination (ca. 50 μE/m^2 s) at 30 °C for 7–10 days. Restreak the colonies if needed onto another AA(N) agar plate containing Sp (10 μg/mL) and Em (10 μg/mL) until the colonies appear to be free of *E. coli*.
7. Perform colony PCR with the conjugation colonies, using one primer matching the DNA insert and one primer matching the vector. To verify colonies containing pZR2239, primers ZR1602 and ZR261 were used. ZR1602 matched the 5′ end of the *ado/aar* insert while ZR261 matched part of the *gfp* encoding sequence contained within the vector [8].
8. Examine the phenotype of at least three independent complementing colonies to verify that the function lost in the double crossover mutant is regained in the complement. For the alkane knockout mutant, complement strains were grown in 100 mL AA/8(N) for 5 days; cultures were incubated at 30 °C, 120 rpm at an aeration rate of 100 mL/min. A resin column inserted into the rubber stopper sealing each flask was used to capture hydrocarbons in the headspace, and hydrocarbons were extracted from the resin on day 5. GC-MS was performed to analyze the hydrocarbon extraction from wild-type *Anabaena* 7120, the double crossover mutant, and the complement. Heptadecane ($C_{17}H_{36}$) was detected in the hydrocarbon extraction from the wild type and complement (*see* **Note 2**). As expected, it was not detected in the mutant. These results suggested that the alkane genes (*ado/aar*), which were knocked out, were responsible for heptadecane production in *Anabaena* 7120 [8].

4 Notes

1. When amplifying the target genes from the chromosome, use a high-fidelity DNA polymerase, such as Phusion (as opposed to a standard Taq polymerase), to ensure that the gene is amplified without error. A standard Taq polymerase may be used when verifying colonies by PCR to check for the double crossover knockout. However, if colonies consistently appear on the conjugation plates and after the 5% sucrose selection step for the double crossover, yet are unable to be verified by PCR using Taq, replace Taq with Phusion polymerase. Also, if colonies appear for the double crossover mutants and yet are unable to be verified by PCR, try gradient PCR with various annealing temperatures to hone in on the right annealing temperature. Verification of the double crossover mutants can be meticulous and require a more fine-tuned PCR.

2. For complementation experiments, the loss of function identified in the knockout mutant will reappear in the complement. However, the complement generally does not regain the full level seen in the wild type. Bear this in mind, as we are not looking to reach the level of function seen in the wild type. To successfully complement the knockout gene, we are merely looking to ascertain that the function absent in the knockout mutant is regained by the complement.
3. The threshold length for efficient homologous crossover/recombination was 330 bp [9], if the coding region of a target gene is shorter than 660 bp, it is impossible to only use the coding region for double crossover knockout. We used the double crossover approach to successfully knocked out a shorter gene *devR* (405 bp) [10] by using the upstream region and downstream region along with *devR* as the flanking sequences for antibiotic cassette, but the antibiotic cassette must be inserted within the coding region of a target gene (*devR*) to be able to knock out it through a double crossover.
4. For conjugation experiment, all plasmids must be transformed into one of four *E. coli* strains. They are *E. coli* HB101, *E. coli* DH10B, *E. coli* NEB 10β, DH5α MCR.
5. For most cyanobacteria, Allen and Arnon medium (AA) can be replaced by BG11 medium (http://www-cyanosite.bio.purdue.edu/media/table/BG11.html) [11].

References

1. Cai YP, Wolk CP (1990) Use of a conditionally lethal gene in Anabaena sp. strain PCC 7120 to select for double recombinants and to entrap insertion sequences. J Bacteriol 172(6): 3138–3145. https://doi.org/10.1128/jb.172.6.3138-3145.1990
2. Heckman KL, Pease LR (2007) Gene splicing and mutagenesis by PCR-driven overlap extension. Nat Protoc 2(4):924–932. https://doi.org/10.1038/nprot.2007.132
3. Elhai J, Vepritskiy A, Muro-Pastor AM, Flores E, Wolk CP (1997) Reduction of conjugal transfer efficiency by three restriction activities of Anabaena sp. strain PCC 7120. J Bacteriol 179(6):1998–2005
4. Elhai J, Wolk CP (1988) Conjugal transfer of DNA to cyanobacteria. Methods Enzymol 167:747–754. https://doi.org/10.1016/0076-6879(88)67086-8
5. Gay P, Le Coq D, Steinmetz M, Berkelman T, Kado CI (1985) Positive selection procedure for entrapment of insertion sequence elements in gram-negative bacteria. J Bacteriol 164(2): 918–921. https://doi.org/10.1128/JB.164.2.918-921.1985
6. Reyrat JM, Pelicic V, Gicquel B, Rappuoli R (1998) Counterselectable markers: untapped tools for bacterial genetics and pathogenesis. Infect Immun 66(9):4011–4017
7. Elhai J, Vepritskiy A, MuroPastor AM, Flores E, Wolk CP (1997) Reduction of conjugal transfer efficiency by three restriction activities of Anabaena sp. strain PCC 7120. J Bacteriol 179(6):1998–2005
8. Gibbons J, Gu L, Zhu H, Gibbons W, Zhou R (2018) Identification of two genes required for heptadecane production in a N2-fixing cyanobacterium Anabaena sp. strain PCC 7120. AMB Express 8(1):167. https://doi.org/10.1186/s13568-018-0700-6
9. Biswas I, Gruss A, Ehrlich SD, Maguin E (1993) High-efficiency gene inactivation and replacement system for gram-positive bacteria. J Bacteriol 175(11):3628–3635. https://doi.org/10.1128/jb.175.11.3628-3635.1993

10. Zhou R, Wolk CP (2003) A two-component system mediates developmental regulation of biosynthesis of a heterocyst polysaccharide. J Biol Chem 278(22):19939–19946. https://doi.org/10.1074/jbc.M300577200

11. Rippka R, Deruelles J, Waterbury J, Herdman M, Stanier R (1979) Generic assignments, strain histories and properties of pure cultures of cyanobacteria. J Gen Microbiol 111(1):1–61

Chapter 17

Expression of Cyanobacterial Biosynthetic Gene Clusters in *Escherichia coli*

Alescia Cullen, Matthew Jordan, and Brett A. Neilan

Abstract

Cyanobacteria represent an attractive source of natural bioactive compounds, ranging from sunscreens to cancer treatments. While many biosynthetic gene clusters (BGCs) that encode cyanobacterial natural products are known, the slow growth and lack of genetic tools in the native producers hampers their modification, characterization, and large-scale production. By engineering heterologous hosts for the expression of cyanobacterial BGCs, sufficient material can be produced for research or industry. Although several hosts have been evaluated for the expression of cyanobacterial natural products, this work details the process of expressing BGCs in *Escherichia coli* via promoter exchange.

Key words Cyanobacteria, Specialized metabolites, Natural products, Heterologous expression, Recombineering, Promoter exchange

1 Introduction

Cyanobacteria are a rich source of natural products, with a 2012 review by Gerwick and Moore determining that 20% of marine drugs currently in use or in clinical trial were isolated from these photosynthetic prokaryotes [1]. There are many different classes of natural products produced by cyanobacteria including polyketides, peptides, alkaloids, and terpenoids, which can range in biological activity from medicinal (antifungal, antibacterial, anticancer) [2], to cyanotoxins that are highly toxic to humans and livestock [3]. These compounds are often synthesized by enzymes that are encoded for by a series of colocated genes, referred to as a biosynthetic gene cluster (BGCs). In cyanobacteria these BGCs can range in size from 2 to 64 kb [4].

The interesting bioactivities that cyanobacterial natural products (also referred to as specialised metabolites) may possess make their encoded biosynthetic gene clusters (BGCs) alluring to characterize. Unfortunately, although a few genetically amiable

Elizabeth Skellam (ed.), *Engineering Natural Product Biosynthesis: Methods and Protocols*,
Methods in Molecular Biology, vol. 2489, https://doi.org/10.1007/978-1-0716-2273-5_17,

cyanobacterial systems exist [5–7], these are largely in unicellular/coccoid species (such as *Synechocystis* spp.) and are less developed in filamentous species [8], which are responsible for the large majority of specialized metabolite production [4, 9, 10]. Filamentous strains are renowned for their difficulty to genetically manipulate due to their multiple chromosome copies and thick sheath surrounding the cell wall, which may harbor heterotrophic symbionts and nucleases [9, 11–13]. Thus, it is clear that a heterologous expression approach must be taken to further characterize cyanobacterial natural product biosynthesis pathways. Furthermore, many cyanobacteria that have a rich ability to produce natural products are prohibitively slow growing (such as those from the genera *Lyngbya* and *Nostoc*) [4, 14–16] or are unable to be grown under lab conditions [17–19]. Therefore, if the large-scale production of these natural products is desired for use commercially, then heterologous production systems are required.

The first attempts to clone cyanobacterial specialized metabolite clusters were of the microcystin [20] and barbamide BGCs [21]. In both cases, fragment(s) of the cluster were generated via a genomic library from the native producer. While heterologous expression of the clones did not occur, the cloned fragments allowed for the elucidation of biosynthetic pathways and insight into the mechanisms of production [20, 21]. At the time, the size of both clusters exceeded what was achievable using traditional cloning methods (restriction endonuclease digest coupled with DNA ligase). However, current cloning techniques, such as recombination engineering (recombineering), have allowed for advances in the production of vectors carrying cyanobacterial specialized metabolite clusters, and since then several have been cloned [22–27].

One consideration when cloning cyanobacterial BGCs is whether the host strain being used contains all the necessary requirements for expression. BGCs harboring polyketides synthases (PKSs) and nonribosomal peptide synthetases (NRPSs) require phosphopantetheinyl transferases (PPTases) to post-translationally modify them for activation. Within cyanobacterial genomes, all PPTases are Family 2 (Sfp-type) PPTases, and are crucial not only for the production of specialized metabolites, but also heterocyst differentiation for nitrogen fixation [28, 29]. Therefore, to successfully express cyanobacterial BGCs heterologously, host strains must either have their own broad range PPTase, such as *Streptomyces coelicolor* [30], or have one inserted into their genome, as observed in the engineered *Escherichia coli* strain BAP1 which contains the gene encoding *Bacillus subtilis* PPTase, *sfp* [31, 32].

Although several expression systems have been trialed to produce cyanobacterial natural products, each has its drawbacks. The focus has primarily been on strains of *Streptomyces* and *E. coli*. *Streptomyces* strains have been used as a host for cyanobacterial natural products due to their native abilities to produce and post-translationally modify specialized metabolites [33, 34]. However, *E. coli* represents an attractive heterologous host due to a more similar GC content to cyanobacteria [18] and better-established genetic tools for the cloning and modification of gene clusters [35]. It is for these reasons that all published accounts of cyanobacterial natural product heterologous production within our lab [18, 24, 36, 37], and the majority of publications from other labs, have used *E. coli* as a heterologous host [17, 23, 25, 26, 38]. While expression of cyanobacterial metabolites in cyanobacterial hosts such as *Synechocystis* [39] and *Anabaena* [27, 40] would be ideal due to genetic relatedness and obviation of the need for genetic modification, including promoter exchange, modified strains of *E. coli* remain the most feasible option for heterologous expression of cyanobacterial specialized metabolites.

The methods outlined below detail the process of promoter exchange to allow gene expression in *E. coli,* and assume the cloning vector containing the entire gene cluster of interest has already been assembled. There are several approaches that can be taken to isolate and assemble the BGC (reviewed by Ongley et al. [41] and Huo et al. [42]), which extend beyond the scope of this work. The examples shown here detail the processes used to exchange the native promoter within the lyngbyatoxin biosynthetic cluster (*ltx*) for the tetracyline inducible promoter P_{tetO}, as published by Ongley et al. [18]. This resulted in the heterologous production of lyngbyatoxin and its precursor indolactam V to titers of 25 and 50 mg/L, respectively [18].

2 Materials

2.1 Equipment

1. Bunsen or laminar flow for sterile steps.
2. Thermal shaker for microcentrifuge tubes.
3. Microcentrifuge (with refrigeration capacities).
4. Electroporator.
5. Incubator (capable of maintaining 37 °C).
6. PCR thermocycler.
7. Electrophoresis system (horizontal electrophoresis tank and power supply).
8. Blue light LED transilluminator.
9. Mass balance.
10. DNA spectrophotometer.
11. Gel documentation system.

12. UV-Vis spectrophotometer.
13. Temperature controlled orbital shaker.

2.2 Consumables

1. 1.5 mL microcentrifuge tubes
2. 18G needle (1.2 mm diameter)
3. 0.1 mm electroporation cuvette
4. Spectrophotometer cuvettes.
5. Sterile spreader.
6. 200 μL PCR tubes
7. Surgical blade.
8. DNA gel extraction kit.
9. Plasmid miniprep kit.
10. Polymeric adsorption resin (XAD-7 or XAD-16).

2.3 Buffers, Solutions, and Media

1. Milli-Q H_2O: Water that has been purified using a Type 1 ultra-pure water system from the Milli-Q line of systems or equivalent.
2. Luria–Bertani (LB) liquid media: 10 g tryptone, 5 g NaCl, 5 g yeast extract made to 1 L in MilliQ H_2O. Media is sterilized via autoclaving and can be stored at room temperature.
3. LB agar (solid media): 10 g tryptone, 5 g NaCl, 5 g yeast extract, 15 g agar made to 1 L in MilliQ H_2O. Media is sterilized via autoclaving and poured into petri dishes when still liquid in a sterile environment. When making LB agar with antibiotics, the media (after autoclaving while still liquid) should be cooled to a maximum of 50 °C before antibiotic is added. Adding antibiotic to the agar at higher temperatures can cause it to degrade. Store plates at 4 °C once cooled and set.
4. Terrific Broth (TB) liquid media: 24 g yeast extract, 20 g tryptone, 4 mL glycerol in 900 mL of MilliQ H_2O. Autoclave to sterilize and allow to cool. Add 100 mL of phosphate buffer (0.17 M KH_2PO_4, 0.72 M K_2HPO_4, autoclaved separately to sterilize) to the media in a sterile environment. Store at room temperature.
5. Chloramphenicol: 30 mg dissolved in 1 mL of ethanol. Store at −20 °C.
6. Gentamicin: 4 mg dissolved in 1 mL of MilliQ H_2O. Store at −20 °C.
7. Tetracycline: 5 mg dissolved in 1 mL of 70% ethanol. Store at −20 °C.
8. PCR reagents: 5× high-fidelity DNA polymerase buffer, 10 mM dNTPs, 50 μM forward primer (ltx-genta5: 5′-**TATT**

CCTTATACCATCAAGATCCTTTGATTTTGAGGGCT AGAAGGCACGAA CCCAGTTGAC-3′—the region in bold being the homology arm for recombineering), 50 μM reverse primer (ltx-tetR3: 5′-**TATCGCCCGCCTACTTCCACT CCAAGGTTGATTCAT**AATCATGTC GATCCTCTTCT CTATCACTGATAGG-3′—the region in bold being the homology arm for recombineering), dimethylsulphoxide (DMSO), high fidelity DNA polymerase, 10× *Taq* DNA polymerase buffer, 10 μM construct specific forward primer (for screening PCRs), 10 μM construct specific reverse primer (for screening PCRs), 50 mM $MgCl_2$, *Taq* DNA polymerase. Store all reagents at −20 °C.

9. 6× DNA loading dye: 25 mg xylene cyanol FF, 25 mg bromophenol blue, made up in 10 mL 30% (v/v) glycerol. Store at room temperature.
10. Tris–acetate–EDTA (TAE) buffer: 242 g of Tris base, 57 mL of glacial acetic acid, 100 mL 0.5 M EDTA (pH 8) made to 1 L in MilliQ H_2O. This makes the 50× stock which is then diluted 20 mL into 1 L MilliQ H_2O for the working stock. Stock and working solutions can be stored at room temperature.
11. 1% agarose: 1 g of molecular grade agarose in 100 mL TAE. Bring solution to a boil in microwave to dissolve agarose and allow to cool to at most 60 °C before use. If a dye is to be added to the agarose before pouring, then agarose must be cool enough to touch before doing so. Liquid agarose solution may be kept in an oven/ incubator at >50 °C to stop from solidifying and have ready to use in liquid form.
12. GelGreen nucleic acid stain kept in the dark at room temperature.
13. GelRed nucleic acid stain kept in the dark at room temperature.
14. L-arabinose: 1 mg (w/v) L-arabinose in 10 mL MilliQ H_2O. Sterilise using syringe filtration (0.22 μM) and store frozen at −20 °C. *See* **Note 1**.
15. Methanol (HPLC grade).

2.4 Biologicals

1. *E. coli* GB05-red: GB2005 with a *lacZ* deletion and P_{BAD}-gbaA inserted at the *ybC* locus [43, 44]. A related strain GB08-red (which can be bought from GeneBridges) may also be used [43].
2. *E. coli* GB05-MtaA: GB2005 modified to contain the *Stigmatella aurantiaca* PPTase, MtaA [44].
3. A plasmid containing the BGC of interest: in this case the *ltx* cluster containing fosmid fos-DE3-86 [22].
4. A template to amplify the antibiotic resistance and inducible promoter cassette from: in this case the plasmid pR6K-Tps-Genta which contains the *aacC1-tetR*-P_{tetO} cassette [45].

3 Methods

In all reports of the heterologous expression of cyanobacterial metabolites by our laboratory, production of said metabolites only occurred when the BGCs were under the control of an *E. coli* inducible promoter instead of the native promoters [18, 24, 36, 37], The methods listed below will detail the steps required to exchange native promoters within a construct for *E. coli* inducible promoters and then ferment these constructs in *E. coli* for the heterologous expression of natural products. Whether or not promoter exchange will be necessary for your construct is dependent on the vector backbone the BGC has been cloned into (*see* **Note 2**).

3.1 Transformation of BGC Construct into E. coli GB05-Red

Before promoter exchange can occur, the plasmid of interest must be transformed into a recombineering proficient strain. In this case GB05-red [44], an *E. coli* strain engineered to contain *recA* and the λ phage *red* operon under the control of pBAD, was used. The transformation and recombineering protocols discussed were adapted from Fu et al. [44]. This method should be performed under sterile conditions.

1. Pick a single colony of *E. coli* GB05-red into 1 mL of LB liquid media in a 1.5 mL microcentrifuge tube. Pierce a hole into the lid of the tube with a hot needle to allow for aeration (*see* **Note 3**). Grow the samples overnight in a shaker at 37 °C, rotating at 950 rpm.
2. Add 40 μL of the overnight culture into a new 1.5 mL tube (with a hole in the lid) containing 1.4 mL of LB liquid media. Grow the subculture in a shaker for 2 h at 37 °C, rotating at 950 rpm.
3. Spin down cultures at 9000 × *g*, for 1 min at 4 °C and discard the supernatant.
4. Resuspend the cell pellet in 1 mL ice cold MilliQ water.
5. Repeat **steps 3** and **4** twice more and after the last step remove the supernatant with a pipette such that the cell pellet remains in between 30 and 50 μL of MilliQ water.
6. Resuspend cells in the remaining liquid and add between 50 and 300 ng of plasmid (in this case a fosmid containing the *ltx* cluster, fos-DE3-86 [22]) to the cell mixture. Incubate on ice for 20 min.
7. Transfer the cell-DNA mixture into a precooled 1 mm electroporation cuvette. Gently tap the cuvette on the bench a few times to ensure the mixture is at the bottom of the well and there are no bubbles. Wipe the sides of the cuvette to remove any condensation and place the cuvette into an electroporator.

8. Electroporate the cells at 1.8 kV and then immediately add 1 mL LB liquid media to the cuvette. Take note of the electroporation time constant. An ideal time constant is around 5 ms, a lower time constant is usually indicative of excess salts in the cell-DNA mixture and will result in lowered transformation efficiencies.
9. Transfer the cells and media from electroporation cuvette into the 1.5 mL microcentrifuge tube from **step 6** and place tube into thermomixer for recovery at 37 °C, 950 rpm for between 1 and 2 h.
10. Plate 30–100 μL of recovered cells onto LB agar plates containing 30 μg/mL chloramphenicol (for the selection of the plasmid) using a sterile spreader and place the plates in a 37 °C incubator overnight.

3.2 PCR Amplification of Promoter Resistance Cassette

Below describes the amplification of a gentamycin resistance, tetracycline repressor and tetracycline promoter (*aacC1-tetR*-P_{tetO}) gene cassette for promoter exchange [18] into the lyngbyatoxin BGC containing fosmid fos-DE3-86 [22]. For more detail on the use of a template to amplify the promoter resistance cassette, please *see* **Note 4**. For greater clarification on the use of homology arms for homologous recombination, please *see* **Note 5** and Fig. 1.

1. To amplify the *aacC1-tetR*-P_{tetO} promoter resistance cassette (*see* **Note 4**), combine the following in 3× 0.2 mL PCR tubes on ice: 30 μL of MilliQ H_2O, 10 μL 5× "Hi-Fi" polymerase buffer, 5 μL 10 mM dNTPs, 1 μL 50 μM forward primer (ltx-genta5), 1 μL 50 μM reverse primer (ltx-tetR3), 1.5 μL DMSO and 0.5 μL high fidelity polymerase (in this case Velocity DNA polymerase), and 1 μL of 10 ng/μL template (a plasmid containing the *aacC1-tetR*-P_{tetO} cassette, pR6K-Tps-Genta [45]).
2. Place the tubes into a thermocycler which has been preheated to 98 °C then subject the tubes to PCR with a 2 min 98 °C initial denaturation, 30 cycles of 30 s at 98 °C, 35 s at 60 °C, and 1 min at 72 °C (*see* **Note 6**). Lastly the tubes are incubated at 72 °C for 10 min for the final elongation and then held at 4 °C before being used or stored.
3. Add 10 μL of 6× loading dye to each PCR product, mix well and then load product into a 1% agarose gel (containing Gel-Green® nucleic acid stain to 1×) in an electrophoresis tank filled with TAE buffer. Run the gel at 80 V for 20 min (*see* **Note 7**).
4. View the gel using blue light box illuminator, and carefully cut out the fluorescent product band using a surgical blade. Trim the excess agarose around the band and place into a premeasured tube.

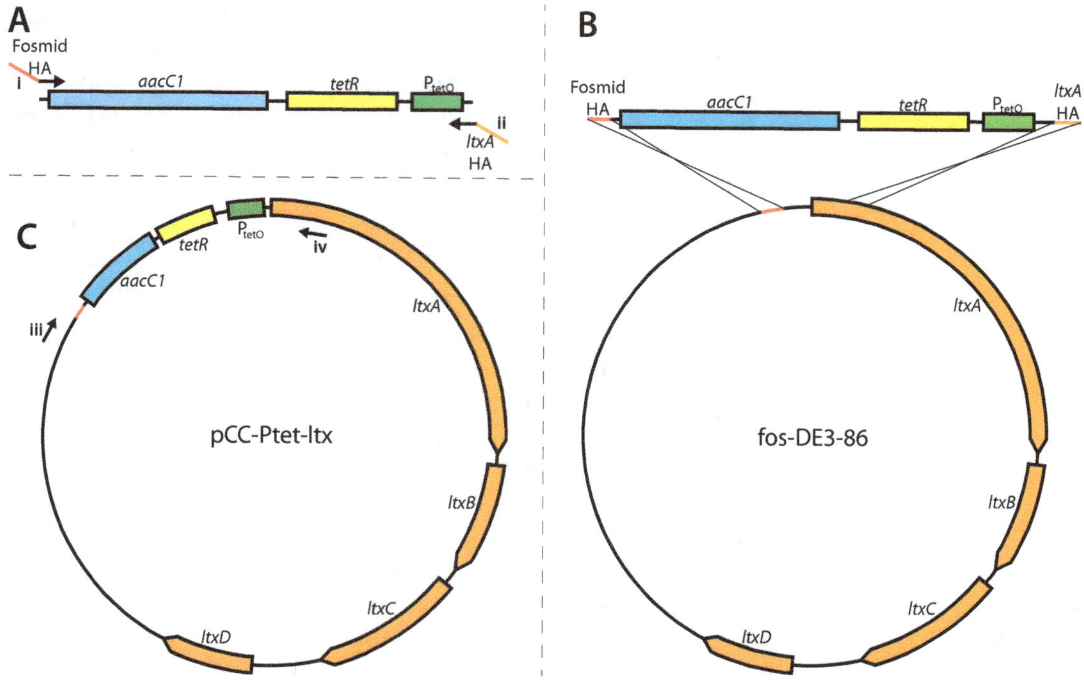

Fig. 1 Schematic of promoter exchange for the *ltx* cluster. (**a**) PCR amplification of promoter cassette and marker. Resistance marker (*aacC1*), tetracycline repressor (*tetR*), and tetracycline inducible promoter (P_{tetO}) are shown. Homology arms (HA) for the fosmid backbone (i) and first gene of the *ltx* cluster (ii) are color-coded. (**b**) Recombination of promoter resistance cassette into the *ltx* containing fosmid (fos-DE3-86). Color coding is as for (**a**). Sites of homologous recombination are indicated by crossed lines. (**c**) *ltx* fosmid following promoter exchange (pCC-Ptet-ltx). Color coding is as per previous panels. The position of potential screening primers is indicated (iii and iv)

5. Weigh the tube again to obtain the band mass and then purify the PCR product out of the agarose using a DNA gel extraction kit (*see* **Note 8**). Elute the sample twice with 6 μL MilliQ H_2O that has been preheated to 50 °C.
6. Measure the concentration of purified product using a spectrophotometer.

3.3 Recombineering of the Promoter Cassette into the BGC Vector

The following steps will result in the in vivo recombineering of the amplified *aacC1*-*tetR*-P_{tetO} cassette (with homology arms) [18] with the fosmid fos-DE3-86 [22], as shown in Fig. 1b. The resulting vector will be called pCC-Ptet-ltx [18] and will allow for the *ltx* gene cluster to be expressed in *E. coli* under the control of an inducible promoter (Fig. 1c). This method should be performed under sterile conditions.

1. Pick a single colony from the transformation above (*E. coli* GB05-red fos-DE3-86) and inoculate 1 mL LB liquid media (containing 30 μg/mL chloramphenicol) in a 1.5 mL

tube with a hole pierced in the lid. Grow transformants overnight in a thermomixer at 37 °C, shaking at 950 rpm.

2. Add 40 μL of the overnight culture to two 1.5 mL tubes (with a hole in the lid) containing 1.4 mL of LB liquid media (containing 30 μg/mL chloramphenicol). Grow the subculture for 1.5 h at 37 °C, rotating at 950 rpm.
3. Take one of the tubes and induce recombinases in GB05-red by adding 40 μL of 10% (w/v) L-arabinose (*see* **Note 1**). The other tube will be the uninduced control.
4. Place the tubes back into thermomixer and continue to grow for 40–45 min at 37 °C, 950 rpm.
5. Spin cultures down at 9000 × *g*, for 1 min at 4 °C and discard supernatant.
6. Resuspend cell pellets in 1 mL ice cold MilliQ water.
7. Repeat **steps 3** and **4** twice more and remove the supernatant after the last step with a pipette such that the cell pellet remains in between 30 and 50 μL of MilliQ water.
8. Resuspend the cells in the remaining liquid and add between 100 and 500 ng of purified promoter resistance cassette with homology arms (*aacC1-tetR*-P_{tetO} cassette) to the cell mixture (both with the recombinases induced and uninduced). Incubate on ice for 20 min.
9. Transfer the cell-DNA mixtures into precooled 1 mm electroporation cuvettes. Gently tap the cuvette on the bench a few times to ensure the mixture is at the bottom of the well and there are no bubbles. Wipe the sides of the cuvette to remove any condensation and place the cuvette into an electroporator.
10. Electroporate the cells at 1.8 kV and then immediately add 1 mL LB liquid media (**without** any antibiotic) to the cuvette. Again, take note of the electroporation time constant, which is ideally 5 ms.
11. Transfer the cells and media from the electroporation cuvette into the 1.5 mL microcentrifuge tube from **step 6** and place the tubes into the thermomixer for recovery at 37 °C, 950 rpm for 75 min.
12. Plate 50–100 μL of recovered cells onto LB agar plates containing 15 μg/mL chloramphenicol (for the selection of the fosmid) and 2 μg/mL gentamycin (for the selection of the resistance antibiotic cassette) using a sterile spreader. *See* **Note 9**.
13. Spin the remaining cells down at 9000 × *g* for 1 min and tip out the supernatant.

14. Resuspend cells in the remaining media and plate entire mixture onto LB agar plates (15 μg/mL chloramphenicol, 4 μg/mL gentamycin) using a sterile spreader (*see* **Note 9**).
15. Place all plates into 37 °C incubator overnight.
16. The ideal result is for the number of colonies on the plates that had the recombinases induced in GB05-red to be far greater than the number of colonies on the plates where the recombinases were uninduced. To improve recombination efficiency *see* **Note 10**.

3.4 Screening

To check if recombination of the promoter cassette into the BGC vector was successful, extraction of the constructs from colonies and screening is required via restriction enzyme digest or PCR. Though restriction digest was used to verify the cloning of pCC-Ptet-ltx [18], our lab will now preferentially PCR screen first due the ease of result interpretation and accuracy of the method. The location of where primers (~20 bp) for this screening would be designed to bind to, is indicated in Fig. 1c.

1. Pick >10 single colonies from the induced recombination above (*E. coli* GB05-red pCC-Ptet-ltx) and inoculate each into separate 1.5 mL tubes (with a hole in the lid) with 1 ml LB liquid media (15 μg/mL chloramphenicol, 4 μg/mL gentamycin). Grow recombinants overnight in a thermomixer at 37 °C, shaking at 950 rpm (*see* **Note 11**).
2. Harvest cells from the media by spinning at 9000 × *g* for 1 min.
3. Discard the supernatant and proceed with fosmid extraction (pCC-Ptet-ltx) from the cells using a plasmid miniprep kit.
4. Quantify the concentration of plasmid minipreps using a spectrophotometer.
5. Combine the following in 0.2 mL PCR tubes on ice with one tube per miniprepped plasmid: 13 μL of MilliQ H_2O, 2 μL 10× polymerase buffer, 1.5 μL 10 mM dNTPs, 1 μL 10 μM forward primer (specific to the site of insertion), 1 μL 10 μM reverse primer (specific to either site of insertion or the inserted construct), 1 μL 50 mM $MgCl_2$, 0.2 μL *Taq* DNA polymerase and 10–50 ng/μL of each miniprepped plasmid as the template.
6. Place the tubes into a thermocycler and subject the tubes to PCR with a 2 min 98 °C initial denaturation, 30 cycles of 10 s at 98 °C, 10 s at an appropriate annealing temperature, and 1 min 30 s at 72 °C (*see* **Note 6**). Lastly, the tubes are incubated at 72 °C for 10 min for the final elongation and then held at 4 °C until ready.
7. Mix 1 μL of 6× loading dye with 5 μL of each PCR product, mix well and then load the product into a 1% agarose gel

(containing GelRed® nucleic acid stain to 1×) in an electrophoresis tank filled with TAE buffer. Run the gel at 80 V for 20 min (*see* **Note 7**).

8. View the gel using UV light in a gel documentation system. Successfully recombined constructs will be identified by having a PCR product of the correct product size.
9. At this step, one may directly continue onto Subheading 3.5 but our group will preferentially retransform the positive minipreps and screen them again (using the protocol in Subheading 3.1 and then Subheading 3.4). The host cell may harbor multiple versions of a plasmid (modified and unmodified) and retransformation will further select for and isolate the modified plasmid from the original.
10. To further confirm the promoter exchange, the plasmid may be digested with a restriction enzyme or sent for sequencing.

3.5 Fermentation

Fermentation of the BGC construct in a strain of *E. coli* containing an appropriate PPTase is required for the production of desired metabolites. In the case of this example, the plasmid pCC-Ptet-ltx will be expressed in the *E. coli* strain GB05-MtaA [44], which has been engineered to contain the broad range *Myxococcus* PPTase, MtaA [18, 46]. For more information on appropriate *E. coli* hosts that may be used for the expression of cyanobacterial BGCs *see* **Note 12**. This method should be performed under sterile conditions.

1. The promoter exchanged plasmids need to be transformed into an expression strain of *E. coli*. In this example the construct (pCC-Ptet-ltx) will be transformed into *E. coli* GB05-MtaA using the methods in Subheading 3.1.
2. Grow transformants in 1 mL LB liquid media (with 15 μg/mL chloramphenicol, 4 μg/mL gentamycin) in a 1.5 mL tube (with a hole in the lid) overnight at 37 °C, rotating at 950 rpm.
3. Use 500 μL of the overnight cultures to inoculate 50 mL of TB liquid media (with 15 μg/mL chloramphenicol, 4 μg/mL gentamycin) in a 250 mL flask. Grow cultures in a temperature controlled orbital shaker at 30 °C, 200 rpm.
4. Continually monitor the absorbance of the culture at 600 nm (OD_{600}) using a spectrophotometer until the density is 0.03–0.1 lower than the induction density (ideally OD_{600} between 0.2 and 0.6). In this example the OD_{600} was 0.17 [18].
5. Transfer the cultures to an incubator set at 18 °C and continue to grow for ~30 min until the culture reaches induction density (in this case 0.2). Induce the P_{tetO} promoter with tetracycline to a final concentration of 0.5 μg/mL.

6. Continue fermentation at 18 °C, 200 rpm for between 19 h to 5 days (in this example 5 days were used [18]). *See* **Note 13** for greater information on the choice of fermentation conditions.
7. Depending on the method of metabolite extraction, a resin such as amberlite XAD-7 or XAD-16 may be added to the culture to sequester extracellular metabolites that are expelled into the growth media. If this step is taken the resin should be added 1–2 days before the harvest of the culture.
8. Harvest cells (and the resin if used) by centrifugation at 4000 × *g* for 30 min. If a resin is used, then the supernatant can be discarded. If a resin is not used but the metabolite of interest may be extracellular then the media should be stored separately from the cell pellet.
9. Store the cells (and media) at −30 °C until extraction.
10. To extract metabolites from cell pellets, resuspend pellets in HPLC-grade methanol and analyze samples using ELISA, HPLC, or LC-MS. Specific methodology used to extract each type of specialized metabolite will vary greatly and are dependent on the chemistry of the compounds. For greater detail please *see* **Note 14**.

4 Notes

1. Ensure that the correct form of arabinose is being used—the recombinases will only be induced if L-arabinose is used. For difficult promoter exchanges, freshly dissolving the arabinose on the day of the experiment may help.
2. In two of our publications detailing the heterologous expression of cyanobacterial toxins, cloning of their large BGC was possible through the creation of gDNA fosmid libraries using the Epicentre copy control fosmid library kit [18, 24]. pCC1FOS is an appealing vector as it is able to harbor extremely large DNA inserts; however, the cloning of a cluster in its native orientation necessitates promoter exchange to facilitate expression. By using other methods of cloning to capture BGCs [35, 42], such as restriction-ligation cloning (for small BGCS) and TAR, Golden Gate or DiPaC (for larger BGCs), backbones that already include an inducible promoter (such as pET28b) may be used. In these cases, promoter exchange is not necessary, and the reader should skip to Subheading 3.5.
3. The pierced hole at the top of microcentrifuge tubes is to promote aeration of the cultures despite the small headspace above them. Though some may be concerned about the risks these holes play in contamination of the culture, this is unlikely

to be due to the speed at which the tubes are rotating and the fast growth rates of *E. coli*. Even in tubes containing media that have not been inoculated, growth of a contaminant in the first 24 h has not been observed.

4. The resistance marker is required so that the new promoter can be selected for when recombineered into the vector. If lacking access to a template with the promoter (and its regulator) and an antibiotic resistance gene, such a vector can be constructed using Gibson Assembly [47], or the entire promoter resistance cassette may be synthesized as a synthetic dsDNA fragment (such as IDT gBlocks™). Figure 1a shows a schematic of the resistance marker-promoter cassette used in this example.
5. Promoter cassettes require flanking homology in order to allow homologous recombination with the BGC vector and should be situated so that no regions of importance are deleted from the cluster (*see* Fig. 1). Homology "arms" can be included on the 5′ ends of the oligonucleotides used to amplify the cassette. A minimum of 20 bp of homology to the region of modification is required, with efficiency of recombination dramatically increasing with a greater homology arm length [44]. For "arms" longer than approximately 50 bp, it is more economical to either get the DNA commercially synthesized already flanking the promoter cassette, or to PCR amplify the homology region from the BGC and assemble with the promoter resistance sequence using overlap-extension PCR. If the promoter resistance cassette has been synthesized with homology arms, then skip to Subheading 3.3.
6. Denaturation temperatures depend on the template complexity and specifications of the polymerase itself. The annealing temperature is calculated to be 5 °C less than the lower melting temperature (T_m) of the binding region (the primer less the homology arm) of the primer. The length of the extension step within the cycling is determined by the predicted size of the PCR product assuming the polymerase amplifies at a rate of 1 kb per 30 s. Using polymerases other than those detailed here may require an adjustment of PCR protocol. A high-fidelity polymerase is recommended for cassette amplification prior to recombineering.
7. The concentration of agarose to use is dependent on the size of the PCR product with higher concentrations used to resolve smaller products. Running time may vary depending on agarose percentage used. Other nucleic acid stains may be used, and the gel can be stained post run instead of pre stained. If using a stain that is UV fluorescent try to excise the bands as quickly as possible as prolonged UV exposure will degrade the sample and reduce cloning efficiency [48]. PCR products

should be loaded into as few wells as possible to minimize excess agarose carryover into next steps. This can be achieved by using combs with wider teeth for larger well sizes or by using tape to join multiple teeth together for a wider well.

8. The PCR products cannot be purified using methods such as ethanol precipitation or DNA clean and concentrate kits due to the size of the homology arm primers (~50 bp) being larger than the minimum size for these methods. Thus, the product is purified via a gel extraction.

9. When using antibiotics as a selection marker, one must make sure they are aware of the different classes of antibiotics and their method of resistance. If there is only one resistance marker in the BGC-containing construct when it was created, then it is best to use an antibiotic resistance gene for a completely different family as the selection marker for the promoter exchange. If more than one promoter requires exchange or if there are already many selection markers in the construct, then resistance for an antibiotic within the same family may be used as long as it is a 'stronger' antibiotic. For example, a 'weak' aminoglycoside resistance gene such as for spectinomycin may be cloned into the construct first whereas a 'strong' aminoglycoside resistance gene such for gentamycin may be cloned in later [49]. If more than one resistance marker is being used, then the concentration of each antibiotic used can be lowered up to 50% to reduce stress on the cells while still maintaining selection. This is why in the described method, the concentration of chloramphenicol used was reduced from 30 to 15 μg/mL after promoter exchange. If unsure of which concentration to use when combining two antibiotics, it is best to plate cells on several agar plates with differing antibiotic concentrations.

10. While the recombineering method detailed above is generally adequate, more difficult promoter exchanges can benefit from the slightly altered protocol detailed below.

    (a) Grow an overnight culture of GB05 red containing the cluster of interest, as above.

    (b) Inoculate 50 mL of LB in a 250 mL flask with 500 μL of overnight culture. Selection pressure (in this case chloramphenicol) and 1 mL of freshly made 10% arabinose are added at the time of inoculation. Arabinose is not added to the uninduced control.

    (c) Grow the cultures to an OD_{600} of between 0.4 and 0.6.

    (d) Chill on ice for 10 min before transferring to a chilled 50 mL conical tube and centrifuging at 2900 × *g*, at 4 °C for 10 min.

(e) Remove the supernatant and resuspend the cells in 35 mL of ice-cold milli-Q water.

(f) Repeat the centrifugation and wash steps twice, reducing the volume of water to 25 mL.

(g) Resuspend the cell pellet in 600 μL of ice cold milli-Q water. You will use 60 μL of resuspension per electroporation.

(h) Add DNA to the cell resuspension in a microfuge tube and incubate on ice for 20 min. Transfer to an electroporation cuvette and proceed as per **step 9** of Subheading 3.3.

11. Some constructs may be PCR screened directly from the transformant colonies without growing and prepping plasmids, though there is a risk of contamination from untransformed promoter cassette on the plate. Recombinants may also be screened through restriction digest if appropriate cut sites are present.

12. When considering which *E. coli* host to use it is important to keep in mind the requirements of the BGC cluster itself. If the cluster encodes for a PKS, NRPS, or hybrid of the two, then the cluster can only be expressed in a host that contains an PPTase for post-translational modification. Published *E. coli* strains that contain a PPTase include BAP1 [32] and GB05-MtaA [44]. For other types of BGCs such as RiPPs and alkaloids, protein expression strains such as BL21 (DE3) [50] are commonly used. If the cloned BGC contains any rare codons for *E. coli*, then the plasmid pRARE (Novagen)—encoding for rare tRNAs—may be cotransformed with the cluster to improve expression yields.

13. While the heterologous expression of proteins can occur at higher temperatures, our lab has found that the expression of cyanobacterial natural products is inhibited at physiological temperatures (between 30 and 37 °C) but will occur at low temperatures such as 18 °C [18, 24]. Due to this, the expression of all published cyanobacterial BGCs in our lab were done at 18 °C [18, 24, 36, 37]. The long fermentation periods (a few days instead of overnight) also differentiates the expression of (cyanobacterial) natural products from heterologous protein fermentations. The concentration of natural products in the culture accumulates over time and therefore higher yields of compound can be achieved [51]. Thus, a nutrient rich media such as TB must be used to sustain the growth of the host strain for the entirety of the fermentation period. If a different media is to be used, the duration of fermentation will need to be modified based on nutrient availability and waste accumulation.

14. The method used to extract a metabolite is dependent on its chemistry. Many of the cyanobacterial metabolites our lab has extracted have been very polar compounds so an extraction using methanol is ideal [18, 24, 36, 37]. In some cases, the extract is dried down and is ready to analyze by HPLC after resuspension whereas in other cases extended lysis is required or clean up using a solid-phase extraction (SPE) column. Other solvents that may be used to extract cyanobacterial metabolites include ethyl acetate, chloroform, and acetic acid [52].

References

1. Gerwick WH, Moore BS (2012) Lessons from the past and charting the future of marine natural products drug discovery and chemical biology. Chem Biol 19(1):85–98
2. Mandal S, Rath J (2015) Secondary metabolites of cyanobacteria and drug development. In: Extremophilic cyanobacteria for novel drug development. Springer, New York, pp 23–43
3. Bownik A (2010) Harmful algae: effects of alkaloid cyanotoxins on animal and human health. Toxin Rev 29(3–4):99–114. https://doi.org/10.3109/15569543.2010.516464
4. Dittmann E, Gugger M, Sivonen K, Fewer DP (2015) Natural product biosynthetic diversity and comparative genomics of the cyanobacteria. Trends Microbiol 23(10):642–652
5. Xu Y, Alvey RM, Byrne PO, Graham JE, Shen G, Bryant DA (2011) Expression of genes in cyanobacteria: adaptation of endogenous plasmids as platforms for high-level gene expression in *Synechococcus* sp. PCC 7002. In: Photosynthesis research protocols. Springer, New York, pp 273–293
6. Yu J, Liberton M, Cliften PF, Head RD, Jacobs JM, Smith RD, Koppenaal DW, Brand JJ, Pakrasi HB (2015) *Synechococcus elongatus* UTEX 2973, a fast growing cyanobacterial chassis for biosynthesis using light and CO2. Sci Rep 5: 8132
7. Varman AM, Xiao Y, Pakrasi HB, Tang YJ (2013) Metabolic engineering of *Synechocystis* sp. strain PCC 6803 for isobutanol production. Appl Environ Microbiol 79(3):908–914
8. Sun T, Li S, Song X, Diao J, Chen L, Zhang W (2018) Toolboxes for cyanobacteria: recent advances and future direction. Biotechnol Adv 36(4):1293–1307. https://doi.org/10.1016/j.biotechadv.2018.04.007
9. Welker M, Dittmann E, von Döhren H (2012) Cyanobacteria as a source of natural products. Methods Enzymol 517:23–46
10. Shih PM, Wu D, Latifi A, Axen SD, Fewer DP, Talla E, Calteau A, Cai F, De Marsac NT, Rippka R (2013) Improving the coverage of the cyanobacterial phylum using diversity-driven genome sequencing. Proc Natl Acad Sci 110(3):1053–1058
11. Albertano P, Urzì C (1999) Structural interactions among epilithic cyanobacteria and heterotrophic microorganisms in Roman hypogea. Microb Ecol 38(3):244–252
12. Hu B, Yang G, Zhao W, Zhang Y, Zhao J (2007) MreB is important for cell shape but not for chromosome segregation of the filamentous cyanobacterium *Anabaena* sp. PCC 7120. Mol Microbiol 63(6):1640–1652
13. Engene N, Rottacker EC, Kaštovský J, Byrum T, Choi H, Ellisman MH, Komárek J, Gerwick WH (2012) *Moorea producens* gen. nov., sp. nov. and *Moorea bouillonii* comb. nov., tropical marine cyanobacteria rich in bioactive secondary metabolites. Int J Syst Evol Microbiol 62(5):1171–1178
14. Yin Q, Carmichael WW, Evans WR (1997) Factors influencing growth and toxin production by cultures of the freshwater cyanobacterium *Lyngbya wollei* Farlow ex Gomont. J Appl Phycol 9(1):55–63
15. Burja AM, Abou-Mansour E, Banaigs B, Payri C, Burgess JG, Wright PC (2002) Culture of the marine cyanobacterium, *Lyngbya majuscula* (Oscillatoriaceae), for bioprocess intensified production of cyclic and linear lipopeptides. J Microbiol Methods 48(2): 207–219. https://doi.org/10.1016/S0167-7012(01)00324-4
16. Volk R-B (2005) Screening of microalgal culture media for the presence of algicidal compounds and isolation and identification of two bioactive metabolites, excreted by the cyanobacteria *Nostoc insulare* and *Nodularia harveyana*. J Appl Phycol 17(4):339–347
17. Donia MS, Hathaway BJ, Sudek S, Haygood MG, Rosovitz MJ, Ravel J, Schmidt EW

(2006) Natural combinatorial peptide libraries in cyanobacterial symbionts of marine ascidians. Nat Chem Biol 2(12):729–735. https://doi.org/10.1038/nchembio829

18. Ongley SE, Bian X, Zhang Y, Chau R, Gerwick WH, Müller R, Neilan BA (2013) High titer heterologous production of Lyngbyatoxin in *E. coli*, a protein kinase C activator from an uncultured marine Cyanobacterium. ACS Chem Biol 8(9):1888
19. Schorn MA, Jordan PA, Podell S, Blanton JM, Agarwal V, Biggs JS, Allen EE, Moore BS (2019) Comparative genomics of cyanobacterial symbionts reveals distinct, specialized metabolism in tropical Dysideidae sponges. MBio 10(3):e00821
20. Tillett D, Dittmann E, Erhard M, von Döhren H, Börner T, Neilan BA (2000) Structural organization of microcystin biosynthesis in *Microcystis aeruginosa* PCC7806: an integrated peptide–polyketide synthetase system. Chem Biol 7(10):753–764
21. Chang Z, Flatt P, Gerwick WH, Nguyen V-A, Willis CL, Sherman DH (2002) The barbamide biosynthetic gene cluster: a novel marine cyanobacterial system of mixed polyketide synthase (PKS)-nonribosomal peptide synthetase (NRPS) origin involving an unusual trichloroleucyl starter unit. Gene 296(1): 235–247
22. Edwards DJ, Gerwick WH (2004) Lyngbyatoxin biosynthesis: sequence of biosynthetic gene cluster and identification of a novel aromatic prenyltransferase. J Am Chem Soc 126(37):11432–11433
23. Schmidt EW, Nelson JT, Rasko DA, Sudek S, Eisen JA, Haygood MG, Ravel J (2005) Patellamide A and C biosynthesis by a microcin-like pathway in *Prochloron didemni*, the cyanobacterial symbiont of *Lissoclinum patella*. Proc Natl Acad Sci 102(20):7315–7320
24. Liu T, Mazmouz R, Ongley SE, Chau R, Pickford R, Woodhouse JN, Neilan BA (2017) Directing the heterologous production of specific cyanobacterial toxin variants. ACS Chem Biol 12:2021
25. Greunke C, Duell ER, D'Agostino PM, Glöckle A, Lamm K, Gulder TAM (2018) Direct pathway cloning (DiPaC) to unlock natural product biosynthetic potential. Metab Eng 47:334–345. https://doi.org/10.1016/j.ymben.2018.03.010
26. D'Agostino PM, Gulder TA (2018) Direct pathway cloning combined with sequence-and ligation-independent cloning for fast biosynthetic gene cluster refactoring and heterologous expression. ACS Synth Biol 7(7): 1702–1708
27. Taton A, Ecker A, Diaz B, Moss NA, Anderson B, Reher R, Leão TF, Simkovsky R, Dorrestein PC, Gerwick L (2020) Heterologous expression of cryptomaldamide in a cyanobacterial host. ACS Synth Biol 9(12): 3364–3376
28. Copp J, Roberts A, Marahiel M, Neilan B (2007) Characterization of PPTNs, a cyanobacterial phosphopantetheinyl transferase from *Nodularia spumigena* NSOR10. J Bacteriol 189(8):3133–3139
29. Beld J, Sonnenschein EC, Vickery CR, Noel JP, Burkart MD (2014) The phosphopantetheinyl transferases: catalysis of a posttranslational modification crucial for life. Nat Prod Rep 31(1):61–108. https://doi.org/10.1039/c3np70054b
30. Cox RJ, Crosby J, Daltrop O, Glod F, Jarzabek ME, Nicholson TP, Reed M, Simpson TJ, Smith LH, Soulas F (2002) *Streptomyces coelicolor* phosphopantetheinyl transferase: a promiscuous activator of polyketide and fatty acid synthase acyl carrier proteins. J Chem Soc Perkin Trans 1(14):1644–1649
31. Quadri LE, Weinreb PH, Lei M, Nakano MM, Zuber P, Walsh CT (1998) Characterization of Sfp, a *Bacillus subtilis* phosphopantetheinyl transferase for peptidyl carrier protein domains in peptide synthetases. Biochemistry 37(6): 1585–1595
32. Pfeifer BA, Admiraal SJ, Gramajo H, Cane DE, Khosla C (2001) Biosynthesis of complex polyketides in a metabolically engineered strain of *E. coli*. Science 291(5509):1790–1792
33. Jones AC, Ottilie S, Eustáquio AS, Edwards DJ, Gerwick L, Moore BS, Gerwick WH (2012) Evaluation of *Streptomyces coelicolor* A3 (2) as a heterologous expression host for the cyanobacterial protein kinase C activator lyngbyatoxin A. FEBS J 279(7):1243–1251
34. Kim EJ, Lee JH, Choi H, Pereira AR, Ban YH, Yoo YJ, Kim E, Park JW, Sherman DH, Gerwick WH, Yoon YJ (2012) Heterologous production of 4-O-demethylbarbamide, a marine cyanobacterial natural product. Org Lett 14(23):5824–5827. https://doi.org/10.1021/ol302575h
35. Zhang JJ, Tang X, Moore BS (2019) Genetic platforms for heterologous expression of microbial natural products. Nat Prod Rep 36(9):1313–1332
36. Katoch M, Mazmouz R, Chau R, Pearson LA, Pickford R, Neilan BA (2016) Heterologous production of cyanobacterial mycosporine-like amino acids mycosporine-ornithine and mycosporine-lysine in *Escherichia coli*. Appl Environ Microbiol 82(20):6167–6173

37. D'Agostino PM, Javalkote VS, Mazmouz R, Pickford R, Puranik PR, Neilan BA (2016) Comparative profiling and discovery of novel glycosylated mycosporine-like amino acids in two strains of the cyanobacterium *Scytonema* cf. *crispum*. Appl Environ Microbiol 82(19): 5951–5959
38. Ruffner DE, Schmidt EW, Heemstra JR (2015) Assessing the combinatorial potential of the RiPP Cyanobactin tru pathway. ACS Synth Biol 4(4):482–492. https://doi.org/10.1021/sb500267d
39. Yang G, Cozad MA, Holland DA, Zhang Y, Luesch H, Ding Y (2018) Photosynthetic production of sunscreen shinorine using an engineered cyanobacterium. ACS Synth Biol 7(2): 664–671
40. Videau P, Wells KN, Singh AJ, Eiting J, Proteau PJ, Philmus B (2019) Expanding the natural products heterologous expression repertoire in the model Cyanobacterium *Anabaena* sp. strain PCC 7120: production of pendolmycin and teleocidin B-4. ACS Synth Biol 9(1):63–75
41. Ongley SE, Bian X, Neilan BA, Müller R (2013) Recent advances in the heterologous expression of microbial natural product biosynthetic pathways. Nat Prod Rep 30(8): 1121–1138
42. Huo L, Hug JJ, Fu C, Bian X, Zhang Y, Müller R (2019) Heterologous expression of bacterial natural product biosynthetic pathways. Nat Prod Rep 36(10):1412–1436
43. Fu J, Teucher M, Anastassiadis K, Skarnes W, Stewart AF (2010) Chapter eight - a Recombineering pipeline to make conditional targeting constructs. In: Wassarman PM, Soriano PM (eds) Methods in enzymology, vol 477. Academic Press, Boston, pp 125–144. https://doi.org/10.1016/S0076-6879(10)77008-7
44. Fu J, Bian X, Hu S, Wang H, Huang F, Seibert PM, Plaza A, Xia L, Müller R, Stewart AF, Zhang Y (2012) Full-length RecE enhances linear-linear homologous recombination and facilitates direct cloning for bioprospecting. Nat Biotechnol 30(5):440–446. https://doi.org/10.1038/nbt.2183
45. Bian X, Huang F, Stewart FA, Xia L, Zhang Y, Müller R (2012) Direct cloning, genetic engineering, and heterologous expression of the syringolin biosynthetic gene cluster in *E. coli* through Red/ET recombineering. Chembiochem 13(13):1946–1952. https://doi.org/10.1002/cbic.201200310
46. Gaitatzis N, Hans A, Müller R, Beyer S (2001) The *mtaA* gene of the myxothiazol biosynthetic gene cluster from *Stigmatella aurantiaca* DW4/3-1 encodes a phosphopantetheinyl transferase that activates polyketide synthases and polypeptide synthetases. J Biochem 129(1):119–124
47. Gibson DG, Young L, Chuang R-Y, Venter JC, Hutchison CA III, Smith HO (2009) Enzymatic assembly of DNA molecules up to several hundred kilobases. Nat Methods 6(5):343
48. Sharafi Z, Adrangi S (2019) Replacing UV with blue light during DNA purification increases the efficiency of ligation-independent cloning. Trends Pept Protein Sci 3(e1):1–8
49. Mahbub Alam M, Kobayashi N, Ishino M, Sumi A, Kobayashi K-I, Uehara N, Watanabe N (2005) Detection of a novel *aph(2″)* allele *(aph[2″]-Ie)* conferring high-level gentamicin resistance and a spectinomycin resistance gene *ant(9)-Ia* (*aad9*) in clinical isolates of enterococci. Microb Drug Resist 11(3):239–247
50. Studier FW, Moffatt BA (1986) Use of bacteriophage T7 RNA polymerase to direct selective high-level expression of cloned genes. J Mol Biol 189(1):113–130. https://doi.org/10.1016/0022-2836(86)90385-2
51. Liu T, Mazmouz R, Pearson LA, Neilan BA (2019) Mutagenesis of the microcystin tailoring and transport proteins in a heterologous cyanotoxin expression system. ACS Synth Biol 8(5):1187–1194. https://doi.org/10.1021/acssynbio.9b00068
52. Haque F, Banayan S, Yee J, Chiang YW (2017) Extraction and applications of cyanotoxins and other cyanobacterial secondary metabolites. Chemosphere 183:164–175. https://doi.org/10.1016/j.chemosphere.2017.05.106

Chapter 18

Saccharomyces cerevisiae as a Heterologous Host for Natural Products

Maximilian Otto, Dany Liu, and Verena Siewers

Abstract

Cell factories can provide a sustainable supply of natural products with applications as pharmaceuticals, food-additives or biofuels. Besides being an important model organism for eukaryotic systems, *Saccharomyces cerevisiae* is used as a chassis for the heterologous production of natural products. Its success as a cell factory can be attributed to the vast knowledge accumulated over decades of research, its overall ease of engineering and its robustness. Many methods and toolkits have been developed by the yeast metabolic engineering community with the aim of simplifying and accelerating the engineering process.

In this chapter, a range of methodologies are highlighted, which can be used to develop novel natural product cell factories or to improve titer, rate and yields of an existing cell factory with the goal of developing an industrially relevant strain. The addressed topics are applicable for different stages of a cell factory engineering project and include the choice of a natural product platform strain, expression cassette design for heterologous or native genes, basic and advanced genetic engineering strategies, and library screening methods using biosensors. The many engineering methods available and the examples of yeast cell factories underline the importance and future potential of this host for industrial production of natural products.

Key words *Saccharomyces cerevisiae*, Yeast, Metabolic engineering, Natural products, Platform strains, Biosensors, Engineering toolkits, Expression cassettes, Cell factory engineering, Design–Build–Test–Learn Cycle

1 Introduction

Natural products, here referring mainly to plant secondary metabolites, comprise a large variety of compound classes, including alkaloids, terpenoids and phenylpropanoids. They possess a wide range of applications, for instance as flavors, pigments, and pharmaceuticals [1]. Current production of plant secondary metabolites relies primarily on extraction from natural producers. Due to the low abundance of such compounds in nature, the extraction process is often inefficient and costly. In addition, it threatens the extinction of medicinal plants [2]. Chemical synthesis on the

Elizabeth Skellam (ed.), *Engineering Natural Product Biosynthesis: Methods and Protocols*,
Methods in Molecular Biology, vol. 2489, https://doi.org/10.1007/978-1-0716-2273-5_18,

other hand can be difficult due to the structural complexity of many natural products. The use of microbial cell factories such as *E. coli* or *S. cerevisiae* thus presents a promising alternative [3]. It enables controlled fermentative production, independent of external factors such as climate conditions or geographic location. This allows for a more consistent production process. Unlike plants, which must be harvested seasonally, microorganisms can be cultured in a matter of days. The ease of genetic engineering allows for direction of flux toward desired compounds and decreased loss of intermediates to competing pathways [4]. Natural products could thus be obtained at higher productivities, yields, and titers. Heterologous synthesis also facilitates the discovery of novel synthetic bioactive derivatives [5].

S. cerevisiae, commonly known as baker's yeast, presents several advantages over other organisms. Genomic integration of heterologous genes is very efficient in baker's yeast due to the high rate of homologous recombination (HR). This is especially important for natural product biosynthesis, which often involves long, complex pathways, and therefore requires numerous genomic modifications in order to implement the heterologous pathway and to fine-tune endogenous metabolism. As a eukaryotic organism, yeast carries many organelles that are also present in plants. This allows functional expression of enzymes such as cytochrome P450 monooxygenases, which require localization to specific endomembrane systems. Additionally, many plant biosynthetic pathways are compartmentalized into different organelles. This is more easily reconstituted in eukaryotes compared to bacteria [5]. From an industrial point of view, yeast is considered a very robust organism—it is tolerant toward low pH and osmotic stress and furthermore resistant toward phage contaminations [6]. Many biotechnological processes using *S. cerevisiae* as a production host are designated as "GRAS" (generally recognized as safe) by the U.S. Food and Drug Administration [7]. In addition, sophisticated genome-scale metabolic models (GEMs) have been developed for *S. cerevisiae*, which can accelerate and simplify the cell factory design process [8].

Several examples have demonstrated the suitability of baker's yeast as an industrial cell factory for natural products. Paddon et al. [9] developed a semisynthetic process for high-level production of the antimalarial drug artemisinin, where the complete biosynthetic pathway of artemisinic acid was implemented in yeast, which was further processed to artemisinin by chemical conversion. This resulted in 25 g/L artemisinic acid titers in fed-batch bioreactor cultivations [9]. In 2018, Li et al. successfully reconstructed the complete biosynthesis pathway of noscapine by functionally expressing 25 heterologous genes as well as making 6 modifications to yeast's native metabolism [10]. This constitutes the longest example of a plant biosynthesis pathway implemented in a heterologous host to date and exemplifies the potential of *S. cerevisiae* to serve as a host organism for producing complex natural products.

The aim of this book chapter is to provide an overview of metabolic engineering possibilities in *S. cerevisiae*. We describe the most essential considerations for cell factory design, construction and optimization with a focus on methods that might not be feasible in other organisms due to differences in biology and lack of knowledge or genetic tools. We hope that this will help the reader to find the most promising strains and tools for developing a novel natural product cell factory.

2 Platform Strains

To produce natural products in any given host, a basic requirement is that the biosynthetic pathway can be linked to the host's native metabolism via suitable precursor molecules. As central metabolism is highly conserved among different organisms, this is generally feasible [11].

To enable natural product synthesis in heterologous hosts, platform strains with efficient production of such precursors are needed. In many cases, a single platform strain allows the synthesis of a diverse range of products [12]. Moreover, researchers aiming to produce a specific compound can build on existing platform strains, that produce high levels of an appropriate precursor. The knowledge of previously engineered high producers can also facilitate the strain development for related platforms.

Terpenoids are the largest group of plant natural products. Some well-known terpenoids include the anticancer drug paclitaxel and the antimalarial drug artemisinin. Terpenoids are further classified into monoterpenoids, sesquiterpenoids, diterpenoids, triterpenoids, and carotenoids [13]. Their precursors isopentenyl diphosphate (IPP) and dimethylallyl pyrophosphate (DMAPP) as well as geranyl diphosphate (GPP) and farnesyl diphosphate (FPP) are part of the yeast-endogenous mevalonate pathway and are derived from acetyl coenzyme A (acetyl-CoA). Terpenoids can thus be produced by upregulation of the precursor supply and introduction of the downstream heterologous pathway. Meadows et al. [14] succeeded in producing over 100 g/L of the sesquiterpene farnesene in bioreactor fermentations by introducing heterologous reactions to optimize ATP consumption, carbon loss, and redox balance for improved acetyl-CoA supply.

Alkaloids are a class of nitrogen containing secondary metabolites and can be further grouped into benzylisoquinoline alkaloids (BIAs), monoterpene indole alkaloids (MIAs), and tropane alkaloids (TAs) [15]. BIAs possess a wide range of bioactive properties. Morphine and codeine for instance are commonly used analgesics [16]. BIAs are derived from the aromatic amino acid L-tyrosine, which is the last yeast native metabolite within the BIA biosynthesis pathway. A key branchpoint toward various BIAs is (*S*)-reticuline. A high-level (*S*)-reticuline producing strain was developed by Pyne et al. [17] and provides a highly demanded platform for various downstream BIAs. De novo production of several BIAs has been successfully established in yeast [10, 18, 19].

MIAs are a group of over 2000 compounds, several of which are applied as anticancer, antimalarial, and antiarrhythmic agents [20]. They are derived from the aromatic amino acid L-tryptophan, which is further converted to tryptamine. Tryptamine undergoes a condensation reaction with secologanin to form strictosidine, which constitutes the first committed step in the MIA biosynthesis pathway. The monoterpene secologanin is a derivative of the mevalonate pathway. L-tryptophan and IPP/DMAPP thus represent the last endogenous yeast metabolites within the MIA pathway [20, 21]. The complete strictosidine synthesis pathway has been reconstructed in yeast [21].

TAs comprise around 200 tropane ring-based alkaloids, which display a wide range of pharmacological effects [22]. The last yeast-endogenous precursor involved in TA biosynthesis are the amino acids L-arginine and L-ornithine. Overproduction of L-ornithine has previously been engineered, reaching titers of over 5 g/L in fed-batch cultivations [23]. Srinivasan and Smolke [24] were able to produce tropine, a key building block for TA synthesis, as well as its noncanonical derivative cinnamoyltropine, from simple sugars.

In recent times, cannabinoids have become another interesting target for heterologous biosynthesis due to their bioactive properties and potential medical applications [25]. Cannabinoids are derived from GPP of the mevalonate pathway and olivetolic acid, which is synthesized from hexanoyl-CoA and malonyl-CoA. De novo production of cannabinoids tetrahydrocannabinolic acid (THCA) and cannabidiolic acid (CBDA) has recently been realized in yeast [26].

Phenylpropanoids such as flavonoids and stilbenoids are derived from two aromatic amino acids, L-tyrosine and L-phenylalanine. A common precursor is *p*-coumaric acid (pCA), which after ligation with coenzyme A to *p*-coumaroyl-CoA, can be further cyclized with three malonyl-CoA moieties to form flavonoids and stilbenoids. A pCA platform strain has previously been developed in *S. cerevisiae*, that enables production of 12.5 g/L pCA in fed-batch fermentations [27]. Some successful phenylpropanoid pathway reconstructions include the production of the stilbenoid resveratrol [28] and the central flavonoid naringenin [29, 30]. Another important engineering target is the supply of malonyl-CoA, which concerns not only flavonoids and stilbenoids, but also polyketides and fatty acid derivatives. A comprehensive summary of strategies for malonyl-CoA overproduction was given by Milke and Marienhagen [31].

Table 1 lists a collection of platform strains that can be used to produce alkaloids, terpenoids and phenylpropanoids.

3 Design of Expression Cassettes

While the choice of genes to be expressed in a cell factory might not be trivial, the more challenging aspect of cell factory design is the balancing of finite cellular resources between cell maintenance, cell proliferation and product formation. This is an inherently difficult task due to the tight regulation of the central carbon metabolism that the metabolic engineer aims to interfere with [33]. The metabolic flux should ideally be redirected as much as possible toward the product, while at the same time metabolic imbalances need to be avoided, for example, the accumulation of intermediates or the drainage of central metabolites resulting in growth retardation. Thanks to our increasingly better understanding of basic biological processes and the progress in the field of systems biology, cells are no longer regarded as black boxes or as mere "bags of enzymes" [8, 34]. Nonetheless, optimal allocation of cellular resources remains one of the major challenges in metabolic engineering and the design of gene expression cassettes is central to this challenge. In this section of the chapter the most critical elements of expression cassettes are discussed.

Besides the coding sequences, promoters are usually the first element that is taken into consideration when designing an expression cassette due to their crucial role in determining expression levels and expression patterns. Promoter characteristics have been studied on a genome level [35–38] and also more focused on subsets of promoters relevant for metabolic engineering applications [39–44]. Using the same promoter for multiple genes should be avoided since repeated sequences cause genome instability [45]. A caveat of promoter characterization studies is that there is no standardized way of determining promoter strength. Depending on how and under which conditions expression levels are measured (e.g., plasmid-based expression vs. genomic integration, GFP signal vs. qPCR as output, shake-flask experiments vs. chemostat), different conclusions are drawn on relative promoter strength or how constitutive a given promoter is. Here we categorize promoters into three different groups according to their expression pattern: (1) constitutive promoters, transcribed in virtually all growth conditions; (2) small-molecule inducible promoters, activated or repressed through a specific inducer molecule; (3) condition-dependent promoters, dependent on growth phase, cellular stress or carbon source. Table 2 describes a range of commonly used native promoters in yeast metabolic engineering and Table 3 highlights promoter engineering approaches and heterologous promoters.

Table 1
Collection of *S. cerevisiae* platform strains for alkaloids, terpenoids, and phenylpropanoids

Product class	Precursor(s)	Genotype description	Description	References
Terpenoids	Acetyl-CoA (farnesene as final product)	*ADA- PK- PTA- ald4Δ ald6Δ rhr2Δ NADH-HMGr-* *No detailed genotype information available*	Substitution of native PDH-bypass reactions with heterologous acetaldehyde dehydrogenase (acetylating) (EutE from *D. zeae*) to lower the ATP cost; replacement of native NADPH-dependent HMG-CoA reductase (HMGr) with NADH-specific one (from *S. pomeroyi)* to mitigate change in cofactor usage; implementation of nonoxidative glycolysis pathway (phosphoketolase PK from *L. mesenteroides* and phosphotransacetylase PTA from *C. kluyveri*) to generate acetyl-CoA from acetyl phosphate; complementation of ADA/NADH-HMGr pathway (excess ATP and reducing power) and PK/PTA pathway (deficient in ATP and reducing power)	[14]
Benzylisoquinoline alkaloids, phenylpropanoids	Tyrosine	*aro10Δ zwf1Δ RsTAL ARO4^{FBR} ZmTYRC*	Reduced loss of tyrosine to Ehrlich pathway (*aro10Δ*); feedback-resistant DAHP synthase (Aro4^{K229L}); feedback-resistant cyclohexadienyl dehydrogenase TYRC from *Z. mobilis*	[32]

Benzylisoquinoline alkaloids	(*S*)-Reticuline	*ari1Δ adh6Δ ypr1Δ ydr541cΔ aad3Δ gre2Δ hfd1Δ ARO4^{FBR}- ARO2- ARO7^{FBR}- TYR1- ARO10- CYP76AD5 PpDODC CjNCSΔN$_{35}$ (2x) Ps6OMT (2x) PsCNMT EcNMCH AtATR2 Ps4'OMT2(2x) CYP76AD5*	Deletion of several oxidoreductases and dehydrogenases (*ari1Δ adh6Δ ypr1Δ ydr541cΔ aad3Δ gre2Δ hfd1Δ*) to eliminate 4-hydroxyphenylacetaldehyde (4-HPAA) consumption; overexpression of genes involved in precursor (L-tyrosine) supply; feedback-resistant DAHP synthase (Aro4^{K229L}) and chorismate mutase (Aro7^{G141S})	[17]
Monoterpene indole alkaloids	Strictosidine	*tHMGR- AgGPPS2 IDI1- CrCPR CrCYB5* *CrGES CrG8H (4x) CrGOR CrISY CrIO Cr7-DLGT Cr7-DLH CrLAMT CrSLS CrSTR CrTDC CrADH2* *MAF1- GgFPS^{N144W} SAM2- ZWF1- atf1Δ oye2Δ*	High copy number of geraniol 8-hydroxylase (G8H); deletion of *ATF1* and *OYE2* to eliminate consumption of geraniol; expression of CrADH2 to assist cytochrome P450-catalyzed reactions	[21]
Tropane alkaloids	L-Ornithine	*car2Δ CAR1- ARG3⁻ EcArgA EcArgB CgArgC CgArgD CgArgJ ORT1- AGC1- GDH1- MTH1-ΔT-*	Deletion of *CAR2* and downregulation of *ARG3* to reduce ornithine degradation or consumption; engineering of mitochondrial trafficking of L-ornithine and L-glutamate; internal deletion in the hexose transporter regulation protein Mth1p to alleviate Crabtree effect and improve carbon flux to TCA cycle (α-ketoglutarate)	[23]
Tropane alkaloids	Tropine	*meu1Δ oaz1Δ hfd1Δ ald2-5Δ ARG2- CAR1- FMS1- SPE1- AsADC speB AbPMT1(2x) DsPMT1 DmMPO1$^{ΔC-PTS1}$ DsTR1 AtATR1 AbPYKS (2x) AbCYP82M3*	Deletion of *MEU1/OAZ1* to deregulate polyamine biosynthesis regulatory mechanisms; deletion of aldehyde dehydrogenases to lower 4-methylaminobutanal oxidation	[24]
Cannabinoids	Cannabigerolic acid (CBGA) tetrahydrocannabinolic acid (THCA)	*CsTKS (2x) CsOAC (2x) CsAAE1 CsPT4-T CsTHCAS (2x)* *EfMvaE EfMvaS ERG12- ERG8- ERG19-*	Implementation of heterologous hexanoyl-CoA biosynthesis pathway (ReBktB, CnPaaH1, CaCrt, TdTer); implementation of heterologous	[26]

(continued)

Table 1 (continued)

Product class	Precursor(s)	Genotype description	Description	References
		IDI1- ERG20 erg9$^-$ ReBktB CnPaaH1 CaCrt TdTer*	olivetolic acid biosynthesis pathway (CsTKS, CsOAC, CsAAE1); GPP overproduction by upregulation of native mevalonate pathway (*ERG12*, *ERG8*, *ERG19*, *IDI1*) and heterologous expression of EfMvaE, EfMvaS; use farnesyl pyrophosphate synthase mutant (Erg20$^{F96W/N127W}$) that preferentially produces GPP over FPP; introduction of CsPT4 for CBGA production; introduction of THCA synthase for THCA production	
Phenylpropanoids	*p*-Coumaric acid	*gpp1Δ pdc5Δ aro10Δ gal80Δ ARO1- ARO2- ARO3- ARO4^{FBR}- ARO7^{FBR}- EcaroL PHA2- CYB5- AtPAL2 AtC4H AtATR2 FjTAL MtPDH1 Bbxfpk Ckpta PFK1$^-$ PFK2$^-$ PYK1$^-$*	Implementation of heterologous phosphoketolase (PHK) pathway (*Bbxfpk, Ckpta*) to divert carbon flux from glycolysis to erythrose-4-phosphate (E4P); feedback-resistant DAHP synthase (Aro4^{K229L}) and chorismate mutase (Aro7^{G141S}); deletion of *GPP1* to reduce acetate accumulation due to acetyl phosphate degradation from PHK pathway implementation; dynamic regulation of TAL/PAL branch and PHK pathway using p*GAL*-controlled expression; downregulation of *PFK1/2*, *PYK1* to increase flux through PHK pathway; deletion of *ARO10* and *PDC5* to reduce aromatic acid degradation	[27]

Downstream flavonoids	Naringenin	*FjTAL At4CL AtCHS AtCHI* $ARO4^{FBR}$- $ARO7^{FBR}$- *aro10Δ pdc1Δ pdc5Δ pdc6Δ tsc13Δ::MdECR EcLpdA EcAceE EcAceF ACC1- ERG9⁻*	Feedback-resistant DAHP synthase ($Aro4^{K229L}$) and chorismate mutase ($Aro7^{G141S}$); deletion of *ARO10* and *PDC5* to reduce aromatic acid degradation; complementation of endogenous enoyl reductase (*TSC13*) with heterologous MdECR to eliminate phloretic acid accumulation; deletion of *PDC1/6* to reduce pyruvate consumption (to acetaldehyde); upregulation of malonyl-CoA supply by introduction of heterologous PDH complex (LpdA, AceE, AceF from *E. coli*); downregulation of *ERG9* to reduce ergosterol content in cell membrane and increase membrane permeability for naringenin flux out of the cell	[29]

Table 2
Selection of native *S. cerevisiae* promoters commonly used in metabolic engineering

Group	Description	Promoter	References
Constitutive	Strong promoters	p*TEF1*, p*PGK1*, p*TDH3*, p*CCW12*, p*ENO2*	[40, 43, 44]
	Medium promoters	p*RPL18B*, p*HHF1*, p*ALD6*	[40]
	Weak promoters	p*REV1*, p*CYC1*, p*CIT2*, p*ACT1*	[40, 43]
Inducible	Induction by galactose	p*GAL1*/p*GAL10* (bidirectional), p*GAL3* and p*GAL7*	[9, 46–48]
	Repression by inorganic phosphate	p*PHO5*	[49]
	Induction by Cu(II)	p*CUP1*	[46, 50–52]
	Repression by methionine	p*MET3*, p*MET25*	[48, 53–56]
	Repression by thiamine	p*THI4*	[57]
Condition-dependent	Active predominantly in glucose phase	p*ADH1* (full length), p*HXT1*, p*TPO2*	[36, 41, 58–60]
	Active predominantly in ethanol phase	p*ADH2*, p*HXT7*, p*HXK1*, p*ICL1*	[36, 41, 61–63]
	Active predominantly in stationary phase	p*HSP26*, p*HSP104*, p*SSA1*	[64]
	Active under anaerobic conditions	p*DAN1*	[65, 66]

Strong, constitutive promoters like p*TEF1*, p*PGK1*, and p*TDH3* are often a first choice to drive heterologous gene expression [39, 44, 85, 86]. For one, this brute-force approach helps to overcome the tight regulation of the central carbon metabolism and can redirect the flux toward the product. In addition, secondary metabolite pathway enzymes mostly have lower catalytic efficiencies compared to enzymes involved in the central carbon metabolism possibly due to lower selection pressure and physiochemical constraints [87], subsequently requiring high, constant expression to overcome this limitation. Furthermore, heterologous enzymes might not work as efficiently as in the native host, for example, due to physiochemical differences in the cellular milieu. Medium or weak promoters are beneficial for potentially toxic enzymes or to fine-tune pathways containing enzymes with widely different catalytic efficiencies, thereby avoiding accumulation of pathway intermediates. It is worth noting that there is no precise definition of "constitutive" and every promoter changes expression strength in different growth phases and culture conditions [38, 41, 43]. Interestingly, Keren and coworkers [38] found that 60–80% of *S. cerevisiae* and *E. coli* promoters appear to be constitutive in the sense that their expression levels in different conditions depend on a global scaling factor instead of on specific regulation.

An often used strategy in metabolic engineering is to divide a process into growth phase and production phase. In the growth

Table 3
Selection of heterologous promoters, promoter engineering and transcription factor engineering approaches in *S. cerevisiae*

Description	Features	References
Synthetic hybrid promoters	Library of constitutive and galactose-inducible promoters by combining enhancer elements and core promoters; mRNA levels increased by ≈2.5-fold compared to wt p*TDH3*	[67]
Synthetic hybrid promoters	Library of constitutive or diauxic shift induced promoters by combining upstream activating sequence and core promoter elements; expression increased ≈2-fold compared to p*TEF1*	[68]
Library of p*TEF1* promoter	Error-prone PCR used to construct library of p*TEF1* variants; expression levels ranging from 8 to 120% compared to wt p*TEF1*	[69, 70]
Engineering based on nucleosome architecture predictions	Redesign to increase expression of native promoters and design of purely synthetic promoter	[71]
Chimeric intron promoters	Introns in the 5′ UTR or directly after initiation codon increase gene expression; p*TDH3* expression levels increased ≈50-fold for some proteins; also enhanced expression for modified p*GAL10* and p*CUP1*	[72]
Minimal promoters	Engineering of constitutive and inducible minimal promoters up to 80% reduced in length	[73, 74]
aTc-inducible and IPTG-inducible promoters	Based on bacterial repressor TetR and its operator sequence	[75]
Dox-repressible promoter	Using synthetic TetR-VP16 activator; comparable induction to p*GAL1*	[76]
Camphor repressible promoter	Using CamR repressor from *Pseudomonas putida*	[77]
Heterologous *GAL* promoters	*GAL* promoters from other *Saccharomyces* species characterized in *S. cerevisiae*	[78]
Gal4 protein engineering	Hyperactive $Gal4^{T406A/V413A}$ mutant for increased expression from p*GAL* promoters	[79]
p*ADH1* engineering	Shift expression pattern by reducing promoter length; truncated p*ADH1* has more constitutive behavior	[80, 81]
p*DAN1* engineering	p*DAN1* mutants with higher expression levels and induction under less stringent anaerobiosis	[82]
Synthetic stress-responsive promoters	Promoters induced under acidic conditions; engineering strategy based on modifying TF-binding sites of p*YGP1*	[83]
Plant-derived synthetic TFs	Large library of orthogonal synthetic promoters; library based on combination of plant TFs, activation domains and synthetic promoters; up to ten-fold higher expression than p*TDH3*	[84]

(continued)

Table 3
(continued)

Description	Features	References
Bacteria-derived synthetic TFs	Toolkit containing synthetic activators and synthetic repressors, both based on bacterial repressors; toolkit also contains synthetic TF-dependant promoter library with varying expression range; more than three-fold higher expression than p*TDH3*	[57]

aTc anhydrotetracycline, *IPTG* isopropyl β-D-1-thiogalactopyranoside, *Dox* doxycycline

phase the expression of natural product pathway enzymes is kept to a minimum, cellular resources are utilized for cell maintenance and cell proliferation. After a sufficient cell density is reached, the process is switched to the production phase, in which cellular resources are allocated to cell maintenance and product formation. The goal is to achieve maximum production capacities at high cell densities and avoid growth retardation due to sequestration of resources or (by-)product toxicity. To establish a two-phase process either inducible promoters or condition-dependent promoters are utilized.

The most prominent example of endogenous inducible promoters are the galactose-responsive p*GAL1*/p*GAL10* (bidirectional), p*GAL3*, and p*GAL7*, which are strongly repressed in glucose and induced up to 1000-fold in galactose [86, 88, 89]. They have been extensively studied [89–91] and are widely used in metabolic engineering. Examples include natural product cell factories producing artemisinic acid [9, 92] and cannabinoids [26]. Deletion or regulation of the *GAL2* gene was shown to change the normally ON/OFF behavior of *pGAL1* to a galactose concentration-dependent behavior [93] which enables fine-tuning of expression levels. *See* Tables 2 and 3 for more inducible promoters.

The described small-molecule inducible promoters enable precise and often dose-dependent control of expression via human intervention. However, the use of even simple molecules like galactose in large-scale fermentation processes can result in substantial additional costs [94]. Condition-dependent promoters allow the separation of growth and production phase without requiring this external stimulation. *S. cerevisiae* has a diauxic growth, first consuming glucose and producing ethanol (glucose phase), followed by consuming ethanol as energy and carbon source after the diauxic shift (ethanol phase). This biology led to promoters that are up- or down-regulated depending on glucose concentration in the culture. These promoters can be exploited to favour cell proliferation during the glucose phase and favour product formation in the ethanol phase. For example, by using p*HXT1* for regulating the expression of *ERG9*, an essential enzyme in the ergosterol biosynthetic pathway, carbon flux was redirected toward α-santalene production during cultivation in low glucose concentration. The expression patterns of condition-dependent promoters can be very dependent on the experimental setup, for example, batch or fed-batch cultivation [41]. More condition-dependent promoters are described in Tables 2 and 3.

Promoter engineering has been undergone to shorten promoter sequences, tweak expression strength or change promoter behavior, sometimes using artificial or heterologous transcription factors (TFs) [85, 95] (Table 3). Many of these (partially) synthetic promoters have the advantage of being more orthogonal (less involvement of the native yeast regulatory network) [57, 96, 97], which should result in a more predictable and programmable behavior. In addition, compared to native promoters, synthetic promoters have shorter sequence homology to native genomic sequences where unwanted HR could occur [85]. While only a minority of *S. cerevisiae* promoters contain a TATA box [98], its sequence was shown to affect expression levels [99] and even render a promoter nonfunctional [100]. Besides the TATA box, the preinitiation complex (PIC) region and the 5′UTR directly influence expression strength [101, 102] and should undergo special scrutiny in promoter engineering approaches.

Even though terminators often do not get the same attention as promoters since their dynamic range is much smaller [103], different terminators can affect protein expression levels by multiple folds [40, 104–106]. Table 4 summarizes terminators of varying strength as well as terminator engineering strategies.

Expression cassette design can also encompass changes to the coding region of a gene of interest. This can include codon optimization [110] or a variety of strategies to improve substrate channeling such as, enzyme-fusions (fusing enzymes that catalyze sequential reactions), enzyme scaffolding (recruitment of multiple pathway enzymes onto a scaffold) and organelle targeting (localizing pathway enzymes in a separate compartment). Besides increasing the local concentration of enzymes, organelle targeting brings the additional benefit of enabling utilization of the precursor and cofactor pool of a specific organelle (e.g., the acetyl-CoA in the mitochondria). We refer to a recent review from Yocum et al. [111] in which substrate channelling approaches and their usefulness in metabolic engineering are well summarized. Heterologous genes are commonly synthesized as codon-optimized variants for metabolic engineering applications. However, some studies suggest that codon-optimized enzymes do not always perform better, which might depend on the specific protein or on the optimization algorithm used [112–115].

A noteworthy addition in the context of expression cassette design are modular cloning toolkits that have been developed for

Table 4
Selection of native *S. cerevisiae* terminators and terminator engineering approaches

Description	Examples/features	References
Strong terminators	t*RPL41B*, t*RPL15A*, t*DIT1*, t*PRM9*, t*CPS1*, t*TDH1*	[40, 104, 105, 107]
Medium terminators	t*ADH1*, t*CYC1*, t*PGK1*	[40, 105, 107]
Weak terminators	t*GIC1*, t*ICY2*, t*ECM13*, t*ENO1*, t*TDH3*	[40, 104, 106, 107]
Short synthetic terminators	35–70 bp long	[108]
Engineering terminators based on nucleosome occupancy	Algorithm-aided designing of synthetic terminators	[109]

multiple model organisms including *S. cerevisiae* [40, 116, 117]. The toolkit from Lee and coworkers [40] provides an opportunity to speed up the build process, reduce costs, simplify the sharing of genetic parts and improve reproducibility. It combines a library of well-characterized, standardized DNA parts (promoters, terminators, yeast markers, etc.) with an efficient Golden-Gate assembly strategy. From these standardized parts multigene cassettes for plasmid expression or genomic integration can be assembled and the toolkit can easily be expanded to include new genetic parts or applications.

While this section covered a more rational engineering approach for expression cassette design, sophisticated methods have been developed to optimize gene expression levels of multienzyme pathways. These methods either use combinatorial libraries or an iterative, systematic approach. A selection of expression optimization methods and other advanced engineering strategies is highlighted in the following cell factory engineering section.

4 Cell Factory Engineering

The iterative process of the Design–Build–Test–Learn cycle makes metabolic engineering inherently time-consuming. While the Design, Test, and Learn phase can be greatly accelerated by combining GEMs and machine learning (ML) with high-throughput methods and -omics data collection [8, 118], the Build phase has become a major time-constraint in the engineering cycle, despite the advances in DNA synthesis [119] and sequencing [120]. Progress was made in large-scale liquid-handling systems and cultivation robotics, but acquisition and maintenance costs often make these systems unattainable for academic laboratories. However, in an attempt to simplify and standardize strain engineering, many methods and toolboxes have been developed in the last decades especially for synthetic biology and metabolic engineering purposes. The most recent and paradigm-shifting innovation in biological engineering is the CRISPR-Cas technology, which has been successfully transferred to model and nonmodel organisms alike and made genetic engineering more efficient and less time-consuming [121]. Compared to *S. cerevisiae* other yeasts and fungi could arguably be more promising chassis candidates for specific natural products, because of, for example, superior precursor and cofactor supply [122], but the vast amounts of genetic tools available and the overall ease of engineering in *S. cerevisiae* are unprecedented in other yeasts and fungi.

In the following section we highlight methodologies and toolkits for *S. cerevisiae* engineering that are useful either for constructing a natural product proof-of-concept strain or for developing a proof-of-concept strain into a high-performance natural product cell factory. Production of different natural products will provide different challenges for the metabolic engineer. The aim of this section is not to provide an exhaustive list of all published engineering efforts, but to focus on a variety of possible engineering techniques.

Genomic integrations are usually favorable for cell factories over plasmid-based expression systems, mostly due to the number of genes that need to be expressed in cell factories. Natural product biosynthetic pathways can consist of dozens of genes and additional copies of heterologous and native genes are likely necessary to increase production. In addition, expression of transporters to avoid product toxicity and to simplify extraction procedures or chaperons to improve protein folding capabilities might also be desirable. Further advantages of genomic integrations are increased strain stability, decreased strain heterogeneity [40, 123], and no reliance on selection markers, which can change cell physiology [124, 125]. The efficient HR machinery in *S. cerevisiae* makes genome engineering a straightforward and simple procedure. With CRISPR-Cas technology [126], it is possible to induce DNA double-strand breaks (DSB) at virtually any locus in the genome (only requiring the presence of a PAM site), simply by expressing a guide RNA (gRNA) with complementary sequence to the target site. CRISPR-Cas technology enabled marker-less genome modifications with near 100% efficiency [126, 127] and furthermore makes it possible to multiplex genome modifications, for example perform multiple integrations in one transformation step [128]. The choice of which genome engineering method to use depends on several factors, for example: the number of genes that need to be integrated, the desired genomic locus (e.g., a well-characterized integration site or a native gene that should be replaced or knocked out) and the number of cassette copies that

Table 5
Genome engineering tools for *S. cerevisiae*

Name	Application	Features	References
EasyClone-MarkerFree	Marker-free integration in well-characterized loci	Toolkit for simultaneous integration of 1 or 3 cassettes (2 genes per cassette); choice of 11 well-characterized integration sites (high expression levels, high genetic stability; no effect on cell growth)	[127]
CRISPR-Cas9 Swiss Army Knife	Marker-free integrations, KOs and point mutations	Toolkit for multiplexing of up to six modifications; includes a design tool for custom gRNAs and repair fragments; double-gRNA vectors assembled in vitro; single-gRNA vectors assembled in vivo	[129]
CasEMBLR	Marker-free integrations and KIs, potential use for combinatorial libraries	Method for simultaneous multiloci integrations; integration cassettes are assembled in vivo; includes a design tool for gRNAs	[128]
Synthetic landing pad platform strains for multiple integrations	Marker-free integration in well-characterized loci	Platform strain containing synthetic landing pads; loci are targeted by a single gRNA and allow simultaneous integration of up to five cassettes	[130]
Di-CRIPSR (-integration CRISPR)	Multicopy integrations	Multicopy marker-free integration in site of Ty retrotransposon; copy number ranges from 4 to 18 copies per clone with an average of 10; demonstrated on a 24-kb biosynthetic pathway	[131]
Synthetic landing pad platform strain for multicopy gene integrations	Multicopy integrations	Platform strain with synthetic landing pads in genome for integrations, allows precise control of copy number; maximum of 4 copies	[132]
High-copy genome integration via *in vivo* DNA assembly	Multicopy integrations	Multicopy marker-free integration in site of Ty retrotransposon; in vivo assembly of a marker-free integration cassette; up to 25 copies with average of 10 copies per clone	[133]

should be integrated. Table 5 summarizes different techniques for genomic integrations, knockouts (KOs), and knockins (KIs).

The RNA polymerase III promoter p*SNR52* was initially used for expression of gRNAs in *S. cerevisiae* [126] and is still used in many CRISPR-Cas applications, yielding satisfactory efficiencies.

Table 6
Strategies for gRNAs expression in *S. cerevisiae*

Description	Features	References
Increased gRNA abundance and expression from tRNA promoter	Higher gRNA abundance and CRISPR-Cas efficiencies were achieved by including a 5′ self-cleaving hepatitis delta virus ribozyme in the transcript; besides p*SNR52* tRNA promoters can be used for gRNA expression	[134]
gRNA design enabling expression from RNA Pol II promoters	gRNA is flanked by two self-cleaving ribozymes; design allows the use of any promoter for gRNA expression; can potentially be used for expressing multiple gRNAs from one transcript	[135]
Csy4 for cleavage of gRNA arrays	Cleavage of single CRISPR array transcript into individual gRNAs by bacterial endoribonuclease Csy4	[136, 137]
Cas12a for cleavage of gRNA arrays	Cas12a is capable of cleaving CRISPR arrays into individual crRNAs and inducing a crRNA-targeted DSB	[138]
(Lightning) GTR-CRISPR (gRNA-tRNA array CRISPR)	Utilizing the endogenous tRNA processing machinery for cleavage of CRISPR array into individual gRNAs	[139]

Relying on a single promoter especially for expression of multiple gRNAs is problematic and several strategies have been developed to widen the possibilities for gRNA expression. Ryan and coworkers [134] demonstrated that gRNA expression levels are critical for CRIPSR-Cas targeting and attention should be paid to the choice of promoter and overall gRNA design. Table 6 summarizes techniques for the expression of single or multiple gRNAs in *S. cerevisiae*.

The yeast metabolic engineering community has developed creative tools for cell factory construction and optimization that go well beyond genomic integrations and KOs. The tools are either based on (semi)rational design or on directed evolution to explore the solution space of a given cell factory. The methods are validated, mostly standardized and can be adapted for different products. Table 7 lists a variety of more sophisticated engineering techniques covering combinatorial libraries, targeted mutagenesis, gene regulation by deactivated/dead Cas9 (dCas9), and more. Biosensors, as an indispensable tool for high-throughput screening or selection, are discussed in the following text section.

5 Biosensors

Although heterologous synthesis of natural products presents a promising alternative to other methods, the large-scale production of many compounds is still hindered by low titers, rates, and yields

Table 7
Advanced strategies for metabolic engineering in *S. cerevisiae*

Name	Application	Features	References
CRISPR/Cas9-RNA interference system for combinatorial metabolic engineering	Library construction for simultaneous KD and overexpression of multiple genes	Method for cell factory optimization that combines CRISPR-Cas9-mediated genomic integration, in vivo assembly and RNA interference; integration cassettes consist of multiple transcription units (overexpression or KD cassettes) and are assembled *in vivo*	[140]
CasPER (Cas9-mediated protein evolution reaction)	Library construction by integrating error-prone PCR fragments	CRISPR-Cas9-mediated genomic integrations of error-prone PCR fragments to create targeted and stable mutant libraries, multiple genes can be targeted simultaneously	[141]
STEPS (Systematically Test Enzyme Perturbation Sensitivities)	Optimization of BGC expression levels	Method to identify production-limiting reactions in multigene metabolic pathways in a systematic and iterative manner; expression levels of individual pathway genes is modulated by CRISPRa or CRISPRi	[142]
COMPASS (Combinatorial Pathway Assembly)	Library construction to optimize BGC expression levels	Toolkit for construction and integration of combinatorial libraries; libraries are based on combining pathway genes with promoters responsive to artificial transcription factors; toolkit includes a detailed workflow for constructing the library from part vectors	[143]
MiYA (Machine-learning workflow in conjunction with YeastFab Assembly strategy)	Library construction to optimize BGC expression levels	Machine-learning-assisted workflow for construction of combinatorial libraries; libraries are assembled from pool of standardized parts	[144]
Engineering transcriptional programs using dCas9	Simultaneous CRISPRa and CRISPRi of different genes	Activation or repression of gene is dependent on gRNA sequence, which contains scaffold that recruits either	[145]

(continued)

Table 7 (continued)

Name	Application	Features	References
		RNA-binding protein fused to activator or RNA-binding protein fused to repressor	
Inducible dCas9 gene regulation	Inducible CRISPRa and CRISPRi systems	Multiple native genes can be up or down regulated in a dose-dependent manner; studies highlight importance of gRNA design for regulation	[146, 147]
CRISPR-AID (CRISPR activation, inhibition and deletion)	Trifunctional CRISPR-system for simultaneous deletion, gene activation and gene repression	Uses three orthogonal Cas proteins only recognizing their cognate gRNA; combinatorial gRNA libraries possible	[148]
RAGE (RNAi-assisted genome evolution)	Construction of a genome-wide KD library	Plasmid-based expression of RNAi cassettes; genomic integration of beneficial KD cassette enables multigene KD and iterative screening cycles	[149]
mCRISTAR (multiplexed CRISPR transformation-associated recombination)	Transferring whole BGC to yeast and exchanging promoters	Whole BGC can be transferred into yeast via transformation-associated recombination; Cas9 cuts gene cluster into operon fragments, which are reassembled into expression cassettes with synthetic promoters; resulting gene cluster can then be shuttled into another heterologous host (or potentially be directly expressed in yeast); tool can be used to activate silent BGCs or to optimize expression levels of BGC enzymes	[150]
Automated multiplex genome-scale engineering in yeast	Library construction for simultaneous KD and overexpression of multiple genes	RNAi KD cassettes and overexpression cassettes are being integrated in delta site of Ty retrotransposon, standardized cassette architecture enables multiplexing and iterative strain engineering using robotics	[151]

(continued)

Table 7 (continued)

Name	Application	Features	References
POTAC (Pathway Optimization by Tuning Antibiotic Concentration)	Optimization of BGC expression levels	Plasmid system for optimizing expression levels of pathway genes, copy number of plasmids is tunable by changing antibiotic concentration	[152]
CHAnGE (CRISPR–Cas9-and homology-directed-repair-assisted genome-scale engineering	Library construction for genome-wide KOs or mutagenesis of genes with single-nucleotide resolution	Yeast library is based on synthesized oligo library that include both, HR repair template and gRNA for permanent genome modifications; oligos are assembled into plasmids and transformed to yeast; iterative cycles of selection and engineering	[153]
OrthoRep	*In vivo* mutagenesis of user-defined gene	Orthogonal DNA polymerase-plasmid pair; mutation rate increased about 100,000-fold compared to wt mutation rate	[154]
Target-AID (target-activation-induced cytidine deaminase)	*In vivo* mutagenesis of user-defined loci	Approach utilizes fusion of nCas9 mutant and activation-induced cytidine deaminase (AID); mutating mostly cytidines in a 5 bp range	[155]
yEvolvR	*In vivo* mutagenesis of user-defined loci	Approach utilizes nCas9 fused to error-prone DNA polymerase; mutation rate increased about 10,000-fold compared to wt mutation rate; mutations occur mostly in 40-bp window	[156]
eMAGE	*In vivo* mutagenesis using synthetic oligonucleotides	Combinatorial genome modifications based on annealing of synthetic oligonucleotides during DNA replication; allowing single-bp precision and iterative cycles of engineering	[157]

BGC biosynthetic gene cluster, *CRISPRa* CRISPR activation, *CRISPRi* CRISPR interference, *KD* gene knockdown, *nCas9* Cas9 nickase

[158]. The introduction of heterologous pathways can create a metabolic burden on the host organism leading to growth impediment, which in turn will hinder production of the target compound. Furthermore, producer strains are typically evaluated by analytical methods, such as chromatography and mass spectrometry, which are relatively low in throughput. The application of in vivo biosensors may expedite the development of high-producer strains. Not only can biosensors be used to regulate heterologous gene expression in order to improve balance between cell growth and product synthesis, they also enable high-throughput screening of enzymes, pathways and strain libraries, which may accelerate the "Test" phase of the Design-Build-Test-Learn cycle [159–161]. Although most biosensors are implemented in prokaryotes, the development of biosensors in eukaryotes is increasing [162].

Biosensors enable the transduction of a metabolite's concentration into an expression change of a reporter gene. If the reporter gene encodes a fluorescent protein, methods such as flow cytometry or fluorescence-activated cell sorting (FACS) can be used to screen enzyme or strain libraries for promising candidates. If the metabolite is linked to antibiotic resistance or auxotrophic complementation, high producer strains can be selected based on cell growth [163]. A summary of biosensors for natural products and their intermediates is given in Table 8. They include transcription factor-, RNA-, enzyme-, and G-protein coupled receptor-based (GPCR-based) biosensors.

Most natural product-responsive biosensors are based on heterologous transcription factors (TF), that have a ligand binding domain with high affinity for the compound of interest, and a DNA binding domain that allows the TF to regulate transcription of a reporter gene. The TF thus converts the input signal (concentration of the compound of interest) into an easily measurable output signal (e.g., fluorescence). Many prokaryotic transcription factors have been shown to function in *S. cerevisiae*. Prokaryotic promoters however cannot be directly applied in eukaryotes. Therefore, TF-DNA binding sites must be embedded into native promoters instead. Although this is a nontrivial task, some generalized strategies and methods have been reported [181, 182].

Suitable TFs may not be available for certain molecules of interest. Moreover, TFs can be difficult to reprogram to alter specificity toward structurally similar compounds. Riboswitch-based biosensors present some advantages in this regard, as they are more easily adaptable to new ligands [176, 179]. In addition, unlike TFs, riboswitch biosensors do not require the incorporation of DNA-binding sites into the reporter promoter. In riboswitches, the RNA aptamer acts as the ligand binding domain. Upon binding, the RNA undergoes a conformational rearrangement, leading to a measurable change in expression levels of the reporter gene.

Table 8
Biosensors developed for *S. cerevisiae* for natural products and precursors

Target compound	Compound class	Biosensor type	Sensing element	Source	References
Malonyl-CoA	Thioester of malonic acid and coenzyme A, precursor of flavonoids and polyketides	TF (repressor)	FapR	*Bacillus subtilis*	[164–166]
Naringenin	Flavonoid	TF (activator)	FdeR	*Herbaspirillum seropedicae*	[167, 168]
p-Coumaric acid	Hydroxycinnamic acid, precursor of phenylpropanoids	TF (repressor)	PadR	*Bacillus subtilis*	[169]
Protocatechuic acid	Plant phenolic acid	TF (activator)	PcaQ	*Sinorhizobium meliloti*	[167]
4-Hydroxybenzoic acid (pHBA), anthranilic acid	Benzoic acid derivatives: pHBA (precursor of arbutin, resveratrol, shikonin etc.), anthranilic acid (precursor of some alkaloids)	TF (activator)	sBAD (LexA_DBD-HbaR_LBD-B112_AD)	Chimeric, HbaR from *Rhodopseudomonas palustris*	[170]
Vanillin, vanillic acid	Plant phenolic aldehyde, phenolic acid	TF (repressor)	VanR	*Caulobacter crescentus*	[171]
Digoxin, digoxigenin	Plant steroid glycoside, plant steroid aglycone	TF (activator)	Dimerization of Gal4_DBD-DIG and VP16_AD-DIG [172] Gal4_DBD-DIG-VP16_AD [173]	Chimeric, computationally designed ligand binding domain DIG10.3	[172–174]
Isopentenyl pyrophosphate	Precursor of terpenoids	TF (activator)	Dimerization of Gal4_DBD-Idi/Idi1/Erg20 and Gal4_AD-Idi/Idi1/Erg20	Chimeric, *idi* from *E. coli* or *IDI1* and *ERG20* from *S. cerevisiae*	[175]
Theophylline	Purine alkaloid	RNA switch	RNA aptamer	Synthetic	[176]
L-3,4-dihydroxyphenylalanine (L-DOPA)	Precursor of benzylisoquinoline alkaloids	Enzyme-coupled	DOPA dioxygenase	*Mirabilis jalapa*	[177]

Paromomycin	Aminoglycoside antibiotic	RNA switch	RNA aptamer	Synthetic	[178]
Neomycin	Aminoglycoside antibiotic	RNA switch	RNA aptamer + Hammerhead ribozyme (HHR) + Gal4 mRNA	HHR from *Schistosoma mansoni*	[179]
2,4-Diacetylphloroglucinol (DAPG)	Natural phenol	TF (repressor)	PhlF (ON switch), PhlF-3xVP16 (OFF switch)	*Pseudomonas protegens*	[97]
Camphor	Terpenoid	TF (repressor)	CamR-3xVP16 (OFF switch)	*Pseudomonas putida*	[77]
Serotonin	Precursor of hydroxystrictosidine (modified monoterpene indole alkaloids)	GPCR	Human GPCR 5-HT_4 isoform b	*Homo sapiens*	[180]

Enzyme-coupled biosensors are based on enzymes that can convert the target compound into either a colorimetric or fluorometric product [177, 183]. This approach allows simple detection methods, it is however limited to a smaller number of effector molecules.

The above described biosensors are typically used to sense intracellular levels of the target molecule. GPCR biosensors are useful in the case that extracellular concentrations are of interest. Here, a heterologous GPCR known to bind the target molecule, is linked to the yeast native GPCR pathway, typically the pheromone response pathway. The signaling transduction can then be coupled to expression of, for example, a fluorescent protein [162, 184].

6 Outlook

Cell factories enable sustainable production of high-value natural products and the demand for efficient cell factories is expected to rise during the shift away from a petroleum-based economy. Historically, *S. cerevisiae* has been at the forefront of basic biological research, metabolic engineering, systems and synthetic biology. *S. cerevisiae* has been the first eukaryote to be fully sequenced [185] and is likely to be the first eukaryote carrying a synthetic genome [186]. We expect that cell factory design and construction will greatly benefit from the synergies between high-throughput methods developed in metabolic engineering, more predictive GEMs developed in systems biology and sophisticated gene circuit designs developed in synthetic biology. The aim is to minimize the number of DBTL-cycle iterations required to construct an industrially viable strain and to overall improve titers, rates, and yields via precise fine-tuning of the yeast metabolism. ML algorithms are likely to be an essential part of this development and the algorithms will benefit from the vast knowledge available thanks to decades of *S. cerevisiae* research. ML can potentially be applied in all three mentioned research fields [118, 187], possibly creating an overarching framework that can help to combine them.

The *S. cerevisiae* synthetic genome project, Sc2.0 [186], allows for speculation on how cell factories might look like in the future. The synthetic chromosomes contain SCRaMbLE sites at regular intervals, which allow the reshuffling of chromosomes, creating a large pool of genetic diversity [188]. The technique has already been applied to natural product cell factories [189–191]. Once the yeast synthetic genome project is completed applying the SCRaMbLE method in strain Sc2.0 could lead to genotypes that are hard or virtually impossible to obtain with other engineering techniques and pave the way toward a minimal or minimized cell chassis that can be customized to precisely meet the requirements of a given industrial production process.

This book highlights a variety of domesticated host organisms that can be exploited to produce natural products. Each host brings different capabilities and engineering options, making them more or less suitable for specific natural products. In this chapter, we highlighted some features that *S. cerevisiae* excels at, specifically the ease of genome engineering and the creative engineering and screening tools developed by the yeast metabolic engineering and synthetic biology community. The many examples of yeast natural product cell factories underline its importance in the field.

References

1. Facchini PJ, Bohlmann J, Covello PS, De Luca V, Mahadevan R, Page JE, Ro DK, Sensen CW, Storms R, Martin VJ (2012) Synthetic biosystems for the production of high-value plant metabolites. Trends Biotechnol 30(3):127–131. https://doi.org/10.1016/j.tibtech.2011.10.001
2. Brower V (2008) Back to nature: extinction of medicinal plants threatens drug discovery. J Natl Cancer Inst 100(12):838–839. https://doi.org/10.1093/jnci/djn199
3. Huang B, Guo J, Yi B, Yu X, Sun L, Chen W (2008) Heterologous production of secondary metabolites as pharmaceuticals in *Saccharomyces cerevisiae*. Biotechnol Lett 30(7):1121–1137. https://doi.org/10.1007/s10529-008-9663-z
4. Krivoruchko A, Nielsen J (2015) Production of natural products through metabolic engineering of *Saccharomyces cerevisiae*. Curr Opin Biotechnol 35:7–15. https://doi.org/10.1016/j.copbio.2014.12.004
5. Cravens A, Payne J, Smolke CD (2019) Synthetic biology strategies for microbial biosynthesis of plant natural products. Nat Commun 10(1):2142. https://doi.org/10.1038/s41467-019-09848-w
6. Zhang Y, Nielsen J, Liu Z (2017) Engineering yeast metabolism for production of terpenoids for use as perfume ingredients, pharmaceuticals and biofuels. FEMS Yeast Res 17(8). https://doi.org/10.1093/femsyr/fox080
7. Sewalt V, Shanahan D, Gregg L, La Marta J, Carrillo R (2016) The Generally Recognized as Safe (GRAS) process for industrial microbial enzymes. Ind Biotechnol 12(5):295–302. https://doi.org/10.1089/ind.2016.0011
8. Yu R, Nielsen J (2019) Big data in yeast systems biology. FEMS Yeast Res 19(7):foz070. https://doi.org/10.1093/femsyr/foz070
9. Paddon CJ, Westfall PJ, Pitera DJ, Benjamin K, Fisher K, McPhee D, Leavell MD, Tai A, Main A, Eng D, Polichuk DR, Teoh KH, Reed DW, Treynor T, Lenihan J, Fleck M, Bajad S, Dang G, Dengrove D, Diola D, Dorin G, Ellens KW, Fickes S, Galazzo J, Gaucher SP, Geistlinger T, Henry R, Hepp M, Horning T, Iqbal T, Jiang H, Kizer L, Lieu B, Melis D, Moss N, Regentin R, Secrest S, Tsuruta H, Vazquez R, Westblade LF, Xu L, Yu M, Zhang Y, Zhao L, Lievense J, Covello PS, Keasling JD, Reiling KK, Renninger NS, Newman JD (2013) High-level semi-synthetic production of the potent antimalarial artemisinin. Nature 496(7446):528–532. https://doi.org/10.1038/nature12051
10. Li Y, Li S, Thodey K, Trenchard I, Cravens A, Smolke CD (2018) Complete biosynthesis of noscapine and halogenated alkaloids in yeast. Proc Natl Acad Sci U S A 115(17):E3922–E3931. https://doi.org/10.1073/pnas.1721469115
11. Nielsen J (2015) BIOENGINEERING. Yeast cell factories on the horizon. Science 349(6252):1050–1051. https://doi.org/10.1126/science.aad2081
12. Billingsley JM, DeNicola AB, Tang Y (2016) Technology development for natural product biosynthesis in *Saccharomyces cerevisiae*. Curr Opin Biotechnol 42:74–83. https://doi.org/10.1016/j.copbio.2016.02.033
13. Paramasivan K, Mutturi S (2017) Progress in terpene synthesis strategies through engineering of *Saccharomyces cerevisiae*. Crit Rev Biotechnol 37(8):974–989. https://doi.org/10.1080/07388551.2017.1299679
14. Meadows AL, Hawkins KM, Tsegaye Y, Antipov E, Kim Y, Raetz L, Dahl RH, Tai A, Mahatdejkul-Meadows T, Xu L, Zhao L, Dasika MS, Murarka A, Lenihan J, Eng D, Leng JS, Liu C-L, Wenger JW, Jiang H, Chao L, Westfall P, Lai J, Ganesan S, Jackson P, Mans R, Platt D, Reeves CD, Saija PR, Wichmann G, Holmes VF,

Benjamin K, Hill PW, Gardner TS, Tsong AE (2016) Rewriting yeast central carbon metabolism for industrial isoprenoid production. Nature 537:694. https://doi.org/10.1038/nature19769

15. Chen R, Yang S, Zhang L, Zhou YJ (2020) Advanced strategies for production of natural products in yeast. iScience 23(3):100879. https://doi.org/10.1016/j.isci.2020.100879
16. Minami H (2013) Fermentative production of plant benzylisoquinoline alkaloids in microbes. Biosci Biotechnol Biochem 77(8): 1617–1622. https://doi.org/10.1271/bbb.130106
17. Pyne ME, Kevvai K, Grewal PS, Narcross L, Choi B, Bourgeois L, Dueber JE, Martin VJJ (2020) A yeast platform for high-level synthesis of tetrahydroisoquinoline alkaloids. Nat Commun 11(1):3337. https://doi.org/10.1038/s41467-020-17172-x
18. Galanie S, Thodey K, Trenchard IJ, Filsinger Interrante M, Smolke CD (2015) Complete biosynthesis of opioids in yeast. Science 349(6252):1095. https://doi.org/10.1126/science.aac9373
19. Hawkins KM, Smolke CD (2008) Production of benzylisoquinoline alkaloids in *Saccharomyces cerevisiae*. Nat Chem Biol 4(9): 564–573. https://doi.org/10.1038/nchembio.105
20. O'Connor SE, Maresh JJ (2006) Chemistry and biology of monoterpene indole alkaloid biosynthesis. Nat Prod Rep 23(4):532–547. https://doi.org/10.1039/B512615K
21. Brown S, Clastre M, Courdavault V, O'Connor SE (2015) De novo production of the plant-derived alkaloid strictosidine in yeast. Proc Natl Acad Sci U S A 112(11): 3205–3210. https://doi.org/10.1073/pnas.1423555112
22. Kohnen-Johannsen KL, Kayser O (2019) Tropane alkaloids: chemistry, pharmacology, biosynthesis and production. Molecules 24(4):796. https://doi.org/10.3390/molecules24040796
23. Qin J, Zhou YJ, Krivoruchko A, Huang M, Liu L, Khoomrung S, Siewers V, Jiang B, Nielsen J (2015) Modular pathway rewiring of *Saccharomyces cerevisiae* enables high-level production of L-ornithine. Nat Commun 6: 8224. https://doi.org/10.1038/ncomms9224
24. Srinivasan P, Smolke CD (2019) Engineering a microbial biosynthesis platform for *de novo* production of tropane alkaloids. Nat Commun 10(1):3634. https://doi.org/10.1038/s41467-019-11588-w
25. Ozber N, Watkins JL, Facchini PJ (2020) Back to the plant: overcoming roadblocks to the microbial production of pharmaceutically important plant natural products. J Ind Microbiol Biotechnol 47(9-10):815–828. https://doi.org/10.1007/s10295-020-02300-9
26. Luo X, Reiter MA, d'Espaux L, Wong J, Denby CM, Lechner A, Zhang Y, Grzybowski AT, Harth S, Lin W, Lee H, Yu C, Shin J, Deng K, Benites VT, Wang G, Baidoo EEK, Chen Y, Dev I, Petzold CJ, Keasling JD (2019) Complete biosynthesis of cannabinoids and their unnatural analogues in yeast. Nature 567(7746):123–126. https://doi.org/10.1038/s41586-019-0978-9
27. Liu Q, Yu T, Li X, Chen Y, Campbell K, Nielsen J, Chen Y (2019) Rewiring carbon metabolism in yeast for high level production of aromatic chemicals. Nat Commun 10(1): 4976. https://doi.org/10.1038/s41467-019-12961-5
28. Li M, Schneider K, Kristensen M, Borodina I, Nielsen J (2016) Engineering yeast for high-level production of stilbenoid antioxidants. Sci Rep 6:36827. https://doi.org/10.1038/srep36827
29. Du Y, Yang B, Yi Z, Hu L, Li M (2020) Engineering *Saccharomyces cerevisiae* coculture platform for the production of flavonoids. J Agric Food Chem 68(7): 2146–2154. https://doi.org/10.1021/acs.jafc.9b07916
30. Li H, Gao S, Zhang S, Zeng W, Zhou J (2021) Effects of metabolic pathway gene copy numbers on the biosynthesis of (2S)-naringenin in *Saccharomyces cerevisiae*. J Biotechnol 325:119–127. https://doi.org/10.1016/j.jbiotec.2020.11.009
31. Milke L, Marienhagen J (2020) Engineering intracellular malonyl-CoA availability in microbial hosts and its impact on polyketide and fatty acid synthesis. Appl Microbiol Biotechnol 104(14):6057–6065. https://doi.org/10.1007/s00253-020-10643-7
32. Gold ND, Gowen CM, Lussier FX, Cautha SC, Mahadevan R, Martin VJ (2015) Metabolic engineering of a tyrosine-overproducing yeast platform using targeted metabolomics. Microb Cell Fact 14:73. https://doi.org/10.1186/s12934-015-0252-2
33. Nielsen J, Keasling Jay D (2016) Engineering cellular metabolism. Cell 164(6):1185–1197. https://doi.org/10.1016/j.cell.2016.02.004

34. Duran L, López JM, Avalos JL (2020) ¡Viva la mitochondria!: harnessing yeast mitochondria for chemical production. FEMS Yeast Res 20(6):foaa037. https://doi.org/10.1093/femsyr/foaa037
35. Zhu J, Zhang MQ (1999) SCPD: a promoter database of the yeast *Saccharomyces cerevisiae*. Bioinformatics 15(7):607–611. https://doi.org/10.1093/bioinformatics/15.7.607
36. Boer VM, de Winde JH, Pronk JT, Piper MDW (2003) The genome-wide transcriptional responses of *Saccharomyces cerevisiae* grown on glucose in aerobic chemostat cultures limited for carbon, nitrogen, phosphorus, or sulfur. J Biol Chem 278(5):3265–3274. https://doi.org/10.1074/jbc.M209759200
37. Chang DT-H, Huang C-Y, Wu C-Y, Wu W-S (2011) YPA: an integrated repository of promoter features in *Saccharomyces cerevisiae*. Nucleic Acids Res 39(suppl_1):D647–D652. https://doi.org/10.1093/nar/gkq1086
38. Keren L, Zackay O, Lotan-Pompan M, Barenholz U, Dekel E, Sasson V, Aidelberg G, Bren A, Zeevi D, Weinberger A, Alon U, Milo R, Segal E (2013) Promoters maintain their relative activity levels under different growth conditions. Mol Syst Biol 9(1):701. https://doi.org/10.1038/msb.2013.59
39. Sun J, Shao Z, Zhao H, Nair N, Wen F, Xu J-H, Zhao H (2012) Cloning and characterization of a panel of constitutive promoters for applications in pathway engineering in *Saccharomyces cerevisiae*. Biotechnol Bioeng 109(8):2082–2092. https://doi.org/10.1002/bit.24481
40. Lee ME, DeLoache WC, Cervantes B, Dueber JE (2015) A highly characterized yeast toolkit for modular, multipart assembly. ACS Synth Biol 4(9):975–986. https://doi.org/10.1021/sb500366v. PMID: 25871405
41. Maury J, Kannan S, Jensen NB, Oberg FK, Kildegaard KR, Forster J, Nielsen J, Workman CT, Borodina I (2018) Glucose-dependent promoters for dynamic regulation of metabolic pathways. Front Bioeng Biotechnol 6:63. https://doi.org/10.3389/fbioe.2018.00063
42. Mumberg D, Müller R, Funk M (1995) Yeast vectors for the controlled expression of heterologous proteins in different genetic backgrounds. Gene 156(1):119–122. https://doi.org/10.1016/0378-1119(95)00037-7. PMID: 7737504
43. Reider Apel A, d'Espaux L, Wehrs M, Sachs D, Li RA, Tong GJ, Garber M, Nnadi O, Zhuang W, Hillson NJ, Keasling JD, Mukhopadhyay A (2017) A Cas9-based toolkit to program gene expression in *Saccharomyces cerevisiae*. Nucleic Acids Res 45(1):496–508. https://doi.org/10.1093/nar/gkw1023
44. Partow S, Siewers V, Bjorn S, Nielsen J, Maury J (2010) Characterization of different promoters for designing a new expression vector in *Saccharomyces cerevisiae*. Yeast 27(11):955–964. https://doi.org/10.1002/yea.1806
45. Peng B, Plan MR, Carpenter A, Nielsen LK, Vickers CE (2017) Coupling gene regulatory patterns to bioprocess conditions to optimize synthetic metabolic modules for improved sesquiterpene production in yeast. Biotechnol Biofuels 10(1):43. https://doi.org/10.1186/s13068-017-0728-x
46. Peng B, Williams TC, Henry M, Nielsen LK, Vickers CE (2015) Controlling heterologous gene expression in yeast cell factories on different carbon substrates and across the diauxic shift: a comparison of yeast promoter activities. Microb Cell Fact 14:91–91. https://doi.org/10.1186/s12934-015-0278-5
47. West RW Jr, Chen SM, Putz H, Butler G, Banerjee M (1987) GAL1-GAL10 divergent promoter region of *Saccharomyces cerevisiae* contains negative control elements in addition to functionally separate and possibly overlapping upstream activating sequences. Genes Dev 1(10):1118–1131. https://doi.org/10.1101/gad.1.10.1118
48. Mumberg D, Müller R, Funk M (1994) Regulatable promoters of *Saccharomyces cerevisiae*: comparison of transcriptional activity and their use for heterologous expression. Nucleic Acids Res 22(25):5767–5768. https://doi.org/10.1093/nar/22.25.5767
49. Kramer RA, DeChiara TM, Schaber MD, Hilliker S (1984) Regulated expression of a human interferon gene in yeast: control by phosphate concentration or temperature. Proc Natl Acad Sci 81(2):367. https://doi.org/10.1073/pnas.81.2.367
50. Labbé S, Thiele DJ (1999) [8] Copper ion inducible and repressible promoter systems in yeast. In: Methods in enzymology, vol 306. Academic Press, Boston, pp 145–153. https://doi.org/10.1016/S0076-6879(99)06010-3
51. Etcheverry T (1990) [26] Induced expression using yeast copper metallothionein promoter. In: Methods in enzymology, vol 185. Academic Press, Boston, pp 319–329. https://doi.org/10.1016/0076-6879(90)85028-M

52. Butt TR, Ecker DJ (1987) Yeast metallothionein and applications in biotechnology. Microbiol Rev 51(3):351–364
53. Mao X, Hu Y, Liang C, Lu C (2002) MET3 promoter: a tightly regulated promoter and its application in construction of conditional lethal strain. Curr Microbiol 45(1):37–40. https://doi.org/10.1007/s00284-001-0046-0
54. Sangsoda S, Cherest H, Surdin-Kerjan Y (1985) The expression of the MET25 gene of *Saccharomyces cerevisiae* is regulated transcriptionally. Mol Gen Genet 200(3): 407–414. https://doi.org/10.1007/BF00425724
55. Kerjan P, Cherest H, Surdin-Kerjan Y (1986) Nucleotide sequence of the *Saccharomyces cerevisiae* MET25 gene. Nucleic Acids Res 14(20):7861–7871. https://doi.org/10.1093/nar/14.20.7861
56. Cherest H, Kerjan P, Surdin-Kerjan Y (1987) The *Saccharomyces cerevisiae* MET3 gene: nucleotide sequence and relationship of the 5′ non-coding region to that of MET25. Mol Gen Genet 210(2):307–313. https://doi.org/10.1007/BF00325699
57. Rantasalo A, Kuivanen J, Penttilä M, Jäntti J, Mojzita D (2018) Synthetic toolkit for complex genetic circuit engineering in *Saccharomyces cerevisiae*. ACS Synth Biol 7(6): 1573–1587. https://doi.org/10.1021/acssynbio.8b00076
58. Denis CL, Ferguson J, Young ET (1983) mRNA levels for the fermentative alcohol dehydrogenase of *Saccharomyces cerevisiae* decrease upon growth on a nonfermentable carbon source. J Biol Chem 258(2): 1165–1171
59. Ozcan S, Johnston M (1995) Three different regulatory mechanisms enable yeast hexose transporter (HXT) genes to be induced by different levels of glucose. Mol Cell Biol 15(3):1564. https://doi.org/10.1128/MCB.15.3.1564
60. Scalcinati G, Knuf C, Partow S, Chen Y, Maury J, Schalk M, Daviet L, Nielsen J, Siewers V (2012) Dynamic control of gene expression in *Saccharomyces cerevisiae* engineered for the production of plant sesquitepene α-santalene in a fed-batch mode. Metab Eng 14(2):91–103. https://doi.org/10.1016/j.ymben.2012.01.007
61. Shen MWY, Fang F, Sandmeyer S, Da Silva NA (2012) Development and characterization of a vector set with regulated promoters for systematic metabolic engineering in *Saccharomyces cerevisiae*. Yeast 29(12):495–503. https://doi.org/10.1002/yea.2930
62. Fang F, Salmon K, Shen MWY, Aeling KA, Ito E, Irwin B, Tran UPC, Hatfield GW, Da Silva NA, Sandmeyer S (2011) A vector set for systematic metabolic engineering in *Saccharomyces cerevisiae*. Yeast (Chichester, England) 28(2):123–136. https://doi.org/10.1002/yea.1824
63. Ye L, Berden JA, van Dam K, Kruckeberg AL (2001) Expression and activity of the Hxt7 high-affinity hexose transporter of Saccharomyces cerevisiae. Yeast 18(13):1257–1267. https://doi.org/10.1002/yea.771
64. Apel AR, d'Espaux L, Wehrs M, Sachs D, Li RA, Tong GJ, Garber M, Nnadi O, Zhuang W, Hillson NJ (2016) A Cas9-based toolkit to program gene expression in *Saccharomyces cerevisiae*. Nucleic Acids Res 45:496
65. ter Linde JJ, Liang H, Davis RW, Steensma HY, van Dijken JP, Pronk JT (1999) Genome-wide transcriptional analysis of aerobic and anaerobic chemostat cultures of *Saccharomyces cerevisiae*. J Bacteriol 181(24): 7409–7413. https://doi.org/10.1128/jb.181.24.7409-7413.1999
66. Kwast KE, Lai L-C, Menda N, James DT 3rd, Aref S, Burke PV (2002) Genomic analyses of anaerobically induced genes in *Saccharomyces cerevisiae*: functional roles of Rox1 and other factors in mediating the anoxic response. J Bacteriol 184(1):250–265. https://doi.org/10.1128/jb.184.1.250-265.2002
67. Blazeck J, Garg R, Reed B, Alper HS (2012) Controlling promoter strength and regulation in *Saccharomyces cerevisiae* using synthetic hybrid promoters. Biotechnol Bioeng 109(11):2884–2895. https://doi.org/10.1002/bit.24552
68. Wang J, Zhai H, Rexida R, Shen Y, Hou J, Bao X (2018) Developing synthetic hybrid promoters to increase constitutive or diauxic shift-induced expression in *Saccharomyces cerevisiae*. FEMS Yeast Res 18(8). https://doi.org/10.1093/femsyr/foy098
69. Alper H, Fischer C, Nevoigt E, Stephanopoulos G (2005) Tuning genetic control through promoter engineering. Proc Natl Acad Sci U S A 102(36):12678. https://doi.org/10.1073/pnas.0504604102
70. Nevoigt E, Kohnke J, Fischer CR, Alper H, Stahl U, Stephanopoulos G (2006) Engineering of promoter replacement cassettes for fine-tuning of gene expression in *Saccharomyces cerevisiae*. Appl Environ Microbiol 72(8): 5266. https://doi.org/10.1128/AEM.00530-06
71. Curran KA, Crook NC, Karim AS, Gupta A, Wagman AM, Alper HS (2014) Design of synthetic yeast promoters via tuning of

nucleosome architecture. Nat Commun 5: 4002. https://doi.org/10.1038/ncomms5002

72. Hoshida H, Kondo M, Kobayashi T, Yarimizu T, Akada R (2017) 5′-UTR introns enhance protein expression in the yeast *Saccharomyces cerevisiae*. Appl Microbiol Biotechnol 101(1):241–251. https://doi.org/10.1007/s00253-016-7891-z
73. Redden H, Alper HS (2015) The development and characterization of synthetic minimal yeast promoters. Nat Commun 6(1): 7810. https://doi.org/10.1038/ncomms8810
74. Decoene T, De Maeseneire SL, De Mey M (2019) Modulating transcription through development of semi-synthetic yeast core promoters. PLoS One 14(11):e0224476. https://doi.org/10.1371/journal.pone.0224476
75. Ellis T, Wang X, Collins JJ (2009) Diversity-based, model-guided construction of synthetic gene networks with predicted functions. Nat Biotechnol 27(5):465–471. https://doi.org/10.1038/nbt.1536
76. Garí E, Piedrafita L, Aldea M, Herrero E (1997) A set of vectors with a tetracycline-regulatable promoter system for modulated gene expression in *Saccharomyces cerevisiae*. Yeast 13(9):837–848. https://doi.org/10.1002/(sici)1097-0061(199707)13:9<837::Aid-yea145>3.0.Co;2-t
77. Ikushima S, Zhao Y, Boeke JD (2015) Development of a tightly controlled off switch for *Saccharomyces cerevisiae* regulated by camphor, a low-cost natural product. G3 (Bethesda) 5(10):1983–1990. https://doi.org/10.1534/g3.114.012765
78. Peng B, Wood RJ, Nielsen LK, Vickers CE (2018) An expanded heterologous GAL promoter collection for Diauxie-inducible expression in *Saccharomyces cerevisiae*. ACS Synth Biol 7(2):748–751. https://doi.org/10.1021/acssynbio.7b00355
79. Zhou P, Xu N, Yang Z, Du Y, Yue C, Xu N, Ye L (2020) Directed evolution of the transcription factor Gal4 for development of an improved transcriptional regulation system in *Saccharomyces cerevisiae*. Enzyme Microb Technol 142:109675. https://doi.org/10.1016/j.enzmictec.2020.109675
80. Ruohonen L, Penttilä M, Keränen S (1991) Optimization of Bacillus alpha-amylase production by *Saccharomyces cerevisiae*. Yeast 7(4):337–346. https://doi.org/10.1002/yea.320070404
81. Ruohonen L, Aalto MK, Keränen S (1995) Modifications to the ADH1 promoter of *Saccharomyces cerevisiae* for efficient production of heterologous proteins. J Biotechnol 39(3): 193–203. https://doi.org/10.1016/0168-1656(95)00024-k
82. Nevoigt E, Fischer C, Mucha O, Matthäus F, Stahl U, Stephanopoulos G (2007) Engineering promoter regulation. Biotechnol Bioeng 96(3):550–558. https://doi.org/10.1002/bit.21129
83. Rajkumar AS, Liu G, Bergenholm D, Arsovska D, Kristensen M, Nielsen J, Jensen MK, Keasling JD (2016) Engineering of synthetic, stress-responsive yeast promoters. Nucleic Acids Res 44(17):e136–e136. https://doi.org/10.1093/nar/gkw553. PMID: 27325743
84. Naseri G, Balazadeh S, Machens F, Kamranfar I, Messerschmidt K, Mueller-Roeber B (2017) Plant-derived transcription factors for orthologous regulation of gene expression in the yeast *Saccharomyces cerevisiae*. ACS Synth Biol 6(9):1742–1756. https://doi.org/10.1021/acssynbio.7b00094
85. Tang H, Wu Y, Deng J, Chen N, Zheng Z, Wei Y, Luo X, Keasling JD (2020) Promoter architecture and promoter engineering in *Saccharomyces cerevisiae*. Metabolites 10(8): 1–20. https://doi.org/10.3390/metabo10080320
86. Da Silva NA, Srikrishnan S (2012) Introduction and expression of genes for metabolic engineering applications in *Saccharomyces cerevisiae*. FEMS Yeast Res 12(2):197–214. https://doi.org/10.1111/j.1567-1364.2011.00769.x
87. Bar-Even A, Noor E, Savir Y, Liebermeister W, Davidi D, Tawfik DS, Milo R (2011) The moderately efficient enzyme: evolutionary and physicochemical trends shaping enzyme parameters. Biochemistry 50(21):4402–4410. https://doi.org/10.1021/bi2002289
88. Adams BG (1972) Induction of galactokinase in *Saccharomyces cerevisiae*: kinetics of induction and glucose effects. J Bacteriol 111(2): 308–315
89. St John TP, Davis RW (1981) The organization and transcription of the galactose gene cluster of *Saccharomyces*. J Mol Biol 152(2): 285–315. https://doi.org/10.1016/0022-2836(81)90244-8
90. Lohr D, Venkov P, Zlatanova J (1995) Transcriptional regulation in the yeast GAL gene family: a complex genetic network. FASEB J

9(9):777–787. https://doi.org/10.1096/fasebj.9.9.7601342

91. Traven A, Jelicic B, Sopta M (2006) Yeast Gal4: a transcriptional paradigm revisited. EMBO Rep 7(5):496–499. https://doi.org/10.1038/sj.embor.7400679. PMID: 16670683
92. Ro D-K, Paradise EM, Ouellet M, Fisher KJ, Newman KL, Ndungu JM, Ho KA, Eachus RA, Ham TS, Kirby J, Chang MCY, Withers ST, Shiba Y, Sarpong R, Keasling JD (2006) Production of the antimalarial drug precursor artemisinic acid in engineered yeast. Nature 440(7086):940. https://doi.org/10.1038/nature04640
93. Hawkins KM, Smolke CD (2006) The regulatory roles of the galactose permease and kinase in the induction response of the GAL network in *Saccharomyces cerevisiae*. J Biol Chem 281(19):13485–13492. https://doi.org/10.1074/jbc.M512317200
94. Nevoigt E (2008) Progress in metabolic engineering of *Saccharomyces cerevisiae*. Microbiol Mol Biol Rev 72(3):379–412. https://doi.org/10.1128/MMBR.00025-07
95. Xu N, Wei L, Liu J (2019) Recent advances in the applications of promoter engineering for the optimization of metabolite biosynthesis. World J Microbiol Biotechnol 35(2):33. https://doi.org/10.1007/s11274-019-2606-0
96. Khalil Ahmad S, Lu Timothy K, Bashor Caleb J, Ramirez Cherie L, Pyenson Nora C, Joung JK, Collins James J (2012) A synthetic biology framework for programming eukaryotic transcription functions. Cell 150(3):647–658. https://doi.org/10.1016/j.cell.2012.05.045
97. Ikushima S, Boeke JD (2017) New orthogonal transcriptional switches derived from Tet repressor homologues for *Saccharomyces cerevisiae* regulated by 2,4-diacetylphloroglucinol and other ligands. ACS Synth Biol 6(3):497–506. https://doi.org/10.1021/acssynbio.6b00205
98. Basehoar AD, Zanton SJ, Pugh BF (2004) Identification and distinct regulation of yeast TATA box-containing genes. Cell 116(5):699–709. https://doi.org/10.1016/s0092-8674(04)00205-3
99. Mogno I, Vallania F, Mitra RD, Cohen BA (2010) TATA is a modular component of synthetic promoters. Genome Res 20(10):1391–1397. https://doi.org/10.1101/gr.106732.110
100. Chen W, Struhl K (1988) Saturation mutagenesis of a yeast his3 "TATA element": genetic evidence for a specific TATA-binding protein. Proc Natl Acad Sci U S A 85(8):2691–2695. https://doi.org/10.1073/pnas.85.8.2691
101. Lubliner S, Regev I, Lotan-Pompan M, Edelheit S, Weinberger A, Segal E (2015) Core promoter sequence in yeast is a major determinant of expression level. Genome Res 25(7):1008–1017. https://doi.org/10.1101/gr.188193.114
102. Cuperus JT, Groves B, Kuchina A, Rosenberg AB, Jojic N, Fields S, Seelig G (2017) Deep learning of the regulatory grammar of yeast 5′ untranslated regions from 500,000 random sequences. Genome Res 27(12):2015–2024. https://doi.org/10.1101/gr.224964.117
103. Matsuyama T (2019) Recent developments in terminator technology in *Saccharomyces cerevisiae*. J Biosci Bioeng 128(6):655–661. https://doi.org/10.1016/j.jbiosc.2019.06.006
104. Yamanishi M, Ito Y, Kintaka R, Imamura C, Katahira S, Ikeuchi A, Moriya H, Matsuyama T (2013) A genome-wide activity assessment of terminator regions in *Saccharomyces cerevisiae* provides a "Terminatome" toolbox. ACS Synth Biol 2(6):337–347. https://doi.org/10.1021/sb300116y
105. Curran KA, Karim AS, Gupta A, Alper HS (2013) Use of expression-enhancing terminators in *Saccharomyces cerevisiae* to increase mRNA half-life and improve gene expression control for metabolic engineering applications. Metab Eng 19:88–97. https://doi.org/10.1016/j.ymben.2013.07.001
106. Wei L, Wang Z, Zhang G, Ye B (2017) Characterization of terminators in Saccharomyces cerevisiae and an exploration of factors affecting their strength. ChemBioChem 18(24):2422–2427. https://doi.org/10.1002/cbic.201700516
107. Ito Y, Yamanishi M, Ikeuchi A, Imamura C, Tokuhiro K, Kitagawa T, Matsuyama T (2013) Characterization of five terminator regions that increase the protein yield of a transgene in *Saccharomyces cerevisiae*. J Biotechnol 168(4):486–492. https://doi.org/10.1016/j.jbiotec.2013.09.024
108. Curran KA, Morse NJ, Markham KA, Wagman AM, Gupta A, Alper HS (2015) Short synthetic terminators for improved heterologous gene expression in yeast. ACS Synth Biol 4(7):824–832. https://doi.org/10.1021/sb5003357
109. Morse NJ, Gopal MR, Wagner JM, Alper HS (2017) Yeast terminator function can be modulated and designed on the basis of predictions of nucleosome occupancy. ACS

Synth Biol 6(11):2086–2095. https://doi.org/10.1021/acssynbio.7b00138

110. Lanza AM, Curran KA, Rey LG, Alper HS (2014) A condition-specific codon optimization approach for improved heterologous gene expression in *Saccharomyces cerevisiae*. BMC Syst Biol 8(1):33. https://doi.org/10.1186/1752-0509-8-33

111. Yocum HC, Pham A, Da Silva NA (2021) Successful enzyme colocalization strategies in yeast for increased synthesis of non-native products. Front Bioeng Biotechnol 9:37

112. Cripwell RA, Rose SH, van Zyl WH (2017) Expression and comparison of codon optimised *Aspergillus tubingensis* amylase variants in *Saccharomyces cerevisiae*. FEMS Yeast Res 17(4). https://doi.org/10.1093/femsyr/fox040

113. Agashe D, Martinez-Gomez NC, Drummond DA, Marx CJ (2013) Good codons, bad transcript: large reductions in gene expression and fitness arising from synonymous mutations in a key enzyme. Mol Biol Evol 30(3):549–560. https://doi.org/10.1093/molbev/mss273

114. Curran KA, Leavitt JM, Karim AS, Alper HS (2013) Metabolic engineering of muconic acid production in *Saccharomyces cerevisiae*. Metab Eng 15:55–66. https://doi.org/10.1016/j.ymben.2012.10.003

115. Alexeyev MF, Winkler HH (1999) Gene synthesis, bacterial expression and purification of the *Rickettsia prowazekii* ATP/ADP translocase. Biochim Biophys Acta Biomembr 1419(2):299–306. https://doi.org/10.1016/S0005-2736(99)00078-4

116. Weber E, Engler C, Gruetzner R, Werner S, Marillonnet S (2011) A modular cloning system for standardized assembly of multigene constructs. PLoS One 6(2):e16765. https://doi.org/10.1371/journal.pone.0016765. PMID: 21364738

117. Moore SJ, Lai H-E, Kelwick RJR, Chee SM, Bell DJ, Polizzi KM, Freemont PS (2016) EcoFlex: a multifunctional MoClo Kit for *E. coli* synthetic biology. ACS Synth Biol 5(10):1059–1069. https://doi.org/10.1021/acssynbio.6b00031

118. Lawson CE, Martí JM, Radivojevic T, Jonnalagadda SVR, Gentz R, Hillson NJ, Peisert S, Kim J, Simmons BA, Petzold CJ, Singer SW, Mukhopadhyay A, Tanjore D, Dunn JG, Garcia Martin H (2021) Machine learning for metabolic engineering: A review. Metab Eng 63:34–60. https://doi.org/10.1016/j.ymben.2020.10.005

119. Kosuri S, Church GM (2014) Large-scale de novo DNA synthesis: technologies and applications. Nat Methods 11(5):499–507. https://doi.org/10.1038/nmeth.2918

120. Heather JM, Chain B (2016) The sequence of sequencers: the history of sequencing DNA. Genomics 107(1):1–8. https://doi.org/10.1016/j.ygeno.2015.11.003

121. Donohoue PD, Barrangou R, May AP (2018) Advances in industrial biotechnology using CRISPR-Cas systems. Trends Biotechnol 36(2):134–146. https://doi.org/10.1016/j.tibtech.2017.07.007

122. Lacerda MP, Oh EJ, Eckert C (2020) The model system *Saccharomyces cerevisiae* versus emerging non-model yeasts for the production of biofuels. Life 10(11):1–20. https://doi.org/10.3390/life10110299

123. Jensen NB, Strucko T, Kildegaard KR, David F, Maury J, Mortensen UH, Forster J, Nielsen J, Borodina I (2014) EasyClone: method for iterative chromosomal integration of multiple genes in *Saccharomyces cerevisiae*. FEMS Yeast Res 14(2):238–248. https://doi.org/10.1111/1567-1364.12118

124. Pronk JT (2002) Auxotrophic yeast strains in fundamental and applied research. Appl Environ Microbiol 68(5):2095–2100. https://doi.org/10.1128/aem.68.5.2095-2100.2002

125. Alam MT, Zelezniak A, Mülleder M, Shliaha P, Schwarz R, Capuano F, Vowinckel J, Radmanesfahar E, Krüger A, Calvani E, Michel S, Börno S, Christen S, Patil KR, Timmermann B, Lilley KS, Ralser M (2016) The metabolic background is a global player in *Saccharomyces* gene expression epistasis. Nat Microbiol 1: 15030–15030. https://doi.org/10.1038/nmicrobiol.2015.30

126. Dicarlo JE, Norville JE, Mali P, Rios X, Aach J, Church GM (2013) Genome engineering in *Saccharomyces cerevisiae* using CRISPR-Cas systems. Nucleic Acids Res 41: 4336

127. Jessop-Fabre MM, Jakociunas T, Stovicek V, Dai Z, Jensen MK, Keasling JD, Borodina I (2016) EasyClone-MarkerFree: A vector toolkit for marker-less integration of genes into *Saccharomyces cerevisiae* via CRISPR-Cas9. Biotechnol J 11(8):1110–1117. https://doi.org/10.1002/biot.201600147

128. Jakočiunas T, Rajkumar AS, Zhang J, Arsovska D, Rodriguez A, Jendresen C, Skjødt ML, Nielsen AT, Borodina I, Jensen MK, Keasling JD (2015) CasEMBLR: Cas9-facilitated multiloci genomic integration of *in vivo* assembled DNA parts in *Saccharomyces*

cerevisiae. ACS Synth Biol 4(11):1226–1234. https://doi.org/10.1021/acssynbio.5b00007. PMID: 25781611

129. Mans R, van Rossum HM, Wijsman M, Backx A, Kuijpers N, van den Broek M, Daran-Lapujade P, Pronk JT, van Maris A, Daran J-MG (2015) CRISPR/Cas9: a molecular Swiss army knife for simultaneous introduction of multiple genetic modifications in *Saccharomyces cerevisiae*. FEMS Yeast Res 15(2):fov004. https://doi.org/10.1093/femsyr/fov004
130. Baek S, Utomo JC, Lee JY, Dalal K, Yoon YJ, Ro D-K (2021) The yeast platform engineered for synthetic gRNA-landing pads enables multiple gene integrations by a single gRNA/Cas9 system. Metab Eng 64:111–121. https://doi.org/10.1016/j.ymben.2021.01.011
131. Shi S, Liang Y, Zhang MM, Ang EL, Zhao H (2016) A highly efficient single-step, markerless strategy for multi-copy chromosomal integration of large biochemical pathways in *Saccharomyces cerevisiae*. Metab Eng 33:19–27. https://doi.org/10.1016/j.ymben.2015.10.011. PMID: 26546089
132. Bourgeois L, Pyne ME, Martin VJJ (2018) A highly characterized synthetic landing pad system for precise multicopy gene integration in yeast. ACS Synth Biol 7(11):2675–2685. https://doi.org/10.1021/acssynbio.8b00339
133. Huang S, Geng A (2020) High-copy genome integration of 2,3-butanediol biosynthesis pathway in Saccharomyces cerevisiae via in vivo DNA assembly and replicative CRISPR-Cas9 mediated delta integration. J Biotechnol 310:13–20. https://doi.org/10.1016/j.jbiotec.2020.01.014
134. Ryan OW, Skerker JM, Maurer MJ, Li X, Tsai JC, Poddar S, Lee ME, DeLoache W, Dueber JE, Arkin AP, Cate JHD (2014) Selection of chromosomal DNA libraries using a multiplex CRISPR system. eLife 3:e03703. https://doi.org/10.7554/eLife.03703
135. Gao Y, Zhao Y (2014) Self-processing of ribozyme-flanked RNAs into guide RNAs *in vitro* and *in vivo* for CRISPR-mediated genome editing. J Integr Plant Biol 56(4):343–349. https://doi.org/10.1111/jipb.12152
136. Ferreira R, Skrekas C, Nielsen J, David F (2018) Multiplexed CRISPR/Cas9 genome editing and gene regulation using Csy4 in *Saccharomyces cerevisiae*. ACS Synth Biol 7(1):10–15. https://doi.org/10.1021/acssynbio.7b00259
137. McCarty NS, Shaw WM, Ellis T, Ledesma-Amaro R (2019) Rapid assembly of gRNA arrays via modular cloning in yeast. ACS Synth Biol 8(4):906–910. https://doi.org/10.1021/acssynbio.9b00041
138. Verwaal R, Buiting-Wiessenhaan N, Dalhuijsen S, Roubos JA (2018) CRISPR/Cpf1 enables fast and simple genome editing of *Saccharomyces cerevisiae*. Yeast 35(2):201–211. https://doi.org/10.1002/yea.3278
139. Zhang Y, Wang J, Wang Z, Zhang Y, Shi S, Nielsen J, Liu Z (2019) A gRNA-tRNA array for CRISPR-Cas9 based rapid multiplexed genome editing in *Saccharomyces cerevisiae*. Nat Commun 10(1):1053. https://doi.org/10.1038/s41467-019-09005-3. PMID: 30837474
140. Kildegaard KR, Tramontin LRR, Chekina K, Li M, Goedecke TJ, Kristensen M, Borodina I (2019) CRISPR/Cas9-RNA interference system for combinatorial metabolic engineering of *Saccharomyces cerevisiae*. Yeast 36(5):237–247. https://doi.org/10.1002/yea.3390
141. Jakočiūnas T, Pedersen LE, Lis AV, Jensen MK, Keasling JD (2018) CasPER, a method for directed evolution in genomic contexts using mutagenesis and CRISPR/Cas9. Metab Eng 48:288–296. https://doi.org/10.1016/j.ymben.2018.07.001
142. Deaner M, Alper HS (2017) Systematic testing of enzyme perturbation sensitivities via graded dCas9 modulation in *Saccharomyces cerevisiae*. Metab Eng 40:14–22. https://doi.org/10.1016/j.ymben.2017.01.012
143. Naseri G, Behrend J, Rieper L, Mueller-Roeber B (2019) COMPASS for rapid combinatorial optimization of biochemical pathways based on artificial transcription factors. Nat Commun 10(1):2615. https://doi.org/10.1038/s41467-019-10224-x
144. Zhou Y, Li G, Dong J, Xing X-h, Dai J, Zhang C (2018) MiYA, an efficient machine-learning workflow in conjunction with the YeastFab assembly strategy for combinatorial optimization of heterologous metabolic pathways in *Saccharomyces cerevisiae*. Metab Eng 47:294–302. https://doi.org/10.1016/j.ymben.2018.03.020
145. Zalatan JG, Lee ME, Almeida R, Gilbert LA, Whitehead EH, La Russa M, Tsai JC, Weissman JS, Dueber JE, Qi LS, Lim WA (2015) Engineering complex synthetic transcriptional programs with CRISPR RNA scaffolds. Cell 160(1–2):339–350. https://doi.org/10.1016/j.cell.2014.11.052

146. Jensen ED, Ferreira R, Jakočiūnas T, Arsovska D, Zhang J, Ding L, Smith JD, David F, Nielsen J, Jensen MK, Keasling JD (2017) Transcriptional reprogramming in yeast using dCas9 and combinatorial gRNA strategies. Microb Cell Fact 16(1):46. https://doi.org/10.1186/s12934-017-0664-2
147. Smith JD, Suresh S, Schlecht U, Wu M, Wagih O, Peltz G, Davis RW, Steinmetz LM, Parts L, St. Onge RP (2016) Quantitative CRISPR interference screens in yeast identify chemical-genetic interactions and new rules for guide RNA design. Genome Biol 17(1):45. https://doi.org/10.1186/s13059-016-0900-9
148. Lian J, HamediRad M, Hu S, Zhao H (2017) Combinatorial metabolic engineering using an orthogonal tri-functional CRISPR system. Nat Commun 8(1):1688. https://doi.org/10.1038/s41467-017-01695-x
149. Si T, Luo Y, Bao Z, Zhao H (2015) RNAi-assisted genome evolution in *Saccharomyces cerevisiae* for complex phenotype engineering. ACS Synth Biol 4(3):283–291. https://doi.org/10.1021/sb500074a
150. Kang H-S, Charlop-Powers Z, Brady SF (2016) Multiplexed CRISPR/Cas9- and TAR-mediated promoter engineering of natural product biosynthetic gene clusters in yeast. ACS Synth Biol 5(9):1002–1010. https://doi.org/10.1021/acssynbio.6b00080
151. Si T, Chao R, Min Y, Wu Y, Ren W, Zhao H (2017) Automated multiplex genome-scale engineering in yeast. Nat Commun 8:15187. https://doi.org/10.1038/ncomms15187
152. Lian J, Jin R, Zhao H (2016) Construction of plasmids with tunable copy numbers in *Saccharomyces cerevisiae* and their applications in pathway optimization and multiplex genome integration. Biotechnol Bioeng 113(11):2462–2473. https://doi.org/10.1002/bit.26004
153. Bao Z, HamediRad M, Xue P, Xiao H, Tasan I, Chao R, Liang J, Zhao H (2018) Genome-scale engineering of *Saccharomyces cerevisiae* with single-nucleotide precision. Nat Biotechnol 36(6):505. https://doi.org/10.1038/nbt.4132. PMID: 29734295
154. Ravikumar A, Arzumanyan GA, Obadi MKA, Javanpour AA, Liu CC (2018) Scalable, continuous evolution of genes at mutation rates above genomic error thresholds. Cell 175(7):1946–1957.e1913. https://doi.org/10.1016/j.cell.2018.10.021
155. Nishida K, Arazoe T, Yachie N, Banno S, Kakimoto M, Tabata M, Mochizuki M, Miyabe A, Araki M, Hara KY, Shimatani Z, Kondo A (2016) Targeted nucleotide editing using hybrid prokaryotic and vertebrate adaptive immune systems. Science 353(6305):aaf8729. https://doi.org/10.1126/science.aaf8729
156. Tou CJ, Schaffer DV, Dueber JE (2020) Targeted diversification in the *S. cerevisiae* genome with CRISPR-guided DNA polymerase I. ACS Synth Biol 9(7):1911–1916. https://doi.org/10.1021/acssynbio.0c00149
157. Barbieri EM, Muir P, Akhuetie-Oni BO, Yellman CM, Isaacs FJ (2017) Precise editing at DNA replication forks enables multiplex genome engineering in eukaryotes. Cell 171(6):1453–1467.e1413. https://doi.org/10.1016/j.cell.2017.10.034
158. Liu X, Ding W, Jiang H (2017) Engineering microbial cell factories for the production of plant natural products: from design principles to industrial-scale production. Microb Cell Fact 16(1):125. https://doi.org/10.1186/s12934-017-0732-7
159. Hossain GS, Saini M, Miyake R, Ling H, Chang MW (2020) Genetic biosensor design for natural product biosynthesis in microorganisms. Trends Biotechnol 38(7):797–810. https://doi.org/10.1016/j.tibtech.2020.03.013
160. Tan SZ, Prather KL (2017) Dynamic pathway regulation: recent advances and methods of construction. Curr Opin Chem Biol 41:28–35. https://doi.org/10.1016/j.cbpa.2017.10.004
161. Cheng F, Tang XL, Kardashliev T (2018) Transcription factor-based biosensors in high-throughput screening: advances and applications. Biotechnol J 13(7):e1700648. https://doi.org/10.1002/biot.201700648
162. Qiu C, Zhai H, Hou J (2019) Biosensors design in yeast and applications in metabolic engineering. FEMS Yeast Res 19(8):foz082. https://doi.org/10.1093/femsyr/foz082
163. Rogers JK, Taylor ND, Church GM (2016) Biosensor-based engineering of biosynthetic pathways. Curr Opin Biotechnol 42:84–91. https://doi.org/10.1016/j.copbio.2016.03.005
164. Li S, Si T, Wang M, Zhao H (2015) Development of a synthetic malonyl-CoA sensor in *Saccharomyces cerevisiae* for intracellular metabolite monitoring and genetic screening. ACS Synth Biol 4(12):1308–1315. https://doi.org/10.1021/acssynbio.5b00069
165. David F, Nielsen J, Siewers V (2016) Flux control at the malonyl-CoA node through

hierarchical dynamic pathway regulation in *Saccharomyces cerevisiae*. ACS Synth Biol 5(3):224–233. https://doi.org/10.1021/acssynbio.5b00161

166. Dabirian Y, Li X, Chen Y, David F, Nielsen J, Siewers V (2019) Expanding the dynamic range of a transcription factor-based biosensor in *Saccharomyces cerevisiae*. ACS Synth Biol 8(9):1968–1975. https://doi.org/10.1021/acssynbio.9b00144
167. Skjoedt ML, Snoek T, Kildegaard KR, Arsovska D, Eichenberger M, Goedecke TJ, Rajkumar AS, Zhang J, Kristensen M, Lehka BJ, Siedler S, Borodina I, Jensen MK, Keasling JD (2016) Engineering prokaryotic transcriptional activators as metabolite biosensors in yeast. Nat Chem Biol 12(11):951–958. https://doi.org/10.1038/nchembio.2177
168. Wang R, Cress BF, Yang Z, Hordines JC 3rd, Zhao S, Jung GY, Wang Z, Koffas MAG (2019) Design and characterization of biosensors for the screening of modular assembled naringenin biosynthetic library in *Saccharomyces cerevisiae*. ACS Synth Biol 8(9):2121–2130. https://doi.org/10.1021/acssynbio.9b00212
169. Siedler S, Khatri NK, Zsohar A, Kjaerbolling I, Vogt M, Hammar P, Nielsen CF, Marienhagen J, Sommer MOA, Joensson HN (2017) Development of a bacterial biosensor for rapid screening of yeast p-coumaric acid production. ACS Synth Biol 6(10):1860–1869. https://doi.org/10.1021/acssynbio.7b00009
170. Castano-Cerezo S, Fournie M, Urban P, Faulon JL, Truan G (2019) Development of a biosensor for detection of benzoic acid derivatives in *Saccharomyces cerevisiae*. Front Bioeng Biotechnol 7:372. https://doi.org/10.3389/fbioe.2019.00372
171. D'Ambrosio V, Pramanik S, Goroncy K, Jakočiūnas T, Schönauer D, Davari MD, Schwaneberg U, Keasling JD, Jensen MK (2020) Directed evolution of VanR biosensor specificity in yeast. Biotechnol Notes 1:9–15. https://doi.org/10.1016/j.biotno.2020.01.002
172. Jester BW, Tinberg CE, Rich MS, Baker D, Fields S (2018) Engineered biosensors from dimeric ligand-binding domains. ACS Synth Biol 7(10):2457–2467. https://doi.org/10.1021/acssynbio.8b00242
173. Feng J, Jester BW, Tinberg CE, Mandell DJ, Antunes MS, Chari R, Morey KJ, Rios X, Medford JI, Church GM, Fields S, Baker D (2015) A general strategy to construct small molecule biosensors in eukaryotes. Elife 4:e10606. https://doi.org/10.7554/eLife.10606
174. Tinberg CE, Khare SD, Dou J, Doyle L, Nelson JW, Schena A, Jankowski W, Kalodimos CG, Johnsson K, Stoddard BL, Baker D (2013) Computational design of ligand-binding proteins with high affinity and selectivity. Nature 501(7466):212–216. https://doi.org/10.1038/nature12443
175. Chou HH, Keasling JD (2013) Programming adaptive control to evolve increased metabolite production. Nat Commun 4:2595. https://doi.org/10.1038/ncomms3595
176. Michener JK, Smolke CD (2012) High-throughput enzyme evolution in *Saccharomyces cerevisiae* using a synthetic RNA switch. Metab Eng 14(4):306–316. https://doi.org/10.1016/j.ymben.2012.04.004
177. DeLoache WC, Russ ZN, Narcross L, Gonzales AM, Martin VJ, Dueber JE (2015) An enzyme-coupled biosensor enables (S)-reticuline production in yeast from glucose. Nat Chem Biol 11(7):465–471. https://doi.org/10.1038/nchembio.1816
178. Boussebayle A, Torka D, Ollivaud S, Braun J, Bofill-Bosch C, Dombrowski M, Groher F, Hamacher K, Suess B (2019) Next-level riboswitch development-implementation of Capture-SELEX facilitates identification of a new synthetic riboswitch. Nucleic Acids Res 47(9):4883–4895. https://doi.org/10.1093/nar/gkz216
179. Klauser B, Atanasov J, Siewert LK, Hartig JS (2015) Ribozyme-based aminoglycoside switches of gene expression engineered by genetic selection in *S. cerevisiae*. ACS Synth Biol 4(5):516–525. https://doi.org/10.1021/sb500062p
180. Ehrenworth AM, Claiborne T, Peralta-Yahya P (2017) Medium-throughput screen of microbially produced serotonin via a G-protein-coupled receptor-based sensor. Biochemistry 56(41):5471–5475. https://doi.org/10.1021/acs.biochem.7b00605
181. Mannan AA, Liu D, Zhang F, Oyarzun DA (2017) Fundamental design principles for transcription-factor-based metabolite biosensors. ACS Synth Biol 6(10):1851–1859. https://doi.org/10.1021/acssynbio.7b00172
182. Ambri F, D'Ambrosio V, Di Blasi R, Maury J, Jacobsen SAB, McCloskey D, Jensen MK, Keasling JD (2020) High-resolution scanning of optimal biosensor reporter promoters in yeast. ACS Synth Biol 9(2):218–226. https://doi.org/10.1021/acssynbio.9b00333

183. Yang D, Kim WJ, Yoo SM, Choi JH, Ha SH, Lee MH, Lee SY (2018) Repurposing type III polyketide synthase as a malonyl-CoA biosensor for metabolic engineering in bacteria. Proc Natl Acad Sci U S A 115(40): 9835–9844. https://doi.org/10.1073/pnas.1808567115
184. Lengger B, Jensen MK (2020) Engineering G protein-coupled receptor signalling in yeast for biotechnological and medical purposes. FEMS Yeast Res 20(1):foz087. https://doi.org/10.1093/femsyr/foz087
185. Goffeau A, Barrell BG, Bussey H, Davis RW, Dujon B, Feldmann H, Galibert F, Hoheisel JD, Jacq C, Johnston M, Louis EJ, Mewes HW, Murakami Y, Philippsen P, Tettelin H, Oliver SG (1996) Life with 6000 Genes. Science 274(5287):546. https://doi.org/10.1126/science.274.5287.546
186. Zhang W, Mitchell LA, Bader JS, Boeke JD (2020) Synthetic genomes. Annu Rev Biochem 89(1):77–101. https://doi.org/10.1146/annurev-biochem-013118-110704
187. Helmy M, Smith D, Selvarajoo K (2020) Systems biology approaches integrated with artificial intelligence for optimized metabolic engineering. Metab Eng Commun 11: e00149. https://doi.org/10.1016/j.mec.2020.e00149
188. Dymond J, Boeke J (2012) The *Saccharomyces cerevisiae* SCRaMbLE system and genome minimization. Bioeng Bugs 3(3):168–171. https://doi.org/10.4161/bbug.19543
189. Jia B, Wu Y, Li B-Z, Mitchell LA, Liu H, Pan S, Wang J, Zhang H-R, Jia N, Li B, Shen M, Xie Z-X, Liu D, Cao Y-X, Li X, Zhou X, Qi H, Boeke JD, Yuan Y-J (2018) Precise control of SCRaMbLE in synthetic haploid and diploid yeast. Nat Commun 9(1):1933. https://doi.org/10.1038/s41467-018-03084-4
190. Blount BA, Gowers GOF, Ho JCH, Ledesma-Amaro R, Jovicevic D, McKiernan RM, Xie ZX, Li BZ, Yuan YJ, Ellis T (2018) Rapid host strain improvement by *in vivo* rearrangement of a synthetic yeast chromosome. Nat Commun 9(1):1932. https://doi.org/10.1038/s41467-018-03143-w
191. Liu W, Luo Z, Wang Y, Pham NT, Tuck L, Pérez-Pi I, Liu L, Shen Y, French C, Auer M, Marles-Wright J, Dai J, Cai Y (2018) Rapid pathway prototyping and engineering using in vitro and in vivo synthetic genome SCRaMbLE-in methods. Nat Commun 9(1):1936. https://doi.org/10.1038/s41467-018-04254-0

Chapter 19

Investigating Plant Biosynthetic Pathways Using Heterologous Gene Expression: Yeast as a Heterologous Host

Shanhui Xu, Sheng Wu, and Yanran Li

Abstract

Plant natural products (PNPs) are valuable resources for the development of pharmaceuticals and agrochemicals, yet the biosynthesis and metabolism of PNPs are largely unknown. Heterologous pathway reconstitution is a heavily adopted strategy in secondary metabolism characterization. Yeast systems have been broadly utilized in the heterologous production of PNPs and have been considered as a promising platform to investigate plant biosynthetic pathways. Here, we describe the reconstitution and verification of the upstream part of brassinolide biosynthesis in *S. cerevisiae* using this method.

Key words *Saccharomyces cerevisiae*, Yeast, Biosynthesis, Natural products, Heterologous gene expression, Metabolic engineering

1 Introduction

Plant natural products (PNPs) are of great pharmaceutical significance and exhibit a wide range of bioactivities [1]. One major challenge in developing and using PNPs for medicinal and agricultural purposes is their limited accessibility because of the low abundance in nature and their difficult chemical synthesis due to the highly complex structures. Bioproduction from engineered plants or microbes provides a promising alternative for the sourcing of these valuable molecules, yet the biosynthetic routes of most PNPs are largely unknown. Heterologous expression of biosynthetic pathways is a well-received strategy to characterize natural product biosynthesis. For example, heterologous reconstitution of phytochemical biosynthesis in model plants such as *Arabidopsis thaliana* and *Nicotiana benthamiana* has been demonstrated as a highly

Shanhui Xu and Sheng Wu contributed equally to this chapter.

Elizabeth Skellam (ed.), *Engineering Natural Product Biosynthesis: Methods and Protocols*, Methods in Molecular Biology, vol. 2489, https://doi.org/10.1007/978-1-0716-2273-5_19,

promising and successful strategy to elucidate plant secondary metabolism [2]. Biosynthesis of a number of PNPs has been successfully elucidated recently by employing this heterologous reconstitution strategy, including but not limited to bietoposide aglycone from mayapple [3], vinblastine from Madagascar periwinkle [4, 5], diosgenin from Himalayan Paris and fenugreek [6].

Recently, yeast systems have been frequently utilized in the heterologous production of PNPs because they are fermentable, fast-growing, and metabolically robust [7–10]. It is thus believed that the bioproduction and biosynthesis could be achieved and elucidated at the same time through reconstituting the putative plant biosynthetic pathways in *S. cerevisiae* in a step-wise fashion [11]. *S. cerevisiae* as a heterologous microbial host: (1) enables more efficient enzyme functional characterization and stable genomic modification because it is easier to genetically engineer; (2) allows high-throughput screening because yeast is a unicellular microorganism; and (3) provides a clean background free of related enzymes and metabolites, which results in clearer characterization of the enzyme function. In addition, such pathway reconstitution strategies allow greater access to related pathway intermediates, which are normally low in abundance in the native producing hosts, thereby greatly aiding in the characterization of pathway enzymes.

To simultaneously reconstitute and investigate the putative plant biosynthetic pathway in *S. cerevisiae* in a stepwise fashion, four major steps are performed iteratively (Fig. 1). The genes putatively involved in the PNP biosynthesis are generally proposed through functional genomics in plants [2, 14, 15] (Fig. 1, step I). The candidate genes will then be expressed in yeast strains producing the corresponding precursors (Fig. 1, step II; methods as described in Subheading 3.1). If desired functions are detected, the identified genes will be integrated into the corresponding strain to extend the biosynthetic pathway in yeast (Fig. 1, step III; methods as described in Subheading 3.2), which can be used to characterize downstream enzymes. Often, the titer of the pathway intermediates in yeast needs to be enhanced to provide enough substrates for the functional identification of enzymes catalyzing the subsequent step, and this is generally achieved through metabolic engineering and protein engineering of the corresponding plant enzymes (Fig. 1, step IV; methods as described in Subheading 3.3). The engineering strategies to optimize the performance of heterologous plant pathways in yeast have been carefully discussed in several recent reviews [16–21]. Throughout the heterologous gene expression and engineering in yeast, metabolites synthesized from the engineered yeast strains are always analyzed to identify the function of candidate genes and quantify the production of the target compounds (Fig. 1, step V; methods as described in Subheading 3.4).

Here, we use the reconstitution and characterization of the upstream part of brassinolide (BL) biosynthesis as an example to illustrate this procedure in detail [22]. We engineered yeast to

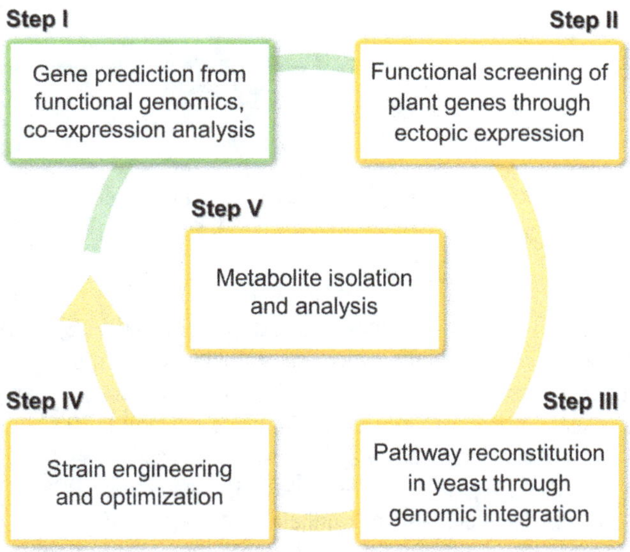

Fig. 1 Flowchart depicting steps to investigate plant biosynthetic pathways using yeast-based bioproduction platforms. Here, we mainly demonstrate the methods of steps highlighted in yellow

produce campesterol, a major phytosterol and the precursor of BL [12, 13]. Δ^7-sterol-C5-desaturase (DWF7), 7-dehydrocholesterol reductase (DWF5), and $\Delta^{24(28)}$-sterol reductase (DWF1) were first introduced into *S. cerevisiae* to convert episterol (a native precursor to ergosterol in yeast) to campesterol. The mevalonate (MVA) pathway was then upregulated to enhance the supply of squalene, the general precursor of the sterol biosynthetic pathway. Subsequently, acyltransferases (ARE1, ARE2) and $\Delta^{24(28)}$-sterol reductase (ERG4) were inactivated from the campesterol-producing strain so that campesterol produced is accessible to downstream synthetic enzymes. The resulting yeast strain enabled us to reconstitute the activities of CYP90A1 and CYP90B1, two BL-synthetic enzymes functioning after the synthesis of campesterol and validate their reaction sequence in BL biosynthesis (Fig. 2).

2 Materials

2.1 Ectopic Gene Expression

2.1.1 Strains and Plasmids (Table 1)

2.1.2 Reagents and Reagent Preparation

1. Polymerases used for polymerase chain reaction (PCR): Q5 High-Fidelity DNA polymerase and buffers; Expand High Fidelity PCR System; Taq DNA polymerase; Phusion.
2. dNTP mix (10 mM).

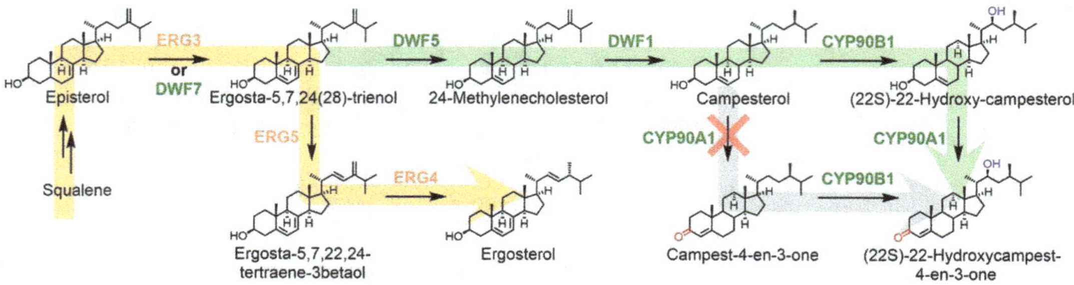

Fig. 2 Engineered biosynthetic pathway of campesterol and (*22S*)-22-Hydroxycampest-4-en-3-one in yeast. The yellow arrow represents the native ergosterol biosynthetic pathway in yeast; the green arrow represents the plant pathway of campesterol and (*22S*)-22-Hydroxycampest-4-en-3-one through campesterol; the grey arrow represents an alternative pathway from campesterol to (*22S*)-22-Hydroxycampest-4-en-3-one that is excluded according to yeast pathway reconstitution results. Yeast and plant enzymes are marked in yellow and green, respectively. ERG3: Δ^7-sterol-C5(6)-desaturase; ERG5: C22-sterol desaturase; ERG4: $\Delta^{24(28)}$-sterol reductase (in ergosterol biosynthetic pathway); DWF7: Δ^7-sterol-C5(6)-desaturase; DWF5: C7(8)-reductase; DWF1: $\Delta^{24(28)}$-sterol reductase (in campesterol biosynthetic pathway); CYP90B1: C22-hydroxylase; CYP90A1: C3-oxidase; FPP: Farnesyl diphosphate

Table 1
Strains and plasmids used in ectopic gene expression

Strain	Genotype	References
S. cerevisiae CEN.PK2-1D	*MATα, ura3-52, trp1-289, leu2-3, 112 his3Δ1, MAL2-8C, SUC2*	[23]
E. coli Top10	*mcrA, Δ(mrr-hsdRMS-mcrBC), Phi80lacZ(del)M15, ΔlacX74, deoR, recA1, araD139, Δ(ara-leu)7697, galU, galK, rpsL(SmR), endA1, nupG*	Invitrogen
pDONR221	*attP1-ccdB-attP2, kanR, pUC ori*	Invitrogen
pAG414GPD-ccdB	*Centromeric CEN/ARS, TRP1,* P_{GPD}*-attR1-ccdB-attR2-*T_{CYC1}*, ampR, ColE1 ori, F1 ori*	[24]
pCS2814 (pAG416-ccdB)	*Centromeric CEN/ARS, URA3, attR1-ccdB-attR2, ampR, ColE1 ori, F1 ori*	[11]
pCS3129 (pTPI1-CYP82Y1)	*attL1-*P_{TPI1}*-CYP82Y1-*T_{STE2}*-attL2, kanR, pUC_origin*	[11]
pYL571 (pAG416TPI1-CYP82Y1)	*Centromeric CEN/ARS, URA3,* P_{TPI1}*-CYP82Y1-*T_{STE2}*, ampR, ColE1 ori, F1 ori*	This work
pYL624 (pAG415GPD-DWF1)	*Centromeric CEN/ARS, LEU2,* P_{GPD}*-DWF1-*T_{CYC1}*, ampR, ColE1 ori, F1 ori*	[22]
pYL625 (pAG414GPD-DWF5)	*Centromeric CEN/ARS, TRP1,* P_{GPD}*-DWF5-*T_{CYC1}*, ampR, ColE1 ori, F1 ori*	[22]

3. DNA ladder.
4. Gateway™ pDONR221 Vector.
5. Gateway™ BP Clonase ™ II Enzyme mix.
6. Gateway™ LR Clonase ™ II Enzyme mix.
7. Gibson one-pot, isothermal DNA assembly mixture: 320 μL 5× ISO buffer [25], 0.64 μL T5 Exonuclease, 20 μL Phusion, 40 μL Taq ligase, 820 μL water.
8. LB broth (Luria–Bertani): 25 g/L in water, autoclave before use.
9. LB agar: 40 g/L in water, autoclave and pour plates.
10. Agarose: 0.8 g/100 mL in water, microwave to dissolve.
11. 10 mg/mL ethidium bromide.
12. Kanamycin: stock concentration 35 mg/mL in water, final working solution 35 μg/mL. Filter to sterilize.
13. Ampicillin: stock concentration 50 mg/mL in water, final working solution 50 μg/mL. Filter to sterilize.
14. Competent *E. coli* TOP 10 cells.
15. QIAprep Spin Miniprep Kit.
16. Zymoclean Gel DNA Recovery Kit.
17. *S. cerevisiae* Advanced Gateway Destination Vector Kit (Addgene) [24].
18. 1 M lithium acetate (LiAc): Filter to sterilize.
19. 50% Polyethylene glycol 4000 (PEG 4000) solution: 50 g/100 mL in water (w/v). Filter to sterilize.
20. 10 mg/mL salmon sperm DNA solution (SSD).
21. 10× TE buffer: 1 mL of 1 M EDTA buffer, 10 mL of 1 M Tris-HCl, 89 mL of water, pH 8. Filter to sterilize.
22. 0.1 M LiAc-TE buffer: 10 mL 1 M LiAc, 10 mL 10× TE buffer, 80 mL water.
23. 40% PEG-LiAc solution: 80 mL 50% PEG 4000, 10 mL 10× TE buffer, 10 mL 1 M LiAc. Store at room temperature and avoid light.
24. Yeast media, amino acid dropout (D.O.) mixes: 10× D.O. solutions are made based on the instruction from the vendor. Filter to sterilize.
25. Synthetic defined medium (SDM): 1.7 g yeast nitrogen base without amino acids and ammonium sulfate, 5.0 g $(NH_4)_2SO_4$, 20 g agar (do not add agar if making liquid medium), *quantum satis* (q.s.) to 800 mL with deionized (D.I.) water, then autoclave. Add 100 mL of 20% (w/v) glucose or other carbon sources, 100 mL of 10× D.O. solution.

Table 2
Primers needed for cloning of ectopic gene expression

Primer name	Sequence
CYP90B1-*attB1*-F	5′-ggggacaagtttgtacaaaaaagcaggctTCTATACAatgttcgaaacagagcatcatactc-3′
CYP90B1-*attB2*-R	5′-ggggaccactttgtacaagaaagctgggtcTTACAGAATACGAGAAACCCTAATAGGC-3′
CPY90B1-499 bp-F	5′-gtgggtggaaagtgttaccgg-3′
M13R	5′-caggaaacagctatgac-3′
pTPI1-yCYP90A1-F	5′-aaaaaacacatacataaactaaaatatacaATGGCTTTCACTGCTTTCTTG-3′
pTPI1-yCYP90A1-R	5′-ttacttttcaaagccgtaaattttgatcaTTAGGTAGCGAAATCCCTTCTCT-3′
TPI1-STE2-F	5′-tgatcaaaatttacggctttgaaaaagtaa-3′
TPI1-STE2-R	5′-tgtatatttagtttatgtatgtgtttttgtag-3′

26. YPD complex medium (1 L): 10 g yeast extract, 20 g peptone, 80 mg adenine, 20 g agar (do not add agar if making liquid medium), q.s. to 900 mL with D.I. water and autoclave. Add 100 mL of 20% (w/v) glucose solution.
27. Primers (Table 2).

2.1.3 Equipment

1. Thermal cycler.
2. Horizontal Electrophoresis Systems.
3. Gel imaging system.
4. Microvolume spectrophotometer.
5. Vortex mixer.
6. Incubator and shaker with temperature setting.
7. Microcentrifuge.
8. Water bath.

2.2 Pathway Reconstruction Through Genomic Integration

2.2.1 Strains and Plasmids (Table 3)

2.2.2 Reagents and Reagent Preparation

1. G418 disulfate salt: stock concentration 200 mg/mL in water, final working solution 200 μg/mL. Filter to sterilize.
2. Lyticase buffer: 22.5 μL of Tween 20, 22.5 μL of NP-40, 83.3 μL of 3 M KCl, 50 μL of 1 M Tris-HCl at pH 8.3, 150 μL of 50 mM $MgCl_2$ and 5 mg lyticase. Make up to 5 mL total volume with water (4.67 mL). Store at −20 °C.
3. Primers (Table 4).

Table 3
Strains and plasmids used in genomic integration-based pathway reconstruction

Strain	Genotype	Reference
CEN.PK2-1D	*MATα, ura3-52, trp1-289, leu2-3, 112 his3Δ1, MAL2-8C, SUC2*	[23]
pYL215 (pPGK1-DWF5)	*attL1-*P_{PGK1}*-dwf5-*T_{PHO5}*-attL2, kanR, pUC ori*	[22]
pYL228 (pTPI1-ATR1)	*attL1-*P_{TPI1}*-atr1-*T_{STE2}*-attL2, kanR, pUC ori*	[22]
pYL218 (pPYK1-DWF1)	*attL1-*P_{PYK1}*-dwf1-*T_{MFA1}*-attL2, kanR, pUC ori*	[22]
pYL212 (p416GPD-DWF7)	*Centromeric CEN/ARS, URA3,* P_{GPD}*-DWF7-*T_{CYC1}*, ampR, ColE1 ori, F1 ori*	[22]
pCS1056 (pTEF1-ATR1-KanMX)	*loxP-*P_{AgTEF1}*-kanmx-*T_{AgTEF1}*-loxP, ampR, ColE1 ori*	[11]
pYL223 (pTEF1-ERG10)	*attL1-*P_{TEF1}*-erg10-*T_{CYC1}*-attL2, kanR, pUC ori*	[22]
pYL216 (pPGK1-ERG12)	*attL1-*P_{PGK1}*-erg12-*T_{PHO5}*-attL2, kanR, pUC ori*	[22]
pYL219 (pPYK1-tHMG1)	*attL1-*P_{PYK1}*-tHmg1-*T_{MFA1}*-attL2, kanR, pUC ori*	[22]
pYL226 (pTPI1-ERG13)	*attL1-*P_{TPI1}*-erg13-*T_{STE2}*-attL2, kanR, pUC ori*	[22]
pUG72	*loxP-*P_{KlURA3}*-KlURA3-*T_{KlURA3}*-loxP, ampR, ColE1 ori*	[26]

2.3 Strain Engineering

2.3.1 Strains and Plasmids (Table 5)

2.3.2 Reagents and Reagent Preparation (Same as Subheading 2.2.2)

1. Yeast genomic DNA: prepared using lyticase (procedure as Subheading 3.2.3, **step 3**) or commercial kit (e.g., YeaStar Genomic DNA Kit™).
2. Primers (Table 6).

2.4 Metabolite Isolation and Analysis

2.4.1 Strains and Plasmids (Table 7)

Table 4
Primer sequences needed to clone DNA fragments for YYL55 construction

DNA fragments	Template	Primer sequence
leu2 locus-upstream	Yeast gDNA	F: 5′-tcgagaccaagaagaacattgc-3′ R: 5′-catggcggctattttcattagaatggtatatccttgaaa-3′
P_{PYK1}-*dwf1*-T_{MFA1}	pYL218 (pPYK1-DWF1)	F: 5′-tataccattctaatgaaaatagccgccatgaccccggat-3′ R: 5′-gcttttacgagttcagaattctcttaggattcgattcac-3′
P_{TPI1}-*atr1*-T_{STE2}	pYL228 (pTPI1-ATR1)	F: 5′-atcctaagagaattctgaactcgtaaaagcaaaggtggt-3′ R: 5′-ttcttgcaaatgcctcggaccttaatacattcagacact-3′
P_{PGK1}-*dwf5*-T_{PHO5}	pYL215 (pPGK1-DWF5)	F: 5′-atgtattaaggtccgaggcatttgcaagaattactcgtg-3′ R: 5′-gatccactagtggccttcactcttcctttttttcatttt-3′
loxP-P_{AgTEF1}-*kanmx*-T_{AgTEF1}-*loxP*	pCS1056 (pTEF1-ATR1-KanMX)	F: 5′-aaaaggaagagtgaaggccactagtggatctgatatcac-3′ R: 5′-gagtattgataatgagcggccgtcgacaacccttaatat-3′
P_{GPD}-*dwf7*-T_{CYC1}	pYL212 (p416GPD-DWF7)	F: 5′-gttgtcgacggccgctcattatcaatactcgccatttca-3′ R: 5′-aagagaatctttttagcaaattaaagccttcgagcgtcc-3′
leu2 locus-downstream	Yeast gDNA	F: 5′-aaggctttaatttgctaaaaagattctcttttttatga-3′ R: 5′-cggtccaaagggaataggtg-3′

2.4.2 Reagents and Reagent Preparation

1. Campesterol (purity >98%).
2. Organic solvents used for extraction: dimethylformamide (DMF; Molecular biology grade); absolute ethanol (Molecular biology grade), acetone.
3. Solvents used as mobile solvents on liquid chromatography–mass spectrometry (LC-MS): LC-MS–grade methanol, water, and formic acid.
4. 0.5 mm glass beads.

2.4.3 Equipment

1. Eppendorf concentrator plus.
2. Cell disruptor, for example, FastPrep-24 MP Biomedicals Sample Preparation System.
3. Poroshell 120 EC-C18 (100 mm × 3.0 mm, 100 Å, particle size 2.7 μm; Agilent) or an equivalent.
4. Shimadzu LC-MS 2020 high-performance liquid chromatography–mass spectrometry or an equivalent.
5. High-resolution mass spectrometer (HRMS) and associated software (optional).

Table 5
Strains and plasmids used in strain engineering

Strain	Genotype	Reference
YYL55	*leu2Δ::P_{PYK1}-dwfl-T_{MFA1}, P_{TPI1}-atr1-T_{STE2}, P_{PGK1}-dwf5-T_{PHO5}, loxP-P_{AgTEF1}-kanmx-T_{AgTEF1}-loxP, P_{GPD}-dwf7-T_{CYC1}*	[22]
YYL56	*leu2Δ::P_{PYK1}-dwfl-T_{MFA1}, P_{TPI1}-atr1-T_{STE2}, P_{PGK1}-dwf5-T_{PHO5}, P_{GPD}-dwf7-T_{CYC1};* *ybl059wΔ::P_{PGK1}-erg12-T_{PHO5}, P_{TEF1}-erg10-T_{CYC1}, P_{PYK1}-tHmg1-T_{MFA1}, P_{TPI1}-erg13-T_{STE2}*	[22]
YYL64	*leu2Δ::P_{PYK1}-dwfl-T_{MFA1}, P_{TPI1}-atr1-T_{STE2}, P_{PGK1}-dwf5-T_{PHO5}, P_{GPD}-dwf7-T_{CYC1};* *ybl059wΔ::P_{PGK1}-erg12-T_{PHO5}, P_{TEF1}-erg10-T_{CYC1}, P_{PYK1}-tHmg1-T_{MFA1}, P_{TPI1}-erg13-T_{STE2};* *are1Δ; are2Δ*	[22]
pYL223 (pTEF1-ERG10)	*attL1-P_{TEF1}-erg10-T_{CYC1}-attL2, kanR, pUC ori*	[22]
pYL216 (pPGK1-ERG12)	*attL1-P_{PGK1}-erg12-T_{PHO5}-attL2, kanR, pUC ori*	[22]
pYL219 (pPYK1-tHMG1)	*attL1-P_{PYK1}-tHmg1-T_{MFA1}-attL2, kanR, pUC ori*	[22]
pYL226 (pTPI1-ERG13)	*attL1-P_{TPI1}-erg13-T_{STE2}-attL2, kanR, pUC ori*	[22]
pSH63	*Centromeric CEN/ARS, TRP1,* P_{GAL1}-*Cre*-T_{CYC1}, *ampR, ColE1 ori*	[26]

3 Methods

3.1 Ectopic Gene Expression

The biosynthetic genes can be introduced and expressed in yeast through genomic integration or from plasmids. Generally, plasmids are more convenient and flexible to handle than integration, but have to be maintained in the presence of certain selection pressure. Auxotrophic selectable markers (e.g., *LEU2*, *TRP1*, *HIS3*, *URA3*) are commonly utilized in yeast ectopic expression. Due to the rapid turnaround rate, ectopic expression is generally utilized for functional screening of candidate genes of unknown functions. Ectopic expression is also more convenient when the expression condition (i.e., expression level and induction timing) is unknown and needs to be examined by using different promoters or copy numbers. The natural promoters from *S. cerevisiae* and other yeast strains are currently the most widely utilized and can be divided into several

Table 6
Primer sequences needed to clone DNA fragments for YYL56 construction

DNA fragments	Template	Primer sequence
ybl059w-upstream	Yeast gDNA	F: 5′-cacaattcgaagccgtatctaacgcag-3′ R: 5′-ttcttgcaaatgcctcattcactgatttacaagtgcagt-3′
P_{PGK1}-*erg12*-T_{PHO5}	pYL216 (pPGK1-ERG12)	F: 5′-aaaaggaagagtgaacaaattaaagccttcgagcgtccc-3′ R: 5′-cagatccactagtggatatgcatagcttcaaaatgtttc-3′
P_{TEF1}-*erg10*-T_{CYC1}	pYL223 (pTEF1-ERG10)	F: 5′-gtaaatcagtgaatgaggcatttgcaagaattactcgtg-3′ R: 5′-gaaggctttaatttgttcactcttcctttttttcatttt-3′
loxP-p_{KlURA3}-*KlURA3*-t_{KlURA3}-*loxP*	pUG72	F: 5′-tgaagctatgcatatccactagtggatctgatatcacct-3′ R: 5′-atcctaagagaattcgccctgcaggtcgacaacccttaa-3′
P_{PYK1}-*tHmg1*-T_{MFA1}	pYL219 (pPYK1-tHMG1)	F: 5′-gtcgacctgcagggcgaattctcttaggattcgattcac-3′ R: 5′-atgtattaaggtccgaaaatagccgccatgacccc-3′
P_{TPI1}-*erg13*-T_{STE2}	pYL226 (pTPI1-ERG13)	F: 5′-catggcggctattttcggaccttaatacattcagacac-3′ R: 5′-tatggtataacactatgaactcgtaaaagcaaaggtgg-3′
ybl059w-downstream	Yeast gDNA	F: 5′-gcttttacgagttcatagtgttataccatagtagtagtttcaataata tattccactacttatatgtg-3′ R: 5′-catcatagcgataaagcggaccatcgg-3′

categories: constitutive promoters of different expression strengths (e.g., strong: *GPD* promoter, *TEF1* promoter; weak: *CYC1* promoter) [27], inducible promoters (e.g., *GAL1* promoter is induced by galactose) [27], late-stage promoters (e.g., *HXT7* promoter is induced when glucose is consumed [28], *ADH2* promoter is induced when ethanol is produced [28]). More recently, libraries of synthetic promoters and terminators of minimal length and/or various expression levels have been developed [29–32] and enabled more versatile tuning on the expression level. The expression level can also be tuned by using different types of plasmids (e.g., the high-copy number 2 μ plasmid, the low-copy number centromere plasmid). The low-copy number centromere plasmid is more often utilized due to its higher stability [33, 34]. Yeast expression plasmids can be constructed using various cloning strategies, and here we mainly use Gateway™ cloning [24] and Gibson Assembly [25].

3.1.1 Construction of Yeast Expression Plasmid Using Gateway® Cloning

The construction of the yeast expression plasmid pYL621 (pAG414GPD-yCYP90B1) is utilized as the example to demonstrate this method.

1. The *CYP90B1* gene can be obtained from PCR. Design the forward and reverse primers with the following overhang including *attB* and the Kozak sequences (forward: 5-′-ggggacaagtttgtacaaaaaagcaggcttcTATACA-3′) and (reverse: 5′-ggggacaagtttgtacaaaaaagcaggcttc-3′) before the binding sequences of *CYP90B1*, respectively (primers are CYP90B1-

Table 7
Strains and plasmids used in metabolite isolation and analysis

Strain	Genotype	Reference
YYL67	*leu2Δ::P_{PYK1}-dwf1-T_{MFA1}, P_{TPI1}-atr1-T_{STE2}, P_{PGK1}-dwf5-T_{PHO5}, P_{GPD}-dwf7-T_{CYC1}* *ybl059wΔ::P_{PGK1}-erg12-T_{PHO5}, P_{TEF1}-erg10-T_{CYC1}, P_{PYK1}-tHmg1-T_{MFA1}, P_{TPI1}-erg13-T_{STE2}* *are1Δ; are2Δ; erg4Δ*	[22]
pYL621 (pAG414GPD-yCYP90B1)	*Centromeric CEN/ARS, TRP1, attR1-P_{GPD}-yCYP90B1-T_{CYC1}-attR2, ampR, ColE1 ori, f1 ori*	[22]
pYL622 (pAG416TPI1-yCYP90A1)	*Centromeric CEN/ARS, URA3, attR1-P_{TPI1}-yCYP90A1-T_{STE2}-attR2, ampR, ColE1 ori, f1 ori*	[22]
pCS2812 (pAG414-ccdB)	*Centromeric CEN/ARS, TRP1, attR1-ccdB-attR2, ampR, ColE1 ori, f1 ori*	[11]
pCS2814 (pAG416-ccdB)	*Centromeric CEN/ARS, URA3, attR1-ccdB-attR2, ampR, ColE1 ori, f1 ori*	[11]

attB1-F and CYP90B1-attB2-R as listed in Table 2). Use *A. thaliana* cDNA library as the template to conduct regular PCR using high fidelity DNA polymerase (Q5 or Expand) to obtain the *attB*-flanked DNA fragment encoding the gene of interest.

2. Construct the entry clone through a BP reaction. Set up the BP reaction mixture to include: 50 ng of the *attB*-flanked DNA fragment, 50 ng of pDONR221 (containing *attP* flankings), 0.5 μL of Gateway™ BP Clonase™ II Enzyme mix, q.s. to 2.5 μL with 1× TE buffer (pH 8.0). Incubate the reaction mixture at 25 °C for 1–2 h. Transform into competent *E. coli* TOP 10 cells with the reaction mixture. Plate on an LB plate with 35 μg/mL of kanamycin, then incubate at 37 °C for 12–16 h, and colonies will appear.
3. Pick six colonies and conduct colony PCR to confirm the presence of *CYP90B1* in the plasmid construct. Design the forward primer to include 15–20 bp sequence from *CYP90B1* and use M13R as the reverse primer. The expected PCR product size is preferred to be within 1 kb. Then, make the master solution of 6.4 × PCR reactions using Taq polymerase following the recipes provided by the vendor (without adding templates). Aliquot 15 μL of the master solution into each PCR tube. Pick six colonies from the LB plate from the previous step with the pipette tip, restreak each colony on an LB plate containing kanamycin, and then resuspend the

remaining colonies on the tips (could be invisible) to the PCR mixture. Perform the PCR reactions using Taq polymerase on the thermal cycler, and then check the size of PCR products on the gel (primers are CYP90B1-499 bp-F and M13R as shown in Table 2). The positive one with pENTR-CYP90B1 should exhibit a band of around 500 bp.

4. Inoculate one positive colony into 5 mL LB medium with 35 mg/L of kanamycin and grow at 37 °C, 250 rpm for 12–16 h. Prepare plasmids from the overnight culture by using commercial columns from Qiagen and send for DNA sequencing to verify the entry clone pENTR-CYP90B1. Entry plasmids have *CYP90B1* flanked by *attL* sequences.
5. Construct the yeast expression plasmids through an LR reaction between the entry clone and the destination plasmids (containing *attR* sequences) from the Yeast Gateway Kit [24] with the desired selectable marker (i.e., *LEU2*, *TRP1*, *HIS3*, *URA3*), promoter (i.e., *GPD* promoter, *GAL1* promoter), N-terminal or C-terminal tag (i.e., EGFP, ECFP, EYFP, Cerulean, DsRed, HA, or TAP), and copy number (i.e., 2 μ, centromere). Set up the LR reaction mixture to include 40 ng of the entry plasmid (e.g., pENTR-CYP90B1), 40 ng of pDEST (e.g., pAG414GPD-ccdB), 0.5 μL of Gateway™ LR Clonase™ II Enzyme mix and q.s. to 2.0 μL with 1× TE buffer (pH 8.0). Incubate the reaction mixture at 25 °C for 1–2 h. Transform competent *E. coli* TOP 10 cells with the reaction mixture. Plate on an LB plate containing 50 μg/mL of ampicillin, incubate at 37 °C for 12–16 h, then colonies will appear.
6. Due to the high accuracy of LR reaction, colony PCR is not necessary. Inoculate one colony into 5 mL LB medium with 50 μg/mL of ampicillin and incubate at 37 °C, 250 rpm for 12–16 h. Prepare plasmid from the overnight culture and send for DNA sequencing to verify the yeast expression plasmid pAG414GPD-yCYP90B1.

3.1.2 Construction of Yeast Expression Plasmid Using Gibson Assembly

The construction of the yeast expression plasmid pYL622 (pAG416TPI1-yCYP90A1, Table 7) is utilized as an example to demonstrate this method.

Gibson Assembly Followed by Gateway LR Cloning

1. CYP90A1 can be obtained from PCR. CYP90A1 was *S. cerevisiae* codon-optimized (yCYP90A1) through gene synthesis. Design the forward and reverse primers to include 30 bp overhang (in lower case) from the vector sequence at the 5′ end of the 20–25 bp yCYP90A1 specific sequence (primers are pTPI1-yCYP90A1-F and pTPI1-yCYP90A1-R as shown in Table 2).

2. Obtain the DNA fragment of the entry vector harboring both promoter and terminator from pCS3129 (pTPI1-CYP82Y1) using high-fidelity DNA polymerase (Q5 or Expand) with primers TPI1-STE2-F and TPI1-STE2-R as shown in Table 2, followed by gel purification using Zymoclean Gel DNA Recovery Kit.
3. Construct the entry clone through Gibson assembly. To set up Gibson reaction, add 10–50 ng of each DNA fragment, preferentially making them equimolar, to 7.5 μL of 1.33× Gibson matrix (recipe can be found in Subheading 2.1.2), and q.s. to 10 μL with D.I. water. Incubate the reaction mixture at 50 °C for 50 min. Transform the competent *E. coli* TOP 10 cells with the reaction mix, and plate on the LB plate with appropriate antibiotics for selection (in this case is 35 μg/mL kanamycin), incubate at 37 °C for 12–16 h, and colonies will appear.
4. Colony PCR is conducted to confirm the presence of gene of interests in the plasmid construct (Same procedure as Subheading 3.1.1, **step 3**).
5. Inoculate one positive colony into 5 mL LB medium with 35 mg/L of kanamycin and grow at 37 °C, 250 rpm for 12–16 h. Purify plasmids from the overnight culture by using commercial columns from Qiagen and send for DNA sequencing to verify the entry clone pTPI1-yCYP90A1.
6. Construct the yeast expression plasmids through LR reaction between the entry clone pTPI1-yCYP90A1 (which has *attL* sequences before *TPI1* promoter and after *STE2* terminator) and the destination plasmids pCS2814 (pAG416-ccdB) [11]. The setup of LR reaction, *E. coli* transformation, and plasmid prep, and verification are the same as Subheading 3.1.1.

Direct Gibson Assembly (Alternative to Above)

1. Obtain the DNA fragment of the expression vector from pYL571 (pAG416TPI1-CYP82Y1) using high-fidelity DNA polymerase (Q5 or Expand) and primers TPI1-STE2-F and TPI1-STE2-R as listed in Table 2 followed by gel purification using Zymoclean Gel DNA Recovery Kit.
2. Construct the expression plasmid through Gibson assembly. The Gibson reaction setup and *E. coli* transformation are the same as **step 3** above.
3. Colony PCR is conducted to confirm the presence of *CYP90A1* in the resulting plasmid construct (Same procedure as Subheading 3.1.1, **step 3**).

4. Inoculate one positive colony into 5 mL LB medium with 50 mg/L of ampicillin and grow at 37 °C, 250 rpm for 12–16 h. Purify plasmids from the overnight culture by using commercial columns from Qiagen and send for DNA sequencing to verify the clone pAG416TPI1-yCYP90A1.

3.1.3 Yeast Plasmid Transformation

The protocol presented here is for one yeast plasmid transformation. It should be scaled up in accordance with the number of transformations needed. This protocol is based on the previously reported protocol [35] with minor modifications. Materials used in this procedure should be all sterilized. We use the transformation of CEN.PK2-1D with pAG415GPD-DWF1 and pAG414GPD-DWF5 for the synthesis of campesterol as the example.

1. Inoculate one single colony of *S. cerevisiae* CEN.PK2-1D from a fresh YPD plate into 0.5 mL of YPD liquid medium and incubate at 30 °C, 250 rpm overnight (~12 to 24 h) until stationary phase (e.g., for CEN.PK2-1D, OD_{600} ~ 20 in YPD).
2. Inoculate 25 μL of the overnight yeast culture into 5 mL fresh YPD medium (OD_{600} ~ 0.1) in a 50 mL Erlenmeyer flask, and incubate at 30 °C, 250 rpm to OD_{600} ~ 0.4–1.0, which usually takes 4 h.
3. Prepare 5 μL of SSD (5 μL per transformation) by denaturing it at 95 °C for 5 min and chill immediately on ice.
4. Prewarm the SDM plates with appropriate selections at 30 °C.
5. Harvest the cells by centrifuging the yeast culture at 3000 rpm (1711 × *g*) for 3 min. Resuspend the cells with 2.5 mL sterile D.I. water and collect the cells again through centrifugation. Repeat this step with 2.5 mL of sterile D.I. water.
6. Resuspend the cells with 1 mL of 0.1 M LiAc-TE buffer and transfer to a microcentrifuge tube. Centrifuge to remove the supernatant.
7. Add 50 μL of 0.1 M LiAc-TE buffer to resuspend the cell pellets.
8. Make the transformation mix.

   For one single-plasmid transformation (items should be added following the order from top to bottom as listed below).

Transformation mixture component	Volume (μL)
0.1 M LiAc-TE buffer resuspended cells	50
SSD	5
DNA (~400 ng/ μL)	1
40% PEG-LiAc	300

For one multiple-plasmid transformation (items should be added following the order from top to bottom as listed below).

Transformation mixture component	Volume (μL)
0.1 M LiAc-TE buffer resuspended cells	50
SSD	5
DNA (~400 ng/μL)	2/plasmid
40% PEG-LiAc	300

For multiple-plasmid transformation, the total volume of DNA should be less than 5 μL. To ensure the success of transformation, ≥500 ng of each plasmid is recommended. More cells may be used to enhance the efficiency of multiple-plasmid transformation (e.g., for a 3- or 4-plasmid transformation, use 7–10 mL of yeast culture). Mix well by vortexing.

9. Incubate at 30 °C on a rotor for 15–30 min.
10. Place the tubes at 42 °C water bath for 30–60 min.
11. Take out the transformation mix. Centrifuge to the mixture at 13,000 rpm (17,000 × *g*) for 30 s to pellet the cells and remove the supernatant.
12. Plate the cells. For auxotrophic selection: resuspend the cells with 100 μL of water and plate on the appropriate SDM selection plate. For eukaryotic antibiotic selection: resuspend the cells with 1 mL of YPD. Incubate the cells in the tube at 30 °C on a rotor for 2 h. Then centrifuge the mixture at 13,000 rpm (17,000 × *g*) for 10–30 s to pellet the cells. Resuspend the cells with 100 μL of water and plate on the appropriate plate with antibiotics.
13. Incubate the plate at 30 °C for 3–4 days until colonies appear. (Go to Subheading 3.4 for strain culturing and metabolite analysis.)

3.2 Pathway Reconstruction Through Genomic Integration

Once the gene function and expression condition are confirmed through ectopic expression, the expression cassette can be integrated into the yeast genome. One major limitation of ectopic expression is the limited number of selectable markers—normally, we can introduce up to four plasmids of different selectable markers to one single yeast strain. To express more genes, chromosomal integration is needed. In addition, compared with ectopic expression, chromosomal integration also exhibits higher genetic stability and uniform expression across yeast culture [36]. We use the establishment of the campesterol-producing pathway in *S. cerevisiae* CEN.PK2-1D to construct YYL55 (*leu2Δ::*P_{PYK1}-*dwf1*-T_{MFA1}, P_{TPI1}-*atr1*-T_{STE2}, P_{PGK1}-*dwf5*-T_{PHO5}, P_{AgTEF1}-*kanmx*-T_{AgTEF1}, P_{GPD}-*dwf7*-T_{CYC1}) as an example to demonstrate this method.

Although DWF7 has the same catalytic function as ERG3 (Fig. 2), introducing one copy of DWF7 significantly enhanced the production of campesterol in yeast expressing DWF1 and DWF5. A selectable marker is normally included to identify the clones with the gene expression cassettes successfully integrated into the chromosome. The selectable markers can be rescued using the Cre-loxP system [37].

3.2.1 Preparation of DNA Fragments Encoding Gene Expression Cassettes

1. Define the locus to incorporate the expression cassettes (e.g., *leu2*). Prepare expression cassettes for each gene flanked with promoters and terminators through PCR using high fidelity DNA polymerase (Q5 or Expand) and primers as listed in Table 4. A selectable marker (e.g., KanMX) is required for the selection of positive colonies. Make sure the promoter and terminator are not natively paired on the *S. cerevisiae* genome to avoid unexpected homologous integrations. Design primers so that each fragment contains 30 bp end homology to each other (Fig. 3, Table 4). DNA fragments containing 500–1000 bp upstream and downstream of the locus of choice are used to enhance the integration efficiency (Fig. 3, Table 4).
2. Purify each fragment by gel electrophoresis and Zymoclean Gel DNA Recovery Kit (Zymo Research). Use a microvolume spectrophotometer to determine the concentration of DNA (at A260 nm). Ensure each DNA fragment is over 600 ng for a seven-piece transformation and adjust the molar ratio of each DNA fragment to similar level.

3.2.2 Yeast Transformation with the DNA Fragments

1. Inoculate one single colony of *S. cerevisiae* CEN.PK2-1D from a fresh YPD plate into 0.5 mL YPD liquid medium and incubate at 30 °C, 250 rpm overnight (~ 12 to 24 h) to stationary phase (e.g., for CEN.PK2-1D, OD_{600} ~ 20 in YPD).
2. Inoculate 75 μL of the overnight yeast culture into 15 mL of YPD liquid medium (OD_{600} ~ 0.1) in a 50 mL Erlenmeyer flask, and incubate at 30 °C, 250 rpm to OD_{600} ~ 0.4–1.0, which usually takes 4 h.
3. Prepare 10 μL of SSD by denaturing it at 95 °C for 5 min and chill immediately on ice.
4. Prewarm the SDM plates with appropriate selections at 30 °C.
5. Harvest the cells by centrifuging the yeast culture at 3000 rpm (1711 × *g*) for 3 min. Resuspend the cells with 10 mL of sterile D.I. water and collect the cells through centrifugation. Repeat this step with 10 mL of sterile D.I. water twice.
6. Resuspend the cells with 1 mL of 0.1 M LiAc-TE buffer. Centrifuge to remove the supernatant.
7. Add 50 μL of 1 M LiAc buffer to resuspend the cell pellets.

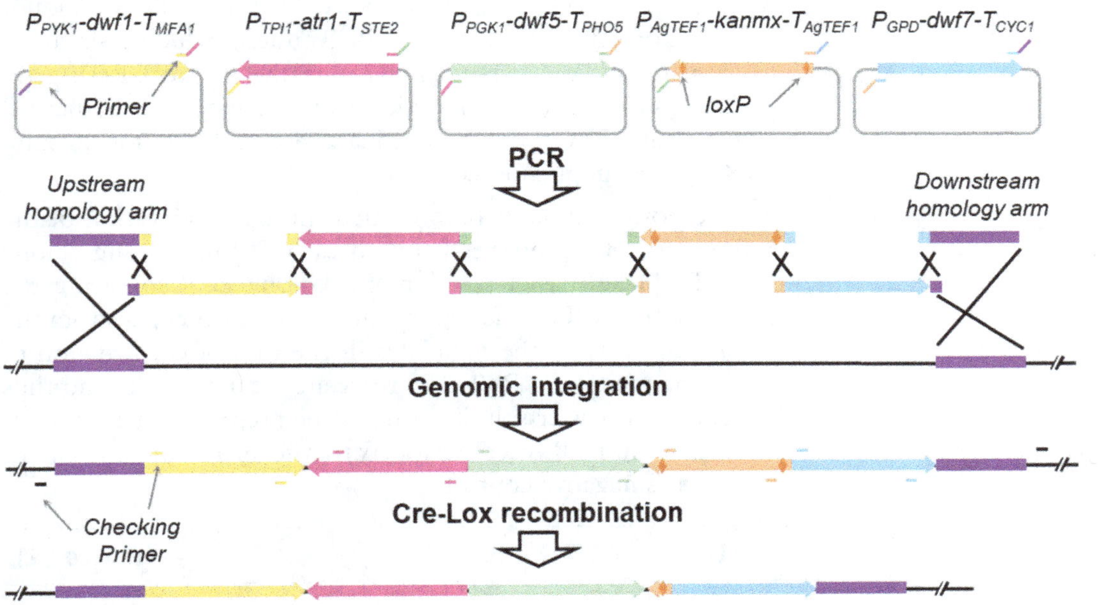

Fig. 3 Diagram illustrating multiple cassettes assembly onto yeast genome through homologous recombination followed by removal of the selectable marker through Cre-lox recombination

8. Make the transformation mixture as described below (items should be added following the order from top to bottom as listed below).

Transformation mixture component	Amount
1 M LiAc buffer resuspended cells	50 μL
SSD	10 μL
[a]DNA (~400 ng/μL)	600 ng/fragment
50% PEG-LiAc	300 μL

[a] Concentrate the DNA fragment mix to reduce the volume lower than 5 μL

9. Incubation, heat-shock and yeast plating are the same procedure as **steps 9–13** in Subheading 3.1.3.

3.2.3 Screening for Positive Clones

1. Design primers to examine if all the DNA fragments are integrated into the yeast genome. Design the forward primers to include 15–20 bp of the sequence from the fragment upstream of the boundary to be checked, and the reverse primer to include 15–20 bp reverse compliment sequence from the fragment downstream of the boundary (Fig. 3). Ensure the primers do not have multiple binding sites. The expected PCR product size is recommended to be within 1 kb.
2. Pick six colonies from the plate and restreak on YPD plate and incubate at 30 °C for 2–3 days.

3. Prepare the crude genomic DNA for colony PCR. Aliquot 15 μL of lyticase buffer into 7 PCR tubes (including six transformants and one parent strain as negative control). Resuspend cells (~10^6 cells) in the lyticase buffer, incubate in thermal cycler at 37 °C for 1 h, and inactivate at 95 °C for 20 min. Keep the samples on ice before use.
4. The positive mutant is confirmed through verification of the presence of boundaries between each fragment using colony PCR. Make master solutions for colony PCR following the recipe below. Use the crude genomic DNA prepped from the previous step as the template. In the case of construction of YYL55 from CENPK2. 1D, there are six boundaries and thus six sets of master solutions should be prepared (each set contains seven PCR reactions for six transformants and one parent strain as negative control).

Yeast colony PCR (7.4 × 15 μL)	Volume (μL)
10× Taq buffer	11.1
dNTPs	2.22
[a]Primer forward	2.22
[a]Primer reverse	2.22
Taq polymerase	0.56
H_2O	81.59

[a] Primers are designed to generate DNA fragments of size ~500–1000 bp

Aliquot 13.5 μL of the Taq polymerase master solution as described in **step 3** of Subheading 3.1.1 to each PCR tube and add 1.5 μL of crude genomic DNA prepped from the previous step. There are seven samples for each set of colony PCR, including 6 colonies and 1 negative control (e.g., CEN.PK2-1D). Perform the PCR reactions using Taq polymerase on the thermal cycler. A lower annealing temperature (usually 50 °C) is recommended.

5. Verify the size of the PCR products by agarose gel electrophoresis.
6. Perform a functional test if applicable: Inoculate 3 positive colonies in 0.5 mL YPD liquid medium, and incubate at 30 °C, 250 rpm for 12–16 h. Inoculate 10 μL of the yeast overnight culture from the previous step into 1 mL of SDM/synthetic completed. Incubate them at 30 °C, 250 rpm for 3 days. Extract the sterols from the yeast cells and conduct LC-MS analysis to confirm the function of the integrated genes as described in Subheadings 3.4.2 and 3.4.3.
7. The colony exhibiting the best performance should be selected, designated (e.g., YYL55), and stored.

3.2.4 Removing the Selectable Marker

1. Transform YYL55 with pSH63 (pGal-CRE plasmid) and plated on SDM/-Trp plate (with 2% glucose), incubate at 30 °C for 2 days.
2. Inoculate 0.5 mL of SDM/-Trp liquid medium with 2% (w/v) glucose using one colony from the previous step, and incubate at 30 °C, 250 rpm overnight or till stationary phase.
3. Inoculate 10 μL of yeast culture from the previous step into 1 mL of SDM/-Trp liquid medium with 2% (w/v) galactose (galactose is added because Cre recombinase is regulated by *GAL1* promoter), and incubate the culture at 30 °C, 250 rpm for 2 days. This step will enable the removal of the selectable marker-encoding cassette flanked by *loxP* sites.
4. Inoculate 10 μL of yeast culture from the previous step into 1 mL of YPD liquid medium, and incubate at 30 °C, 250 rpm for 2 days to remove the plasmid from the yeast strain.
5. Take 1 μL of the yeast culture (~2×10^5 cells) from the previous step and dilute it in 1 mL sterile D.I. water as "serial dilution 1". Take 1 μL of "serial dilution 1" (~200 cells) and dilute it again in 1 mL sterile D.I. water as "serial dilution 2". Take 100 μL from "serial dilution 2" and spread on YPD plate. Incubate the plate at 30 °C for 2 days. There should be around 20 colonies shown on the plate.
6. Replicate the YPD plate on YPD plate with antibiotics (e.g., G418), SDM/-Trp plate, and YPD plate. Incubate the plates at 30 °C for 24 h. The colonies that can grow on YPD but not on the other SDM plates or YPD plate with antibiotics are selectable marker free.

3.3 Strain Engineering

3.3.1 Overexpression of Precursor Biosynthetic Pathway to Enhance Product Titer

To improve campesterol production, the yeast endogenous MVA pathway has been upregulated to enhance the supply of squalene by introducing an extra copy of MVA pathway genes on the pathway. Thus, expression cassettes encoding genes involved in MVA pathway and squalene synthesis (*erg13*, *thmg1*, *erg10*, *erg12*, *erg8*, *erg19*, *erg20*, *erg2*, *erg7*, *erg9*, and *idi1*) were integrated onto yeast genome. Here, we use the construction of YYL56 (*leu2Δ*::P_{PYK1}-*dwf1*-T_{MFA1}, P_{TPI1}-*atr1*-T_{STE2}, P_{PGK1}-*dwf5*-T_{PHO5}, P_{GPD}-*dwf7*-T_{CYC1}; *ybl059wΔ*:: P_{PGK1}-*erg12*-T_{PHO5}, P_{TEF1}-*erg10*-T_{CYC1}, P_{PYK1}-*tHmg1*-T_{MFA1}, P_{TPI1}-*erg13*-T_{STE2}) from YYL55 (Table 5) as an example to demonstrate the method.

1. Select an integration locus (e.g., *ybl059w*), the disruption of which will not affect the function of yeast.
2. Prepare DNA fragments as listed in Table 6 through PCR using high fidelity DNA polymerase (Q5 or Expand), purify the DNA fragments, transform to YYL55, and screen for the positive mutants follow the same procedure as described in Subheadings 3.2.1–3.2.3.

3.3.2 Inactivation of Competing Enzymes to Redirect the Flux Toward Desired Product

Though campesterol can be synthesized efficiently from YYL56 (Table 5), no product can be detected upon the introduction of *CYP90B1*. Through metabolite analysis, we found that *are1*, *are2*, and *erg4* need to be inactivated to make campesterol accessible to CYP90B1 [22]. Here, we use the inactivation of *erg4* from YYL64 (Table 5) to generate YYL67 (*leu2Δ::*P_{PYK1}*-dwf1-*T_{MFA1}, P_{TPI1}*-atr1-*T_{STE2}, P_{PGK1}*-dwf5-*T_{PHO5}, P_{GPD}*-dwf7-*T_{CYC1}; *ybl059wΔ::* P_{PGK1}*-erg12-*T_{PHO5}, P_{TEF1}*-erg10-*T_{CYC1}, P_{PYK1}*-tHmg1-*T_{MFA1}, P_{TPI1}*-erg13-*T_{STE2}, *are1Δ*, *are2Δ*, *erg4Δ*) as an example to demonstrate the method. In this protocol, we exploit yeast homologous recombination to replace *erg4* with a selectable marker *URA3* that can be rescued by Cre recombinase as described in Subheading 3.2.4.

1. Select a selectable marker that is suitable for the strain (e.g., *URA3*).
2. Design primers to obtain *loxP-*p_{KlURA3}*-KlURA3-*t_{KlURA3}*-loxP* flanked with 80–100 bp homology arm (HA) for the integration into *erg4* (5′-erg4 upstream HA-*loxP-*p_{KlURA3}*-KlURA3-*t_{KlURA3}*-loxP*-downstream HA-3′).
3. Amplify the DNA fragment from yeast genomic DNA by PCR using high-fidelity DNA polymerase (Q5 or Expand) followed by gel purification to obtain >600 ng of DNA.
4. Transform the DNA fragment from the previous step into yeast using the method described in Subheading 3.2.2.
5. Screening for the positive colonies and rescue the selectable marker using the methods described in Subheadings 3.2.3 and 3.2.4.

3.4 Metabolite Isolation and Analysis

With the inactivation of *are1*, *are2*, and *erg4*, campesterol produced by the engineered yeast strain (YYL67) can be used to identify the sequence of CYP90A1 and CYP90B1 in the synthesis of (*22S*)-22-hydroxycampest-4-en-3-one from campesterol. Here, we use the functional verification of CYP90A1 and CYP90B1 in YYL67 to describe how to conduct yeast metabolite analysis.

3.4.1 Yeast Culturing and Fermentation

1. Transform YYL67 with pAG414GPD-yCYP90B1, pAG416TPI1-yCYP90A1, and/or the empty vector to obtain the following strains.

Strain	Plasmid	Selection
YYL67	pAG414GPD-yCYP90B1	-TRP
	pAG414-ccdB (negative control)	
	pAG416TPI1-yCYP90A1	-URA
	pAG416-ccdB (negative control)	
	pAG414GPD-yCYP90B1, pAG416TPI1-yCYP90A1	-TRP/URA
	pAG414-ccdB, pAG416-ccdB (negative control)	

2. Pick three different colonies of each strain from the corresponding SDM selective plates, resuspend in SDM liquid medium with the corresponding selection, and incubate at 30 °C and 250 rpm overnight or till the stationary phase (OD_{600} ~ 9).
3. Inoculate 10 μL of yeast culture from the previous step in 1 mL of fresh SDM liquid medium with the corresponding selection, and incubate at 30 °C, 250 rpm for 3 days.

3.4.2 Metabolite Extraction from Cell Culture

1. Transfer 1 mL fermentation broth from the previous step to 1.5 mL microcentrifuge tubes, centrifuge at 13,000 rpm (17,000 × *g*) for 30 s to pellet the cells. The supernatant of culture is then collected in different microcentrifuge tubes.
2. Extract the target metabolites from cell pellets. Yeast pellets are mixed with 100 μL of acetone, 400 μL of methanol and 50 μL of 0.5 mm glass beads. Use cell disruptor to break cells. Centrifuge at 13,000 rpm (17,000 × *g*) for 10 min to pellet the cell debris and beads. Collect the supernatant. Use an addition 400 μL of methanol to vortex the pellets for 5 min. Centrifuge the mixture at 13,000 rpm (17,000 × *g*) for 10 min to pellet the cell debris and beads again. Combine the supernatant and evaporate to dryness using Eppendorf concentrator plus.
3. Redissolve the dried compounds in 40 μL of ethanol.

3.4.3 Metabolite Analysis and Characterization

1. LC method setting. The metabolites are analyzed using Shimadzu LC-MS 2020 on Agilent HPLC column Poroshell 120, EC-C18, 3.0 × 100 mm, 2.7 μm. The target compounds were separated under the linear gradient started from 80% (v/v) methanol in water (0.1% formic acid) to 100% (v/v) methanol over 8 min followed by isocratic wash with 100% (v/v) methanol for 12 min at the flow rate of 0.5 mL/min.
2. Mass spectrum setting. The mass spectrometer is operated in positive ionization mode. Set up selected ion monitoring (SIM) using the characteristic m/z^+ of the target compounds (campesterol: $m/z^+ = [C_{28}H_{49}O\text{-}H_2O]^+ = 383.3$, and 22-OH-campesterol: $m/z^+ = [C_{28}H_{49}O_2\text{-}H_2O]^+ = 399.3$) to enhance the detection sensitivity.
3. The presence of campesterol in the extracts of yeast is confirmed through comparison of the retention time and mass spectrum to those of the authentic campesterol standard.
4. Standards of 22-OH-campesterol and (22S)-22-Hydroxycampest-4-en-3-one are not commercially available. With the detection of the putative compounds using Shimadzu LC-MS 2020, the extracts of the corresponding yeast strains are then analyzed by high-resolution tandem mass spectrometry (HRMS/MS) using Waters Synapt G2-Si Q-TOF system.

Yeast sample preparation and LC chromatograph are the same as described in **step 2** "LC method setting". The mass spectrometer is operated in the positive ionization mode. MS/MS fragmentation data is obtained using Collision Induced Dissociation (CID) fragmentation, with collision voltage at 25 V.

3.4.4 Compound Quantification

Quantification of campesterol in the engineered yeast strains (e.g., YYL64, YYL67) is necessary to evaluate different engineering strategy in enhancing campesterol production.

1. Yeast metabolite extraction is the same as described in Subheading 3.4.2. At least three biological replicates should be used to ensure the reliability of quantification results.
2. Prepare the serial dilution of standard campesterol as 100, 200, 400, 600, 800 mg/L. Inject 10 μL of each standard solution, equivalent to campesterol injected at 1, 2, 4, 6, 8 μg.
3. Yeast extracts are analyzed using the same method as described in Subheading 3.4.3. The peak area of campesterol is calculated by LabSolutions (Shimadzu).
4. Standard curve of campesterol is plotted as the amount of campesterol (μg) injected against peak area (mass abundance). Use the standard curve to estimate the quantity of campesterol of samples from the peak area of the yeast extracts.

4 Notes

1. Genes of interests can also be codon-optimized and obtained through gene synthesis. Codon optimization of synthetic genes using GeneOptimizer™ software (GeneArt/Fisher) may increase the level of expression [38]. For an example of Gateway™ cloning, CYP90B1 from *A. thaliana* can be *S. cerevisiae* codon-optimized (yCYP90B1) and designed to include attB flanking and the Kozak sequence (5-′-ggggacaagtttgtacaaaaaagcaggcttcTATACA-yCYP90B1-gacccagctttcttgtacaaagtggtcccc-3′). For an example of Gibson assembly, CYP90A1 from *A. thaliana* can be *S. cerevisiae* codon-optimized (yCYP90A1) and designed to include 30 bp overhang from the vector sequence at 5′ and 3′ end of *yCYP90A1* (5′-aaaaaacacatacataaactaaaatataca-*yCYP90A1*-tgatcaaaatttacggctttgaaaaagtaa-3′).
2. SDM/-Trp/-His medium deteriorate yeast growth.
3. Avoid using promoters and terminators that are natively paired in *S. cerevisiae* when constructing the gene expression cassette for genomic integration.

4. Yeast strains of different genotypes exhibit different sensitivity to heat and LiAc. The heat shock time and Li^+ concentration of yeast transformation methods may need to be fine-tuned to obtain optimized transformation efficiency. For example, 1 M LiAc and 50% PEG 4000 solutions can increase the transformation efficiency for wildtype CEN.PK2-1D. However, the engineered yeast strain YYL67 has low tolerance to LiAc and using 0.1 M LiAc-TE and 50% PEG 4000 solutions will enhance the cell survival rate during transformation procedure.
5. For yeast genomic integration, 2 × YPD can be used to improve yeast growth and increase the competency and transformation efficiency [35].
6. Fresh colonies can enhance the strain performance and are always recommended. Transformants of more than 2 weeks old should not be used.
7. In addition, to integrate multiple gene expression cassettes into specific locus on the genome, the DNA fragments can also be assembled in the BAC/YAC shuttle vector (e.g., linearized GeneArt™ pYES1L Vector from Thermofisher), which is featured with the advantages of both ectopic expression and genomic integration [39].
8. If the cell disruptor is not accessible when extract steroids in yeast, the alternative option is to mix the cell pellets with 100 μL of DMF, followed by the addition of 900 μL of acetone. The mixture is then vortexed for 15 min and spun at 13,000 rpm (17,000 × *g*) for 10 min to pellet the cell debris. The liquid phases are collected and evaporated to dryness using Eppendorf concentrator plus.
9. Since sterols are not exported into the medium, only metabolites within the cell pellets are extracted and analyzed in Subheading 3.4.2. To investigate metabolites that can be secreted into the medium, metabolites from culture medium should also be extracted and analyzed [40]. To extract metabolites from culture medium, use equal volume of suitable organic solvents (target compounds should exhibit higher solubility in the solvent of choice than in water, e.g., ethyl acetate) to the medium. Vortex the mixture vigorously for 15 min. Then centrifuge at 13,000 rpm (17,000 × *g*) for 5 min to separate the organic and aqueous phase. Collect the organic phase carefully using a pipette and evaporate the extract to dryness.

Acknowledgments

We thank Dr. Christina Smolke from Stanford University for kindly providing pCS1056, pCS2812, pCS2814 and pCS3129 as gifts. We thank Anqi Zhou and Tiffany Chiu for valuable feedback in the preparation of the manuscript. This work was supported by Cancer Research Coordinating Committee Research Award (grant to Y.L., CRN-20-634571) and NIH New Innovator Award (grant to Y.L., DP2 AT011445-01).

References

1. Osbourn AE, Lanzotti V (2009) Plant-derived natural products. Springer, Berlin
2. Duge de Bernonville T et al (2020) Identifying missing biosynthesis enzymes of plant natural products. Trends Pharmacol Sci 41(3): 142–146
3. Lau W, Sattely ES (2015) Six enzymes from mayapple that complete the biosynthetic pathway to the etoposide aglycone. Science 349(6253):1224–1228
4. Miettinen K et al (2014) The seco-iridoid pathway from Catharanthus roseus. Nat Commun 5:3606
5. Caputi L et al (2018) Missing enzymes in the biosynthesis of the anticancer drug vinblastine in Madagascar periwinkle. Science 360(6394): 1235–1239
6. Christ B et al (2019) Repeated evolution of cytochrome P450-mediated spiroketal steroid biosynthesis in plants. Nat Commun 10(1): 3206
7. Liu Y, Nielsen J (2019) Recent trends in metabolic engineering of microbial chemical factories. Curr Opin Biotechnol 60:188–197
8. Pyne ME, Narcross L, Martin VJJ (2019) Engineering plant secondary metabolism in microbial systems. Plant Physiol 179(3): 844–861
9. Guerriero G et al (2018) Production of plant secondary metabolites: examples, tips and suggestions for biotechnologists. Genes (Basel) 9(6):309
10. Cravens A, Payne J, Smolke CD (2019) Synthetic biology strategies for microbial biosynthesis of plant natural products. Nat Commun 10(1):2142
11. Li Y, Smolke CD (2016) Engineering biosynthesis of the anticancer alkaloid noscapine in yeast. Nat Commun 7:12137
12. Piironen V et al (2000) Plant sterols: biosynthesis, biological function and their importance to human nutrition. J Sci Food Agric 80(7): 939–966
13. Peres ALGL et al (2019) Brassinosteroids, the sixth class of phytohormones: a molecular view from the discovery to hormonal interactions in plant development and stress adaptation. Int J Mol Sci 20(2):331
14. Jacobowitz JR, Weng JK (2020) Exploring uncharted territories of plant specialized metabolism in the postgenomic era. Annu Rev Plant Biol 71:631–658
15. Kautsar SA et al (2017) plantiSMASH: automated identification, annotation and expression analysis of plant biosynthetic gene clusters. Nucleic Acids Res 45(W1):W55–W63
16. Smolke C et al (2018) Synthetic biology: parts, devices and applications. Wiley, Hoboken
17. Li SJ, Li YR, Smolke CD (2018) Strategies for microbial synthesis of high-value phytochemicals. Nat Chem 10(4):395–404
18. Nielsen J (2019) Yeast systems biology: model organism and cell factory. Biotechnol J 14(9): 1800421
19. Redden H, Morse N, Alper HS (2015) The synthetic biology toolbox for tuning gene expression in yeast. FEMS Yeast Res 15(1): 1–10
20. Smanski MJ et al (2016) Synthetic biology to access and expand nature's chemical diversity. Nat Rev Microbiol 14(3):135–149
21. Jensen MK, Keasling JD (2015) Recent applications of synthetic biology tools for yeast metabolic engineering. FEMS Yeast Res 15(1): 1–10
22. Xu S, Chen C, Li Y (2020) Engineering of phytosterol-producing yeast platforms for functional reconstitution of downstream biosynthetic pathways. ACS Synth Biol 9(11): 3157–3170
23. van Dijken JP et al (2000) An interlaboratory comparison of physiological and genetic properties of four Saccharomyces cerevisiae strains. Enzym Microb Technol 26(9–10):706–714
24. Alberti S, Gitler AD, Lindquist S (2007) A suite of gateway cloning vectors for high-throughput genetic analysis in Saccharomyces cerevisiae. Yeast 24(10):913–919
25. Gibson DG et al (2009) Enzymatic assembly of DNA molecules up to several hundred kilobases. Nat Methods 6(5):343–345
26. Gueldener U et al (2002) A second set of loxP marker cassettes for Cre-mediated multiple gene knockouts in budding yeast. Nucleic Acids Res 30(6):e23

27. Partow S et al (2010) Characterization of different promoters for designing a new expression vector in Saccharomyces cerevisiae. Yeast 27(11):955–964
28. Peng B et al (2015) Controlling heterologous gene expression in yeast cell factories on different carbon substrates and across the diauxic shift: a comparison of yeast promoter activities. Microb Cell Factories 14:91
29. Redden H, Alper HS (2015) The development and characterization of synthetic minimal yeast promoters. Nat Commun 6:7810
30. Kotopka BJ, Smolke CD (2020) Model-driven generation of artificial yeast promoters. Nat Commun 11(1):2113
31. Curran KA et al (2015) Short synthetic terminators for improved heterologous gene expression in yeast. ACS Synth Biol 4(7):824–832
32. Wang Z et al (2019) Yeast synthetic terminators: fine regulation of strength through linker sequences. Chembiochem 20(18):2383–2389
33. Tschumper G, Carbon J (1983) Copy number control by a yeast centromere. Gene 23(2): 221–232
34. Trenchard IJ, Smolke CD (2015) Engineering strategies for the fermentative production of plant alkaloids in yeast. Metab Eng 30:96–104
35. Gietz RD, Schiestl RH (2007) High-efficiency yeast transformation using the LiAc/SS carrier DNA/PEG method. Nat Protoc 2(1):31–34
36. Jensen NB et al (2014) EasyClone: method for iterative chromosomal integration of multiple genes in Saccharomyces cerevisiae. FEMS Yeast Res 14(2):238–248
37. Sauer B (1994) Recycling selectable markers in yeast. BioTechniques 16(6):1086–1088
38. Zhou Z et al (2016) Codon usage is an important determinant of gene expression levels largely through its effects on transcription. Proc Natl Acad Sci 113(41):E6117–E6125
39. Bruschi C, Gjuracic K, Tosato V (2006) Yeast artificial chromosomes. Encyclopedia of Life Sciences. ChichesterJohn Wiley & Sons Ltd
40. Wu S, Ma X, Zhou A, Valenzuela A, Zhou K, Li Y (2021) Establishment of strigolactone-producing bacteriumyeast consortium. Sci Adv 7: eabh4048. https://doi.org/10.1126/sciadv.abh4048

Chapter 20

Rapid Combinatorial Coexpression of Biosynthetic Genes by Transient Expression in the Plant Host *Nicotiana benthamiana*

Ling Chuang and Jakob Franke

Abstract

Metabolic engineering of heterologous hosts requires coexpression of multiple genes, often more than ten for a single pathway. Traditional approaches to create genetic constructs for this purpose are highly inflexible and suffer from very low throughput. In this book chapter, we describe a powerful method to overcome this bottleneck, namely, combinatorial co-expression in the Australian tobacco plant *Nicotiana benthamiana*. This system is based on *Agrobacterium tumefaciens*-mediated transient gene expression, often called agroinfiltration. Herein, instead of creating complex multigenic constructs, coexpression is achieved by simply mixing different plasmid-bearing *Agrobacterium* strains without the need for different selection markers. We present a practical guide for coexpressing multiple biosynthetic genes followed by GC-MS analysis, using production of the plant triterpene β-amyrin as an example. Our chapter provides a guideline to harness the potential of this versatile expression system in the natural product community for studying and engineering metabolic pathways.

Key words Combinatorial coexpression, *Nicotiana benthamiana*, Agroinfiltration, Heterologous expression, Transient expression, *Agrobacterium tumefaciens*, Eukaryotic host, Metabolic engineering

1 Introduction

Rapid and efficient coexpression systems are crucial for discovering, characterizing, and engineering biosynthetic pathways of natural products. Unfortunately, building heterologous organisms that coexpress multiple pathway genes is usually cumbersome and labor-intensive and typically involves the assembly of multigenic constructs which are either plasmid-borne or integrated into the host genome. However, this approach has severely limited flexibility when genes need to be swapped, added, or removed from a coexpression system. As such, it strongly limits the speed with which natural product biosynthetic pathways can be reconstituted and engineered in a heterologous organism.

Elizabeth Skellam (ed.), *Engineering Natural Product Biosynthesis: Methods and Protocols*, Methods in Molecular Biology, vol. 2489, https://doi.org/10.1007/978-1-0716-2273-5_20,

A fast and flexible coexpression system popular within the plant biochemistry community could become a valuable alternative to a wider natural product community. This method relies on *Agrobacterium tumefaciens*-mediated transient expression in the Australian tobacco plant *Nicotiana benthamiana*, often dubbed agroinfiltration. Transient expression refers to the fact that expression occurs only transiently in adult plants for typically 2–7 days [1], thereby avoiding the time-consuming generation and growth of stably transformed plants. Most importantly, this system permits coexpression of multiple genes by simply mixing different *Agrobacterium* strains which each contain a single expression plasmid, without the need for different selection markers. Accordingly, around 10–15 genes can be routinely coexpressed in any combination desired. This versatile strategy is therefore often called mixing-and-matching. In addition, *Nicotiana benthamiana* possesses the cellular machinery for complex posttranslational modifications, making it ideal for expression of eukaryotic genes. Among all conceivable plants, *Nicotiana benthamiana* is an excellent host system for this process, because it is extremely susceptible to plant viruses as a basis for foreign gene expression from virus-based expression systems [2]. In addition, it accumulates biomass very quickly, facilitating production on industrial scale. The only major downside, as in any heterologous host, is that endogenous enzymes can lead to unexpected modifications of nonnative metabolites in vivo [3].

The method described here uses the soil bacterium *Agrobacterium tumefaciens* [4, 5] (synonym *Agrobacterium radiobacter* [6], *Rhizobium radiobacter* [7, 8]) to introduce foreign genes into the plant nuclear genome, leading to foreign gene expression within the infected region. The development of the technique can be traced back to the 1890s when it was noticed that a tumor-like structure at the plant root-shoot junction (gall) is caused by a "tumor inducing principle" from agrobacteria. In the 1970s, with the knowledge of molecular biology, the "tumor inducing principle" was identified as a fragment of DNA (T-DNA, T stands for transfer) that is transferred from the tumor-inducing (Ti) plasmid found in *Agrobacterium tumefaciens* to the nuclear genome of the host plant [9]. Soon after, this natural system was exploited for genetically modifying plants. Because the Ti plasmid is very large and difficult to handle, it is nowadays commonly split into two plasmids [10]. Such binary vector systems consist of a plasmid containing the infection machinery, which is kept in agrobacteria, and a smaller custom T-DNA plasmid containing the gene(s) of interest for insertion.

In the last decade, transient expression in *N. benthamiana* has seen increasing use in the plant biochemistry community, supported by continuous improvements of vector toolkits. Importantly, several binary vector systems have been developed for

Table 1
Examples of natural products produced in *Nicotiana benthamiana* by transient coexpression

Natural product family	Final product	Number of genes coexpressed	Yield	Reference
Alkaloids	*N*-Formyldemecolcine	16	6.3 ± 1.3 μg/g DW	[15]
Sesquiterpenoids	Patchoulol	5	45 μg/g FW	[16]
Diterpenoids	Abietadiene	4–5	100–200 μg/g FW	[16]
Triterpenoids	Melianol	4	30 μg/g DW[a]	[17]
Phenolic compounds	(−)-Deoxypodophyllotoxin	16	4.3 mg/g DW; 0.71 mg/g DW[a]	[18]
Polyketides	C-glucosylated anthraquinone	4	Not provided	[19]

FW fresh weight of leaves, *DW* dry weight of leaves. Yields are calculated from chromatogram data except for the ones marked with an asterisk
[a] Isolated yield

transient expression. An early example was based on the common pCAMBIA vectors [11], which still required coinfiltration of a p19 silencing suppressor gene to reduce post-transcriptional silencing in planta. A detailed agroinfiltration procedure using this system was described previously [12]. A more convenient and very popular vector is pEAQ-HT ("*e*asy *a*nd *q*uick"), which incorporates a hypertranslational cowpea mosaic virus (CPMV) expression system and already includes a silencing suppressor [13]. In 2019, a further improved version termed pHREAC ("*h*igh *r*ecombinant *e*xpression *a*ssociated with *C*PMV") was published [14]. With these tools, several impressive examples involving combinatorial coexpression of more than ten genes have been reported in order to elucidate, as well as engineer, biosynthetic pathways (Table 1). In this chapter, we describe how transient coexpression in *N. benthamiana* can be used to reconstitute and engineer natural product pathways in a eukaryotic host, employing either pEAQ-HT [13] or pHREAC [14]. As an example, we demonstrate production of the well-studied plant triterpene β-amyrin by coexpressing genes encoding β-amyrin synthase and a truncated version of 3-hydroxy-3-methyl-glutaryl-CoA reductase (HMGR) [20] from oat (*Avena sativa*) [21], followed by gas chromatographic analysis. The workflow is summarized in Fig. 1a. While this chapter will focus on coexpression experiments in analytical scale, transfer to preparative scale based on vacuum infiltration is easily possible and well described elsewhere [22]. We hope that this book chapter will contribute to disseminating this powerful eukaryotic combinatorial coexpression system to a wider natural product community.

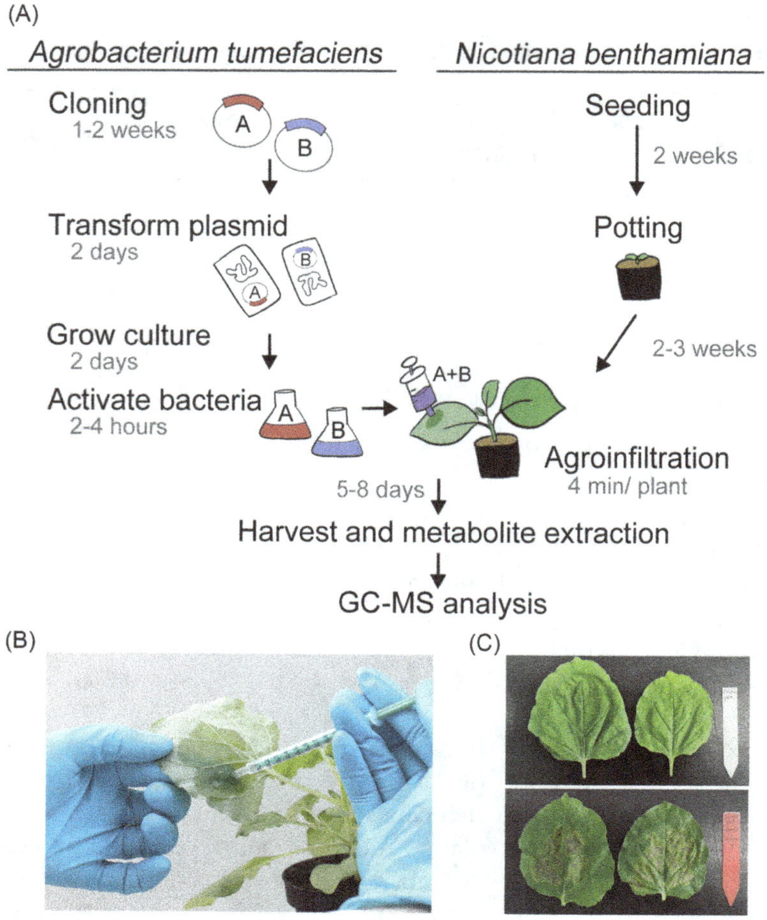

Fig. 1 Combinatorial coexpression for studying and engineering natural product biosynthetic pathways based on *Agrobacterium*-mediated transient expression in the plant *Nicotiana benthamiana* (**a**). Leaves are infiltrated with the soil bacterium *Agrobacterium tumefaciens* ("agroinfiltration") using a needle-less syringe (**b**). *Nicotiana benthamiana* leaves 7 days after infiltration with empty pEAQ-HT vector (upper panel) compared to pHREAC-AsbAS plus pEAQ-AstHMGR (lower panel) (**c**). Note that only the region infiltrated with pHREAC-AsbAS and pEAQ-AstHMGR shows necrotic phenotype caused by local β-amyrin production

2 Materials

2.1 Media and Antibiotics

1. Lysogeny broth (LB) medium: 10 g/L Bacto tryptone, 10 g/L NaCl, and 5 g/L yeast extract (*see* **Note 1**).
2. LB agar with antibiotics: 1 L LB medium with 15 g/L agarose (*see* **Note 2**).
3. Terrific Broth (TB) medium: 24 g/L yeast extract, 20 g/L tryptone, 4 mL/L glycerol and 100 mL/L phosphate buffer (0.17 M KH_2PO_4, 0.72 M K_2HPO_4) (*see* **Note 3**).

4. Super optimal broth with catabolite repression (SOC) medium: 20 g/L tryptone, 5 g/L yeast extract, 0.5 g/L NaCl, 1.86 g/L KCl, 10 mL/L 1 M $MgCl_2$ solution and 20 mL/L 1 M glucose solution (*see* **Note 4**).
5. Kanamycin 1000× stock solution: 50 mg/mL in H_2O (*see* **Note 5**).
6. Gentamicin 1000× stock solution: 25 mg/mL in H_2O (*see* **Note 5**).
7. Rifampicin 1000× stock solution: 50 mg/mL in DMSO (*see* **Note 5**).

2.2 Cloning

1. pEAQ-HT plasmid (GenBank: GQ497234.1), available upon request from Lomonossoff lab [13]. Alternatively, Gateway-ready pEAQ vectors are also available (*see* **Note 6**).
2. pHREAC plasmid (GenBank: MK521430.1), available from Addgene (ID #134908).
3. Thermocycler.
4. Primers for amplifying gene inserts with suitable overhangs for cloning (*see* **Note 7**).
5. DNA polymerase, dNTPs, and compatible buffer.
6. Enzymes for pEAQ-HT cloning: restriction enzymes AgeI and XhoI, and In-Fusion assembly enzyme premix (*see* **Note 8**).
7. Enzymes for pHREAC cloning: BsaI (*see* **Note 9**) and T4 DNA ligase.
8. *E. coli* chemically competent cells.
9. Heat block.
10. Lysogeny broth (LB) medium: the recipe is described in Subheading 2.1.
11. LB agar: the recipe is described in Subheading 2.1
12. Terrific Broth (TB) medium: the recipe is described in Subheading 2.1.
13. Super optimal broth with catabolite repression (SOC) medium: the recipe is described in Subheading 2.1.
14. Antibiotics stock: 50 mg/mL kanamycin as described in Subheading 2.1.
15. Shaker incubator.
16. Gel electrophoresis equipment.
17. TAE buffer 50× stock: 242 g/L Tris base, 57.1 mL/L acetic acid, 100 mL/L 0.5 M EDTA, pH 8.0.
18. Blue light transilluminator.
19. Blue light compatible nucleic acid gel stain.

20. DNA gel loading dye.
21. DNA ladder with a range larger than 10 kb.
22. Microliter spectrophotometer.
23. Gel extraction kit.
24. Plasmid isolation kit.

2.3 Agrobacteria Transformation

1. Electroporator.
2. Electroporation cuvettes with 0.2 cm gap. *See* **Note 10** for cleaning and reusing.
3. LB medium: the recipe is described in Subheading 2.1.
4. LB agar: the recipe is described in Subheading 2.1.
5. SOC medium: the recipe is described in Subheading 2.1.
6. Antibiotic stocks: 25 mg/mL gentamicin, 50 mg/mL rifampicin, and 50 mg/mL kanamycin as described in Subheading 2.1.
7. Spectrophotometer.
8. 1 mM HEPES buffer, *see* **Note 11** for preparation.
9. 1 mM HEPES buffer with 10% (v/v) glycerol, *see* **Note 12** for preparation.

2.4 Growing *Nicotiana benthamiana*

1. Seeds of *Nicotiana benthamiana* LAB strain [2].
2. Green house or phytochamber that can provide 11–16 h light per day and keep the temperature between 18 and 28 °C. *See* **Note 13** for our growth conditions.
3. Seed starting and potting soil: the soil components we use are listed in **Note 14**.
4. Pots and trays: the ones we use are described in Subheading 3.4.
5. Forceps.

2.5 Agrobacteria Infiltration

1. LB medium: the recipe is described in Subheading 2.1.
2. Antibiotic stocks: 25 mg/mL gentamicin, 50 mg/mL rifampicin, and 50 mg/mL kanamycin as described in Subheading 2.1.
3. Bench top centrifuge.
4. Spectrophotometer.
5. MMA infiltration buffer: 10 mM $MgCl_2$, 10 mM 2-(*N*-morpholino)ethanesulfonic acid (MES), 100 μM acetosyringone (*see* **Note 15**).
6. 1 mL syringes.
7. Labeling material for plants: labeling tape or tag, and marker pens.
8. Lab coat, gloves and safety goggles.

2.6 Harvesting and Metabolite Extraction

Adjust according to your target compound; the material that we use for extraction of triterpenes such as β-amyrin is listed below.

1. Cork borer and/or scissors.
2. Freeze dryer with a cold trap.
3. Ball mill and 5 mm steel beads; alternatively, mortar and pestle. *See* **Note 16** for reusing and cleaning steel beads.
4. Saponification solution: 10% (w/v) KOH in 90% (v/v) ethanol.
5. Internal standard: 5α-cholestane.
6. *n*-Hexane (≥99%, GC grade).
7. Rotary vacuum concentrator.
8. Silylation solution: 1:1 pyridine–BSTFA/TMCS (1% in ethyl acetate).
9. Ethyl acetate (≥99.8%, GC grade).

2.7 GC-MS Analysis

1. Gas chromatography mass spectrometry (GC-MS) instrument and software.
2. GC column: Low bleed 5% phenyl dimethylpolysiloxane column (e.g., Optima 5 MS), 30 m length, 0.25 mm inner diameter, 1 μm film thickness. Carrier gas: helium.

3 Methods

Constructs for agroinfiltration can be created either by In-Fusion cloning using the vector pEAQ-HT (Subheading 3.1) or by Golden Gate cloning using the vector pHREAC (Subheading 3.2); only one of these cloning methods is required to follow the complete protocol. The two target genes *AsbAS* and *AstHMGR* used here encode β-amyrin synthase (bAS) and truncated 3-hydroxy-3-methylglutaryl-CoA reductase (tHMGR) from oat (*Avena sativa*) [20, 21].

3.1 Cloning Using pEAQ-HT

3.1.1 Vector Preparation

1. Inoculate 30 mL of TB medium containing 50 μg/mL kanamycin with *E. coli* containing the empty plasmid from plates or cryostock (*see* **Note 17**). Shake at 200 rpm and 37 °C for 16 h. Harvest cells by centrifugation at 13,000 × *g* for 5 min at room temperature. Isolate the plasmid using any commercial kit according to the manufacturer's manual. The yield is around 30–40 μg using a silica membrane–based kit.
2. *See* Fig. 2a for pEAQ-HT vector map and the restriction site sequences. To produce protein without His-tag (*see* **Note 18**), digest 10 μg of pEAQ-HT with 20 U AgeI and 20 U XhoI at 37 °C for 3.5 h in a 50 μL reaction.

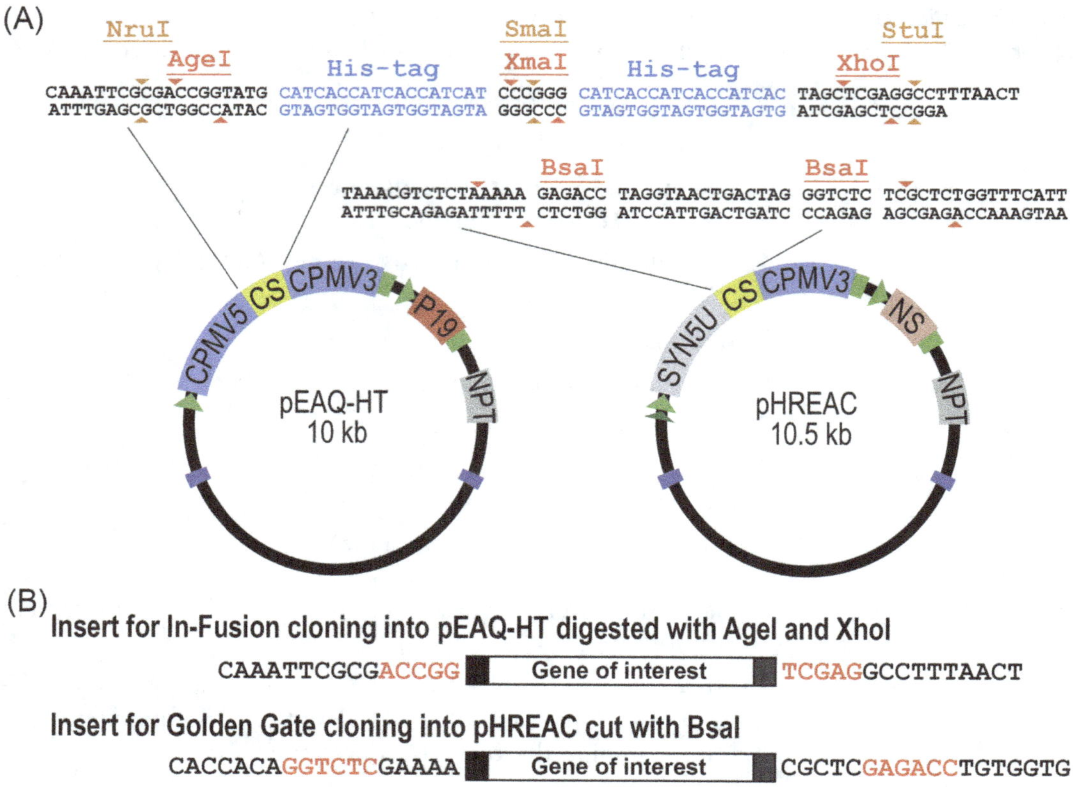

Fig. 2 Vector map (**a**) and insert design (**b**) for pEAQ-HT and pHREAC cloning. The restriction recognition sites are labeled with lines and enzyme name. The cutting sites are indicated with triangles. In the plasmid map, the purple boxes indicate left and right border of T-DNA. Light green arrows indicate 35S promoter from cauliflower mosaic virus (CaMV). Light green boxes indicate NOS terminator. The dark green arrow in pHREAC indicates the enhancer for 35S promoter. *CPMV5* cowpea mosaic virus (CPMV) 5′ untranslated region (UTR), *CS* cloning site, *CPMV3* CPMV 3′ UTR, *P19* RNA silencing suppressor p19 from tomato bushy stunt virus, *SYN5U* synthetic 5′ UTR, *NS* RNA silencing suppressor NS from tomato zonate spot virus, *NPT* neomycin phosphotransferase II that confers kanamycin resistance in *E. coli*

3. Separate the digested plasmid from the residual circular plasmid on a 0.8% agarose gel made with 0.5× TAE buffer. Visualize and cut the 10 kb band using a blue light transilluminator (*see* **Note 19**). Extract the linearized plasmid from the agarose gel using any commercial gel extraction kit (*see* **Note 20**). The yield is around 500–800 ng using a silica membrane–based kit. The purified linearized vector can be stored at −20 °C up to months.

3.1.2 Insert Preparation

1. The insert design for cloning into pEAQ-HT by In-Fusion cloning is shown in Fig. 2b. The insert needs to contain additional 15 bp homology arms at the 5′ and 3′ ends which are complementary to the ends of the linearized vector (*see* **Note 21**). If the insert is amplified from cDNA, include the overlapping sequence CAAATTCGCGACCGG in the forward

Table 2
Insert design of synthetic DNA fragments for cloning *AsbAS* and *AstHMGR* into pEAQ-HT. Homology arms with the pEAQ-HT vector for In-Fusion cloning are indicated in bold

Gene name	NCBI accession	Synthetic sequence
AstHMGR	KY284573.1	**GCCCAAATTCGCGACCGG** ATG—partial coding sequence from 466th bp to 1737th bp of *AstHMGR* (including stop codon)—**TCGAGGCCTTTAACTCTG**
AsbAS1	AJ311789	**GCCCAAATTCGCGACCGG**—full coding sequence of *AsbAS* including start and stop codons–**TCGAGGCCTTTAACTCTG**

primer before the start codon, and include the reverse complement overlapping sequence AGTTAAAGGCCTCGA before the stop codon in the reverse primer (*see* **Note 22**). In this example, we obtained the target coding sequences as synthetic linear DNA fragments (*see* **Note 23**). In this case, directly include the vector homology arms before and after the target gene sequence including start and stop codons (in this example CAAATTCGCGACCGG—gene of interest—TCGAGGCCTTTAACT). Insert design for *AstHMGR* and *AsbAS* is shown in Table 2.

2. If your insert is amplified from cDNA and the results of gel electrophoresis show only one band, the crude PCR reaction can be directly used for In-Fusion assembly without further cleanup. If the results of gel electrophoresis show an additional nonspecific band, purify the PCR product with the expected size by gel extraction according to **step 3** in **Vector preparation**. Measure the insert concentration using a microliter spectrophotometer.

3.1.3 Construct Assembly and Transformation

1. Perform In-Fusion reaction according to the manufacturer's instruction. Transform the assembled product into *E. coli* chemically competent cells by heat shock at 42 °C for 50 s and chill on ice for a minute. Add 200 μL of SOC medium and incubate at 37 °C for 1 h at 200 rpm. Streak 75 μL of recovered cells on an LB plate with 50 μg/mL kanamycin and incubate at 37 °C for 16 h.
2. Perform colony PCR to screen for positive colonies with the vector-specific primers pEAQ_F and pEAQ_R. The sequences of primers are listed in Table 3. Culture the positive colonies overnight in 6 mL TB medium with kanamycin. Isolate the plasmid for Sanger sequencing. Sequence the insert region with vector-specific primers; if this does not cover the full insert length, sequence with an additional internal gene-specific primer (Table 3). After confirming the sequence, the plasmid is ready for transformation to *Agrobacterium tumefaciens*.

Table 3
Primers used in pEAQ-HT cloning

Name	Sequence	Description
pEAQ_F	CAACGTTGTCAGATCGTGCTTC	Forward primer for sequencing and colony PCR
pEAQ_R	CTTTTTGTGTCCTTGC TGAAGGGA	Reverse primer for sequencing and colony PCR
AsbAS_SeqR	CTCTGCTCTTCGATATCGTGG	Internal primer for sequencing *AsbAS*

3.2 Cloning Using pHREAC

3.2.1 Vector Preparation

1. Inoculate 30 mL of TB medium containing 50 μg/mL kanamycin with *E. coli* containing the empty plasmid from plates or cryostock (*see* **Note 17**). Shake at 200 rpm and 37 °C for 16 h. Harvest cells by centrifugation at 13,000 × *g* for 5 min at room temperature. Isolate the plasmid using any commercial kit according to the manufacturer's manual. The yield is around 30–40 μg using a silica membrane–based kit.
2. *See* Fig. 2a for pHREAC vector map and the cloning site sequence. Linearization of pHREAC prior to assembly is not required for Golden Gate cloning.

3.2.2 Insert Preparation

1. If there is any BsaI restriction site in your insert, *see* **Note 24**.
2. *See* Fig. 2b for pHREAC insert design. If the insert is amplified by PCR, include the BsaI recognition site GGTCTC and the homology arm (G)AAAA in the forward primer before the start codon. After BsaI cutting, it will generate a sticky end AAAA that complements the cut pHREAC vector sequence TTTT. Add 7 nucleotides to the 5′ end of the primer before the BsaI recognition site to improve binding of the restriction enzyme. For example, the overhang of the forward primer used here is CACCACAGGTCTCGAAAA before the start codon. The reverse primer also contains the extra 7 nucleotides CACCACA to improve restriction enzyme binding, the BsaI recognition site GGTCTC and a cutting site which in this case generates the directional sticky end AGCG. The overhang for the reverse primer used here is CACCACAGGTCTCGAGCG. All primers are listed in Table 4. In this example, the insert was amplified from the corresponding pEAQ-HT plasmid; alternatively, the insert can be directly obtained as a synthetic gene including the desired overhangs for Golden Gate cloning. For example, ACCACAGGTCTCGAAAA—gene of interest—CGCTCGAGACCTGTGGTG. *See* **Note 23** for handling synthetic DNA fragments.

Table 4
Primers used in pHREAC cloning. Overhangs for Golden Gate cloning are indicated in bold. Constructs should include start and stop codons (underlined)

Name	Sequence	Description
AsbAS_pHREAC_F	**CACCACAGGTCTCGAAAA**ATG TGGAGGCTAACAATAGGTGA	Forward primer for pHREAC-AsbAS cloning
AsbAS_pHREAC_R	**CACCACAGGTCTCGAGCG**TCAGCTC TTAATCGCAAGAAGTC	Reverse primer for pHREAC-AsbAS cloning
pHREAC_F	CTGTCACTTTATTGAGAAGATAGTGG	Forward primer for sequencing and colony PCR
pHREAC_R	CCTTGCTGAAGGGACGACCTG	Reverse primer for sequencing and colony PCR
AsbAS_SeqR	CTCTGCTCTTCGATATCGTGG	Internal primer for sequencing AsbAS

3. If your insert is amplified from cDNA and the results of gel electrophoresis show only one band, clean up the PCR product using any commercial cleanup kit. If the results of gel electrophoresis show any nonspecific band, perform gel extraction according to the next step. Measure the insert concentration using a microliter spectrophotometer.
4. Only in case of multiple PCR products: Separate the insert from the plasmid template or from other nonspecific amplicons on a 0.8% agarose gel made with 0.5× TAE buffer. Visualize and cut the target band using a blue light transilluminator (*see* **Note 19**). Extract the insert from the agarose gel using any commercial gel extraction kit (*see* **Note 20**). Measure the insert concentration using a microliter spectrophotometer.

3.2.3 Construct Assembly and Transformation

1. Perform the Golden Gate reaction by mixing 50 ng of pHREAC plasmid, 2× molar ratio of the insert to the plasmid, 20 U of BsaI (*see* **Note 9**) and 400 U T4 DNA ligase in 1× ligation buffer in a total 20 μL reaction volume for 2 h at 37 °C. Inactivate the enzymes at 80 °C for 10 min.
2. Transform 10 μL of the cloning product into 35 μL of *E. coli* chemically competent cells by heat shock at 42 °C for 50 s and chill on ice for a minute. Add 200 μL of SOC medium and incubate at 37 °C for 1 h at 200 rpm. Streak 75 μL of recovered cells on an LB plate with 50 μg/mL kanamycin and incubate at 37 °C for 16 h.
3. Perform colony PCR to screen for positive colonies with the vector-specific primers pHREAC_F and pHREAC_R (Table 4). Culture the positive colonies overnight in 6 mL TB medium with kanamycin. Isolate the plasmid for Sanger

sequencing. Sequence the insert region with vector-specific primers; if this does not cover the full insert length, sequence with an additional internal gene-specific primer (Table 4). After confirming the sequence, the plasmid is ready for transformation to *Agrobacterium*.

3.3 Transformation of Agrobacteria

The instructions below will yield 20 tubes with 50 μL competent cells each. One tube is required for one plasmid transformation.

3.3.1 Preparation of Agrobacterium tumefaciens *Electrocompetent Cells*

1. Streak out *Agrobacterium tumefaciens* GV3101 (*see* **Note 25**) on an LB agar plate containing 25 μg/mL gentamicin and 50 μg/mL rifampicin (*see* **Note 26**). Incubate at 28 °C for 2 days (*see* **Note 27**).
2. Prepare a 100 mL flask with 20 mL LB medium containing 25 μg/mL gentamicin and 50 μg/mL rifampicin. Inoculate with a single colony from the agar plate. Grow this preculture at 28 °C and 160 rpm for 2 days.
3. To prepare the main culture, inoculate 200 mL of LB medium containing 25 μg/mL gentamicin and 50 μg/mL rifampicin in a 1 L flask with 10 mL of the 2-day preculture. Shake at 28 °C and 160 rpm for 3–4 h until OD_{600} reaches 0.6.
4. Cool 150 mL of 1 mM HEPES buffer and 6 mL of HEPES buffer with 10% glycerol on ice in a laminar flow hood. Conduct all the following steps in a laminar flow hood on ice.
5. Chill the agrobacteria on ice for 15–30 min. Harvest the cells by centrifugation at 4000 rpm (2831 × *g*) at 4 °C for 10 min in two 500 mL centrifuge bottles. Use the same settings also for the following centrifugation steps. If you do not have 500 mL centrifuge bottles or a suitable centrifuge, *see* **Note 28**.
6. Discard the supernatant and resuspend the agrobacteria cells in 50 mL ice-cold HEPES buffer per bottle. Transfer the cell suspension to two 50 mL falcon tubes. Centrifuge to harvest the cells and remove the supernatant.
7. Wash the agrobacteria cells by resuspending in 25 mL ice-cold HEPES buffer per tube. Centrifuge again to harvest the cells and remove the supernatant.
8. Resuspend cells in 2 mL ice-cold HEPES with 10% glycerol per tube. Centrifuge to harvest the cells. Remove the supernatant as much as possible with a 200 μL pipette. Resuspend the cells in 550 μL ice-cold HEPES with 10% glycerol per tube.
9. Chill 20 microcentrifuge tubes (1.5 mL) on ice. Aliquot 55 μL cells to each tube. Snap-freeze the agrobacteria in liquid nitrogen. The competent cells can be stored at −80 °C for months.

3.3.2 Electroporation

1. For one plasmid transformation, thaw one tube of electrocompetent cells on ice. Chill one labeled new 1.5 mL microcentrifuge tube and one electroporation cuvette per transformation on ice.
2. Add 100 ng of the plasmid to be transformed to the cells. Mix by gently pipetting with a 200 μL pipette tip and transfer to the precooled electroporation cuvette. Make sure that the bacterial suspension connects the two electrodes in the cuvette without any air bubbles trapped in between.
3. Use an electroporator to electroshock the cells at 2.2 kV pulse. Typically, this should result in a 5.5 ms long pulse. Add 200 μL of SOC medium to the cuvette and transfer everything to the labeled tube on ice.
4. Recover the cells at 28 °C for 1–2 h at 200 rpm. Streak out 75 μL cells on an LB agar plate containing 25 μg/mL gentamicin, 50 μg/mL rifampicin and 50 μg/mL kanamycin. Incubate at 28 °C for two nights.
5. To confirm successful transformation, select two colonies for each construct to perform colony PCR. The primer sequences are listed in Table 3 for pEAQ-HT and in Table 4 for pHREAC. The transformation efficiency is usually 100%.

3.4 Seeding and Potting of *Nicotiana benthamiana* Plants

1. Prepare space in a greenhouse or a growth chamber. Under our greenhouse conditions (*see* **Note 13**), the resulting plant size is described in **Note 29** for estimating the required space.

3.4.1 Seeding

2. To germinate the seeds, place a tray with drainage holes on a saucer. We use a tray with the dimensions 17 cm width × 20 cm length × 5 cm height for around 100 seeds. Fill it with the seed starting soil. Add water from the bottom.
3. Distribute around 100 *Nicotiana benthamiana* seeds evenly on the soil. Do not cover them with soil.
4. Water them every day to keep the soil moist, but do not let the tray sit in water (*see* **Note 30**). The seeds typically germinate between day 6 and 14. The germination rate is usually more than 90%.

3.4.2 Potting

1. The plants are ready for potting 2 weeks after seeding, when the first pair of true leaves has emerged and expanded (i.e., a total of four leaves including the cotyledons).
2. Prepare pots and fill them up with potting soil. We use pots with 9 cm diameter and 7 cm height. Keep the pots on a tray. Add water to the tray to make the soil moist.
3. Loosen the soil surface in the pots and make a hole with your finger. Pick a healthy-looking seedling with a forceps and transfer it carefully to the new pot. Bury the root gently. Be careful

to hold the seedlings only by the leaves and not by the stem, as this would easily damage the vascular tissue.

4. Water them around twice per week. Increase the frequency of watering when they grow larger.
5. Two weeks after potting, there will be around four true leaves (i.e., six leaves in total). Three to four of them will be suitable for infiltration (i.e., provide low physical resistance to the infiltration). The plants are ready for agroinfiltration at this stage (4 weeks old) until they start to flower (6 weeks old).

3.5 Agroinfiltration

3.5.1 Preparing *Agrobacterium* Strains for Infiltration

1. Calculate the volume of infiltration suspension (i.e., *Agrobacterium* cultures resuspended in MMA buffer for infiltration) needed, choose the infiltration OD_{600} of each strain and prepare the preculture accordingly. For estimating the volume of infiltration suspension, approximately 1–2 mL is required per leaf, and a 4-week-old plant typically has 3–4 leaves suitable for infiltration. The total volume of required infiltration suspension will be: 1–2 (mL/leaf) × 3–4 (leaves/plant) × number of plants to be infiltrated. For example, infiltrating all suitable leaves of 4 plants with the same infiltration suspension would require 12–32 mL. To select the OD_{600} of each bacteria strain, *see* **Note 31** for more detail. We generally use OD_{600} of 0.1 for each strain. The required preculture volume per strain depends on the OD_{600} that the *Agrobacterium* strains reach with your growth conditions. For us 2-day cultures typically reach OD_{600} of ~1 in culture tubes and reach more than OD_{600} of >2 in flasks. Accordingly, 1 mL of a 2-day culture (grown to OD_{600} of ~1) would provide sufficient agrobacteria for infiltrating all suitable leaves of one plant with a final OD_{600} of 0.1.
2. Prepare a preculture with the volume calculated in **step 1** and inoculate the LB medium containing 25 μg/mL gentamicin, 50 μg/mL rifampicin, and 50 μg/mL kanamycin with the *Agrobacterium* strains obtained by transformation. Incubate at 28 °C with shaking at 200 rpm for 2 days (*see* **Note 27**).
3. Prepare fresh MMA infiltration buffer by adding acetosyringone to MM buffer. To the calculated volume of MM buffer freshly add 1:1000 of 100 mM acetosyringone (*see* **Note 32**) and mix by vortexing or by shaking vigorously.
4. Harvest the *A. tumefaciens* cells in centrifuge tubes by centrifugation at 5000 × *g* for 15 min at room temperature. Discard the supernatant by decanting and pipetting.
5. Resuspend each strain in the same volume of MMA infiltration buffer that was previously used for the preculture. Measure OD_{600} with a spectrophotometer, dilute a small amount of culture for accurate measurement if needed.

6. If you only need one agrobacteria strain in the infiltration suspension, dilute the *Agrobacterium* suspension from **step 5** using MMA buffer to the desired infiltration OD_{600} in the required volume. For example, if OD_{600} from **step 5** is 1 and you need 10 mL of infiltration suspension with infiltration OD_{600} of 0.1, take 1 mL of **step 5** culture and add MMA buffer to 10 mL.
7. If you want to coinfiltrate multiple strains, dilute each bacterial strain with MMA buffer to the OD_{600} obtained by this formula: infiltration OD_{600} × number of coinfiltrated strains. For example, if you want to coinfiltrate 3 strains with a final OD_{600} of 0.1, you will need to dilute the individual strains to OD_{600} 0.3. The volume required per strain is calculated as: Total infiltration volume (calculated in **step 1**) ÷ number of coinfiltrated strains. For example, if you plan to infiltrate 12 mL made up from 3 strains, you need 4 mL of every strain with OD_{600} 0.3.
8. Leave the diluted strains at room temperature (not higher than 28 °C) for 2–4 h (*see* **Note 33**).
9. Recheck OD_{600} with a spectrophotometer and adjust with MMA buffer if needed. For single gene infiltration, the infiltration suspension is now ready to use. For multiple gene coinfiltration, mix equal volumes of each strain to be coinfiltrated to bring it to the final volume. Using the example above, mix 4 mL of each strain (each with OD_{600} of 0.3) to bring it to the total volume of 12 mL. The total *Agrobacterium* OD_{600} is now 0.3, but the infiltration OD_{600} of each *Agrobacterium* strain is 0.1.

3.5.2 Infiltration

1. Prepare one 1 mL needleless syringe for each strain combination to be infiltrated, as well as tissue papers for drying the leaves and cleaning, marker pens for marking the infiltrated leaf areas, and plant labels to indicate which strain combination was infiltrated.
2. Wear a lab coat, gloves and safety goggles during the agroinfiltration; bacterial suspension can splash easily when too much pressure is applied during the infiltration.
3. Choose three to four large leaves per plant (*see* **Note 34**). Fill the 1 mL syringe with the bacterial culture. Invert and flip the syringe to remove bubbles. Press it gently on the back of the leaf (abaxial side) with one finger on the other side of the leaf and infiltrate the culture into the leaf (Fig. 1b). If the leaf was successfully infiltrated, the infiltrated area will appear darker. At analytical scale, leaves do not have to be fully infiltrated, as small leaf areas (~1/3 of the total leaf area) already provide sufficient material for downstream analyses.

4. Gently dry the leaf with tissue paper and mark the infiltrated region with red marker pen. Label the pot with the infiltrated strain name(s). Continue to grow the plants as before for 5–8 days before harvesting (*see* **Notes 35** and **36**). In the case of β-amyrin production, necrosis can be observed in the infiltrated leaf region after 5–6 days (Fig. 1c).

3.6 Harvesting and Metabolite Extraction

Adjust the method according to your target compound; the method that we use for extraction of triterpenes such as β-amyrin is described below.

1. Harvest infiltrated leaf areas with scissors or a corkborer. Measure the fresh weight if needed (*see* **Note 37**). We use a corkborer no. 5 (diameter = 10 mm) to collect a total of 5 leaf disks per plant from different infiltrated leaves into a 2 mL microcentrifuge tube for drying and grinding.
2. To dry the leaf samples in tubes, open the tube, cover the opening with tin foil and make small holes with a needle to permit gas exchange. Dry the leaf samples at 0.25 mbar on a freeze dryer with a cold trap overnight or until the weight remains constant.
3. To grind the leaves with a ball mill, add two 5 mm diameter steel beads per 2 mL tube containing the sample and shake at 30 Hz for 20 s. If a ball mill is not available, *see* **Note 38**.
4. To extract triterpenes from leaf powder and to convert esterified steroids and triterpenes into their free alcohol forms, add 250 μL saponification solution to the leaf powder with the steel beads in the 2 mL microcentrifuge tube. Shake vigorously to mix. Use a 1000 μL pipette tip to transfer the mixture with the leaf powder to a 1.5 mL glass vial with screw neck. Some of the leaf powder may remain on the steel beads and at the bottom of the tube. To recover as much as possible, add 250 μL saponification solution to the remaining leaf powder with the steel beads. Shake vigorously to mix. Use a 1000 μL pipette tip to transfer the mixture with the remaining leaf powder to the same glass vial. The leaf powder is now in 500 μL saponification solution in the glass vial. Close the glass vial with the screw cap. Vortex to mix. Incubate at 70 °C for 1 h for efficient saponification. De-coloration of the leaf powder and a transfer of color to the saponification solution will be observed.
5. Let the sample cool to room temperature in a fume hood. Conduct the following procedure in a fume hood.
6. To perform liquid-liquid extraction, add 250 μL H_2O and 500 μL *n*-hexane containing 10 ng/μL internal standard (5-α-cholestane) to the samples in the glass vials. Vortex to mix. Centrifuge at 342 × *g* at 40 °C for 10 min to facilitate phase separation and remove cell debris from the phase boundary (*see*

Note 39). Use a glass pipette or a pipette with a 200 μL filter tip (*see* **Note 40**) to transfer the hexane layer (upper layer) to a new glass vial with a screw neck. Concentrate the extract to dryness in a rotary vacuum concentrator for 1 h at 40 °C and 40 mbar. Store the samples at −20 °C.

3.7 GC-MS Analysis

Adjust the method according to your target compound; the method that we use for analysis of triterpenes such as β-amyrin is described below.

1. To increase the volatility and reduce the polarity of target metabolites, add 100 μL silylation solution to each dried sample. Vortex to mix and incubate at 70 °C for 1 h. Dilute with 250 μL ethyl acetate before injection to GC-MS.
2. For GC-MS analysis, start at 100 °C, increase by 30 °C/min to 275 °C, then increase by 3 °C/min to 300 °C and hold at 300 °C for 15.83 min. Helium is used as carrier gas at a flow rate of 1.5 mL/min. Inject 1 μL sample with a 1:5 split ratio. The injector temperature is set to 300 °C and the detector temperature is set to 250 °C. The mass detector uses a scan range from m/z 43 to 650. The solvent delay is set to 8 min. Triterpenes/sterols typically elute between 11 and 16 min with this method.
3. Use the instrument software to analyze the acquired data. Representative data for leaves infiltrated with pEAQ-AsbAS, pHREAC-AsbAS, and pEAQ-AstHMGR is shown in Fig. 3 (*see* **Note 41**).

4 Notes

1. To prepare 1 L of LB medium, dissolve the components according to the recipe in 800 mL distilled H_2O. Adjust to pH 7.5 with NaOH and add distilled H_2O to 1 L. Autoclave at 121 °C for 20–30 min. Store at room temperature.
2. To prepare 500 mL LB agar, add agarose according to recipe to LB medium. Autoclave in a 500 mL bottle at 121 °C for 20–30 min. Add antibiotics after cooling down to less than 60 °C. Pour 20–25 mL LB agar per 10 cm petri dish. Store the agar plates at 4 °C. If not used immediately, the autoclaved LB agar can be stored at room temperature. Melt the solid LB agar carefully in a regular microwave before use.
3. To prepare 1 L of TB medium, dissolve yeast extract, tryptone and glycerol in 900 mL distilled water and autoclave. After cooling down to less than 60 °C, add 100 mL of sterile phosphate buffer (autoclaved separately). Store at room temperature.

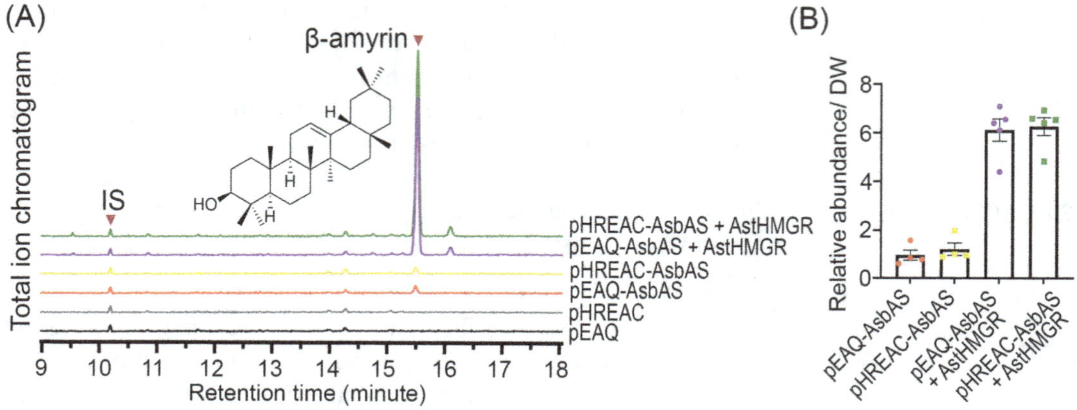

Fig. 3 β-Amyrin production by rapid coexpression in *Nicotiana benthamiana*. GC-MS total ion chromatograms of representative data from two empty vector controls (pEAQ-HT and pHREAC), single gene expression of pEAQ-AsbAS/pHREAC-AsbAS, and two gene coexpression of pEAQ-AsbAS and pHREAC-AsbAS with AstHMGR, respectively (**a**). The vector systems pEAQ-HT and pHREAC result in similar β-amyrin levels in *N. benthamiana* (**b**). Relative ion abundance is calculated as peak area of β-amyrin ($t_R = 15.55$ min) divided by peak area of 5α-cholestane internal standard (IS) ($t_R = 10.33$ min) and calibrated by sample dry weight (mg). The bar graph is shown with mean ± standard error of the mean

4. To prepare 1 L of SOC medium, dissolve the components according to the recipe except for $MgCl_2$ and glucose in 800 mL distilled H_2O. Adjust to pH 7.0 with NaOH before autoclave and add distilled H_2O to 970 mL. Add 10 mL filter-sterilized 1 M $MgCl_2$ solution and filter-sterilized 20 mL 1 M glucose solution after autoclave. Store at 4 °C.
5. To sterilize the kanamycin and gentamicin stock, filtrate the solution after completely dissolving using a 0.2 μm syringe filter. Store at −20 °C. Do not filter the rifampicin stock, as DMSO may damage the filter.
6. The pEAQ-HT vector also has three Gateway-compatible versions, one for use without tags, one including an N-terminal and one with a C-terminal His-tag, respectively. The cloning method for these is described in the original paper [13]. A detailed protocol is also provided in [22].
7. Insert design is described in Subheadings 3.1 and 3.2. The primer sequences for the example used in this chapter are listed in Tables 3 and 4. Alternatively, gene fragments can be obtained directly as synthetic genes (*see* **Note 23**) with suitable overhangs. The synthetic gene sequences for demonstration are listed in Table 2.
8. Restriction enzyme selection is described in Subheading 3.1 and the restriction sites are shown in Fig. 2a. Other Gibson assembly systems can also be used instead of In-Fusion, but require slightly different primer design (*see* **Note 21**).

9. To perform Golden Gate cloning, make sure the BsaI enzyme you have is Golden Gate compatible, that is, it is active in ligation buffer.
10. To reuse electroporation cuvettes, rinse used ones with 70% ethanol for two to three times. Soak in 70% ethanol for 20 min. Rinse with 70% ethanol again. Rinse with 100% ethanol. Leave them upside down on a clean tissue paper for drying at room temperature. We typically reuse our cuvettes until they break during the electroporation; we therefore precool at least one or two extra cuvettes when doing multiple transformations as an emergency backup in case some cuvettes need to be replaced quickly. So far, we have not observed any issues regarding transformation efficiency or contaminations from old reused cuvettes.
11. For 1 mM HEPES buffer preparation, dissolve 238.3 mg of HEPES in 800 mL distilled H_2O. Adjust to pH 7.5 with NaOH and add distilled H_2O to 1 L. Filtrate using 0.2 μm filter in a laminar flow hood.
12. For 1 mM HEPES buffer preparation with 10% (v/v) glycerol, autoclave glycerol separately at 121 °C for 20–30 min. Mix the filtered 1 mM HEPES buffer with autoclaved glycerol in a laminar flow hood.
13. We grow our plants in a greenhouse. Additional artificial light is on from 7 am to 6 pm (11 h per day). The lamps we use provide 60 klm. There is one lamp every 4 m^2 and the lamp is 1.5 m above the plants. When the sunlight exceeds 60 klux on the roof, the artificial light is automatically switched off. The humidity is not controlled; it is usually between 60% and 85%. The temperature is between 21 and 23 °C around the plants. Alternatively, controlled environment rooms (phytochambers) offer more stable and reproducible conditions; typical phytochamber conditions for growing *N. benthamiana* are 16: 8 h day–night cycle, 160 μmol/m^2 s photosynthetic photon flux density (PPFD) [23], 25 °C, 84% humidity for germination, 50% humidity for growing [1].
14. Our soil is purchased from a local supplier. For the seed starting soil, it contains 70% organic material from moderately decomposed bog peat and white peat. The rest consists of clay, calcium carbonate and NPK fertilizer (180 mg/L N, 200 mg/L P_2O_2, 240 mg/L K_2O, 130 mg/L S, 150 mg/L Mg). For the potting soil, it contains 70% organic material from moderately decomposed bog peat, white peat and sod peat. The rest is composed of clay, calcium carbonate and NPK fertilizer (340 mg/L N, 380 mg/L P_2O_2, 450 mg/L K_2O, 130 mg/ L S, 160 mg/L Mg).

15. For infiltration buffer preparation, first prepare a stock without acetosyringone (MM buffer), sterilize using 0.2 μm filter. Prepare 100 mM acetosyringone stock in DMSO, store at −20 °C. Freshly add acetosyringone to MM buffer right before use.
16. To reuse the steel beads, wash them with water to remove the plant debris. Rinse with 70% ethanol three times. Soak in 70% ethanol and ultrasonicate at 40 kHz for 20 min. Rinse with 70% ethanol. Rinse with 100% ethanol. Dry at 70 °C.
17. LB medium for overnight cultures also works, but we observed ~100% higher plasmid yields for pEAQ-HT and pHREAC with TB medium.
18. His-tags are usually included if one needs to purify the target protein by affinity chromatography or track it by western blotting. However, His-tags potentially affect the enzyme production and activity. In this agroinfiltration coexpression system, excluding His-tags is recommended to avoid any tag effect. If a tag is required for protein purification, pEAQ-HT can be digested with XmaI and XhoI for an N-terminal His-tag, or with AgeI and XmaI for a C-terminal His-tag, respectively (*see* Fig. 2a).
19. For gel extraction, cut out bands as small as possible because the residual agarose and the TAE buffer may reduce the binding efficiency of DNA for membrane-based gel extraction kits. Using a blue light transilluminator is preferable for visualizing DNA bands to avoid mutations caused by UV irradiation. Note that not all nucleic acid stains are visible under blue light; we use GelRed nucleic acid gel stain in our lab.
20. If you are using a silica membrane–based gel extraction kit, heat up the elution solution or water to 70 °C before elution. Incubating the tubes with membrane-bound DNA in elution solution at 70 °C before elution improves the yield. We normally use the minimum elution volume the manual recommends (15 μL) and incubate at 70 °C for 15 min before centrifugation for maximum yields.
21. Note that overhang design changes according to the exonuclease used in the seamless homology cloning method and the restriction enzymes you choose. The In-Fusion kit we use here uses a 3′ exonuclease. In contrast, normal Gibson assembly and related kits are based on a 5′ exonuclease; in this case, the overlapping sequence in the forward primer for pEAQ-HT (digested by AgeI and XhoI) will become: TTCTGCCCAAATTCGCGA before the start codon and the reverse primer will become AAACCAGAGTTAAAGGCC before the stop codon. If the cloning method does not include any exonuclease, the required homology sequence is the same as for a 3′ exonuclease. The length of the overlapping sequence

can extend to 20–40 bp to increase the assembly efficiency according to the homology cloning method you use. Design the homology arms according to the manufacturer's instructions. Several kit manufacturers also provide simple online tools for designing primers for In-Fusion cloning and Gibson assembly.

22. Design the gene-specific parts of the primers according to general primer design rules. Remember that the annealing temperature calculation should not include the sequence overlap with the vector.
23. We typically obtain synthetic genes as linear dsDNA gene strings, which are cheaper than standard synthetic genes cloned into a vector. When suitable cloning overhangs are added to the synthetic sequence, it can be directly used in any assembly reaction with high efficiency without additional PCR amplification. Some companies offer gene strings up to 3 kb, which make them suitable for most types of biosynthetic genes. The synthetic DNA strings are usually shipped as dry pellets. To reconstitute the DNA pellet, first briefly centrifuge the tube to make sure the pellet is at the bottom of the tube. Resuspend the pellet in nuclease free water or 1× TE buffer (if your assembly tolerates EDTA) to a concentration of 100 ng/μL. Incubate at 50 °C for 5–15 min and/ or at 4 °C for overnight. Vortex to mix. After dissolving, use the fragment like a normal purified PCR product. Store the resuspended DNA strings at −20 °C.
24. If your insert contains any internal BsaI recognition site, you can alternatively use In-Fusion cloning or any Gibson assembly method for cloning, but the vector will have to be linearized before the assembly. For example, for cloning inserts by In-Fusion into pHREAC linearized with BsaI, the 15 bp homology arms TAAACGTCTCTAAAA and AATGAAAC-CAGAGCG have to be added to the 5′ ends of the forward and reverse primers, respectively. Alternatively, inserts can be domesticated for Golden Gate cloning by introducing silent mutations via gene synthesis or site-directed mutagenesis to remove internal BsaI sites.
25. We use *A. tumefaciens* strain GV3101 (pMP90) in our lab. Other common strains like LBA4404 [10] also work and have been widely used for agroinfiltration in *N. benthamiana*. LBA4404 tends to grow faster than GV3101, but forms clumps in the broth medium which makes it hard to accurately measure the OD_{600}.
26. For GV3101 (pMP90), the purpose of the two antibiotics is as follows: Gentamicin for the helper plasmid pMP90 that contains the infection machinery [24], rifampicin for its genetic

background C58-C1 when it was developed [25]. It is recommended to run an antibiotics test when you first get this strain to make sure that there is no other plasmid in the bacteria. For example, the strain GV3101 (pMP90) itself should not grow on plates containing kanamycin.

27. Very early studies in 1926 showed that *A. tumefaciens*-induced tumor thrives at temperatures between 28 and 30 °C in tomato plants, and no tumor forms above 30 °C [26]. A later study showed that *A. tumefaciens* loses its virulence if it is cultured above 36 °C [27]. As a result, keep your *A. tumefaciens* cultures around 28–30 °C and avoid higher temperature.
28. If you do not have 500 mL centrifuge bottles or a suitable centrifuge, harvest the cells in four 50 mL centrifuge tubes at 4000 rpm (2200 × *g*) for 10 min at 4 °C. Follow the instructions below instead of **step 6**. Discard supernatant and resuspend the cells with 10 mL HEPES buffer per tube. Combine two tubes together. Fill up with HEPES buffer to 50 mL for each tube. You will have two tubes of agrobacteria cells in 50 mL HEPES buffer per tube. Centrifuge again to remove the supernatant. Continue with **step 7**.
29. The plant size depends largely on the growth conditions. A 4-week-old plant grown in our greenhouse conditions (*see* **Note 13**) usually requires 15 cm × 15 cm space. A 6-week-old plant typically requires an area of 29 cm × 29 cm and starts to flower. If the plant growth conditions are optimal, the plants will start to flower later and higher biomass can be obtained [1]. In general, people use 4–6-week-old plants. Infiltrating plants shortly before they start flowering should be avoided as flowering strongly changes their metabolism. In our case, 15 plants require approximately 0.3 m^2, but the size depends on the light and growth conditions. Keep in mind that a full infiltration cycle from potting to harvesting typically takes 6 weeks; maintaining a biweekly or weekly agroinfiltration pattern with 15 plants per batch will therefore roughly require 1 or 2 m^2 growth area, respectively.
30. Avoid overwatering at young stage to prevent moss from outgrowing your plants. In general, if the soil at the bottom of the pot is still wet, they do not need more water.
31. In the publication describing the development of the pEAQ-HT vector, the authors showed that the amount of recombinant protein in a plant protein extract decreased from 2.7 to 1.4 mg/mL when the infiltration bacteria OD_{600} decreased from 1.2 to 0.13 [13]. For metabolite extraction, various groups have tried OD_{600} between 0.2 [22] and 0.6 [15] for single strains. With coexpression, the total OD_{600} can be up to 4.8 (16 genes with OD_{600} 0.3 for each strain) [15]. It is

recommended to optimize the bacteria density when testing new constructs and new *Agrobacterium* strains. In our lab, for triterpene production with GV3101 we always use an OD_{600} of 0.1.

32. It was observed that plant phenolic compounds induce *Agrobacterium tumefaciens* virulence [28]. One of these compounds which is widely used for this purpose is acetosyringone [29, 30].
33. The activation time seems to have limited impact. We have tried 2–4 h and did not observe obvious differences for triterpene production.
34. Avoid leaves which look sick or have been attacked by insects, because these leaves tend to become waxy on the surface, which makes them hard to infiltrate. When too much pressure is applied trying to infiltrate such leaves, bacterial suspensions splash easily.
35. The optimal temperature for *A. tumefaciens* to infect plants varies by plant species. For *Nicotiana* the temperature seems less critical [31]. A study with *Nicotiana tabacum* shows that protein production is optimal at 22 °C and not observable above 27 °C [32]. In general, infiltrated *Nicotiana benthamiana* plants are kept between 23 and 25 °C [12, 13, 22].
36. It is recommended to optimize the harvest time according to the target product. In general, foreign genes are expressed in leaves at least from 2 to 7 days [1]. For metabolite extraction, leaves are typically harvested after 3 [12], 4–5 [15, 22], 6 days [13], or 7–9 days [18].
37. The fresh weight of a big leaf of a 5-week-old plant infiltrated with empty vector is around 500 mg. In this example, five leaf disks collected with a corkborer no. 5 correspond to around 70 mg fresh weight for empty vector control, around 50 mg for AsbAS expressing leaves, and around 10 mg for AsbAS and AstHMGR coexpressing leaves. The dry weight is around 5 mg for all.
38. If a ball mill is not available, grind the leaf with a mortar and pestle, or with a PTFE pestle for microcentrifuge tubes. Freeze the leaf disks in liquid nitrogen to facilitate grinding. For the ball mill, it is not necessary to freeze the leaf disks in liquid nitrogen before grinding.
39. For centrifugation of 1.5 mL glass vials, we use a rotary vacuum concentrator in warmup mode at ambient pressure and maximum speed (1550 rpm, equal to 342 × *g*).
40. It is preferrable to avoid common polypropylene (PP) plasticware as much as possible for GC-MS sample preparation, as PP has slightly limited resistance to the chemicals

used in this protocol. Leaking material such as plasticizers can cause large background peaks or contaminate or even damage GC columns. However, we did not observe obvious negative effects when using PP plasticware for very short times (such as pipette transfer or short centrifugation steps). Consult chemical resistance charts from your supplier for detailed information.

41. For β-amyrin production, pHREAC and pEAQ-HT vectors perform similarly regarding yield, agrobacteria growth rate and plant necrosis phenotype. The resulting β-amyrin levels (relative peak area compared to internal standard and dry weight) between pEAQ-AsbAS ($n = 4$) and pHREAC-AsbAS ($n = 4$) are not significantly different (two-tailed t-test, equal variance (F 1.47 < F critical 9.28), p-value = 0.48, df = 6, t-stat = −0.75). The difference for coexpressing either pEAQ-AsbAS ($n = 5$) or pHREAC-AsbAS ($n = 5$) with pEAQ-AstHMGR is also not significant (two-tailed t-test, equal variance (F 1.56 < F critical 6.38), p-value = 0.81, df = 8, t-stat = −0.25). Previously, Stephenson et al. reported an isolated yield of 3.3 mg/g DW β-amyrin from plants infiltrated with pEAQ-AsbAS and pEAQ-AstHMGR [22].

Acknowledgments

We gratefully acknowledge financial support by the Emmy Noether programme of the Deutsche Forschungsgemeinschaft (FR 3720/3-1) and the SMART BIOTECS alliance between the Technische Universität Braunschweig and the Leibniz Universität Hannover, supported by the Ministry of Science and Culture (MWK) of Lower Saxony. We thank Yvonne Leye and Miriam Fent for excellent horticultural support. We thank Lorenzo Caputi (Max Planck Institute for Chemical Ecology, Jena) and Annika Stein for helpful cloning advice and Dave Biedermann for GC-MS support.

References

1. Leuzinger K, Dent M, Hurtado J, et al (2013) Efficient agroinfiltration of plants for high-level transient expression of recombinant proteins. J Vis Exp (77):e50521. https://doi.org/10.3791/50521
2. Bally J, Jung H, Mortimer C et al (2018) The rise and rise of *Nicotiana benthamiana*: a plant for all reasons. Annu Rev Phytopathol 56: 405–426. https://doi.org/10.1146/annurev-phyto-080417-050141
3. van Herpen TWJM, Cankar K, Nogueira M et al (2010) Nicotiana benthamiana as a production platform for artemisinin precursors. PLoS One 5(12):e14222. https://doi.org/10.1371/journal.pone.0014222
4. Gan HM, Lee MVL, Savka MA (2019) Improved genome of Agrobacterium radiobacter type strain provides new taxonomic insight into Agrobacterium genomospecies 4. PeerJ 7: e6366. https://doi.org/10.7717/peerj.6366
5. Flores-Félix JD, Menéndez E, Peix A et al (2020) History and current taxonomic status of genus Agrobacterium. Syst Appl Microbiol

43:126046. https://doi.org/10.1016/j.syapm.2019.126046

6. Young JM, Pennycook SR, Watson DRW (2006) Proposal that Agrobacterium radiobacter has priority over Agrobacterium tumefaciens. Request for an opinion. Int J Syst Evol Microbiol 56:491–493. https://doi.org/10.1099/ijs.0.64030-0
7. Young JM, Kuykendall LD, Martínez-Romero E et al (2001) A revision of Rhizobium Frank 1889, with an emended description of the genus, and the inclusion of all species of Agrobacterium Conn 1942 and Allorhizobium undicola de Lajudie et al. 1998 as new combinations: Rhizobium radiobacter, R. rhizogenes, R. rubi, R. undicola and R. vitis. Int J Syst Evol Microbiol 51:89–103. https://doi.org/10.1099/00207713-51-1-89
8. White LO (1972) The taxonomy of the crown-gall organism Agrobaterium tumefaciens and its relationship to rhizobia and other agrobacteria. Microbiology 72:565–574. https://doi.org/10.1099/00221287-72-3-565
9. Larebeke NV, Engler G, Holsters M et al (1974) Large plasmid in Agrobacterium tumefaciens essential for crown gall-inducing ability. Nature 252:169–170. https://doi.org/10.1038/252169a0
10. Hoekema A, Hirsch PR, Hooykaas PJJ, Schilperoort RA (1983) A binary plant vector strategy based on separation of vir- and T-region of the Agrobacterium tumefaciens Ti-plasmid. Nature 303:179–180. https://doi.org/10.1038/303179a0
11. Nour-Eldin HH, Hansen BG, Nørholm MHH et al (2006) Advancing uracil-excision based cloning towards an ideal technique for cloning PCR fragments. Nucleic Acids Res 34:e122. https://doi.org/10.1093/nar/gkl635
12. Bach SS, Bassard J-É, Andersen-Ranberg J et al (2014) High-throughput testing of terpenoid biosynthesis candidate genes using transient expression in Nicotiana benthamiana. In: Rodríguez-Concepción M (ed) Plant isoprenoids: methods and protocols. Springer, New York, pp 245–255
13. Sainsbury F, Thuenemann EC, Lomonossoff GP (2009) pEAQ: versatile expression vectors for easy and quick transient expression of heterologous proteins in plants. Plant Biotechnol J 7:682–693. https://doi.org/10.1111/j.1467-7652.2009.00434.x
14. Peyret H, Brown JKM, Lomonossoff GP (2019) Improving plant transient expression through the rational design of synthetic 5′ and 3′ untranslated regions. Plant Methods 15:108. https://doi.org/10.1186/s13007-019-0494-9
15. Nett RS, Lau W, Sattely ES (2020) Discovery and engineering of colchicine alkaloid biosynthesis. Nature 584:148–153. https://doi.org/10.1038/s41586-020-2546-8
16. Sadre R, Kuo P, Chen J et al (2019) Cytosolic lipid droplets as engineered organelles for production and accumulation of terpenoid biomaterials in leaves. Nat Commun 10:1–12. https://doi.org/10.1038/s41467-019-08515-4
17. Hodgson H, Peña RDL, Stephenson MJ et al (2019) Identification of key enzymes responsible for protolimonoid biosynthesis in plants: opening the door to azadirachtin production. Proc Natl Acad Sci U S A 116:17096–17104. https://doi.org/10.1073/pnas.1906083116
18. Schultz BJ, Kim SY, Lau W, Sattely ES (2019) Total biosynthesis for milligram-scale production of etoposide intermediates in a plant chassis. J Am Chem Soc 141:19231–19235. https://doi.org/10.1021/jacs.9b10717
19. Andersen-Ranberg J, Kongstad KT, Nafisi M et al (2017) Synthesis of C-glucosylated octaketide anthraquinones in Nicotiana benthamiana by using a multispecies-based biosynthetic pathway. ChemBioChem 18:1893–1897. https://doi.org/10.1002/cbic.201700331
20. Reed J, Stephenson MJ, Miettinen K et al (2017) A translational synthetic biology platform for rapid access to gram-scale quantities of novel drug-like molecules. Metab Eng 42: 185–193. https://doi.org/10.1016/j.ymben.2017.06.012
21. Haralampidis K, Bryan G, Qi X et al (2001) A new class of oxidosqualene cyclases directs synthesis of antimicrobial phytoprotectants in monocots. Proc Natl Acad Sci U S A 98: 13431–13436. https://doi.org/10.1073/pnas.231324698
22. Stephenson MJ, Reed J, Brouwer B, Osbourn A (2018) Transient expression in Nicotiana benthamiana leaves for triterpene production at a preparative scale. J Vis Exp (138): e58169. https://doi.org/10.3791/58169
23. Goulet M-C, Gaudreau L, Gagné M et al (2019) Production of biopharmaceuticals in Nicotiana benthamiana—axillary stem growth as a key determinant of total protein yield. Front Plant Sci 10:735. https://doi.org/10.3389/fpls.2019.00735
24. Koncz C, Schell J (1986) The promoter of TL-DNA gene 5 controls the tissue-specific expression of chimaeric genes carried by a novel type of Agrobacterium binary vector.

Mol Gen Genet 204:383–396. https://doi.org/10.1007/BF00331014

25. Holsters M, Silva B, Van Vliet F et al (1978) In vivo transfer of the Ti-plasmid of Agrobacterium tumefaciens to Escherichia coli. Mol Gen Genet 163:335–338. https://doi.org/10.1007/BF00271963
26. Riker AJ (1926) Studies on the influence of some environmental factors on the development of crown gall. J Agric Res 32:83–96
27. Hamilton RH, Fall MZ (1971) The loss of tumor-initiating ability inAgrobacterium tumefaciens by incubation at high temperature. Experientia 27:229–230. https://doi.org/10.1007/BF02145913
28. Bolton GW, Nester EW, Gordon MP (1986) Plant phenolic compounds induce expression of the Agrobacterium tumefaciens loci needed for virulence. Science 232:983–985. https://doi.org/10.1126/science.3085219
29. Stachel SE, Timmerman B, Zambryski P (1986) Generation of single-stranded T-DNA molecules during the initial stages of T-DNA transfer from Agrobacterium tumefaciens to plant cells. Nature 322:706–712. https://doi.org/10.1038/322706a0
30. Lee K, Dudley MW, Hess KM et al (1992) Mechanism of activation of Agrobacterium virulence genes: identification of phenol-binding proteins. Proc Natl Acad Sci U S A 89:8666–8670. https://doi.org/10.1073/pnas.89.18.8666
31. Braun AC (1947) Thermal studies on the factors responsible for tumor initiation in crown gall. Am J Bot 34:234–240. https://doi.org/10.2307/2437424
32. Dillen W, De Clercq J, Kapila J et al (1997) The effect of temperature on Agrobacterium tumefaciens-mediated gene transfer to plants. Plant J 12:1459–1463

Chapter 21

Optimized Tools and Methods for Methanotroph Genome Editing

Sreemoye Nath, Jessica M. Henard, and Calvin A. Henard

Abstract

Microbes with the capacity to use methane (CH_4) as a carbon source (methanotrophs) have significant potential for the bioconversion of CH_4-containing natural gas and anaerobic digestion-derived biogas to high value products. These organisms also play a vital role in the biogeochemical cycling of atmospheric CH_4 by serving as the only known biological sink of this gas in terrestrial and aquatic ecosystems. Much is known regarding the enzymes and central metabolic pathways mediating CH_4 utilization in these bacteria. However, large fundamental knowledge gaps exist regarding methanotroph physiology and responses to environmental stimuli, primarily due to a lack of efficient molecular tools to probe gene-function relationships. In this chapter, we describe several recently developed genetic tools and optimized genome editing methods that can be used for methanotroph metabolic engineering and to probe metabolic and physiological governing mechanisms in these unique bacteria.

Key words Methanotroph, Natural gas, Biogas, Biocatalysis, Metabolic engineering, Genetic engineering, Genome editing, CRISPR, Conjugation, Broad host-range plasmid

1 Introduction

Methanotrophs use methane (CH_4) as a carbon and energy source and play a significant role in the biogeochemical carbon cycle. Further, these unique microbes represent promising biocatalysts for use in biomanufacturing processes utilizing CH_4-rich feedstocks, that is, natural gas or biogas. To this end, several research groups have recently introduced improved tools for use in genetic and metabolic engineering of methanotrophs, most based on conjugative broad host range (BHR) plasmids demonstrated in seminal studies to function in diverse methylotrophic and methanotrophic bacteria [1–5]. These include allelic exchange plasmids for the generation of knock-out/knock-in strains [6–10], constitutive and inducible expression plasmids [7, 11–14], promoter-probe plasmids [11, 15, 16], and plasmids for CRISPR-Cas9-based genome editing [11]. These BHR plasmids are routinely

Elizabeth Skellam (ed.), *Engineering Natural Product Biosynthesis: Methods and Protocols*,
Methods in Molecular Biology, vol. 2489, https://doi.org/10.1007/978-1-0716-2273-5_21,

introduced to methanotrophs via biparental mating using *Escherichia coli* to mediate RK2/RP4-based plasmid transfer [17–19]. Successful electroporation-based transformation has been demonstrated in *Methylotuvimicrobium buryatense* [20], but the methodology as described is not consistently successful in most methanotrophs (*see* **Note 1**). In contrast, conjugation offers a reliable method for stable transfer of BHR plasmids to methanotrophic bacteria.

The recent expansion of genetic tools for use in methanotrophs has been instrumental in the development of strains engineered for production of fuels and chemicals (for a review *see* [21] and references therein). Further, these tools are being used to increase fundamental knowledge of methanotroph metabolism and physiology [14, 22–25], of which a detailed understanding is required to realize CH_4 biocatalysis as part of a future biomanufacturing industry. Importantly, advanced genome editing tools, such as CRISPR-Cas9, should enable development of methanotroph genetic variant libraries for high-throughput metabolic engineering and screening of gene–function relationships in these unique bacteria [26, 27].

In this chapter, we outline the plasmids and optimized conjugation method routinely used in our laboratory for CRISPR-Cas9-based genetic manipulation of the Gammaproteobacterial methanotroph *Methylococcus capsulatus*. Notably, the tools and methods described here are founded on previously described methods developed in diverse methylotrophic and methanotrophic bacteria [7, 8, 13, 19, 28]; thus, they likely have utility, with minimal modification, in many methylotrophic bacteria as well as other non-model, industrially promising bacteria.

2 Materials

2.1 Microbial Strains and Plasmids

1. All bacterial strains are available from American Type Culture Collection (ATCC) and/or the German Collection of Microorganisms and Cell Cultures (DSMZ) repositories (Table 1).
2. Plasmids used in the development of the CRISPR-Cas9 genome editing system in *M. capsulatus* are shown in Fig. 1 and include:
   (a) pCAH01.
   (b) pCAH01Sp^RCas9.
   (c) pQCH.
   (d) pBBR1.

2.2 Culture Medium

2.2.1 *E. coli* Strains

1. Lysogeny broth (LB, Lennox) liquid.
2. Lysogeny broth (LB) solid (1.5% agar).

Table 1
Bacterial strains and plasmids

Name	Genotype	Source
Methylococcus capsulatus str. Bath	Wild-type	ATCC
Escherichia coli DH10b	F- *mcrA* Δ(*mrr-hsdRMS-mcrBC*) Φ80*lacZ*ΔM15 Δ*lacX74 recA1 endA1 araD139* Δ(*ara leu*) 7697 *galU galK rpsL* nupG λ-	Zymo research
E. coli S17 λpir	Tp^r Sm^r *recA thi pro hsd*(r^-m^+)RP4–2-Tc::Mu::Km Tn7 λpir	ATCC
Plasmids		*Addgene # or reference*
pCAH01		128162
pCAH01SpRCas9		128160
pQCH		164160
pBBR1		85168

3. ***Antibiotics***: 100 μg/mL carbenicillin; 50 μg/mL spectinomycin.

2.2.2 Methylococcus capsulatus *Nitrate Mineral Salts (NMS) Cultivation and Mating Medium*

1. NMS medium: (Table 2, *see* **Note 2**).
2. Mating medium: 1 g/L KNO_3, 0.2 g/L $MgSO_4{\cdot}7H_2O$), 0.02 g/L $CaCl_2{\cdot}2H_2O$, 2 g/L NaCl, 1.2 g/L nutrient broth, 1 mL/L trace solution, 50 mL/L phosphate solution) 3.10 mg/mL Anhydrotetracycline (aTc) in DMSO (*see* **Note 3**).
3. Oxoid chamber with atmosphere of 20% CH_4 in air.

2.3 Polymerase Chain Reaction and Isothermal Assembly

1. DNA Polymerase, for example, New England Biolabs' Q5 High-Fidelity 2× Master Mix.
2. Molecular grade water (*see* **Note 4**).
3. Forward and Reverse primer pair to amplify pCAH01Sp^{R}Cas9 backbone (*see* **Note 5**).

   Forward (5′) primer oCAH230: GTTCTTTCCTGCGTTAT CCC

   Reverse (3′) primer oCAH231: GTCTGACGCTCAGTGGAA
4. Forward and Reverse primer pair to amplify synthesized cassette containing the gRNA and editing template (lowercase bases indicate homology arms to facilitate isothermal assembly).

   Forward (5′) primer oCAH267: cgttccactgagcgtcagacGAGG TTCAGGCGAAACCG

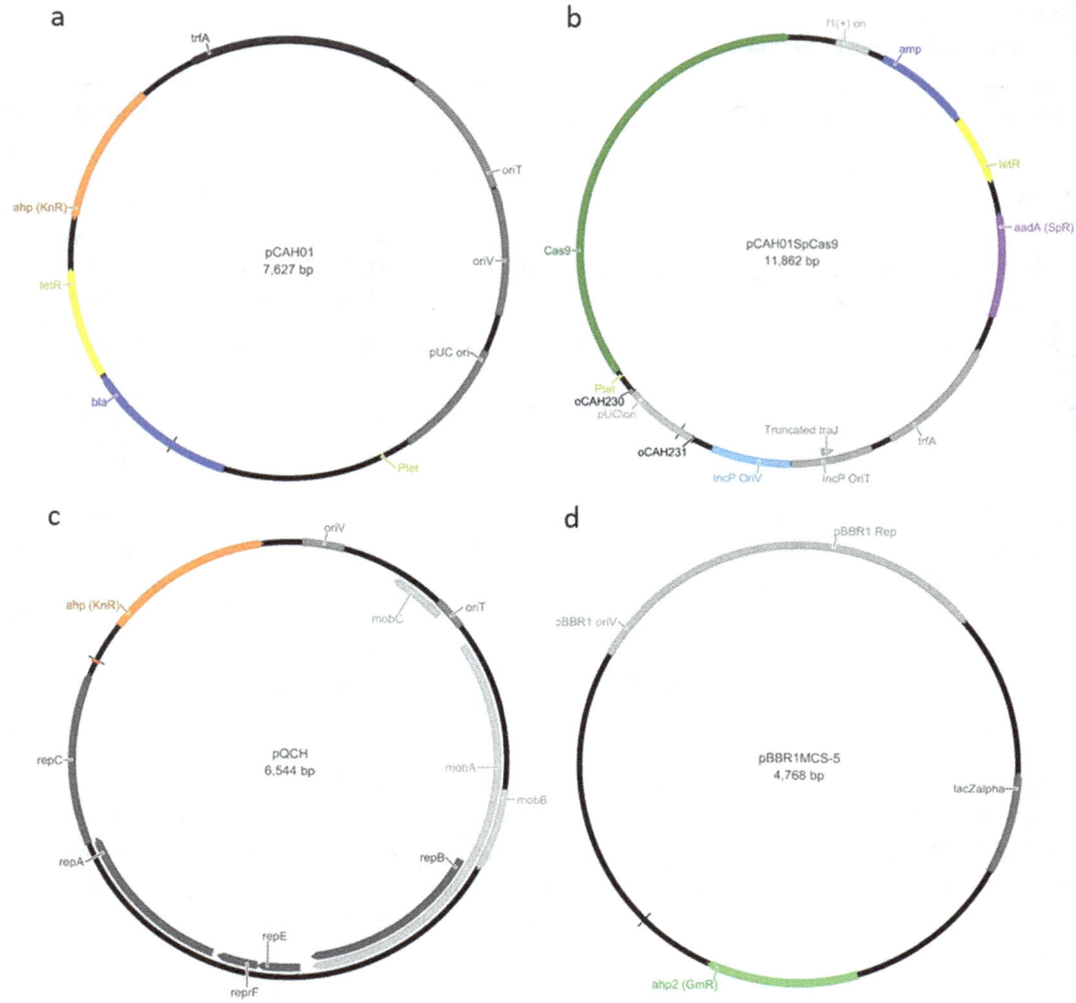

Fig. 1 Plasmid Maps. We previously constructed an Incompatibility group (Inc) P-based inducible plasmid, pCAH01, containing the tetracycline promoter/operator (**a**) [12]. pCAH01 was used to develop a CRISPR-Cas9 genome editing system for use in *M. capsulatus* (**b**) [11]. We also routinely use the IncQ-based plasmid pQCH for homologous/heterologous gene expression and for evaluation of promoter activity (**c**). In addition to IncP and IncQ replicons, *M. capsulatus* can also harbor pBBR1 (**d**), and we have shown that pCAH01, pQCH, and pBBR1 can be simultaneously maintained in this bacterium [11]. These plasmids are available from the Addgene plasmid collection (Addgene.org) or upon reasonable request from the authors

Reverse (3′) primer specific to editing template: gggataacgcaggaaagaacNNNNNNNNNNNNN

5. gRNA and editing template (*see* Fig. 2, **Notes 6** and 7).
6. PCR consumables; thermocycler.
7. DNA agarose gel materials, including agarose and Tris–acetate–EDTA buffer (10× solution: 400 mM Tris–acetate and 10 mM EDTA).
8. PCR cleanup kit.

Table 2
NMS medium composition

Component	NMS
KNO_3	1 g/L
$MgSO_4 \cdot 7H_2O$	0.2 g/L
$CaCl_2 \cdot 2H_2O$	0.02 g/L
Trace solution (1000×)	1 mL/L
Agar[a]	15 g/L
Phosphate solution (pH 6.8)	50 mL/L
Phosphate solution (pH 6.8)	*g/L*
KH_2PO_4	5.44
Na_2HPO_4	5.68
Trace mineral solution (1000×)	*g/L*
Na_2EDTA	5
$FeSO_4 \cdot 7H_2O$	2
$ZnSO_4 \cdot 7H_2O$	0.3
$MnCl_2 \cdot 4H_2O$	0.03
$CoCl_2 \cdot 6H_2O$	0.2
$CuSO_4 \cdot 5H_2O$	1.2
$Na_2O_4W \cdot 2H_2O$	0.3
$NiCl_2 \cdot 6H_2O$	0.05
$Na_2MoO_4 \cdot 2H_2O$	0.05
H_3BO_3	0.03

[a]Only include for solid medium

9. DpnI.
10. Materials for Gibson Assembly.
11. DH10b chemically competent cells (*see* **Note 8**).
12. Sequencing primers to confirm successful CRISPR-Cas9 targeting plasmid construction.

    Forward (5′) primer oCAH232: TTGCTCGGTCGCTATTCG.

    Reverse (3′) primer oCAH3: TTCACTTTTCTCTATCACTGATAG.
13. S17λpir chemically competent cells.
14. Plasmid miniprep kit.

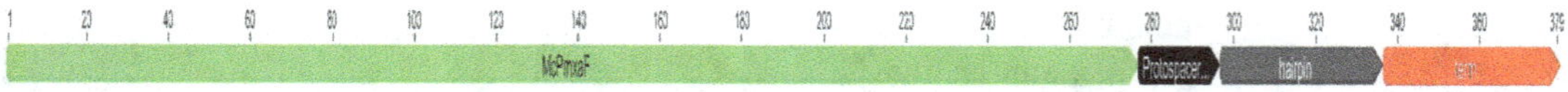

cgttccactgagcgtcagacGAGGTTCAGGCGAAACCGCAGACTCAAGGGCGCTTGCTCCCGGGAAAGA

TCGTATTAGTTTGCCTCGATCGGCGGTCCTTGTGACAGGGAGATATTCCCGACGGATCCGGG

GCATTCGAGCGGAACCGCCCGCCGTGGGAGTTTTTCCAGCGAGCATTCGAGAGTTTTTCAAG

GCGGCTTCGAGGGGTTATTCCGTAACGCCGCCGACATGATCTGTCCCAGAATCTCCGCCGCT

GTTCGTAGAGCGCCGATGCAGGGTCGGCATCAATCATTCTTNNNNNNNNNNNNNNNNNNNN

GTTTTAGAGCTAGAAATAGCAAGTTAAAATAAGGCTAGTCCGTTATCAACTTGAAAAAGTGG

/ \ editing template CACCGAGTCGGTGCTTTTTTTN_{500}gttctttcctgcgttatccc

Fig. 2 The gRNA-containing cassette. The *M. capsulatus* PQQ-dependent methanol dehydrogenase promoter (P_{mxaF}) is used to drive strong, constitutive expression of the synthetic gRNA. The cassette containing the gRNA and editing template is obtained as a synthetic gBlock Gene Fragment from Integrated DNA Technologies

3 Methods

3.1 Design and Construction of CRISPR-Cas9 Plasmid Via Isothermal Assembly for M. capsulatus Gene Targeting

A schematic overview of the CRISPR-Cas9 targeting plasmid assembly is presented in Fig. 3 and detailed below.

1. Set up the following PCR to amplify the pCAH01Sp^{R}Cas9 backbone (*see* **Note 9**).

Reagent	Volume (μL)
Q5 HiFi 2× Master Mix	25
10 μM oCAH230	2.5
10 μM oCAH231	2.5
pCAH01Sp^{R}Cas9 (10 ng/μL)	1
Water	19
Total	**50**

Cycling conditions.

Step	Temperature (°C)	Time	Cycles
Initial denaturation	98	30 s	1
Denaturation Annealing Extension	98 64 72	10 s 15 s 10 min	30
Final extension	72	10 min	1
Hold	4	∞	1

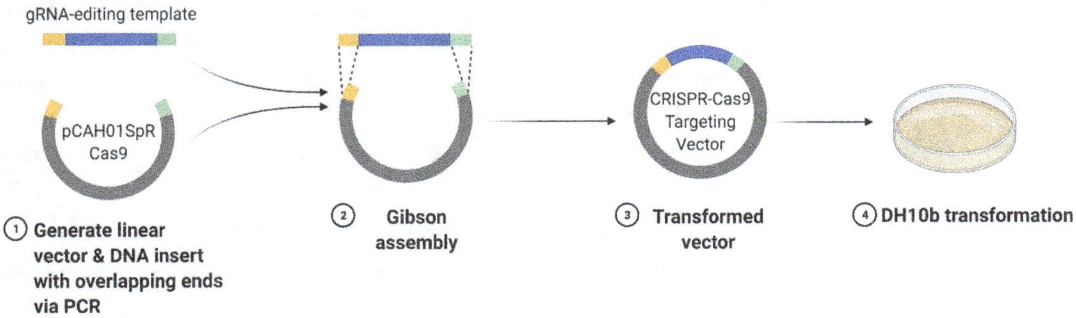

Fig. 3 Overview of isothermal assembly of the pCRISPR targeting plasmid

2. Set up the following PCR to amplify the gRNA-editing template cassette (*see* **Note 7**).

Reagent	Volume (μL)
Q5 HiFi 2× Master Mix	25
10 μM oCAH267	2.5
10 μM editing template reverse primer	2.5
gBlock (10 ng/μL)	1
Water	19
Total	**50**

Cycling conditions.

Step	Temperature (°C)	Time	Cycles
Initial denaturation	98	30 s	1
Denaturation Annealing Extension	98 Variable 72	10 s 15 s 15 s	30
Final extension	72	30 s	1
Hold	4	∞	1

3. Verify that the PCRs were successful and amplified a single product via DNA agarose gel electrophoresis. Mix 5 μL PCR with 1 μL 6× DNA loading dye and compare to a molecular weight marker standard in a 1% agarose gel (*see* **Note 10**).
4. After confirmation of PCR success via electrophoresis, add 1 μL DpnI directly to the pCAH01Sp^{R}Cas9 backbone PCR and incubate 1 h at 37 °C in order to prevent the intact pCAH01Sp^{R}Cas9 plasmid from being carried forward and resulting in background transformants.

5. Purify the DpnI-treated pCAH01Sp^{R}Cas9 backbone and the gRNA-editing template amplicon using a PCR purification kit following the manufacturer's protocol. Elute the DNA in 20 μL nuclease-free water.
6. Determine the DNA fragment concentrations spectrophotometrically.
7. Prepare the isothermal assembly reaction as follows.

Reagent	Volume (μL)
Gibson Assembly Master Mix (2×)	2
Linearized pCAH01Sp^{R}Cas9 (100 ng/μL)	1
gRNA-editing template cassette (20 ng/μL)	1
Total	**4**

Incubate the reaction at 50 °C for 1 h.

8. Thaw a 50 μL aliquot of DH10b chemically competent cells in a 1.5 mL microcentrifuge tube on ice.
9. Add the 4 μL isothermal reaction to the 50 μL cells, flick the tube with your finger five times to mix, and incubate on ice for 30 min.
10. Spread the transformation evenly on LB agar containing 100 μg/mL carbenicillin using a sterile L-rod (*see* **Note 11**).
11. Incubate the selection plate at 37 °C overnight.
12. Screen transformant colonies to confirm presence of the gRNA-editing template cassette by colony PCR using sequencing primers (*see* **Note 12**):

Reagent	Volume
Q5 HiFi 2× Master Mix	12.5 μL
10 μM oCAH232	1.25 μL
10 μM oCAH3	1.25 μL
Bacterial biomass	–
Water	10 μL
Total	**25 μL**

Cycling conditions.

Step	Temperature (°C)	Time	Cycles
Initial denaturation	98	30 s	1
Denaturation Annealing Extension	98 60 72	10 s 15 s 30 s	30
Final extension	72	1 min	1
Hold	4	∞	1

13. Inoculate PCR-positive transformants into 2 mL LB broth containing 100 μg/mL carbenicillin/50 μg/mL spectinomycin in culture tubes and culture overnight at 37 °C in a 200 rpm rotary shaker.
14. Isolate plasmid DNA from PCR-positive transformants by using the plasmid miniprep kit following the manufacturers protocol (*see* **Note 13**).
15. Confirm successful assembly of the CRISPR-Cas9 targeting construct via sanger sequencing using primer oCAH232 (*see* **Note 14**).

3.2 Optimized Methanotroph Conjugation Protocol

A schematic overview of the conjugation protocol is presented in Fig. 4 and detailed below.

1. Thaw a 50 μL aliquot of S17λpir chemically competent cells in a 1.5 mL microcentrifuge tube on ice.
2. Add 1 μL (50–100 ng/ μL) of the sequence-confirmed CRISPR-Cas9 targeting plasmid DNA to the 50 μL cells, flick the tube with your finger five times to mix, and incubate on ice for 30 min.
3. Spread the transformation evenly on LB agar containing 100 μg/mL carbenicillin using a sterile L-rod.
4. Incubate the selection plate at 37 °C overnight.
5. Streak a single CRISPR-Cas9 targeting plasmid-containing S17λpir transformant colony onto LB agar containing 100 μg/mL carbenicillin/50 μg/mL spectinomycin.
6. In a microcentrifuge tube, resuspend an inoculating loopful of plate-derived *M. capsulatus* biomass in 1 mL mating medium.
7. In a separate tube, resuspend an inoculating loopful of plate-derived CRISPR-Cas9 targeting plasmid-containing S17λpir transformant biomass in 1 mL mating medium.
8. Measure the optical density (OD_{600}) of each cell suspension using a spectrophotometer.

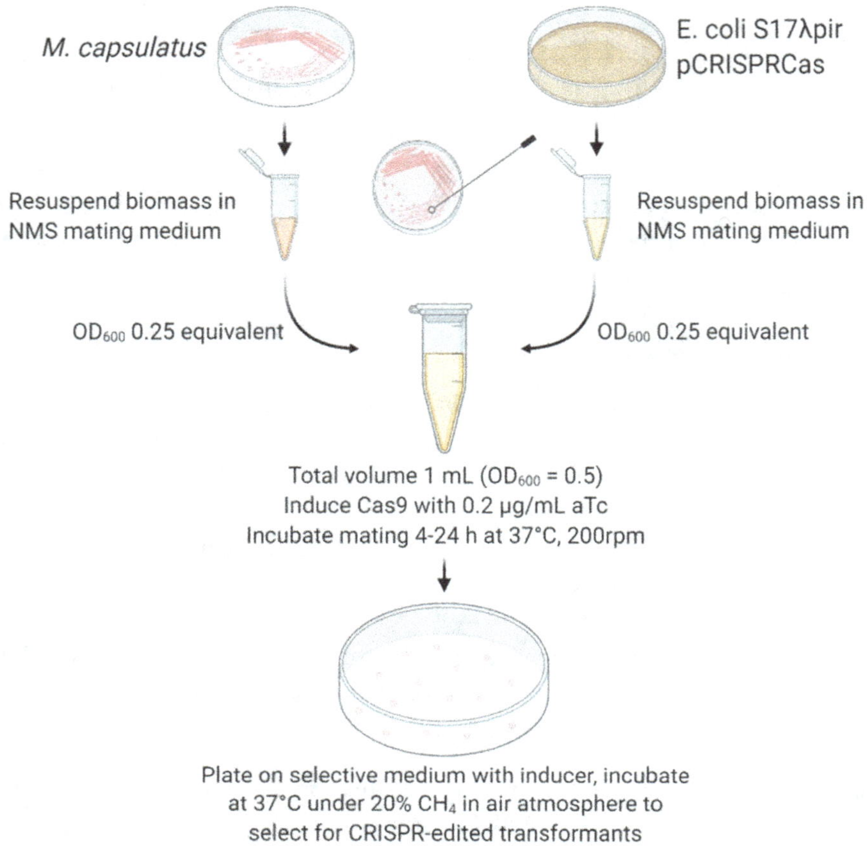

Fig. 4 Overview of mating protocol for *M. capsulatus* genome editing

9. Add an $OD_{600} = 0.25$ equivalent volume of each cell suspension to a 1.5 mL microcentrifuge tube (example: 25 μL of an $OD_{600} = 10$ cell suspension) followed by the addition of mating medium to reach a total volume of 1 mL.
10. Induce Cas9 expression by adding 0.2 μg/mL aTc inducer.
11. Incubate the 1.5 mL microcentrifuge tube/biparental mating in a rotary shaking incubator at 37 °C for 4–24 h at 200 rpm.
12. After incubation, spread the cell mixture onto NMS agar containing 50 μg/mL spectinomycin, 0.2 μg/mL aTc to select for genome edited transformants (*see* **Note 15**).
13. Selection plates should be placed in an oxoid chamber with an atmosphere of 20% CH_4 in air and incubated at 37 °C. Colonies typically appear within 1–2 weeks.
14. Edited transformants can be identified via colony PCR followed by sanger sequencing (*see* **Note 16**).

4 Notes

1. Our laboratory has used the method as described [20] for successful electroporation of *Methylococcus capsulatus* Bath, *M. capsulatus* Texas, *Methylotuvimicrobium alcaliphilum* 20Z, and *Methylosinus trichosporium* OB3b. However, we have found the transformation efficiency associated with electroporation to be significantly lower than conjugation.
2. NMS medium used here is slightly altered from the original formulation previously described [29]. We maintain a single sterile 50× laboratory stock solution of the NMS core components (KNO_3, $MgSO_4$, $CaCl_2$) that is diluted to 1× with autoclaved Milli-Q water prior to the addition of autoclaved phosphate and trace solutions when making NMS liquid medium.
3. Make a stock of aTc by dissolving 10 mg in 1 mL DMSO. aTc is added during the mating to induce Cas9 expression.
4. Autoclaved Milli-Q water can replace molecular grade water but should be tested empirically to confirm it does not inhibit the PCR.
5. We order standard, 25 nmole, single-stranded DNA oligos from Integrated DNA Technologies. Lyophilized oligos are resuspended in nuclease-free water to generate 100 μM stocks. These stocks are diluted ten-fold in nuclease-free water to make the 10 μM working primer stocks to be used in PCRs.
6. 500 bp of homology is required for efficient homologous recombination-dependent repair using the editing template. It is critical that the protospacer adjacent motif (PAM) site is altered or removed during repair to prevent continuous, gRNA-dependent Cas9 targeting and cleavage (*see* [30, 31] on gRNA design). To generate a knockout strain, we alter a PAM site located closest to the 5′ end of the coding sequence of interest to an ochre (TAA) stop codon within the editing template.
7. IDT gRNA-editing template gBlock synthetic DNA fragments can be ordered with pCAH01 homology sequences required for isothermal assembly (as presented in Fig. 2). Thus, editing templates synthesized with the homology arms can be used directly in isothermal assembly reactions without the need for PCR amplification.
8. We use Zymo Research Mix & Go *E. coli* Transformation Buffer Set to produce chemically competent DH10b and S17λpir cells following the manufacturer's protocol.

9. Add the reagents to the 0.2 mL PCR tube in the order listed. After the addition of the water, mix 10× by pipetting, avoiding the generation of bubbles.

10. Occasionally, nonoptimized PCR conditions can lead to nonspecific amplicons. In this case, excise the amplicon corresponding to the correct size and purify using a DNA gel extraction kit.

11. Outgrowth with SOC medium is required for efficient selection of plasmids mediating antibiotic resistance other than penicillin (i.e., kanamycin, gentamicin, spectinomycin) when following the Zymo competent cell transformation protocol. As such, we directly plate transformations on carbenicillin-containing selective medium when possible.

12. Identify 4–6 well isolated colonies, circle and numerically label them with a marker on the underside of the petri dish. Prepare a master mix with all the required PCR reagents and aliquot in 0.2 mL PCR tubes. Using a sterile pipette tip gently touch a colony to be screened, rub the pipette tip on the inside of the PCR tube containing the master mix to introduce the biomass, and mix the reaction by pipetting. DO NOT transfer the whole colony. The vector backbone should be used as a positive control for verification of the gRNA-editing template cassette assembly.

13. For plasmids ≥10 kb, elution buffer should be heated to 50 °C as indicated by the manufacturer. We have observed a significant increase in plasmid yield when heating the elution water compared to room temperature water.

14. Prepare a lab inventory frozen stock of the DH10b and S17 strains harboring a sequence-confirmed construct by mixing 920 μL overnight culture with 80 μL DMSO in a cryogenic vial. Store in −80 °C freezer.

15. After 24 h incubation, the 1.5 mL microcentrifuge tube can be centrifuged at ~16,000 × *g* for 2 min to pellet the cells. Discard 900 μL of the supernatant without disturbing the pellet. Resuspend the pellet in the remaining 100 μL supernatant mating medium and spread on NMS selective medium with an L-rod.

16. The average editing efficiency for the *M. capsulatus* CRISPR-Cas9 system is ~10%, although this varies greatly depending on the target gene, editing template length, recombination distance from the Cas9 restriction site, and gRNA. Users should empirically test different gRNAs to optimize efficiency, especially when developing mutant libraries.

Acknowledgments

This work was supported by University of North Texas institutional start-up funds and College of Science seed funding to CAH. Figures 3 and 4 schematics were created with BioRender.com. Figure 3 was adapted from "Gibson Assembly Cloning 1" retrieved from https://app.biorender.com/biorender-templates.

References

1. Marx CJ, Lidstrom ME (2002) Broad-host-range cre-lox system for antibiotic marker recycling in gram-negative bacteria. BioTechniques 33:1062–1067. https://doi.org/10.2144/02335rr01
2. Marx CJ, Lidstrom ME (2001) Development of improved versatile broad-host-range vectors for use in methylotrophs and other gram-negative bacteria. Microbiology 147:2065–2075. https://doi.org/10.1099/00221287-147-8-2065
3. Lloyd JS, Finch R, Dalton H, Murrell JC (1999) Homologous expression of soluble methane monooxygenase genes in Methylosinus trichosporium OB3b. Microbiology 145 (Pt 2):461–470. https://doi.org/10.1099/13500872-145-2-461
4. Csáki R, Hanczár T, Bodrossy L, Murrell JC, Kovács KL (2001) Molecular characterization of structural genes coding for a membrane bound hydrogenase in Methylococcus capsulatus (Bath). FEMS Microbiol Lett 205:203–207. https://doi.org/10.1111/j.1574-6968.2001.tb10948.x
5. Csáki R, Bodrossy L, Klem J, Murrell JC, Kovács KL (2003) Genes involved in the copper-dependent regulation of soluble methane monooxygenase of Methylococcus capsulatus (Bath): cloning, sequencing and mutational analysis. Microbiology 149:1785–1795. https://doi.org/10.1099/mic.0.26061-0
6. Ishikawa M, Yokoe S, Kato S, Hori K (2018) Efficient Counterselection for Methylococcus capsulatus (Bath) by using a mutated pheS gene. Appl Environ Microbiol 84(23): e01875-18. https://doi.org/10.1128/AEM.01875-18
7. Puri AW, Owen S, Chu F, Chavkin T, Beck DAC, Kalyuzhnaya MG, Lidstrom ME (2015) Genetic tools for the industrially promising methanotroph Methylomicrobium buryatense. Appl Environ Microbiol 81: 1775–1781. https://doi.org/10.1128/AEM.03795-14
8. Ro SY, Rosenzweig AC (2018) Recent advances in the genetic manipulation of Methylosinus trichosporium OB3b. Meth Enzymol 605:335–349. https://doi.org/10.1016/bs.mie.2018.02.011
9. Welander PV, Summons RE (2012) Discovery, taxonomic distribution, and phenotypic characterization of a gene required for 3-methylhopanoid production. Proc Natl Acad Sci U S A 109:12905–12910. https://doi.org/10.1073/pnas.1208255109
10. Marx CJ (2008) Development of a broad-host-range sacB-based vector for unmarked allelic exchange. BMC Res Notes 1:1. https://doi.org/10.1186/1756-0500-1-1
11. Tapscott T, Guarnieri MT, Henard CA (2019) Development of a CRISPR/Cas9 system for Methylococcus capsulatus in vivo gene editing. Appl Environ Microbiol 85(11):e00340-19. https://doi.org/10.1128/AEM.00340-19
12. Henard CA, Smith H, Dowe N, Kalyuzhnaya MG, Pienkos PT, Guarnieri MT (2016) Bioconversion of methane to lactate by an obligate methanotrophic bacterium. Sci Rep 6:21585. https://doi.org/10.1038/srep21585
13. Ojala DS, Beck DAC, Kalyuzhnaya MG (2011) Genetic systems for moderately halo(alkali)-philic bacteria of the genus Methylomicrobium. Meth Enzymol 495:99–118. https://doi.org/10.1016/B978-0-12-386905-0.00007-3
14. He L, Groom JD, Lidstrom ME (2020) The Entner-Doudoroff pathway is an essential metabolic route for Methylotuvimicrobium buryatense 5GB1C. Appl Environ Microbiol 87(3): e02481-20. https://doi.org/10.1128/AEM.02481-20
15. Ali H, Murrell JC (2009) Development and validation of promoter-probe vectors for the study of methane monooxygenase gene expression in Methylococcus capsulatus Bath. Microbiology 155:761–771. https://doi.org/10.1099/mic.0.021816-0
16. Sharpe PL, Dicosimo D, Bosak MD, Knoke K, Tao L, Cheng Q, Ye RW (2007) Use of

transposon promoter-probe vectors in the metabolic engineering of the obligate methanotroph Methylomonas sp. strain 16a for enhanced C40 carotenoid synthesis. Appl Environ Microbiol 73:1721–1728. https://doi.org/10.1128/AEM.01332-06

17. Kalyuzhnaya MG, Puri AW, Lidstrom ME (2015) Metabolic engineering in methanotrophic bacteria. Metab Eng 29:142–152. https://doi.org/10.1016/j.ymben.2015.03.010
18. Henard CA, Guarnieri MT (2018) Metabolic engineering of methanotrophic bacteria for industrial biomanufacturing. In: Kalyuzhnaya MG, Xing X-H (eds) Methane biocatalysis: paving the way to sustainability. Springer International Publishing, Cham, pp 117–132
19. Collins DA, Kalyuzhnaya MG (2018) Navigating methane metabolism: enzymes, compartments, and networks. Meth Enzymol 613: 349–383. https://doi.org/10.1016/bs.mie.2018.10.010
20. Yan X, Chu F, Puri AW, Fu Y, Lidstrom ME (2016) Electroporation-based genetic manipulation in type I Methanotrophs. Appl Environ Microbiol 82:2062–2069. https://doi.org/10.1128/AEM.03724-15
21. Nguyen AD, Lee EY (2020) Engineered Methanotrophy: a sustainable solution for methane-based industrial biomanufacturing. Trends Biotechnol 39(4):381–396. https://doi.org/10.1016/j.tibtech.2020.07.007
22. Nariya S, Kalyuzhnaya MG (2020) Hemerythrins enhance aerobic respiration in Methylomicrobium alcaliphilum 20ZR, a methane-consuming bacterium. FEMS Microbiol Lett 367(2):fnaa003. https://doi.org/10.1093/femsle/fnaa003
23. Groom JD, Ford SM, Pesesky MW, Lidstrom ME (2019) A Mutagenic Screen Identifies a TonB-Dependent Receptor Required for the Lanthanide Metal Switch in the Type I Methanotroph "Methylotuvimicrobium buryatense" 5GB1C. J Bacteriol 201(15):e00120-19. https://doi.org/10.1128/JB.00120-19
24. Fu Y, Li Y, Lidstrom M (2017) The oxidative TCA cycle operates during methanotrophic growth of the type I methanotroph Methylomicrobium buryatense 5GB1. Metab Eng 42: 43–51. https://doi.org/10.1016/j.ymben.2017.05.003
25. Lee JK, Kim S, Kim W, Kim S, Cha S, Moon H, Hur DH, Kim S-Y, Na J-G, Lee JW, Lee EY, Hahn J-S (2019) Efficient production of d-lactate from methane in a lactate-tolerant strain of Methylomonas sp. DH-1 generated by adaptive laboratory evolution. Biotechnol Biofuels 12:234. https://doi.org/10.1186/s13068-019-1574-9
26. Garst AD, Bassalo MC, Pines G, Lynch SA, Halweg-Edwards AL, Liu R, Liang L, Wang Z, Zeitoun R, Alexander WG, Gill RT (2017) Genome-wide mapping of mutations at single-nucleotide resolution for protein, metabolic and genome engineering. Nat Biotechnol 35:48–55. https://doi.org/10.1038/nbt.3718
27. McCarty NS, Graham AE, Studená L, Ledesma-Amaro R (2020) Multiplexed CRISPR technologies for gene editing and transcriptional regulation. Nat Commun 11: 1281. https://doi.org/10.1038/s41467-020-15053-x
28. Crombie A, Murrell JC (2011) Development of a system for genetic manipulation of the facultative methanotroph Methylocella silvestris BL2. Meth Enzymol 495:119–133. https://doi.org/10.1016/B978-0-12-386905-0.00008-5
29. Whittenbury R, Phillips KC, Wilkinson JF (1970) Enrichment, isolation and some properties of methane-utilizing bacteria. J Gen Microbiol 61:205–218. https://doi.org/10.1099/00221287-61-2-205
30. Wang D, Zhang C, Wang B, Li B, Wang Q, Liu D, Wang H, Zhou Y, Shi L, Lan F, Wang Y (2019) Optimized CRISPR guide RNA design for two high-fidelity Cas9 variants by deep learning. Nat Commun 10:4284. https://doi.org/10.1038/s41467-019-12281-8
31. Chuai G, Ma H, Yan J, Chen M, Hong N, Xue D, Zhou C, Zhu C, Chen K, Duan B, Gu F, Qu S, Huang D, Wei J, Liu Q (2018) DeepCRISPR: optimized CRISPR guide RNA design by deep learning. Genome Biol 19:80. https://doi.org/10.1186/s13059-018-1459-4

Chapter 22

Microarray-Based Screening of Putative HSP90 Inhibitors Predicted and Isolated from Microorganisms

Anusha Kishore, Artem Fetter, and Carsten Zeilinger

Abstract

Protein microarrays are useful tools for detecting the presence of a target where different prey and bait combinations exist. Here we describe the extended application for a functional target-oriented screening assay with full length Heat shock proteins (HSPs) for the identification of novel compounds.

Key words Protein microarray, Heat shock proteins, ATP-Cy5, Natural products, Anti HSP90 inhibitor

1 Introduction

Since Mulder and Berzelius, it is known that proteins denature or aggregate under stress, but only an understanding of the molecular causes leads to standard improvements in diagnosis and therapies. How apt the name of the protein is in its ambiguity regarding its chemical nature but also in relation to Greek mythology, we learn today, as we know the variable properties of proteins [1, 2]. Cells try to protect their valuable proteome via various HSPs (chaperones) and to regenerate damages caused by noxious agents such as temperature stress, toxins, oxidative substances, or mutations that impede protein homeostasis and cell survival) [3, 4]. Molecular chaperones correct the resulting protein misfolding. Besides, chaperones are believed to be a possible influencer in evolution, as chaperones can stabilize unstable mutations and thus pave the way for more effective protein forms [5–9]. Many HSPs have been validated as potential drug targets due to their susceptibility for natural products that hinder refolding and enhance the proteotoxic stress [10, 11]. The chaperone's leading influencer is the environmental stress aging eliciting continuously degraded proteins and may be evolutionarily conserved [12–15]. To avoid proteotoxic stress, which hinders protein homeostasis and cell survival, most

Elizabeth Skellam (ed.), *Engineering Natural Product Biosynthesis: Methods and Protocols*,
Methods in Molecular Biology, vol. 2489, https://doi.org/10.1007/978-1-0716-2273-5_22,

proteins are accompanied by HSPs and sequestrated into different routes of renaturation, de novo folding, or denaturation [16–18]. Therefore, the regulation of Hsp90 presence and activity is relevant for most cells and their function. In this position, HSP90 can decide the fate between health and disease by selective refolding of denatured proteins and is a significant player in the evolution.

In recent decades, the HSP90, a protein with a mass of about 90 kDa, has received particular attention [19, 20]. With a proportion of 1–2% of the total cytosolic protein, it is one of the most frequently found proteins within the cell. In recent years, many proteins have been discovered that form a complex with HSP90 and can thus be prevented from misfolding or whose native folding depends on HSP90. The list of proteins interacting with HSP90 also called "clients," has grown steadily since then [21–23]. Many of the HSP90 clients are known oncogenes. If mutations in these oncogenes' appearance and regulation accumulate, malignant growths can occur in the corresponding tissues.

HSP90 and some accompanied (co)chaperones are overexpressed in cancer cells. Since chaperones are known to stabilize defective proteins, it was therefore concluded that HSPs contribute significantly to tumorigenesis by stabilizing the defective oncogenes and being constantly fuelled by transcription factors such as HSF1 of HSP90 levels [24, 25].

HSP90 is a protein with ATPase function and, as a monomer, is composed of three domains that have been strongly conserved during evolution [26, 27]. These domains are called N-terminal domain (NTD), mid-domain (MD), and C-terminal domain (CTD). The binding of the ATP takes place at the NTD. It could be shown that the ATP binding does not take place at the pyrophosphate as is usual for many ATPases, but at the adenine [28, 29]. Thus, the phosphate end of the ATP protrudes from the protein when bound to HSP90, while the Adenine residue penetrates deep into the protein. The hydrolysis of the bound ATP, and the binding of many HSP90 clients takes place at the MD. NTD and MD are connected by a long, flexible, charged, and variable linking region, the influence of which on HSP90 function is not yet fully understood. The CTD has a C-terminal Met-Glu-Glu-Val-Asp (MEEVD) motif that allows cochaperones to bind to HSP90 with a tetratricopeptide repeat (TRP) motif. In eukaryotes, NTD also has a so-called cap region, which acts like a lid over the ATP binding pocket. HSP90 is present in vivo as a homodimer. The CTD binds the monomers together. It is known that HSP90 is modified in vivo posttranslationally [30]. These modifications include phosphorylation, sumoylation, acetylation, and S-nitrosylation. Post-translational modifications can affect the conformational change of HSP90, the ATP affinity, the ATPase reaction, as well as the binding and processing of clients, and the binding of cochaperones [31].

The location and structure of the N-terminal ATP binding pocket in HSP90 of *S. cerevisiae* (scHSP90) have been studied in detail and used as a model for eukaryotic chaperoning, suggesting that existing isoforms and their structural differences gave an advantage in the proteome protection under stress [32–34]. It has been shown that the regions of Amino acids 28–50, 85–94, and 117–124 form the ATP binding pocket that is deeply embedded in the protein. Direct interaction of amino acids with ATP occurs only at position Asp-79. Other important binding sites where the protein is indirectly bound to ATP via hydrogen bonds of water molecules or van der Waal's interactions are positions Leu-34, Asn-37, Gly83, Met-84 Asn-92, Lys-98, Gly-118, Gly-121, Gly-123, Phe-124, and Thr-171. Mg^{2+} is essential for the binding of ATP. The ATP binding pocket is a target for the development and discovery of new HSP inhibitors such as 17-AAG, or Radicicol, or Geldanamycin derivatives [35, 36].

Geldanamycin and its derivatives are soluble in organic solvents, such as Alcohols, Chloroform, Acetone, Benzene, and Ethyl acetate, but hardly soluble in water. Geldanamycin (Gda), a Benzoquinone Ansamycin antibiotic, and also Herbimycin, produced by the soil-dwelling Actinomycetes *Streptomyces hygroscopicus,* is a natural product which was first described to have high potency against mouse leukaemia (L1210) and Oral carcinoma (KB) cells [37–39]. In this time, successful natural products were identified targeting human diseases [40–42]. The target of Gda was identified, which competes with ATP for the N-terminal pocket of Hsp90 and thereby inhibiting the ATPase activity of Hsp90 at deficient concentrations, because the recycling of unfolded proteins was blocked, which occur due to elevated stress in cancer cells, is suppressed [43, 44]. A library has been assembled with Gda and related Benzoquinone and Ansamycin derivatives of fungal-derived HSP90 inhibitors [45]. Interestingly *S. hygroscopicus* had a self-resistance for Gda due to certain amino acid exchanges in the N-terminal ATP binding domain [46, 47].

An ortholog to eukaryotic HSP90 found in Prokaryotes is Heat temperature protein G (HtpG). However, with a sequence homology of about 40% with HSP90 and a similar mode of operation as a chaperone, it serves a different client profile than HSP90 and does not appear to be as essential for the survival of microorganisms as its eukaryotic counterpart. However, bacterial growth under stress conditions is stabilized by HtpG [48]. Differences in the sequences are largely due to sections that are present in HSP90 but are missing in HtpG. An investigation of the hydropathy and the interactome showed that the ATP binding site of the NTD is mostly conserved. Even amino acids that were found to be essential for ATP binding remain conserved. A superposition of the N-terminal ATP-binding pocket of HSP90 and bacterial HtpG is shown in Fig. 1. Thus,

Fig. 1 Sequence alignment of human HSP90 and ortholog Heat shock proteins from microorganisms. High to low homology in the ATP-binding site given dark red to rouge, respectively, while some differences in the conserved motif DALDK of human HSP90 is in *S. hygroscopicus* and *S. iranensis* changed to DALDR (arrow)

HtpG can be used as an HSP90 model with a restriction to the ATP binding pocket.

Despite the strong homology of the ATP binding pocket, it could be shown that affinities to HSP90 inhibitors can differ (Fig. 1). Thus, even small differences in the amino acid sequence or exchange of amino acids between cells can strongly influence the same inhibitor's affinity. It has been shown that the exchange of Lysine to Arginine at position 48 in HtpG from *Streptomyces hygroscopicus* prevents ATP binding [47]. It has also been shown that an impairment of the ATP affinity of XcHtpG can also be caused by allosteric effects [49]. The possibility of exploiting allosteric effects extends the spectrum of potential inhibitors. The different, sometimes species-specific, affinities for the same active compounds have shown promise in selectively controlling plant pests or human pathogens with species-specific HSP90 inhibitors without harming the host. From the findings mentioned above, the possibility still arises to search specifically for active compound producers. Therefore, it is conceivable that organisms that exhibit conspicuous mutations in their HSP90 ortholog, such as in the ATP binding pocket, are potential producers of active substances. A mutation could be evolutionarily justified because the producer prevents poisoning by its active substance. For example, Geldanamycin (Gda) is a potent HSP90 inhibitor. However, the Gda producer *Streptomyces hygroscopicus* itself has Gda resistance. This resistance is due to a mutation in the otherwise highly conserved ATP binding pocket of HtpG. Radicicol (Rad) is also a potent HSP90 inhibitor. The wheel producer *Humicola fuscoatra* shows a significantly lower affinity for Rad. Here, too, a mutation was found in a usually highly conserved region of the ATP binding pocket. Based on this information, alignments in the bacterial database identified several candidates that should give differences in the susceptibility for compounds and serve as sources for new natural products. Several candidates, for example, *Streptomyces iranensis,* related to *Streptomyces hygroscopicus,* were selected and alignment indicates that they are putative producers of substances with anti-HSP90 activity (Fig. 1).

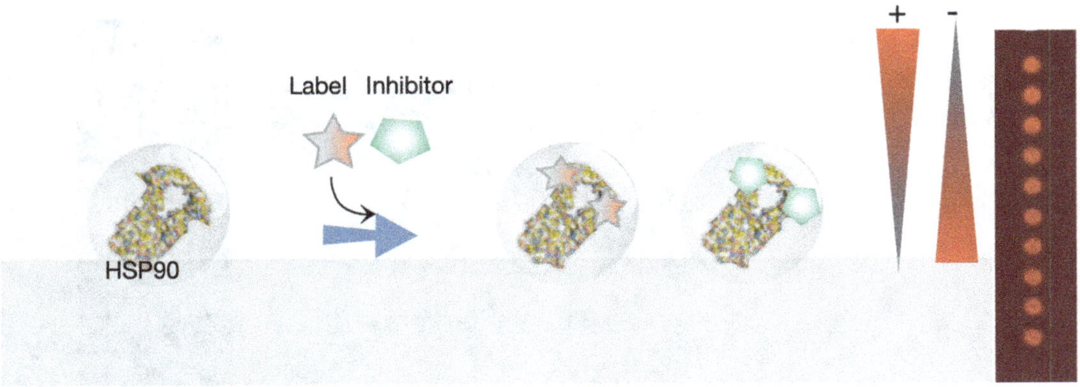

Fig. 2 Scheme of HSP90 microarray-based screening assay for natural products. HSP90 was spotted onto NC microarray slides and incubated with ATP-Cy5 after washing the bound fluorescence intensity of ten spots becomes detected (right side)

A highly miniaturized microarray assay was developed and optimized to investigate ATP binding affinities of HSP and their inhibition using prey and bait configuration for displacement experiments (Fig. 2). The protein to be investigated is placed on a Nitrocellulose (NC) membrane and thus immobilized. Fluorescence-labeled ATP is applied and binds to the protein. If an inhibitory active substance is now added, it hinders the binding of the ATP to the protein. Unbound ATP is washed off. This inhibition can be visualized and quantitatively measured as a decrease in fluorescence intensity on the nitrocellulose membrane. There is a direct correlation between fluorescence intensity and the amount of fluorescence-labeled ATP bound to the protein. The reference inhibitor used to study HSP90-ATP binding is Radicicol, an antifungal antibiotic for which a strong inhibitory effect on HSP90-ATP binding is known. This application was used to screen anti HSP90 libraries (Figs. 3 and 4a) or identify the activity of producers of anti HSP90 substances (Fig. 4b).

2 Materials

Prepare all solutions using double distilled water and analytical grade reagents. Prepare and store all reagents at room temperature (unless indicated otherwise). Diligently follow all waste disposal protocols when disposing of waste materials.

2.1 Contactless Protein Spotting

1. Nitrocellulose glass slides (Unisart® microarray Slide) (*see* **Note 1**).
2. GeSimNanoplotter—The GeSiMNanoplotter is a versatile, noncontact microarray printer for medium to large throughput. Piezo pipettes simultaneously dispense picoliter to nanoliter of the sample (Fig. 3).

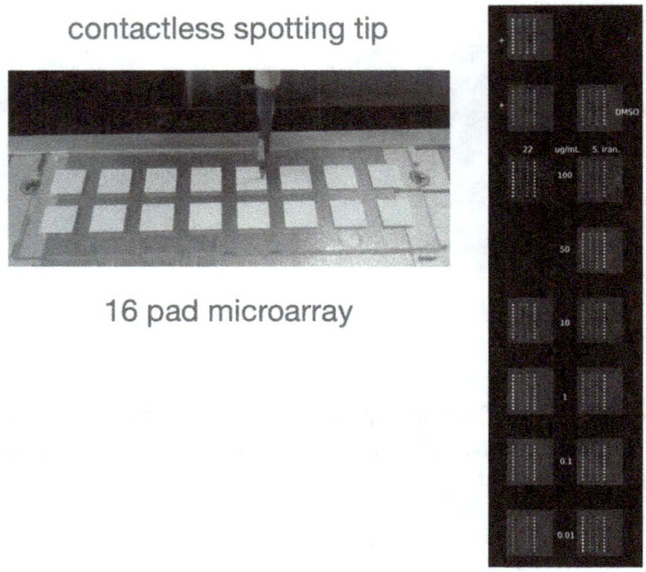

Fig. 3 16 pad NC microarray under spotting process using a contactless spotter. View on the tip and the microarray

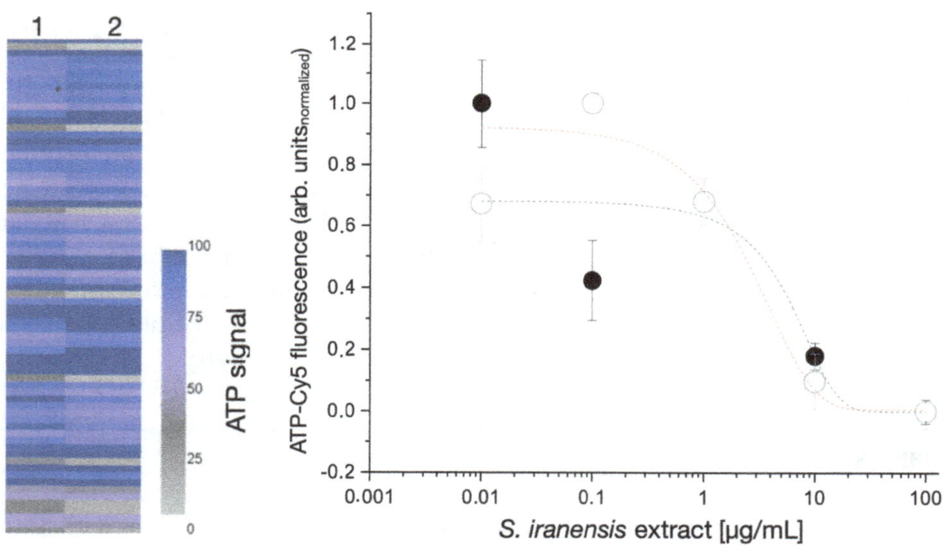

Fig. 4 Heat map of a typical compound screen with 60 compounds on human HSP90 (1) and XcHtpG (2) (**a**) Blue indicates high binding activity of labeled ATP while grey no binding activity. In the first two positions and every tenth, activity was checked repeatedly without inhibitor and with 500 nM Radicicol and IC50 value from identified natural product producer *Streptomyces iranensis* (**b**) tested on HSP90 (open circles) or XcHtpG (filled circles)

2.2 Blocking and Washing of the Membranes

1. HSP90 buffer: 20 mM Tris–HCl, pH-7.5, 50 mM KCl, 6 mM β-mercaptoethanol, 10% v/v Glycerol.
2. Glass container for keeping microarray slides in the buffer.
3. Blocking buffer: 1% BSA in HSP90 buffer.

2.3 Preparation of Incubation Solutions

1. 10× FPI buffer: 20 mM Hepes, 50 mM KCl, 5 mM $MgCl_2{\cdot}6H_2O$, 0.01% Tween 20 and 0.1 mg/mL BSA.
2. 1× FPI buffer: pH 7.3, 1 mM DTT, 2% v/v DMSO, and 20 mM $Na_2MoO_4{\cdot}2H_2O$.
3. Cy5-ATP.
4. Radicicol (*see* **Note 2**).
5. Tested substance (*see* **Note 3**).

2.4 Incubation and Scanning of the Membranes

1. 16-pad hybridization chamber (*see* **Note 4**).
2. PCR sealing foil (*see* **Note 5**).
3. Aluminum foil (*see* **Note 6**).
4. Wet tissues (*see* **Note 7**).
5. GenePix 4300/4400 microarray scanner—Laser slide-based microarray scanner which detects a wide variety of fluorophores.

2.5 Calculation of Binding Activity

1. The binding activity is calculated from fluorescence raw data using Excel and Origin.

3 Methods

FPI buffer should be stored at 4 °C. HSP90 buffer should be kept at room temperature. All steps are performed at RT, unless stated differently. Incubations on a shaker are performed at <50 rpm. Proteins used at a concentration of 3 mg/mL in HSP90 buffer.

3.1 Contactless Protein Spotting

1. The Spotter parameters are adjusted so the spot size and amount of protein on the membrane is as wished. Ten droplets of 100–150 pL per spot have been shown to give the most satisfying results.
2. Fifty microliters of each protein solution is added to the indicated well of a 96-well plate, and the proteins are kept cool, until use.
3. The spotter is initialized, and the activation of the layout is done for the spots on the NC membrane.
4. The 96-well plate is transferred into the spotter for sample uptake.
5. A dummy slide and the slides are placed in the spotter.

6. All slides must align correctly.
7. Each protein is spotted ten times per NC membrane pad (ten spots per column).
8. A maximum of six proteins are spotted on each membrane (per row). More spots may result in an overlapping of the spots and make later evaluation difficult.
9. The spotting procedure is started.
10. If necessary, spotter parameter can be adjusted for desired droplet results (droplet volume, shape, trajectory, and spot size).
11. The exit of droplets from the spotting pipette is ensured.
12. The addition of protein-spots to the membrane is ensured.
13. After spotting, membranes are dried for 15 min at room temperature.

3.2 Blocking and Washing of the Membranes

1. HSP90 buffer is prepared. 100 mL of water is added to a 1 L graduated cylinder. 2.423 g Tris–HCl, 3.728 g KCl, are weighed and 0.412 mL β-mercaptoethanol, and 100 mL Glycerol are transferred to the cylinder. Water is added to a volume of 900 mL. Components are mixed well, and pH is adjusted with HCl. Volume is made up to 1 L with water. The solution is passed through a sterile filter.
2. Blocking buffer is prepared. 1 g of BSA is added in 100 mL of 1× HSP90 buffer and mix well (*see* **Note 8**).
3. The NC membrane slides are placed in a glass container.
4. Per slide, 100 mL of Blocking buffer is added to the container.
5. The container is placed on a shaker for 45 min (blocking).
6. The Blocking buffer is discarded.
7. 100 mL HSP90 buffer per slide is added to the container. The container is placed on a shaker for 20 min. The buffer is then discarded. This washing step is repeated two more times.
8. The NC slides are dried with compressed air. The slides should be kept at an angle so the liquid can flow off easily (*see* **Note 9**).

3.3 Preparation of Incubation Solutions

1. 10× FPI solution is prepared. 20 mL water is added to a 50 mL glass beaker. 2.3831 g Hepes, 1.86 g KCl, 0.51 g $MgCl_2 \cdot 6H_2O$, and 50 mg BSA are weighed and 50 μL Tween 20 are transferred to a beaker. The volume is made 50 mL with water and mixed well.
2. 1× FPI solution is prepared. 0.771 g of DTT is added to 10× FPI mixed well and diluted to 1×. 1 mL DMSO and 0.242 g $Na_2MoO_4 \cdot 2H_2O$ are added. Then pH is adjusted with HCl (*see* **Note 10**). FPI buffer is always kept at 4 °C.

3. Incubation solutions are prepared during blocking or washing steps. All solutions are prepared in 4 °C FPI buffer. Solutions are kept under cool and dark conditions. 50 μL of incubation solution per membrane.
   (a) 0.1 μM Cy5-ATP in FPI buffer (*see* **Note 11**) is the final concentration on the membrane.
   (b) The desired final concentration of the tested substance. Usually, the tested substances are diluted 1:100 times on addition to the membrane.
   (c) Positive control: 50 μL of 0.1 μM Cy5-ATP without inhibitor in 1× FPI.
   (d) Negative control: 50 μL solution of 0.1 μM Cy5-ATP + 1 μM Radicicol in 1× FPI.

3.4 Incubation of the Membranes

1. The dry slides are placed in the incubation chambers.
2. The separation of the membranes can be ensured by applying the buffer solution to a pad and controlling for leakage.
3. 50 μL of incubation solution of different compound concentrations are added to the separated membranes.
4. The chambers are sealed with PCR-sealing foil.
5. The chambers are wrapped with Aluminum foil to maintain dark condition.
6. The chambers are placed in a container on a wet towel.
7. The container is placed at 4 °C on a shaker overnight (*see* **Note 12**).

3.5 Preparation of the Membranes for Scanning

1. On the next day, the incubation solution is discarded.
2. 100 μL of FPI buffer is added to each membrane. The chambers are resealed, rewrapped and placed on a shaker at 4 °C for 5 min. Buffer is discarded. This step is repeated two more times (washing step).
3. The nitrocellulose slides are dried with compressed air. The slides should be kept at an angle so the liquid can flow off easily.

3.6 Scanning of the Membranes

1. The membranes are scanned with a commercially available microarray laser scanner a corresponding software at 635 nm wavelength.
2. The Scanner is calibrated.
3. The dry slides are placed in the Scanner.
4. A quick, low-resolution scan can be used to adjust scanning parameters. Positive control spots should be clearly detectable and be distinguishable from the background. The scan should show no overshooting brightness at any point. If it is the case,

laser intensity or scanner sensitivity should be decreased (*see* **Note 13**).

5. The scan of the slides is saved as a .tiff file.
6. The quantification is done by using Imagene software.
7. The maximum and minimum diameter of spots is determined with the software's measuring tool, used while making a grid.
8. For calculating fluorescence intensity from each spot, a grid is placed on the first pad (Positive control) with the specified number of rows and columns of sample spotted in that pad. Meta grid is placed depending on the rows and columns of the pads used for spotting.
9. After placing the grid each spot has a circle around it, each of which is adjusted separately for the highest signal calculation precision.
10. The evaluation of the program is started.
11. A file with information over the brightness of each spot is received.

3.7 Evaluation of the Fluorescence Intensities

1. With the brightness information for each spot, each protein's average brightness on each membrane is calculated, and the Standard deviation.
2. Data is sorted in the following sequence—Column, Meta Row, Meta Column, Row.

Meta Row, Meta Column, Row, Column	Signal Mean (SM)	Background Mean (BM)
Stdev. BG	SM-BM	SNR:SNR = (SM–BM)/Stdev.BG)

Meta row and Meta column are the numbers of meta row/column in the meta grid where the spot is located. Row and column are the numbers of row/ column in the sub grid where the spot is located. Signal Mean (SM) is pixel intensity averaged over the signal region and Background Mean (BM) is pixel intensity averaged over the local background area.

Mean and Standard deviation are calculated for the ten-fold determination.

c [μM] inhibitor	μ (SM)	σ (SM)	μ (SM-BM)	σ (SM-BM)	μ (SNR)

3. The summary is copied a fitting software, for example, Origin (c, μ (SM-BM), σ (SM-BM) (set as error bar).
4. The average brightness of the positive control is used to normalize the data, fixing the average positive control at a value of

1 and adjusting all other data relative to the control. The negative control is used to normalize for 0.

5. The applied substances' logarithmic concentrations are plotted against relative brightness compared to the positive control (1) and negative control (0).
6. The result is then fitted to the Dose-Response model.
7. IC50 values for the tested substances can now be extracted from the Model-Fit.

4 Notes

1. Make sure to place the slide with the Nitrocellulose side facing upward.
2. Radicicol is a natural product that alters HSP90 function by inhibiting ATP binding.
3. Natural compounds that may inhibit ATP binding to HSP90 or increasing the binding.
4. These are special metal chambers that hold microarray slides and allow separate addition of reagents to each pad on the slide.
5. It is used for stopping evaporation of the reagents. Not placing the chambers in a wet environment can lead to an evaporation of the incubation solution.
6. It is used for maintaining a dark environment to slow down photobleaching of dye.
7. It is used for providing a humid environment when microarray slides are kept for incubation for a long time.
8. Make fresh Blocking buffer.
9. Avoid residues of liquid when drying the slides, or this will result in visible edges forming at the borders of the liquid residues. During drying, keep the Slides at an angle to support the flowing off of all liquid.
10. When 10× FP1 is diluted to 1× FP1 the solution's color changes from bright yellow to light yellow or colorless.
11. Cy5-ATP should be kept in dark conditions. Exposing applied Cy5-solution to direct light might result in a loss of the signal due to degradation of the signaling molecule.
12. Whirling artifacts will occur on the membranes if shaker rpm is too high.
13. Depending on the final concentration and type of solvents in the incubation mixture (such as DMSO or methanol), signal brightness may change.

References

1. Mulder GJ (1839) Ueber die Zusammensetzung einiger thierischen Substanzen. J Prakt Chem 16(1):129–152. https://doi.org/10.1002/prac.18390160137
2. Berzelius JJ (1840) Jahresbericht über die Fortschritte der physischen Wissenschaften. Jahresbericht über die Fortschritte der physischen Wissenschaften. Vol.19 ISBN-10, 0366455516
3. Ritossa F (1996) Discovery of the heat shock response. Cell Stress Chaperones 1(2):97. https://doi.org/10.1379/1466-1268(1996)001<0097:dothsr>2.3.co;2
4. Velazquez JM, Lindquist S (1984) hsp70: nuclear concentration during environmental stress and cytoplasmic storage during recovery. Cell 36(3):655–662. https://doi.org/10.1016/0092-8674(84)90345-3
5. Schopf FH, Biebl MM, Buchner J (2017) The HSP90 chaperone machinery. Nat Rev Mol Cell Biol 18(6):345–360. https://doi.org/10.1038/nrm.2017.20
6. Voth W, Jakob U (2017) Stress-activated chaperones: a first line of defense. Trends Biochem Sci 42(11):899–913. https://doi.org/10.1016/j.tibs.2017.08.006
7. Weber B, Maier A, Buchner J (2020) Peptides in proteins. J Pept Sci 26(2):e3235. https://doi.org/10.1002/psc.3235
8. Draceni Y, Pechmann S (2019) Pervasive convergent evolution and extreme phenotypes define chaperone requirements of protein homeostasis. Proc Natl Acad Sci U S A 116(40):20009–20014. https://doi.org/10.1073/pnas.1904611116
9. Ma P, Xu H, Li J et al (2019) Functionality-independent DNA encoding of complex natural products. Angew Chem Int Ed Engl 58(27):9254–9261. https://doi.org/10.1002/anie.201901485
10. Sontag EM, Vonk WIM, Frydman J (2014) Sorting out the trash: the spatial nature of eukaryotic protein quality control. Curr Opin Cell Biol 26:139–146. https://doi.org/10.1016/j.ceb.2013.12.006
11. Kampinga HH (2006) Chaperones in preventing protein denaturation in living cells and protecting against cellular stress. Handb Exp Pharmacol 172:1–42. https://doi.org/10.1007/3-540-29717-0_1
12. Weids AJ, Ibstedt S, Tamás MJ et al (2016) Distinct stress conditions result in aggregation of proteins with similar properties. Sci Rep 6: 24554. https://doi.org/10.1038/srep24554
13. Mogk A, Bukau B, Kampinga HH (2018) Cellular handling of protein aggregates by disaggregation machines. Mol Cell 69(2):214–226. https://doi.org/10.1016/j.molcel.2018.01.004
14. Määttä TA, Rettel M, Sridharan S et al (2020) Aggregation and disaggregation features of the human proteome. Mol Syst Biol 16(10): e9500. https://doi.org/10.15252/msb.20209500
15. van Dam L, Dansen TB (2020) Cross-talk between redox signalling and protein aggregation. Biochem Soc Trans 48(2):379–397. https://doi.org/10.1042/BST20190054
16. Dai C (2018) The heat-shock, or HSF1-mediated proteotoxic stress, response in cancer: from proteomic stability to oncogenesis. Philos Trans R Soc Lond Ser B Biol Sci 373(1738):20160525. https://doi.org/10.1098/rstb.2016.0525
17. Su K-H, Dai C (2016) Metabolic control of the proteotoxic stress response: implications in diabetes mellitus and neurodegenerative disorders. Cell Mol Life Sci 73(22):4231–4248. https://doi.org/10.1007/s00018-016-2291-1
18. Vihervaara A, Duarte FM, Lis JT (2018) Molecular mechanisms driving transcriptional stress responses. Nat Rev Genet 19(6): 385–397. https://doi.org/10.1038/s41576-018-0001-6
19. Jackson SE (2013) Hsp90: structure and function. Top Curr Chem 328:155–240. https://doi.org/10.1007/128_2012_356
20. Prodromou C (2016) Mechanisms of Hsp90 regulation. Biochem J 473(16):2439–2452. https://doi.org/10.1042/BCJ20160005
21. Radli M, Rüdiger SGD (2018) Dancing with the diva: Hsp90-client interactions. J Mol Biol 430(18 Pt B):3029–3040. https://doi.org/10.1016/j.jmb.2018.05.026
22. Morán Luengo T, Mayer MP, Rüdiger SGD (2019) The Hsp70-Hsp90 chaperone Cascade in protein folding. Trends Cell Biol 29(2): 164–177. https://doi.org/10.1016/j.tcb.2018.10.004
23. Genest O, Wickner S, Doyle SM (2019) Hsp90 and Hsp70 chaperones: collaborators in protein remodeling. J Biol Chem 294(6): 2109–2120. https://doi.org/10.1074/jbc.REV118.002806
24. Calderwood SK, Neckers L (2016) Hsp90 in cancer: transcriptional roles in the nucleus. Adv Cancer Res 129:89–106. https://doi.org/10.1016/bs.acr.2015.08.002

25. Neckers L (2006) Chaperoning oncogenes: Hsp90 as a target of geldanamycin. Handb Exp Pharmacol 172:259–277. https://doi.org/10.1007/3-540-29717-0_11
26. Johnson JL (2012) Evolution and function of diverse Hsp90 homologs and cochaperone proteins. Biochim Biophys Acta 1823(3): 607–613. https://doi.org/10.1016/j.bbamcr.2011.09.020
27. Geller R, Pechmann S, Acevedo A et al (2018) Hsp90 shapes protein and RNA evolution to balance trade-offs between protein stability and aggregation. Nat Commun 9(1):1781. https://doi.org/10.1038/s41467-018-04203-x
28. Biebl MM, Buchner J (2019) Structure, function, and regulation of the Hsp90 machinery. Cold Spring Harb Perspect Biol 11(9). https://doi.org/10.1101/cshperspect.a034017
29. Minari K, de Azevedo ÉC, Kiraly VTR et al (2019) Thermodynamic analysis of interactions of the Hsp90 with adenosine nucleotides: a comparative perspective. Int J Biol Macromol 130:125–138. https://doi.org/10.1016/j.ijbiomac.2019.02.116
30. Prodromou C (2017) Regulatory mechanisms of Hsp90. Biochem Mol Biol J 3(1):2. https://doi.org/10.21767/2471-8084.100030
31. Backe SJ, Sager RA, Woodford MR et al (2020) Post-translational modifications of Hsp90 and translating the chaperone code. J Biol Chem 295(32):11099–11117. https://doi.org/10.1074/jbc.REV120.011833
32. Meyer P, Prodromou C, Liao C et al (2004) Structural basis for recruitment of the ATPase activator Aha1 to the Hsp90 chaperone machinery. EMBO J 23(6):1402–1410. https://doi.org/10.1038/sj.emboj.7600141
33. Ali MMU, Roe SM, Vaughan CK et al (2006) Crystal structure of an Hsp90-nucleotide-p23/Sba1 closed chaperone complex. Nature 440(7087):1013–1017. https://doi.org/10.1038/nature04716
34. Girstmair H, Tippel F, Lopez A et al (2019) The Hsp90 isoforms from S. cerevisiae differ in structure, function and client range. Nat Commun 10(1):3626. https://doi.org/10.1038/s41467-019-11518-w
35. Kitson RRA, Chang C-H, Xiong R et al (2013) Synthesis of 19-substituted geldanamycins with altered conformations and their binding to heat shock protein Hsp90. Nat Chem 5(4): 307–314. https://doi.org/10.1038/nchem.1596
36. Dutton BL, Kitson RRA, Parry-Morris S et al (2014) Synthesis of macrolactam analogues of radicicol and their binding to heat shock protein Hsp90. Org Biomol Chem 12(8): 1328–1340. https://doi.org/10.1039/c3ob42211a
37. Tanida S, Hasegawa T, Hatano K et al (1980) Ansamitocins, maytansinoid antitumor antibiotics. Producing organism, fermentation, and antimicrobial activities. J Antibiot (Tokyo) 33(2):192–198. https://doi.org/10.7164/antibiotics.33.192
38. Omura S, Iwai Y, Takahashi Y et al (1979) Herbimycin, a new antibiotic produced by a strain of Streptomyces. J Antibiot (Tokyo) 32(4):255–261. https://doi.org/10.7164/antibiotics.32.255
39. DeBoer C, Meulman PA, Wnuk RJ et al (1970) Geldanamycin, a new antibiotic. J Antibiot (Tokyo) 23(9):442–447. https://doi.org/10.7164/antibiotics.23.442
40. Sasaki K, Yasuda H, Onodera K (1979) Growth inhibition of virus transformed cells in vitro and antitumor activity in vivo of geldanamycin and its derivatives. J Antibiot (Tokyo) 32(8): 849–851. https://doi.org/10.7164/antibiotics.32.849
41. Patel K, Piagentini M, Rascher A et al (2004) Engineered biosynthesis of geldanamycin analogs for Hsp90 inhibition. Chem Biol 11(12): 1625–1633. https://doi.org/10.1016/j.chembiol.2004.09.012
42. Crump A, Otoguro K (2005) Satoshi Omura: in pursuit of nature's bounty. Trends Parasitol 21(3):126–132. https://doi.org/10.1016/j.pt.2005.01.005
43. Moulin E, Zoete V, Barluenga S et al (2005) Design, synthesis, and biological evaluation of HSP90 inhibitors based on conformational analysis of radicicol and its analogues. J Am Chem Soc 127(19):6999–7004. https://doi.org/10.1021/ja043101w
44. Srivastava BI, DiCioccio RA, Rinehart KL et al (1978) Preferential inhibition of terminal deoxynucleotidyltransferase activity among deoxyribonucleic acid polymerase activities of leukemic and normal cells by geldanamycin, streptoval C, streptovarone, and dapmavarone. Mol Pharmacol 14(3):442–447
45. Winssinger N, Barluenga S (2007) Chemistry and biology of resorcylic acid lactones. Chem Commun 1:22–36. https://doi.org/10.1039/b610344h
46. Piper PW, Millson SH (2012) Spotlight on the microbes that produce heat shock protein 90-targeting antibiotics. Open Biol 2(12): 120138
47. Millson SH, Chua C-S, Roe SM et al (2011) Features of the Streptomyces hygroscopicus

HtpG reveal how partial geldanamycin resistance can arise with mutation to the ATP binding pocket of a eukaryotic Hsp90. FASEB J 25(11):3828–3837

48. Honoré FA, Méjean V, Genest O (2017) Hsp90 is essential under heat stress in the bacterium Shewanella oneidensis. Cell Rep 19(4): 680–687. https://doi.org/10.1016/j.celrep.2017.03.082

49. Yue Q, Stahl F, Plettenburg O et al (2018) The noncompetitive effect of Gambogic acid displaces fluorescence-labeled ATP but requires ATP for binding to Hsp90/HtpG. Biochemistry 57(18):2601–2605. https://doi.org/10.1021/acs.biochem.8b00155

Chapter 23

Isolation of Water-Soluble Metabolites from Marine Invertebrates and Microorganisms

Lamonielli F. Michaliski, Darlon I. Bernardi, and Roberto G. S. Berlinck

Abstract

The isolation of pure organic compounds from biological sources, reaction media, or other complex molecular matrixes is the first step to overcome before further biological and chemical investigations. While the isolation of chemicals soluble in organic solvents is commonly accomplished, the isolation of water-soluble organic compounds is less often addressed. We present here a simple method for the isolation of water-soluble organic compounds, using adsorptive macroporous resins and reversed-phase chromatography.

Key words Hydrophilic, Water-soluble metabolites, Microbial metabolites

1 Introduction

The isolation of pure compounds is the first step to be accomplished for the characterization of unknown structures, the identification of which is essential for further biological and chemical investigations [1, 2]. While a number of reviews, books, and book chapters are devoted to the isolation of compounds soluble in organic solvents, the isolation of water-soluble compounds has been much less discussed. For example, a 2-year (January 1990–December 1991) survey on the literature reporting the isolation of natural products published in the *Journal of Natural Products* demonstrated that 90% of investigations used normal phase chromatography on silica gel as the isolation procedure [3]. An analysis of 2000 articles published in 2012 on the isolation of natural products showed that two-thirds of the articles used normal-phase silica-gel chromatography, while chromatography on C_{18} reversed-phase silica gel or Sephadex LH-20 were much less utilized [4]. These two surveys did not mention the isolation of water-soluble natural metabolites.

Elizabeth Skellam (ed.), *Engineering Natural Product Biosynthesis: Methods and Protocols*,
Methods in Molecular Biology, vol. 2489, https://doi.org/10.1007/978-1-0716-2273-5_23,

A previous review that address the isolation of water-soluble natural products had emphasis on marine metabolites [5]. Few book chapters discuss the isolation of hydrophilic natural products of marine or microbial origin in short sections [6–13]. Nevertheless, the early days of microbial natural products discovery led to the discovery of very polar bioactive metabolites such as aminoglycosides, polar peptides, modified nucleosides, glycosylated polyaromatic polyketides and macrocyclic lactones, which were isolated using ion-exchange, cellulose, and gel-permeation chromatography [14, 15]. This was also true for the isolation of marine natural products, when there was a considerable effort toward the isolation of water-soluble compounds [16].

It seems that in the past there were less concerns on the isolation of hydrophilic natural products than nowadays [16, 17] when the investigation of organic-soluble extracts is largely preferable. However, recent results on the discovery of natural products indicate that water-soluble metabolites represent a valuable class of bioactive compounds [18–21].

We have recently developed a method for the isolation of water-soluble natural products, which has the advantages of (a) minimizing the complexity of metabolite composition by excluding compounds of lower and intermediate polarity, and (b) eliminating "contaminants" such as salts, microbial media components, highly polar pigments, and primary metabolites [17]. This chapter provides the detailed procedure that has been developed and applied for the screening and isolation of hydrophilic natural products.

2 Materials

2.1 Macroporous Adsorptive Resins

1. Amberlite® XAD-2, XAD-4, and XAD-7 resins mixed in a 1: 1:1 ratio (*see* **Note 1**).
2. Diaion® HP-20 resin.

2.2 Macroporous Adsorptive Resin Pretreatment

1. Glass beaker.
2. Glass stick.
3. Ultrasonic bath.
4. Vacuum filtration system with Büchner funnel.
5. Methanol.
6. Acetone.
7. 20% ethanol solution in water.

2.3 Hydrophilic Extract Adsorption on Resins

1. Previously filtered or centrifuged fungal or bacterial fermentation broth or marine invertebrate's aqueous extract, previously partitioned with ethyl acetate.

2. Rotary shaker or rotary incubator shaker.
3. Erlenmeyer flasks or Schott bottles of compatible size to your rotary shaker.

2.4 Macroporous Resin Desorption and Extract Preparation

1. Vacuum filtration system with Büchner funnel.
2. Glass beaker.
3. Glass stick.
4. Methanol.
5. Acetone.
6. Ultrasonic bath.
7. Round bottom flasks.
8. Rotary evaporator.

2.5 Amberlite® XAD-2, XAD-4, and XAD-7 1:1:1 Resin Mixture Cleaning and Storage

1. Glass beaker.
2. Water bath.
3. Heating plate.
4. Glass stick.
5. Methanol.
6. 0.2 M NaOH.
7. 0.2 M HCl.
8. Acetone.
9. 2% Sodium hypochlorite in water.

2.6 Diaion® HP-20 Resin Cleaning and Storage

1. Glass beaker.
2. Ultrasonic bath.
3. Glass stick.
4. Methanol.
5. Acetone.
6. Ethanol: 50% and 20% solutions in water.
7. 2% sodium hypochlorite in water.

2.7 Resin Extract Cleanup

1. Celite® 545.
2. Rotary evaporator.
3. C_{18} derivatized silica gel prepacked column for solid phase extraction (Strata® by Phenomenex).
4. Graduated cylinder.
5. Methanol: 25%, 50%, and 75% solutions in water.
6. Methanol.
7. Water.
8. Vacuum manifold.

3 Methods

3.1 Macroporous Adsorptive Resin Pretreatment

This procedure should be performed with new resins. The procedure for previously used resins is discussed in Subheadings 3.4 and 3.5.

1. Transfer the macroporous adsorptive resin (either Amberlite® XAD-2, XAD-4, and XAD-7 in a 1:1:1 mixture or Diaion® HP-20 resin) to a glass beaker. The amount of resin should correspond to half of the beaker volume.
2. Add methanol (twice the resin volume) and stir gently with a glass stick.
3. Put the beaker in the ultrasonic bath for 20 min (*see* **Note 2**).
4. Remove the methanol by vacuum filtration on a Büchner funnel.
5. Repeat **steps 1–4** using acetone instead of methanol.
6. Transfer the resin back to the beaker and add twice its volume with a 8:2 water–ethanol solution. Stir periodically and gently, using a glass stick until the resin remains in the bottom of the beaker (*see* **Note 3**).
7. Remove floating resin particles by decantation.
8. For an immediate use, remove the water–ethanol mixture by vacuum filtration with a Büchner funnel and proceed to the next section. For storage, keep the resin in the 20% ethanol solution.

3.2 Hydrophilic Extract Adsorption on Resins

This procedure can be performed with a microbial fermentation broth, or with a marine invertebrate aqueous extract, previously filtered and/or centrifuged (*see* **Note 4**). Either Amberlite® XAD-2, XAD-4, and XAD-7 in a 1:1:1 mixture or Diaion® HP-20 resin can be employed (*see* **Note 5**).

1. Transfer the previously filtered or centrifuged fermentation broth or aqueous extract to Erlenmeyer flasks or Schott bottles. Do not exceed half of the container's capacity.
2. Add either Amberlite® XAD-2, XAD-4, and XAD-7 (1:1:1) to a 10% concentration (m/v) or Diaion® HP-20 to a 5% concentration (m/v).
3. Keep the Erlenmeyer flasks or Schott bottles overnight in the rotary shaker or rotary incubator shaker at room temperature and between 120 and 150 rpm.

3.3 Resin Desorption and Extract Preparation

1. After the overnight shaking, collect the resin by vacuum filtration with a Büchner funnel. Rinse the resin in the funnel with MilliQ water. Discard the filtrate and transfer the resin to a glass

beaker. The resin must not occupy more than half of the beaker volume.

2. Add methanol (twice the resin volume) and put the beaker in an ultrasonic bath for 20 min.
3. Remove the methanol by vacuum filtration with a Büchner funnel and collect the organic solvent.
4. Transfer the resin back to the glass beaker.
5. Repeat **steps 2–4** twice more using methanol and three additional times using a 1:1 acetone–methanol mixture.
6. Combine the methanol and 1:1 acetone–methanol 1:1 filtrates and evaporate them in a rotary evaporation. Evaporation provides the hydrophilic resin extract.

3.4 Amberlite® XAD-2, XAD-4, and XAD-7 (1:1:1) Mixture Cleaning and Storage

1. Transfer the used mixture resin to a glass beaker. The resin must not occupy over half of the beaker volume.
2. Add methanol to the glass beaker (twice the resin volume) and put the beaker on a heating plate and let it boil for 1 h.
3. Remove the methanol by vacuum filtration with a Büchner funnel. Discard the filtrate.
4. Transfer the resin to a glass beaker, and add a 0.2 M NaOH solution to the mixture of resins (twice the resin volume). Put the beaker on a water bath and let it boil for 1 h.
5. Repeat **step 3**.
6. Repeat **step 4** using 0.2 M HCl as washing solution.
7. Repeat **step 3**.
8. Transfer the mixture of resins back to a glass beaker and wash with MilliQ (or similar) water until a neutral pH.
9. Filtrate the clean resin in a Büchner funnel, collect and store it in a 20% ethanol solution in water to prevent dryness and microbial contamination (*see* **Note 6**).

3.5 Diaion® HP-20 Resin Cleaning and Storage

Diaion® HP-20 can be reused several times. A good procedure is to clean the resin as soon as the desorption of the hydrophilic material is completed.

1. Transfer the used resin to a glass beaker. The resin should not occupy more than half of the beaker volume.
2. Add a 1:1 ethanol–water mixture (twice the resin volume) and put the beaker in the ultrasonic bath for 20 min.
3. Remove the 1:1 ethanol–water mixture by vacuum filtration with a Büchner funnel. Discard the filtrate.
4. Repeat **steps 1–3** until the filtrate has a pale color or no color at all.

5. Repeat **steps 1–4** using 100% methanol, followed by 100% acetone.
6. Transfer the resin back to the glass beaker and add a 2% sodium hypochlorite solution in water (twice the resin volume). Gently stir the suspension with a glass stick and let it rest overnight.
7. Remove the solution by vacuum filtration with a Büchner funnel. Discard the filtrate.
8. Rinse the resin thoroughly with water, filtrating and discarding the filtrate until the pH of washed water is 7.
9. After filtration, store the resin in a 20% ethanol solution in water to prevent dryness and microbial contamination (*see* **Note 6**).

3.6 Resin Extract Cleanup

Water-soluble extracts can be rapidly generated with our method. Extracts of marine organisms, or from microbial growth media, can be utilized with this procedure (Scheme 1).

1. Exhaustive extraction of marine invertebrates with MeOH results an extract which is concentrated and partitioned between EtOAc and H_2O.
2. The water-soluble fraction is adsorbed on a 1:1:1 mixture of XAD-2, XAD-4, and XAD-7, or on HP-20 macroporous polymeric resins with overnight gentle shaking.
3. Resin desorption of water-soluble organic metabolites is performed with MeOH and with 1:1 MeOH–acetone.
4. The MeOH and 1:1 MeOH–acetone fractions are combined and evaporated.
5. The water-soluble organic fraction is then subjected to a C_{18} reversed-phase (RP) solid-phase extraction (SPE).
6. The C_{18} RP SPE is performed with a gradient of MeOH in H_2O, to give five fractions. The 100% H_2O is discarded (Scheme 1).

Filtered microbial culture media were similarly processed, starting with the EtOAc partition followed by the same procedure. HP-20 resin is used with bacterial culture media instead of the 1:1:1 mixture of XAD-2, -4, and -7. Fractions obtained from the separation of the water-soluble components should be further subjected to bioassays and/or to analysis by HPLC-MS and/or by NMR. Selected fractions will be further evaluated in terms of (a) polarity of compounds; (b) chemical nature (i.e., if acidic, basic, or neutral); (c) chemical class of compounds; (d) mixture complexity; and (e) molecular weight of chemical constituents. In having such information, further separations can be performed by reversed-phase, HILIC, normal-phase, ion-exchange or gel permeation chromatography aiming to separate and purify the desired compounds.

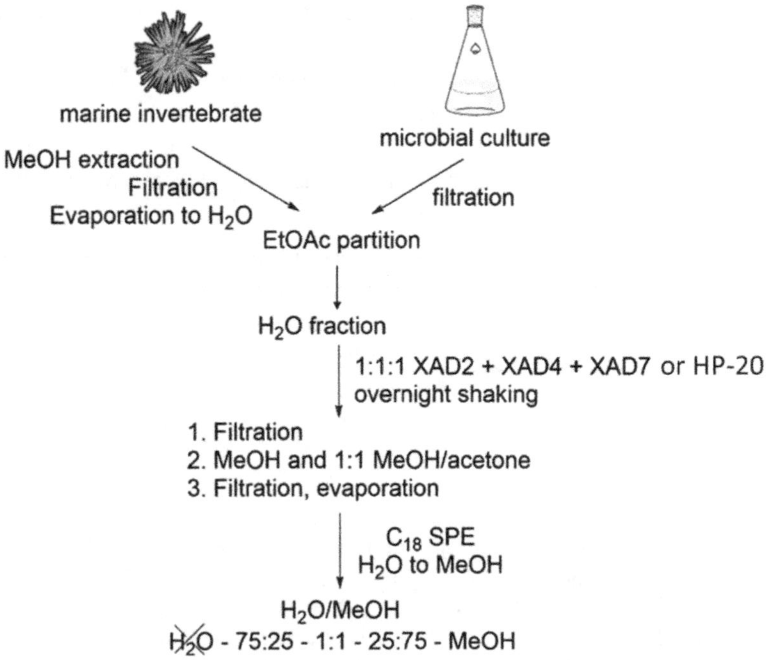

Scheme 1 Procedure for the generation of water-soluble fractions obtained from marine invertebrates or from microbial culture media [14]

4 Notes

1. As each Amberlite® XAD-2, XAD-4, and XAD-7 resin is able to adsorb compounds in different polarity and molecular weight ranges, when using a 1:1:1 mixture of XAD-2, -4, and -7 the majority of the water-soluble organic metabolites may be adsorbed.
2. Since methanol and acetone are volatile, it is advised to avoid excessive exposure by covering the top of the beaker and always working in a fume hood.
3. A magnetic stirrer must not be used for this step since it fragments the resin particles. If the resin has been stored for a long time, it may be dried and float when added to the fermentation broth or aqueous extract, which prevents optimal adsorption. Hence the importance of this step, in order to ascertain that the resin particles are hydrated.
4. When working with microbial cultures, water-soluble metabolites can be concentrated in a H_2O fraction by previously partitioning the culture broth with ethyl acetate. After partition, the residual ethyl acetate should be removed by rotary evaporation before adding any resin to the broth.

5. Experiments with fermentation broths of different bacterial strains indicated similar metabolomic profiles using similar concentrations of either Amberlite® XAD resins or Diaion® HP-20 resin for the same strain in the same growth conditions. However, the yield of fractions obtained after clean-up of extracts was considerably higher when Diaion® HP-20 was utilized.
6. Resins can be stored in its original container, at room temperature, in a dry cabinet, away from direct light. If the storage solution turned colored, it indicates that the resin was not properly cleaned. Thus, the cleaning process should be repeated.

References

1. Bruening RC, Oltz EM, Furukawa J et al (1986) Isolation of tunichrome B-1, a reducing blood pigment of the sea squirt, *Ascidia nigra*. J Nat Prod 49:193–204
2. Nakanishi K (1989) Natural products chemistry—past and future. In: Rowe JW (ed) Natural products of woody plants, 1st edn. Springer-Verlag, Berlin, pp 13–26
3. McAlpine JB, Hochlowski JE (1994) Isolation and purification of secondary metabolites. In: Gullo VP (ed) The discovery of natural products with therapeutic potential. Butterworth-Heinemann, Boston
4. Pauli GF, Chen S-N, Friesen JB, McAlpine JB, Jaki BU (2012) Analysis and purification of bioactive natural products: the AnaPurNa study. J Nat Prod 75:1243–1255
5. Shimizu Y (1985) Bioactive marine natural products, with emphasis on handling of water-soluble compounds. J Nat Prod 48: 223–235
6. Shimizu Y (1998) Purification of water-soluble natural products. In: Cannell RJP (ed) Natural products isolation, 1st edn. Humana Press, Totowa, NJ
7. Wright AE (1998) Isolation of marine natural. In: Cannell RJP (ed) Natural products isolation, 1st edn. Humana Press, Totowa, NJ
8. Shimizu Y, Li B (2005) Purification of water-soluble natural products. In: Sarker SD, Latif Z, Gray AI (eds) Natural products isolation, 2nd edn. Humana Press, Totowa, NJ
9. Dufresne C (1998) Isolation by ion-exchange methods. In: Cannell RJP (ed) Natural products isolation, 1st edn. Humana Press, Totowa, NJ
10. Durham DG (2005) Isolation by ion-exchange methods. In: Sarker SD, Latif Z, Gray AI (eds) Natural products isolation, 2nd edn. Humana Press, Totowa, NJ
11. Dragull K, Beck JJ (2012) Isolation of natural products by ion-exchange methods. In: Sarker SD, Nahar L (eds) Natural products isolation—methods and protocols, 3rd edn. Humana Press, Totowa, NJ
12. Houssen WE, Jaspars M (2012) Isolation of marine natural products. In: Sarker SD, Nahar L (eds) Natural products isolation—methods and protocols, 3rd edn. Humana Press, Totowa, NJ
13. Barrow RA (2005) Isolation of microbial natural products. In: Sarker SD, Latif Z, Gray AI (eds) Natural products isolation, 2nd edn. Humana Press, Totowa, NJ
14. Umezawa H (1988) Low-molecular-weight enzyme inhibitors and immunomodifiers. In: Goodfellow M, Williams ST, Mordarski M (eds) Actinomycetes in biotechnology, 1st edn. Academic, London, pp 284–325
15. Umezawa H, Kondo S (1975) Ion-exchange chromatography of aminoglycoside antibiotics. Method Enzymol 43:263–278
16. Berlinck RGS, Monteiro AF, Bertonha AF et al (2019) Approaches for the isolation and identification of hydrophilic, light-sensitive, volatile and minor natural products. Nat Prod Rep 36: 981–1004
17. Rodríguez JPG, Bernardi DI, Gubiani JR et al (2020) Water-soluble glutamic acid derivatives produced in culture by *Penicillium solitum* IS1-A from King George Island, Maritime Antarctica. J Nat Prod 83:55–65
18. Shi Q-Q, Lu J, Peng X et al (2018) Cimitriteromone A–G, macromolecular triterpenoid–-chromone hybrids from the rhizomes of *Cimicifuga foetida*. J Org Chem 83: 10359–10369

19. Maffioli SI, Fabbretti A, Brandi L et al (2013) Orthoformimycin, a selective inhibitor of bacterial translation elongation from Streptomyces containing an unusual orthoformate. ACS Chem Biol 8:1939–1946
20. Maffioli SI, Zhang Y, Degen D et al (2017) Antibacterial nucleoside-analog inhibitor of bacterial RNA polymerase. Cell 169: 1240–1248
21. Matsui K, Kan Y, Kikuchi J et al (2020) Stalobacin: discovery of novel lipopeptide antibiotics with potent antibacterial activity against multidrug-resistant bacteria. Med Chem 63: 6090–6095

Chapter 24

Natural Product Investigation in Lichens: Extraction and HPLC Analysis of Secondary Compounds in Mycobiont Cultures

Muthukumar Srinivasan, Karthik Shanmugam, and Hariharan Gopalasamudram Neelakantan

Abstract

Chromatography techniques facilitate separation, purification, and identification of secondary compounds (natural products) in lichens and their mycobiont cultures. In particular, high-performance liquid chromatography (HPLC) plays a vital role in the identification of lichen substances because of its high sensitivity, speed, and reliability with the minimal sample. Therefore, we describe the extraction and HPLC protocol for the investigation of secondary compounds with a special focus on lichen mycobiont cultures.

Key words Lichen, Mycobiont culture, Natural product, Lichen compounds, HPLC-PDA

1 Introduction

Lichens are fungi (mycobionts/lichen-forming fungi) which form obligate symbiotic associations with green algae or cyanobacteria (photobionts) to fulfill their nutritional requirements [1]. Lichens are well known for their rich source of secondary metabolites. Most of the lichen compounds belong to aromatic polyketides and are often unique to lichens [2]. Lichen secondary compounds were biosynthesized by acetyl-polymalonyl, shikimic acid and mevalonic acid pathways [3]. However, lichens offer a large number of polyketides through the acetyl-polymalonyl pathway, such as depsides, depsidones, depsones, anthraquinones, xanthones, as well as aliphatic fatty acids, which are derivatives of orcinol or β-orcinol units [4]. These secondary compounds are primarily biosynthesized by the mycobiont in the lichenized condition (lichen thallus found in nature) [3, 5]. Various studies have demonstrated that in response to the nutrients supplemented in the culture media, the aposymbiotically cultured lichen mycobionts also produce diverse

Elizabeth Skellam (ed.), *Engineering Natural Product Biosynthesis: Methods and Protocols*, Methods in Molecular Biology, vol. 2489, https://doi.org/10.1007/978-1-0716-2273-5_24,

secondary compounds; in several cases, one or more than one compound identified in the cultures are found in the lichen thalli [6–10]. On the other hand, several lichen mycobiont cultures biosynthesize secondary compounds whose structures are not similar to the chemical constituents of their lichen thalli [11–15]. More than 1000 compounds characterized from both lichen thalli and mycobiont cultures [16, 17]. However, in recent years, the number of lichen compounds has increased significantly owing to the discovery of many novel compounds from various lichen mycobiont cultures and natural lichen thalli [18–33].

Standardized protocols of natural product investigation in lichens such as spot test (application of specific chemicals on the lichen thallus) [34–36], microcrystallization [36, 37], thin layer chromatography (TLC) [36, 38–42], and high-performance thin-layer chromatography (HPTLC) [43] are available to identify secondary compounds in lichen thalli for many taxa [36, 44–46].

1.1 HPLC Analyses of Lichens and Mycobiont Cultures

Culberson [47] introduced normal phase HPLC to separate lichen extracts as well as compounds found challenging by TLC. The reverse phase with isocratic method was chosen for analyses of some lichens [48, 49]. Later gradient method was found suitable for separation of several lichen genera, Feige et al. [50] developed a standard method with a retention index (I) for the identification of 331 lichen compounds. However, by applying the isocratic method, Yoshimura et al. [51] demonstrated HPLC equipped with a photodiode array detector (PDA) to identify lichen compounds using UV spectral data and relative retention times (RR_t). In recent decades, several chemical studies on lichen mycobiont cultures have been investigated with HPLC-PDA for precise identification of compounds [6, 7, 9, 52–58] (Fig. 1). These studies approached different gradient methods to separate compounds from the mycobiont culture extracts [53, 56, 57]. Therefore, by screening several lichen mycobiont culture extracts (lichen mycobiont cultures available from our laboratory), we optimized an HPLC gradient method for effective and better separation of chemical constituents present in the lichen mycobiont cultures. This chapter presents the extraction and HPLC protocol for the analysis of secondary compounds in lichen mycobiont cultures.

2 Materials

1. Aluminum foil.
2. Scalpels.
3. Forceps.
4. Eppendorf tubes.
5. Glass micropestles.

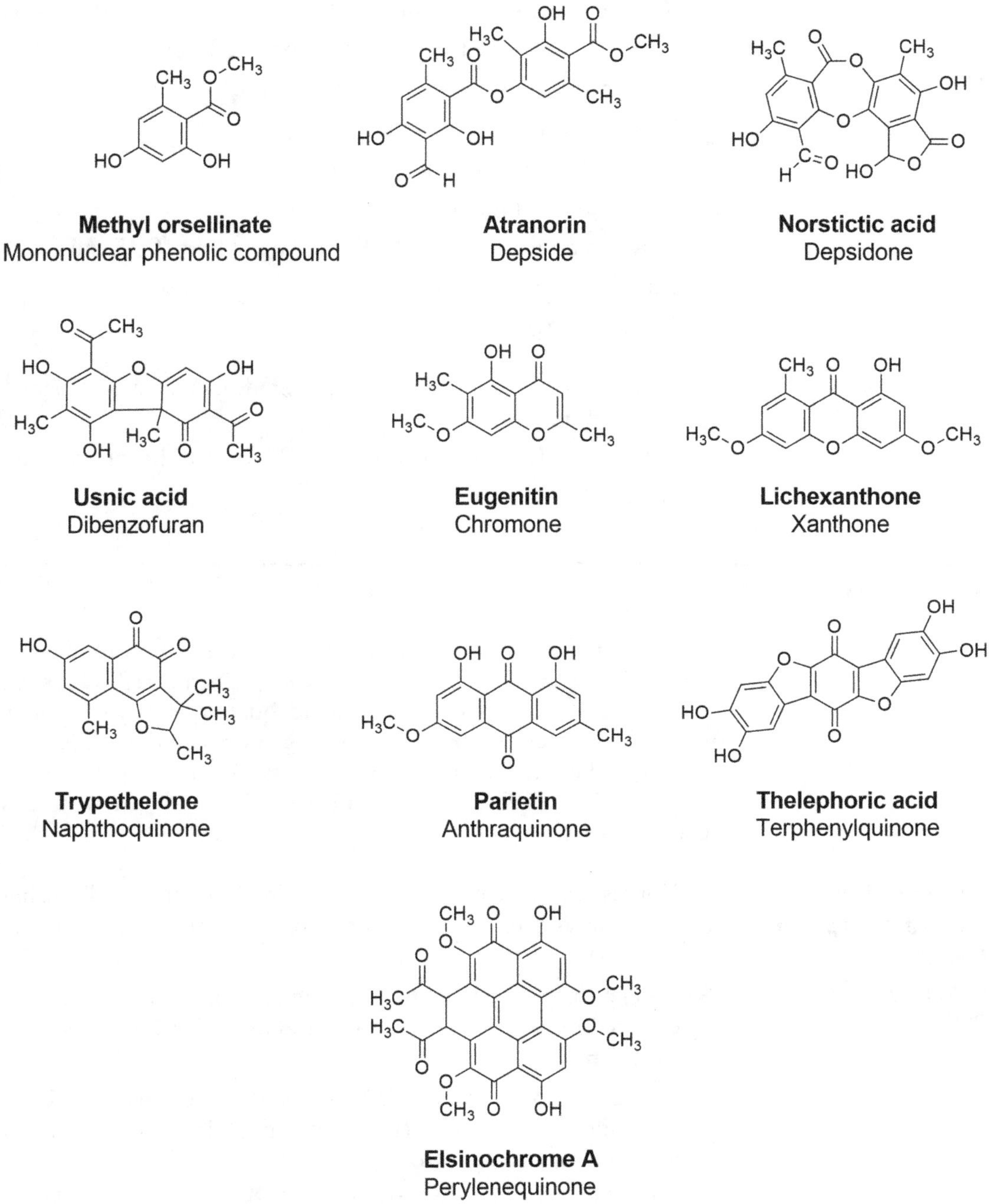

Fig. 1 Diverse classes of secondary compounds biosynthesized by the lichen mycobiont cultures identified using HPLC-PDA

6. Hot air oven.
7. HPLC instrument with photodiode array detector (190–800 nm wavelength range) connected to a computer system and software.

8. Reverse phase C18 column: 250 × 4.6 mm, 5 μm particle size.
9. Screw-cap glass bottles: 500 mL capacity, 2 nos.
10. Glass measuring cylinder: 500 mL, 1 no.
11. Solvent filtration apparatus.
12. Ultrasonic bath sonicator (*see* **Note 1**).
13. Disposable syringes (2 ml).
14. Polytetrafluoroethylene (PTFE) syringe filters (0.45 μm).
15. Micropipettes and tips.
16. HPLC syringe.

2.1 Solvents and Reagents

1. Solvents and chemicals (HPLC grade): Water, methanol, ortho-phosphoric acid.
2. Solvent A: HPLC water containing 1% ortho-phosphoric acid.
3. Solvent B: Methanol.

3 Methods

Carry out the extraction steps and HPLC analysis at room temperature (25 °C). Ensure the preliminary steps in the HPLC system such as degas solvents, checking the adequate quantity of mobile phases in the reservoir, purging, flushing system, and equilibrating column, and check baseline before analysis. Before sample analysis, include a blank HPLC run with a solvent in which the sample is dissolved (*see* **Note 2**).

3.1 Extraction of Secondary Compounds from the Lichen Mycobiont Culture

1. Harvest the lichen mycobiont culture from the media using a scalpel and place it in a small piece of aluminum foil (*see* **Note 3**).
2. To eliminate moisture, transfer the aluminum foil containing culture biomass to a hot air oven and dry it for 2 days at 45 °C (*see* **Note 4**).
3. Using forceps, transfer 100 mg of dried biomass to 2 ml Eppendorf tube and crush the material using a clean glass micropestle (*see* **Note 5**).
4. Add 500 μL of methanol to the crushed biomass and subject it to cold extraction at room temperature for 12 h.

3.2 Filtration

1. Collect the crude extract from the Eppendorf tube using a disposable syringe.
2. By using a syringe filter (PTFE, 0.45 μm), transfer the filtered-crude extract carefully to a 1.5 ml Eppendorf tube (*see* **Note 6**).

3.3 HPLC Analysis

1. Program the run as follows with a total run time of 36 min. Set the flow rate at 1 ml/min.
2. Inject 20 μL of filtered-crude extract into the instrument using an HPLC syringe (*see* **Note 7**).
3. Start the run with 70% solvent B for 3 min and raise to 80% solvent B within 14 min, then to 88% solvent B within 28 min, followed by back to 70% solvent B for 8 min (*see* **Note 8**).
4. Monitor the eluent continuously throughout the run at 200–600 nm.
5. Analyse the eluted chromatogram using the software. Document the UV spectra of each peak eluting in the chromatogram and match those peaks with reference compounds from the spectrum library (*see* **Note 9**).
6. In addition, we also recommend the following approaches to identify the eluted peak.
   (a) Compare the eluted chromatogram/peak with its source lichen thallus chromatogram/peak (*see* **Note 10**).
   (b) Compare the UV spectral data of eluted peaks with previously reported compounds (Fig. 2) from the same lichen mycobiont or its closely related taxa.
   (c) Match the UV spectral data of eluted peaks with the list of lichen substances reported by Huneck and Yoshimura [44].

4 Notes

1. Require when HPLC system lacks online degassers.
2. We strongly recommend a blank run at the beginning of the analysis. The analyst may include between each sample if needed (to check carryover from one sample to the other sample).
3. While harvesting the culture, the biomass should be devoid of media.
4. In some cases, the culture biomass may be mucilaginous. Hence it requires more time to dry. So dry those cultures for around 3–5 days at 50 °C.
5. Some lichen mycobiont cultures may become solid after the drying process. In such cases, soaking dried biomass is sufficient for extraction.
6. While using the syringe filter, some portion of extract settles within the filter. To recover residue, repeat the process two to three times for complete elution of the extract. It is strongly advisable that the crude extract must be filtered using a syringe

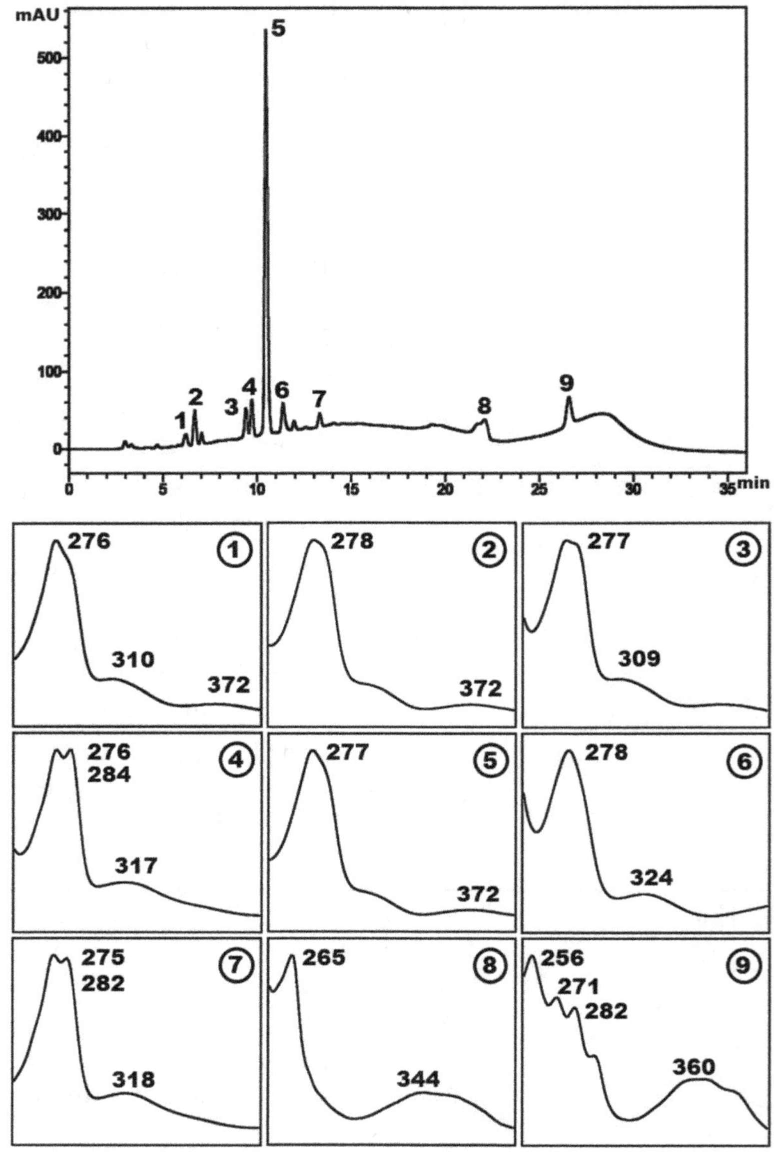

Fig. 2 HPLC chromatogram of *Trypethelium eluteriae* mycobiont cultured on malt-yeast extract medium supplemented with 6% sucrose biosynthesizes trypethelone and phenalenone derivatives. UV spectrum of each peak identified as follows: (1) 5′-hydroxy-8-ethoxytrypethelone; (2) 4′-hydroxy-8-methoxytrypethelone methyl ether; (3) 8-methoxytrypethelone; (4) trypethelone; (5) 8-methoxytrypethelone methyl ether; (6) 8-hydroxytrypethelone methyl ether; (7) trypethelone methyl ether; (8) 7,8-dihydro-6-hydroxy-3-methoxy-1,7,7,8-tetramethyl-5*H*-furo[2′,3′:5,6]-naphtho[1,8-*bc*]furan-5-one (lactone); (9) sclerodin

filter otherwise the HPLC column may get clogged. Do not reuse the filter for other samples.

7. Rinse the HPLC syringe with methanol before and after injection to avoid contamination. While taking extracts, sometimes air bubbles may develop in the syringe. To remove air bubbles, take excess volume and hold the syringe in an upside-down position. Push the plunger to move the air bubble out of the needle. Now the sample is ready for injection.
8. Leave the HPLC system in the same mobile phase for 5–10 min before the new analysis.
9. The spectrum library refers to compound data stored in the HPLC system, that is, the library contains previously analysed compounds, including known, unknown, and novel compounds from various naturally available lichens and lichen mycobiont cultures.
10. Most of the lichen thalli had marker compounds which were helpful for the identification of those compounds in the eluted chromatogram.

Acknowledgments

This work was carried out in the project entitled "Conservation of NTFP Lichens and dependent Paliyar Tribal Livelihoods in Lower Palani Hills—Western Ghats" (Grant number: BT/PR29805/NDB/39/618/2018) supported by the Department of Biotechnology, Government of India.

References

1. Hale ME (1973) Growth. In: Ahmadjian V, Hale ME (eds) The lichens. Academic, New York
2. Elix JA (1996) Biochemistry and secondary metabolites. In: Nash TH (ed) Lichen biology. Cambridge University Press, Cambridge
3. Elix JA, Stocker-Wörgötter E (2008) Biochemistry and secondary metabolites. In: Nash TH III (ed) Lichen Biology, 2nd edn. Cambridge University Press
4. Stocker-Wörgötter E, Cordeiro LMC, Iacomini M (2013) Accumulation of potential pharmaceutically relevant lichen metabolites in lichens and cultured lichen symbionts. In: Atta-ur-Rahman (ed) Studies in natural products chemistry, vol 39. Elsevier, Kidlington, Amsterdam, pp 337–380
5. Honegger R (2009) Lichen-forming fungi and their photobionts. In: Deising HB (ed) Plant relationships. the mycota (a comprehensive treatise on fungi as experimental systems for basic and applied research), vol 5. Springer, Berlin, pp 307–333
6. Stocker-Wörgötter E, Elix JA (2002) Secondary chemistry of cultured mycobionts: formation of a complete chemosyndrome by the lichen fungus of *Lobaria spathulata*. The Lichenologist 34(4):351–359. https://doi.org/10.1006/lich.2002.0395
7. Brunauer G, Hager A, Krautgartner WD et al (2006) Experimental studies on *Lecanora rupicola* (L.) Zahlbr.: chemical and microscopical investigations of the mycobiont and re-synthesis stages. The Lichenologist 38(6): 577–585. https://doi.org/10.1017/S0024282906005895
8. Valarmathi R, Hariharan GN (2007) Soredial culture of *Dirinaria applanata* (Fée) Awasthi: observations on developmental stages and compound production. Symbiosis 43:137–142

9. Fazio AT, Adler MT, Bertoni MD et al (2012) Culture studies on the mycobiont of *Caloplaca erythrantha* (Tuck.) Zahlbr. (Teloschistaceae): high production of major lichen secondary metabolites. The Lichenologist 44(4): 533–542. https://doi.org/10.1017/S0024282912000096
10. Shanmugam K, Srinivasan M, Hariharan GN (2016) Developmental stages and secondary compound biosynthesis of mycobiont and whole thallus cultures of *Buellia subsororioides*. Mycol Progress 15:41. https://doi.org/10.1007/s11557-016-1184-x
11. Hamada N, Ueno T (1990) Lecanoric acid from the mycobiont of the lichen *Stereocaulon curtatum*. Phytochemistry 29(2):678–679. https://doi.org/10.1016/0031-9422(90)85147-8
12. Miyagawa H, Hamada N, Sato M et al (1993) Hypostrepsilic acid, a new dibenzofuran from the cultured lichen mycobiont of *Evernia esorediosa*. Phytochemistry 34(2):589–591
13. Miyagawa H, Yamashita M, Ueno T et al (1997) Hypostrepsilalic acid from a cultured lichen mycobiont of *Stereocaulon japonicum*. Phytochemistry 46(7):1289–1291. https://doi.org/10.1016/S0031-9422(97)80030-1
14. Ernst-Russell MA, Elix JA, Chai CLL et al (1999) Hybocarpone, a novel cytotoxic naphthazarin derivative from mycobiont cultures of the lichen *Lecanora hybocarpa*. Tetrahedron Lett 40(34):6321–6324. https://doi.org/10.1016/S0040-4039(99)01220-4
15. Jarupinthusophon S, Luangsuphabool T, Aree T et al (2019) Naphthoquinones from cultured mycobiont of *Marcelaria cumingii* (Mont.) and their cytotoxicity. Nat Prod Commun 14(12):1–6. https://doi.org/10.1177/1934578X19884383
16. Stocker-Wörgötter E (2008) Metabolic diversity of lichen-forming ascomycetous fungi: culturing, polyketide and shikimate metabolite production, and PKS genes. Nat Prod Rep 25(1):188–200. https://doi.org/10.1039/B606983P
17. Muggia L, Schmitt I, Grube M (2009) Lichens as treasure chests of natural products. Sim News 59:85–97
18. Takenaka Y, Hamada N, Tanahashi T (2011) Aromatic compounds from cultured lichen mycobionts of three *Graphis* species. Heterocycles 83(9):2157–2164. https://doi.org/10.3987/COM-11-12270
19. Takenaka Y, Morimoto N, Hamada N et al (2011) Phenolic compounds from the cultured mycobionts of *Graphis proserpens*. Phytochemistry 72(11–12):1431–1435. https://doi.org/10.1016/j.phytochem.2011.04.017
20. Takenaka Y, Taguchi S, Le DH et al (2013) 3,4-Dihydro-2H,5H-pyrano[3,2-c]chromene and benzophenone derivatives from cultured lichen mycobionts of Vietnamese *Graphis* sp. Heterocycles 87(12):2651–2657. https://doi.org/10.3987/COM-13-12853
21. Millot M, Kaouadji M, Champavier Y et al (2013) Usnic acid derivatives from *Leprocaulon microscopicum*. Phytochem Lett 6(1):31–35. https://doi.org/10.1016/j.phytol.2012.10.009
22. Le DH, Takenaka Y, Hamada N et al (2013) A 14-membered macrolide and isocoumarin derivatives from the cultured lichen mycobionts of *Graphis vestitoides*. Chem Pharm Bull 61(3):358–362. https://doi.org/10.1248/cpb.c12-00853
23. Ly HD, Vo TN, Duong TH et al (2015) A new depside and two new diphenyl ether compounds from the lichen *Ramalina farinacea* (L.) Ach. Phytochem Lett 11:146–150. https://doi.org/10.1016/j.phytol.2014.12.003
24. Huynh BLC, Duong TH, Do TML et al (2016) New γ-lactone carboxylic acids from the lichen *Parmotrema praesorediosum* (Nyl.) Hale, Parmeliaceae. Rec Nat Prod 10(3):332–340
25. Pogam PL, Lamer A-CL, Siva B et al (2016) Minor pyranonaphthoquinones from the apothecia of the lichen *Ophioparma ventosa*. J Nat Prod 79(4):1005–1011. https://doi.org/10.1021/acs.jnatprod.5b01073
26. Duong TH, Huynh BLC, Chavasiri W et al (2017) New erythritol derivatives from the fertile form of *Roccella montagnei*. Phytochemistry 137:156–164. https://doi.org/10.1016/j.phytochem.2017.02.012
27. Duong T-H, Beniddir MA, Genta-Jouve G et al (2018) Tsavoenones A–C: unprecedented polyketides with a 1,7-dioxadispiro[4.0.4.4]tetradecane core from the lichen *Parmotrema tsavoense*. Org Biomol Chem 16:5913–5919. https://doi.org/10.1039/C8OB01280F
28. Mallavadhani UV, Sudhakar AVS (2018) Roccellatol, a new β-orcinol based metabolite from the lichen *Roccella montagnei*. Nat Prod Res 32(3):268–274. https://doi.org/10.1080/14786419.2017.1353508
29. Khader SZA, Ahmed SSZ, Arunachalam T et al (2018) Radical scavenging potential, antiinflammatory and antiarthritic activity of isolated isomer methyl-γ-Orsellinate and roccellatol from *Roccella montagnei* Bel. Bull Fac Pharm

Cairo Univ 56(1):39–45. https://doi.org/10.1016/j.bfopcu.2018.02.001

30. Phan H-V-T, Devi AP, Le H-D et al (2020) Dilatatone, a new chlorinated compound from *Parmotrema dilatatum*. Nat Prod Res 35(21): 3608–3612. https://doi.org/10.1080/14786419.2020.1716348
31. Vo VG, Le H-D, Tran T-N et al (2020) A new eremophilane-sesquiterpene from the cultured lichen mycobiont of *Graphis* sp. Nat Prod Res. https://doi.org/10.1080/14786419.2020.1779717
32. Srinivasan M, Shanmugam K, Kedike B et al (2020) Trypethelone and phenalenone derivatives isolated from the mycobiont culture of *Trypethelium eluteriae* Spreng. and their antimycobacterial properties. Nat Prod Res 34(23):3320–3327. https://doi.org/10.1080/14786419.2019.1566823
33. Bui VM, Huynh BLC, Pham NKT et al (2021) Usneaceratins A and B, two new secondary metabolites from the lichen *Usnea ceratina*. Nat Prod Res. https://doi.org/10.1080/14786419.2021.1901288
34. Nylander W (1866) Hypochlorite of lime and hydrate of potash, two new criteria in the study of lichens. Bot J Linn Soc 9(38):358–365. https://doi.org/10.1111/j.1095-8339.1866.tb01301.x
35. Vitikainen O (2001) William Nylander (1822–1899) and lichen chemotaxonomy. The Bryologist 104(2):263–267. https://doi.org/10.1639/0007-2745(2001)104[0263:WNALC]2.0.CO;2
36. Orange A, James PW, White FJ (2001) Microchemical methods for the identification of lichens. British Lichen Society, London
37. Shibata S (2000) Yasuhiko Asahina (1880–1975) and his studies on lichenology and chemistry of lichen metabolites. The Bryologist 103(4):710–719. https://doi.org/10.1639/0007-2745(2000)103[0710:YAAHSO]2.0.CO;2
38. Culberson CF, Kristinsson H-D (1970) A standardized method for the identification of lichen products. J Chromatogr A 46:85–93. https://doi.org/10.1016/S0021-9673(00)83967-9
39. Culberson CF (1972) Improved conditions and new data for identification of lichen products by standardized thin-layer chromatographic method. J Chromatogr A 72(1): 113–125. https://doi.org/10.1016/0021-9673(72)80013-X
40. Culberson CF (1974) Conditions for the use of Merck Silica Gel 60 F_{254} plates in the standardized thin-layer technique for lichen products. J Chromatogr 97:107–108
41. Culberson CF, Culberson WL, Johnson A (1981) A standardized TLC analysis of β-orcinol depsidones. The Bryologist 84(1):16–29. https://doi.org/10.2307/3242974
42. Culberson CF, Johnson A (1982) Substitution of methyl *tert.*-butyl ether for diethyl ether in the standardized thin-layer chromatographic method for lichen products. J Chromatogr A 238(2):483–487. https://doi.org/10.1016/S0021-9673(00)81336-9
43. Arup U, Ekman S, Lindblom L et al (1993) High performance thin layer chromatography (HPTLC), an improved technique for screening lichen substances. The Lichenologist 25(1):61–71. https://doi.org/10.1006/lich.1993.1018
44. Huneck S, Yoshimura I (1996) Identification of lichen substances. Springer-Verlag, Berlin
45. Elix JA (2014) A catalogue of standardized chromatographic data and biosynthetic relationships for lichen substances. Published by the author, Canberra
46. Schumm F, Elix JA (2015) Atlas of images of thin layer chromatograms. Books on Demand GmbH, Norderstedt
47. Culberson CF (1972) High-speed liquid chromatography of lichen extracts. The Bryologist 75(1):54–62. https://doi.org/10.2307/3241522
48. Culberson WL, Culberson CF (1978) *Cetrelia cetrarioides* and *C. monachorum* (Parmeliaceae) in the new world. The Bryologist 81(4): 517–523. https://doi.org/10.2307/3242338
49. Lumbsch HT, Elix JA (1985) A new species of the lichen genus *Diploschistes* from Australia. Pl Syst Evol 150:275–279. https://doi.org/10.1007/BF00984201
50. Feige GB, Lumbsch HT, Huneck S et al (1993) Identification of lichen substances by a standardized high-performance liquid chromatographic method. J Chromatogr A 646(2):417–427. https://doi.org/10.1016/0021-9673(93)83356-W
51. Yoshimura I, Kinoshita Y, Yamamoto Y et al (1994) Analysis of secondary metabolites from Lichen by high performance liquid chromatography with a photodiode array detector. Phytochem Anal 5(4):197–205. https://doi.org/10.1002/pca.2800050405
52. Stocker-Wörgötter E, Elix JA, Grube M (2004) Secondary chemistry of lichen-forming fungi: chemosyndromic variation and DNA-analyses of cultures and chemotypes in the *Ramalina farinacea* complex. The Bryologist 107(2):

152–162. https://doi.org/10.1639/0007-2745(2004)107[0152:SCOLFC]2.0.CO;2

53. Stocker-Wörgötter E, Hager A, Elix JA (2009) Intraspecific chemical variation within the crustose lichen genus *Haematomma*: anthraquinone production in selected cultured mycobionts as a response to stress and nutrient supply. Phytochem Rev 8:561. https://doi.org/10.1007/s11101-009-9149-1
54. Brunauer G, Hager A, Grube M et al (2007) Alterations in secondary metabolism of aposymbiotically grown mycobionts of *Xanthoria elegans* and cultured resynthesis stages. Plant Physiol Biochem 45(2):146–151. https://doi.org/10.1016/j.plaphy.2007.01.004
55. Cordeiro LMC, Iacomini M, Stocker-Wörgötter E (2004) Culture studies and secondary compounds of six *Ramalina* species. Mycol Res 108(5):489–497. https://doi.org/10.1017/S0953756204009402
56. Fazio AT, Bertoni MD, Adler MT et al (2009) Culture studies on the mycobiont isolated from *Parmotrema reticulatum* (Taylor) Choisy: metabolite production under different conditions. Mycol Prog 8:359. https://doi.org/10.1007/s11557-009-0609-1
57. Fazio AT, Adler MT, Parnmen S et al (2018) Production of the bioactive pigment elsinochrome A by a cultured mycobiont strain of the lichen *Graphis elongata*. Mycol Prog 17:479–487. https://doi.org/10.1007/s11557-017-1374-1
58. Stocker-Wörgötter E, Elix JA (2009) Experimental studies of lichen-forming fungi: formation of depsidones and shikimic-acid derivatives by the cultured mycobionts of three selected species of *Rhizocarpon* (Lecideaceae, lichenized Ascomycota). In: Thell A, Seaward MRD, Feuerer T (eds) The diversity of lichenology: Jubilee volume, Bibliotheca lichenologica, vol 100. J. Cramer in der Gebrüder Borntraeger Verlagsbuchhandlung, Berlin, pp 513–531

Index

A

B

C

Elizabeth Skellam (ed.), *Engineering Natural Product Biosynthesis: Methods and Protocols*, Methods in Molecular Biology, vol. 2489, https://doi.org/10.1007/978-1-0716-2273-5,

U

W

Y

GPSR Compliance
The European Union's (EU) General Product Safety Regulation (GPSR) is a set of rules that requires consumer products to be safe and our obligations to ensure this.

If you have any concerns about our products, you can contact us on

ProductSafety@springernature.com

In case Publisher is established outside the EU, the EU authorized representative is:

Springer Nature Customer Service Center GmbH
Europaplatz 3
69115 Heidelberg, Germany

www.ingramcontent.com/pod-product-compliance
Lightning Source LLC
Chambersburg PA
CBHW082051210726
48526CB00010B/2791
9781071622728